HIGHLY EXCITED ATOMS

CAMBRIDGE MONOGRAPHS ON ATOMIC, MOLECULAR
AND CHEMICAL PHYSICS

This book is an introduction to the physics of highly excited, easily perturbed or interacting atoms.

The book begins with a brief introduction to the traditional view of electron shells and their properties, and then goes on to discuss Rydberg states, quantum defect theory, atomic f values, centrifugal barrier effects, giant resonances, autoionisation, inner-shell and double-excitation spectra, **K**-matrix theory, atoms in high laser fields, statistical methods, quantum chaos and atomic effects in solids. A full chapter is devoted to the properties of atomic clusters. The emphasis is throughout on radial properties, orbital collapse, many-body effects, the breakdown of the independent particle approach, the emergence of chaos and the behaviour of atoms inside clusters and solids. A very full account of autoionisation includes not only the standard treatment for isolated resonances, but also several alternative approaches. The book discusses many experimental examples and has many diagrams and a comprehensive reference list.

This book will be of interest to atomic, molecular, chemical, optical and laser physicists, as well as to researchers in the emerging field of cluster physics.

CAMBRIDGE MONOGRAPHS ON ATOMIC, MOLECULAR AND CHEMICAL PHYSICS: 9

General Editors: A. Dalgarno, P. L. Knight, F. H. Read, R. N. Zare

Highly Excited Atoms

JEAN-PATRICK CONNERADE

Imperial College of Science and Technology

CAMBRIDGE
UNIVERSITY PRESS

CAMBRIDGE UNIVERSITY PRESS
Cambridge, New York, Melbourne, Madrid, Cape Town, Singapore, São Paulo

Cambridge University Press
The Edinburgh Building, Cambridge CB2 2RU, UK

Published in the United States of America by Cambridge University Press, New York

www.cambridge.org
Information on this title: www.cambridge.org/9780521432320

First published 1998
This digitally printed first paperback version 2005

A catalogue record for this publication is available from the British Library

Library of Congress Cataloguing in Publication data

Connerade, J.-P.
Highly excited atoms / Jean-Patrick Connerade.
p.   cm. – (Cambridge monographs on atomic, molecular, and chemical physics: 9)
Includes bibliographical references and index.
ISBN 0 521 43232 4
1. Atomic spectroscopy.   2. Rydberg states.   3. Auger effect.
I. Title.   II. Series.
QC454.A8C66   1997
539.7–dc21   96–39294 CIP

ISBN-13 978-0-521-43232-0 hardback
ISBN-10 0-521-43232-4 hardback

ISBN-13 978-0-521-01788-6 paperback
ISBN-10 0-521-01788-2 paperback

*To Jocelyne,
who allowed me to spend so many
hours on this book, and urged me on
when I might well have given it up.*

Contents

Preface

Motivation for this book

Atomic physics is a well-established subject with a distinguished history, and many books are available which cover its traditional applications. However, it is also a rapidly developing research area, and it is perhaps not surprising that most of the classic texts on which undergraduate courses are usually based no longer reflect its evolution. When the early texts on the subject were written, the prime concern was to demonstrate by many beautiful examples how the principles of quantum mechanics find application in atomic physics. Since numerical methods for solving the radial Schrödinger equation were known in principle but were not generally available, the emphasis was on angular momentum algebra and on formal developments involving electron spin, while the radial integrals were treated as parameters.

Such tools are of course essential in the armoury of any practising atomic physicist, but, in the author's view, lengthy developments in angular momentum algebra no longer form the best introduction to the subject. With ready access to fast computers, solving the radial equation is now a straightforward matter, and there exist many excellent codes for this purpose. Thus, a significant change of attitude has occurred amongst researchers: it is no longer sensible to concentrate on the angular part of the central field equation. Indeed, one can argue that the opposite approach is the correct one. The properties of spherical harmonics need only to be determined once, and can then be used to model all central field atoms. On the other hand, the radial equation must be solved afresh for each system, and sometimes for each excited state. Thus, the behaviour of radial functions is crucial. While the methods of Racah's algebra are indispensable tools, they are in themselves no longer so clearly at the heart of the subject. In fact, the most important recent discoveries in the physics of many-electron atoms are arguably connected with the subtleties of the radial equation and with the study of electron–electron correlations.

The 'traditional' atomic physics courses, which are taught in support of an undergraduate course in elementary quantum mechanics, treat hydrogen, helium and the first three rows of the periodic table in progressively diminishing detail as the atomic number increases. Atoms are regarded as vehicles for introducing the quantisation of angular momentum, demonstrating the existence of spin and establishing the central field model. This approach no longer suffices to provide an up-to-date background in the subject. In fact, if young physicists embark on research involving excited or interacting atoms, they will encounter entirely different subject material, and run the danger of not recognising it as a direct extension of an undergraduate course.

To make matters worse, when a more complete syllabus is attempted at undergraduate level (usually taught by a solid state physicist, who requires information on the transition and rare-earth elements, often omitted from the atomic physics curriculum), the standard textbooks on atomic physics provide little support. A specific example is atomic orbital collapse: many texts (and indeed many teachers) use the outdated semi-classical screening model to account for the long periods, rather than the modern explanation, due to the pioneering work of Fermi and Göppert-Mayer. This is surprising: Fermi pointed out long ago that the semi-classical screening model is inconsistent. The correct explanation is far easier to understand, accounts properly for the facts, and provides a very fine example of how to use the elementary principles of quantum wells. Indeed, the double valley model and the *full* explanation of the periodic table should be in any undergraduate course.

Similarly, quantum defect theory plays a very important role in modern descriptions of atomic physics, and should be included at a less rudimentary level than is found in most texts. Again, its modern developments provide an excellent illustration of many fundamental principles of scattering theory. The principles underlying the Lu–Fano graph are easily grasped, and provide excellent insight into an important aspect of the many-body problem, namely interchannel coupling. Likewise double- and inner-shell excitation are hardly discussed in textbooks, structure in the continuum receives little attention, etc, etc.

In fact, students often gain the impression from an undergraduate course that atoms are extraordinarily simple systems, which can be solved exactly and are fully understood. They seem to appear in a physics course only as pedagogical examples. This mistaken perspective sits uncomfortably beside the equally prevalent notion (not only amongst students) that 'heavy' atoms contain 'too many' particles at which point the physics becomes messy and hard to understand. Only hydrogen and helium, one is told, can really be worked out. Indeed, both views are sometimes expressed in the same breath!

The present monograph is an attempt to redress the situation just described. First, its basic motivation is *not* to provide illustrations of the principles of elementary quantum mechanics. In the present book, quantum mechanics is treated entirely as a tool, and its developments will not be discussed. It is assumed that the reader is already familiar with the material in an intoductory undergraduate course, leading via the solution of the Schrödinger equation for hydrogen to the static mean field and central field approximations for many-electron atoms. We concentrate on the properties which make many-electron atoms interesting, different from each other and from other physical systems. Second, we are interested in highly excited atoms, by which we mean atoms in Rydberg states or in states above the first ionisation threshold, because these are the situations in which many-body effects come to the fore.

In respect of motivation, atomic physics has experienced the same transformation as physics itself in recent years. The problem of the nature of the forces between just two particles is no longer the central preoccupation. As the unification of forces has been pursued to higher and higher energies, eventually out of reach of even the highest energy laboratory accelerators, it has become apparent that the unsolved problem at achievable energies, and indeed at all energies remains the many-body problem (which should perhaps more aptly in our context be named the few-body problem).

If we proceed systematically from small to large systems, then even in elementary quantum mechanics[1] the few-body problem, beginning with just three particles, is unsolved. Many recent developments, including current studies of chaos and quantum chaos, serve to underline the fundamental significance of this fact, even for as simple an atom as helium.

Intriguing questions arise concerning the applicability of the correspondence principle in any quantum system whose underlying classical dynamics becomes chaotic. Also, the Pauli principle (which possesses no classical analogue) somehow contributes to making the quantum few-body problem simpler to handle than the classical one.

For many-electron atoms, since Coulomb interactions between just two particles are well understood, complexities arising from the nature of the forces do not arise. One can thus concentrate entirely on the few-body problem. The very simplicity of the Coulomb interaction, which made atoms seem uninteresting to many physicists twenty years ago, now serves to place atomic physics, once again, in a central position.

---

[1] I.e. excluding quantum electrodynamics, in which the situation is even worse.

$$\star \qquad \star \qquad \star$$

In our definition, a 'highly excited' atom is one for which an electron has been raised to an energy comparable to or higher than the ionisation potential. Thus, the properties of highly excited atoms include the behaviour of high Rydberg states, of the ionisation continuum, and of any further excitations or structures which may lie embedded within the continuum towards higher energies. For hydrogen, the concept of a 'highly excited' atom is not particularly fruitful, since only one electron can anyway be involved. For many-electron atoms, it becomes important, because alternative excitation channels become available: only for highly excited atoms does one enter the range where more than one electron, or an electron other than the valence electron, can become involved in the excitation process.

Ideally, one should probe the many-electron atom using sufficient energy for its many-body properties to become apparent. In the *optical range*, only the valence electron is excited, and so the bound Rydberg spectra leading to the first ionisation potential only provide information about many-body effects when a perturbation happens to occur. For this reason, the study of irregularities in Rydberg spectra assumes a special significance as one enters the 'highly excited' spectral range; if radiation is used to excite the atom, then the most favourable transition energy lies in the vacuum ultraviolet or soft X-ray ranges of the spectrum for a single photon transition.

A many-electron atom is the simplest quantum system in which the N-body problem can be studied, where N is the number of electrons (described as 'few' or 'many' according to the system). In many-electron atoms, the electrostatic force dominates the interaction between electrons and the nucleus, and also the interactions between electrons. Its laws are amongst the best-known in physics, and so the nature of the interaction between the particles introduces no uncertainty of understanding. Complexities may be expected to arise from the many-body nature of the system, and the subtleties of behaviour which distinguish N-electron atoms from hydrogen, or more generally from *independent particle* models (such as the one-configuration Hartree–Fock atom) assume a special significance.

The Hartree–Fock procedure adds to the methods of elementary quantum mechanics a series of approximations which turn the problem of the N-electron atom into an 'equivalent' set of coupled one-electron problems to be solved self-consistently. Despite its simplicity and success, this procedure is far from being obvious, as can be appreciated if we move from quantum to classical mechanics. The possible occurrence of chaos in the three-particle problem implies that its 'solution' would be essentially different from the solution of the two-body problem. Chaos does not occur in quantum mechanics, but its disappearance is not as sudden and complete as this statement implies. Recent experiments on Rydberg atoms in high magnetic fields demonstrate that traces of the order-to-chaos transition (the so-called 'scars') persist in real quantum systems. The question whether some regimes of parameter space exist where isolated many-electron atoms might exhibit similar behaviour remains unanswered. This question, and many others besides, are all hidden from view in the independent particle model. It is therefore of great significance to investigate departures from this model (electron–electron *correlations*) experimentally. We might even ask the question: can correlations trigger an order-to-chaos transition in highly excited atoms?

To the 'quantum chaologist', the highly excited atom is a versatile probe to study the statistical properties of spectra as well as dynamical effects. The long range Coulomb force allows very high quantum numbers to be reached, so that the semiclassical limit can be closely approached. Atoms in strong magnetic fields provide a most beautiful example of the transition from order to chaos in a quantised system, analogous to the classical nonlinear problem of a pendulum in a magnetic field.

⋆   ⋆   ⋆

Observations of highly excited atoms provide a whole catalogue of specific many-body effects, some of which do, and some of which do not, find analogues in other branches of physics (e.g. nuclear or solid state physics). The explorations of atomic physicists in the realm of many-body interactions have been aided by new sources of radiation – mainly the electron accelerators (synchrotron radiation) and tunable dye lasers. Generally, experiment has led the way in the discovery of new effects, but powerful and new theoretical methods were developed, or borrowed from other fields, and refined beyond recognition. Many-electron atoms serve as a testing ground for these new methods of calculation. New theoretical problems have even emerged, such as the subtle interplay between relativity and correlations, which are the subject of much discussion, because of the possible effects involving negative energy states, whose role in self-consistent field computations from the many-electron Dirac equation remains a matter of interest to researchers.

In the laboratory, the development of short pulse, high power lasers has turned conventional approximations of atomic physics inside out. No longer can one assert that the electromagnetic field always acts as a small perturbation: indeed the question of the interaction between the radiation field and matter is now an open one again. It is yet another facet of the strong field problem, and also involves quantum chaos. The strong AC external field introduces the notion of the 'dressed atom,' but also the need to understand the dynamics of short pulse excitation. Several previously unsuspected phenomena have been discovered experimentally, and a complete theory remains to be formulated.

Finally, a subject of fundamental importance in atomic physics is the study of how electronic properties are modified by the atomic environment, as in molecules or in the solid state. New situations have been found at the frontier between atomic physics, molecular physics and the physics of condensed matter. This area has grown considerably with the discovery of giant resonances which, though atomic in origin, were first observed in the soft X-ray spectra of solids. Since then, resonant photoemission has become a well-established experimental technique in solid state physics, and valence fluctuations and intermediate valence effects in solids have been shown to involve localised orbitals which are partly atomic in character.

A novel system has been found, which is ideally suited to explore this frontier: clusters containing small or large numbers of atoms can now be made. These are new objects whose properties evolve from the free atom limit to that of the solid as a function of size, or of the number of atoms they contain. Such objects are of quantum scale when they are small, but achieve macroscopic dimensions as their size increases. Thus, one can study the evolution of properties which persist from quantum to macroscopic sizes, or else search for the earliest appearance of solid state properties, for example plasmon oscillations in solids, as a function of the number of atoms in the cluster.

Clusters also demonstrate the ubiquity and generality of the basic principles of physics: the stability of metal clusters is governed by a shell closure closely related to that of nuclear physics. Indeed, the collective, giant dipole resonances in clusters and in nuclei obey the same laws over changes of fourteen decades in scale size.

In short, much remains to be investigated and understood in atomic physics. Not only is it an open subject in the fundamental sense, but it is also open in its numerous connections to and applications in other fields of science. Atoms, in this context, are not to be regarded as isolated and pure, but rather as interacting objects, whose properties undergo a rich variety of amazingly subtle transformations.

The present monograph attempts to fill some of the gaps which exist in present texts without duplicating too much of the traditional material. It is aimed at young researchers, or at researchers from fields other than pure atomic physics who may be interested in some of the aspects just described. Complexities of detail have been avoided wherever possible in order to stress the important basic principles, and a large number of fresh examples drawn from contemporary research papers have been used to illustrate the text.

Again, it is assumed that the reader is familiar with the usual content of an undergraduate course in atomic physics, *viz.* the principles of quantum mechanics, the hydrogen atom, elementary treatments of angular momentum, of spin-orbit interaction, the Pauli principle, static mean fields, the central field model, the building-up principle, spectroscopic notation and the working of dipole selection rules.

⋆ ⋆ ⋆

The present work is not a textbook. It is written in the style of a monograph, from a particular perspective. Nonetheless, it is hoped that it will fill a gap in the literature, being both a complement to a modern introductory atomic physics course and a preliminary reference for an investigator who actually needs to use atomic physics in his research.

J.-P. Connerade

April 1997
London

1
Closed shells, sphericity, stability and 'magic numbers'

1.1 'Magic numbers'

It is a consequence of the regularity of the periods in the first few rows of the periodic table that we can think of individual bound electrons in the atom each occupying one of the available orbitals. In reality, this is only an approximation, but it usually holds very well. As more and more electrons are added to the system, the lowest energy orbitals are filled in order, in accordance with the Pauli principle and as described by the *aufbau* or building-up principle of Bohr [1] and Stoner[2].[1] Eventually, as filling progresses, stable shells are formed for certain 'magic numbers'. We may regard the existence of these closed shells as an experimental fact (the evidence for which came first from chemistry) or as a property of the independent particle model and central field approximation, which allows us to use as solutions of the angular Schrödinger equation, a single set of spherical harmonics, whatever atom is involved.

The great merit of this approximation is that we need solve the angular equation only once in order to establish the general solution. Thereafter, *all* bound electrons in *all* atoms possess angular functions

$$\Theta(\theta)\Phi(\phi) = Y_m^\ell(\theta, \phi) = P_{|m|}^\ell(\theta) \exp \pm im\phi \qquad (1.1)$$

in the usual notation for spherical harmonics and Legendre polynomials. The difference between atoms, in this model, depends:(i) on the occupation numbers of spherical harmonics (i.e. how many electrons are in each) and (ii) on radial solutions, which can only be obtained numerically for many-electron atoms and must be computed separately for each case.

Thus, the existence of closed shells provides experimental evidence that

---

[1] The *aufbau* principle is due to Bohr, who did not, however, obtain the correct values of the magic numbers. The correct values were obtained for the first time by Stoner.

we may indeed treat electrons as *independent particles* each one with a given value of the angular momentum quantum number ℓ. Unfortunately, this evidence is confined to those atoms which do have closed shells and, as will become apparent in later chapters, it is actually the reemergence of spherical symmetry as each period is completed which restores the validity of the independent particle model, after the breakdown which occurs within each period.

In the present chapter, we describe how the concept of closed shells arises within the central field model, and what consequences flow from it.

1.2   The biaxial theorem

It is a mathematical property of spherical harmonic functions $Y_{\ell m}(\theta, \phi)$ that they obey an *addition rule* which is known as the *biaxial theorem*: if $\mathbf{r}_1$ and $\mathbf{r}_2$ are two vectors, with directions described by the polar angles (θ_1, ϕ_1) and (θ_2, ϕ_2), and if α is the included angle between the two vectors, then

$$P_\ell(\cos \alpha) = \frac{4\pi}{2\ell + 1} \sum_{m=-\ell}^{+\ell} Y_{\ell m}(\theta_1, \phi_1)^* Y_{\ell m}(\theta_2, \phi_2) \qquad (1.2)$$

This result is proved in textbooks on mathematics – see for example Jeffreys and Jeffreys [3].

Now, if we choose that the two vectors point in the *same* direction, i.e. that the angles $\theta_1 = \theta_2 \equiv \theta$ say, and $\phi_1 = \phi_2 \equiv \phi$ then $\alpha = 0$ and, since $P_\ell(1) = 1$, we have that

$$\sum_{m=-\ell}^{+\ell} |Y_{\ell m}(\theta, \phi)|^2 = \frac{2\ell + 1}{4\pi} \qquad (1.3)$$

The significance of this last equation is that, although it contains a full summation over spherical harmonics on the left hand side, each one of which individually contains a complicated angular dependence, the right hand side is *completely independent of either θ or ϕ*.

The Pauli principle allows each orbital $Y_{\ell m}(\theta, \phi)$ to be occupied at most by two electrons, one with spin up and the other with spin down. If we fill all the individual angular wavefunctions which are solutions of the independent electron central field equations for a given value of ℓ, by putting all $2(2\ell + 1)$ (the factor of 2 arises because there are two spin states) electrons into a given subshell, then the resulting charge shell, given by

$$4\pi r^2 \rho(r) dr = 8\pi r^2 |R_{n\ell}(r)|^2 \sum_{m=-\ell}^{+\ell} |Y_{\ell m}(\theta, \phi)|^2 \, dr \qquad (1.4)$$

is spherically symmetric.

It is also possible to produce a spherically symmetric *half-filled* subshell, in which each orbital contains only one electron, and the spins are all aligned parallel to each other. This situation arises, for example, in the element Mn, which has five electrons in the d subshell, all five of which have their spins pointing in the same direction. The charge density for the half-filled subshell is spherically symmetric, and it therefore has zero total orbital angular momentum $L = 0$ and a total spin $S = 5/2$ (its multiplicity is $2S + 1 = 6$). Its ground state is therefore ${}^6S_{5/2}$.

The five electrons in Mn do not, however, fill the subshell, because there are five other empty orbitals very close in energy, namely those with spins antiparallel to the first five. These are filled for the element Zn, which has ten electrons in the $3d$ subshell, at which point this subshell is *closed* because no further electrons can be added to it, in accordance with the Pauli principle.

We thus have a simple model (the *aufbau* or building-up principle of Bohr [1] and Stoner [2]) which correctly predicts the periodic structure of Mendeleev's table of the elements. More precisely, one should state that Mendeleev's table is the experimental evidence which allows us to use an independent electron central field model and to associate each electron in a closed shell with a spherical harmonic of given n and ℓ, because there is no physical reason why a particular ℓ for an individual electron should be a valid quantum number: angular momentum in classical mechanics is only conserved when there is spherical symmetry.

Within the central field model, it is useful to define for each value of ℓ an *effective radial potential*

$$V_{eff}(r) = V_{el}(r) + \frac{\ell(\ell+1)\hbar^2}{2mr^2} \tag{1.5}$$

which includes both the electrostatic central field and the centrifugal term, and is therefore different for electrons of differing angular momenta. This idea will reappear many times, in particular in chapter 5.

Note that if the subshell is *not* either filled or half-filled, equation (1.3) does not apply. Strictly, the conservation of angular momentum can only be valid: (a) for atoms with filled or half-filled outermost subshells in the ground state; (b) for alkali atoms, which have a compact ionic core and one electron orbiting well outside this core; and (c) for rather special excited states, e.g. $2s^2$ of He. Even in such cases, states may turn out not to be pure, so the breakdown of ℓ characterisation for individual electrons is a frequent occurrence in atomic spectra.

It is a consequence of such breakdown, when it does occur, that the association of each electron with an independent particle orbital ceases to be valid.

It is implicit in the *aufbau* principle that closed subshells are 'nested' in order within one another, forming a spherical core of electrons outside of which the outermost electron (known as the valence or optical electron) is added as the atom is 'built up'.

Unfortunately, the simple *aufbau principle*, which only involves angular momentum together with a fixed scheme for the ordering of the energy levels $E_{n\ell}$ derived from observations of the alkali spectra (see chapter 2) fails beyond the first three periods, and straightforward ordering of the subshells breaks down. The reasons for this failure are interesting and important, and will be taken up in detail in chapter 5.

There is a danger in interpreting the filling of shells entirely on the basis of angular functions. In fact (as we shall see below) it is really necessary to distinguish between *subshells*, which arise by filling all the m-states of a given ℓ and *shells*, which arise by filling all the ℓ-states of a given n. The difference between the two is controlled by the radial rather than the angular part of the solution, which determines $E_{n\ell}$, the energy of a level of given n and ℓ: states of different n are usually well separated in energy, while states of the same n but different ℓ are usually closer together. Consequently, a filled shell tends to be very stable, but a filled subshell may not be, a question which we discuss below in more detail.

1.3   Chemical valence

The subject of chemical valence is a vast one, and is not covered in the present book. Excellent introductions to it exist in other monographs [4] to which the interested reader may refer. A few brief comments are made here for completeness, since the connection with the periodic table has been mentioned above.

One of the reasons why closed shells are so important is that they provide the first and simplest model of chemical valence, namely the notion that shells must be 'completed' to form ionic bonds, which yields the valence merely by counting vacancies or else by counting the number of electrons in a less than half-filled shell. Of course, valence is really much more complicated than this idea suggests: the fact that identical neutral atoms can attract each other strongly, as evidenced by the formation of homopolar molecules such as H_2 or N_2 cannot be explained on the Bohr model and requires the introduction of exchange forces or the pairing of electrons with antiparallel spins. Then there are the complexities of molecular hybridisation, etc. In all these situations, angular momentum closure may still appear relevant, since it suggests that no electron is available for pairing when the total angular momentum is zero. However, the case is frequently overstated: it turns out that $J = 0$ is a necessary but not a sufficient condition for chemical stability. Of equal significance is the radial equation, which determines the energy splittings between subshells and shells, and how easy it is to break them open.

Even if we do accept the simple Bohr theory of valence, there is still the difficulty of how to handle chemical valence when the *aufbau* principle breaks down for free atoms, or when the n and ℓ of individual electrons are poorly defined. In some cases, instabilities of valence can be expected. Nonintegral valences are indeed observed for many elements of the long periods in the condensed phase. This aspect of chemical valence will be further discussed in chapter 11, where it will also be related to properties of the radial equation.

1.4 Ionisation potentials

Closed shells are of fundamental importance, but have so little spectral signature that they are characterised essentially by one number: the binding energy. It is customary to plot ionisation potentials V_i against atomic number A in order to display the regularities which underpin the periodic table: in many ways, this is simpler and more direct than a detailed study of the chemical properties of the elements. In the present section, we discuss the screening effect of closed shells and suggest other ways of plotting such information.

Ionisation potentials are displayed as a function of atomic number in fig. 1.1 The plot of ionisation potentials immediately brings out some interesting features: one sees that the most stable (i.e. the most compact) atoms are the rare gases. In fact, the smallest atom is He. For the element following a rare gas (an alkali atom), the ionisation potential is particularly low. This can be understood as a consequence of the excellent screening of the nuclear charge by the compact closed shell of electrons. The same, however, is not true of closed subshells, or at least the effect is then much less pronounced.

It is interesting to consider this from another point of view: the difference ΔV_i between the ionisation potential V_i^{RG} of a rare gas and the ionisation potential of the alkali atom adjacent to it is clearly larger when the screening is more efficient, because a more compact closed shell has a higher ionisation potential and also screens the nuclear charge best. Therefore, ΔV_i should itself be a function of the ionisation potential V_i^{RG} of the rare gas, which is a direct measure of how compact it is. We show such a plot in fig. 1.2: the remarkable result is that a straight line is obtained, showing that the dependence is even simpler than one might anticipate.

If we plot single configuration Hartree–Fock calculations on the same

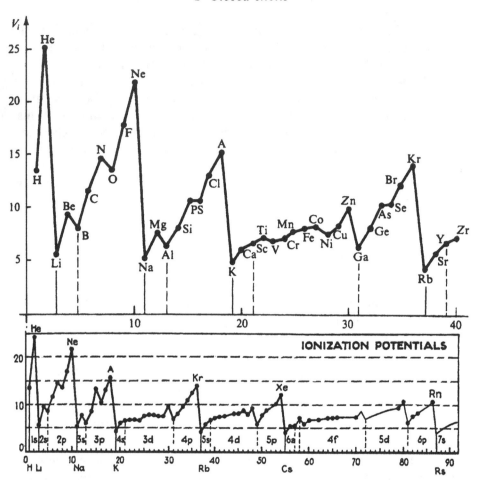

Fig. 1.1. Ionisation potentials of the elements plotted as a function of atomic number. Note the very high values for elements with closed shells and the very low values immediately following them. Note also that closed *subshells* give a much less pronounced peak, and in some cases a barely visible change. The top of the figure shows the first portion of the graph at a greater magnification.

graph, the theoretical points lie rather close to the same straight line (especially for the light atoms), despite the fact that there are significant discrepancies between Hartree–Fock and experimental ionisation potentials. Thus, this linear dependence is a general rule for atomic central fields *provided the core is compact* and lies well within the valence shell.

This simple behaviour does *not* occur for closed subshells (the corresponding plot would show a good deal of scatter). There are a number of reasons for which the core ceases to be compact when only the outermost subshell is closed. They will be discussed in chapter 5.

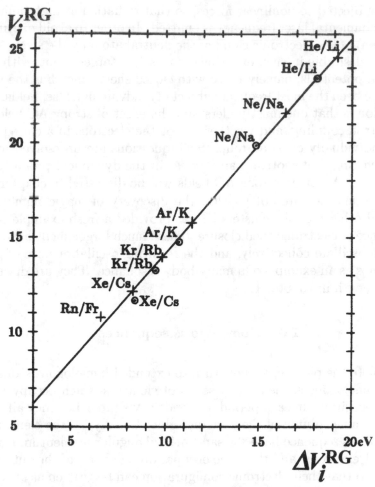

Fig. 1.2. Plot of the drop in ionisation potential between a rare gas and the following alkali against the ionisation potential of the rare gas: X – experimental data; O – Hartree–Fock calculations. Even though experiment differs considerably from theory, all points lie close to the same straight line.

1.5 Collective motion in closed shells

Current interest in the stability of closed shells stems from several different areas of research. The subject of giant resonances [5] will be described later in the present book. It involves collective motion of all the electrons in a closed shell or subshell, and their cooperative response is therefore of interest [6].

In high harmonic generation experiments [9], atoms are subjected to superintense laser fields, in which the motions of the driven electrons become subjected to nonlinear forces, so that radiation at high multiples of the fundamental laser frequency is emitted. In order to avoid disruption (ionisation) of the electronic cloud of the neutral atom by the strong laser field, it is desirable in such experiments to select target atoms with high ionisation potentials, namely those with closed shells, and thus the noble gases have been the most frequent subject of study in high laser fields. The implication is that one must understand the effect of strong AC fields on interelectron coupling, and whether or not the electrons in a closed shell quiver individually or as a group. Such questions are unresolved at the time of writing, but motivate an interest in the dynamical properties of closed shells. Atoms in strong AC fields will be discussed in chapter 9.

In quite another area of physics, the discovery of magic numbers in alkali and other metallic clusters [7] has provided a fresh example of the significance of electronic shell closure. These much larger shells have been shown to oscillate collectively, and the resulting oscillations are of great significance as an example of a many-body resonance. They are discussed at some length in chapter 12.

1.6   Homologous sequences

To complete the picture, we now turn to extended homologous sequences [8]. A homologous sequence is a series of elements which occupy corresponding positions in each period of Mendeleev's table, i.e. they all occur down the same column of the periodic table. Usually, all the elements in a homologous sequence have the same orbital angular momentum ℓ of the outermost electrons, and the same occupation number q of the outermost subshell, so that their electronic configuration can be written as ℓ^q, where the principal quantum number n of the outermost electrons is of course different for each element of the sequence. A homologous sequence, however, does not always exhaust the list of possible elements of this type. Thus, if we form a list of all the elements with a closed s^2 outer subshell, we must include He, which is a rare gas, then Zn, Cd and Hg, which are all metals despite possessing closed subshells and fairly high ionisation potentials, and also Ca, Sr, Ba and Ra, which are alkaline earths with rather low ionisation potentials, despite their closed subshells. We may also include the elements Yb and No, which possess filled $4f$ and $5f$ subshells respectively. Such a list is called an *extended* homologous sequence.

Although closed subshells differ fundamentally from closed shells, they do share some of their properties, in particular as regards the structure of excited configurations, and this allows comparisons between elements belonging to such sequences to be useful. An example will be found in chapter 5.

Here we merely note that, for some elements of the sequence (alkaline earths) which have closed outer subshells but low ionisation potentials, the next element in the table can have a *higher* ionisation potential, i.e. ΔV_i in fig. 1.2 can actually be negative. What this means is that screening by the outer subshell is not at all effective, and that the outermost electron of the next element penetrates deeply into the s^2 subshell. In other words, it is *not* forming a compact core, as we see also from the fact that alkaline-earth elements possess a low ionisation potential. Thus, a plot analogous to fig. 1.2 is unsuccessful for the extended homologous sequence. Again, we stress that *spherical symmetry alone (determined by angular momenta) does not suffice to produce stability and effective screening*. It *must* be accompanied by compactness of the core, which is a property governed by the *radial equation*. Perhaps the best illustration of this point is the fact that He, the smallest atom, is also the most permanent of all permanent gases, while Be, which is about 2.6 times larger, is a stable solid, despite the fact that both possess outer s^2 configurations.

1.7 The Hartree–Fock equations

The central field approximation and the simplifications which result from it allow one to construct a highly successful quantum-mechanical model for the N-electron atom, by using Hartree's principle of the *self-consistent field* (SCF). In this method, one equation is obtained for each radial function, and the system is solved iteratively until convergence is obtained, which leaves the total energy stationary with respect to variations of all the functions (the variational principle). The Hartree–Fock equations for an N-electron system are equivalent to several one electron radial Schrödinger equations (see equation (2.2)), with terms which make the solution for one orbital dependent on all the others. In essence, the full N-electron problem is approximated by a smaller number of coupled one-electron problems. This scheme is sometimes (somewhat inappropriately) referred to as a one-electron model: in fact, the Hartree–Fock equations are a genuine N-electron theory, but describe an *independent particle* system.

The simplest SCF system is based on Hartree's original model, which treats the many-electron wavefunction Ψ as a simple product of one-electron functions ψ_i. Since this is also the rule for combining independent probabilities, this is clearly an independent particle approach, but violates the Pauli principle, because no account is taken of exchange. The next stage beyond this is to write the wavefunction as a properly antisymmetrised determinant of products thus:

$$\Psi \equiv \frac{1}{N^{1/2}} \begin{vmatrix} \psi_1(x_1)\psi_1(x_2)...\psi_1(x_N) \\ \psi_2(x_1)\psi_2(x_2)...\psi_2(x_N) \\ \vdots \qquad \vdots \qquad \vdots \\ \psi_N(x_3)\psi_N(x_3)...\psi_N(x_N) \end{vmatrix} \qquad (1.6)$$

known as a Slater determinant . This is the simplest combination which automatically satisfies the Pauli principle. By minimising the expectation value of the total energy with respect to variation of the radial parts ϕ_i of the wavefunctions, where $\psi_i(x) = \phi_i(r)\xi_i(\eta)$ and $\xi_i(\eta)$ is a spin function, one obtains the Hartree–Fock equations for N electrons in an external central potential $V_{ext}(r)$:

$$-\tfrac{1}{2}\nabla_i^2\phi_i(r) + V_{ext}(r)\phi_i(r) + \left\{ \sum_j^N \int \frac{\phi_j^\star(r')\phi_j(r')}{|\,r - r'\,|}dr' \right\}\phi_i(r)$$

$$\qquad (1.7)$$

$$- \sum_{j\,spins//} \left\{ \int \frac{\phi_j^\star(r')\phi_i(r')}{|\,r - r'\,|}dr' \right\}\phi_j(r) = \epsilon_i\phi_i(r)$$

(For a derivation of the Hartree–Fock equations, see, e.g., [10].) The first term is the kinetic energy of electron i. The second term is the potential energy in V_{ext}, which is the Coulomb potential due to the nucleus. The third term is a correction to the Coulomb potential due to the centrally symmetric average of the electrostatic repulsion between the electrons. The fourth term is an effective *exchange potential*, due to the Pauli Principle. The sum of the external, Coulomb and electrostatic repulsion potentials is often referred to as the Hartree potential.

Note that the external, Coulomb and electrostatic repulsion terms are all three multiplicative on ϕ_i (as is the Hartree potential). For this reason, they are called *local* operators. The exchange term contains ϕ_i in the kernel of an integral and is referred to as a *nonlocal* operator. Much of the complexity of solving the Hartree–Fock equations arises from the nonlocality of the exchange term. Various simplifications have therefore been devised so as to replace it by an effective local operator.

Independent electron models are only an approximation. Any effect not included within a particular independent electron model is called a *correlation*. Note that electron correlations are defined with respect to a specific model and therefore depend on the model used. Thus exchange forces appear as a *Pauli correlation* in Hartree's model. The main effect of Pauli correlations is to reduce the probability of electrons with parallel spins approaching each other. Owing to this reduction, each electron seems to be surrounded by a 'hole' or a space devoid of other electrons.

This is called a *Fermi hole* and is the first example we encounter of a particle being *dressed* (i.e. having its properties modified) by many-body forces. Strictly speaking, the Fermi hole differs for each electron, but the interaction can be made local by averaging it over different orbitals, and this is referred to as the Hartree–Slater approximation .

The other correlations which are neglected in the Hartree–Fock model are the *Coulomb correlations*, due to the approximate treatment implied by using an averaged central field. Often, they are small. The Hartree–Fock model is fairly robust, because the next higher order contribution to the many-body perturbation series is zero (Brillouin's theorem).

In order to solve the Hartree–Fock equations explicitly, one can write them as:

$$\frac{-\hbar^2}{2m}\frac{d^2 u_{n\ell}}{dr^2} + \left(\frac{\ell(\ell+1)\hbar^2}{2mr^2} - \frac{e^2}{4\pi\epsilon_0 r}[\mathcal{Z} - Y_{n\ell}(r)] + \varepsilon_{n\ell,n\ell}\right)u_{n\ell}$$

$$= \frac{e^2}{4\pi\epsilon_0 r}X_{n\ell}(r) + \sum_{n'}\varepsilon_{n'\ell,n\ell}u_{n\ell} \tag{1.8}$$

where the functions $Y_{n\ell}(r)$ and $X_{n\ell}(r)$ are, respectively, corrections to the Coulomb potential and the exchange function. They depend on the radial eigenfunctions $u_{n\ell}$ through the definitions

$$\left.\begin{array}{l} Y_{n\ell}(r) \equiv \sum_{n'\ell';k} A_{n\ell n'\ell' k}Y^k_{n'\ell',n'\ell}(r) \\ X_{n\ell}(r) \equiv \sum_{n'\ell'=n\ell;k} B^k_{n\ell n'\ell'}Y^k_{n\ell,n'\ell'}(r)u_{n'\ell'} \end{array}\right\} \tag{1.9}$$

where

$$Y^k_{n\ell,n'\ell'}(r) = \int_0^r \frac{s^k}{r}u_{n\ell}(s)u_{n'\ell'}(s)ds + \int_r^\infty \frac{r^{k+1}}{s}u_{n'\ell'}(s)ds$$

are radial integrals, while $A^k_{n\ell n'\ell'}$, $B^k_{n\ell n'\ell'}$ are numerical coefficients which depend on the occupation of the orbitals and the $\varepsilon_{n\ell,n\ell}$ are energy matrix elements which yield the theoretical removal energies from each orbital. For a derivation of the Hartree–Fock equations, see e.g. [10]. The effective potential in the Hartree–Fock equations may look as though it plays the same role as the potential in the H atom, but this is deceptive: in a two-particle problem, the potential is always the same, and so the dynamical properties of the system are completely determined by it. In an *N*-particle problem, not only is the potential different for each one of the orbitals in the problem, but *the potential depends on the state of the system*: for example, the potential obtained for an electron in the ground state of an atom may be quite different from the potential experienced by even a corresponding electron in an excited configuration. In other words, there is no single determination of the effective potential which can fully represent the dynamical properties of the atom. Note that, although an infinite summation over the index k is implied, only a few terms are actually required, depending on the electronic configuration: this is proved in texts on multiplet structure theory [17].

1.8   The SCF procedure

The system of equations (1.8) is based on the central field approximation, and therefore its application to real atoms is entirely dependent on the existence of closed shells, which restore spherical symmetry in each successive row of the periodic table. For spherically symmetric atoms with closed shells, the Hartree–Fock equations do not involve neglecting noncentral electrostatic interactions and are therefore said to apply exactly. This does *not* mean that they are expected to yield exact values for the experimental energies, but merely that they will apply better than for atoms which are not centrally symmetric. One should bear in mind that, in any real atom, there are many excited configurations, which mix in even with the ground state and which are not spherically symmetric. Even if one could include all of them in a Hartree–Fock multiconfigurational calculation, they would not be exactly represented. Consequently, there is no such thing as an exact solution for any many-electron atom, even under the most favourable assumptions of spherical symmetry.

To solve equations (1.8), which connect the effective potential (i.e. the radial equation) for one orbital $n\ell$ to the *solutions* of the equations (i.e. the $u_{n'\ell'}$ of all the other orbitals, it is necessary to break into the circle by assuming some (reasonably realistic) approximate eigenfunctions $u_{n'\ell'}$, from which the potential and exchange functions $Y_{n\ell}(r)$ and $X_{n\ell}(r)$ can be computed. Once this is done, equations (1.8) for the orbital $u_{n\ell}$ are determined, and can be integrated numerically. This is done in turn for each orbital, starting from the approximate set of orbitals and yielding a new set of functions, which are hopefully a better approximation to the correct solution than the previous choice. If the functions are indeed improved, then in turn they should yield better potential and exchange functions, and a new set of improved equations which can again be integrated numerically. This procedure is continued until no further improvement is achieved by going round the loop, at which point the wavefunctions are said to be 'self-consistent' with each other. Since the 'correct' set of orbitals would be expected to have the same property, the equations are then said to be solved (if we assume that there is only one 'correct' solution to the problem, and that no other will be self-consistent). Of course, it is not obvious that self-consistency will be achieved. It is possible that a bad initial approximation may have been made. In this case, the solutions might not improve and, after a certain number of iterations, the procedure will diverge. This is indeed, precisely what can occur: picking the starting solutions in such a way as to obtain convergence is something of an art. For example, one may perform a calculation for a highly charged ion which is isoelectronic with the problem one is really interested in, and gradually reduce the charge of the ion, scaling the converged solutions of one stage of ionisation to use them as starting approximations for the next lower stage of ionisation, so as to approach rather slowly a case for which convergence is difficult. The reason why this procedure works well is that solutions tend to become more and more hydrogenic with increasing charge in an isolectronic sequence.

1.9 The output of an SCF calculation

The output of an SCF calculation using one of the standard codes normally contains: (i) the binding energy of the state, referred to the completely stripped nucleus (at least two calculations are therefore required to obtain the *binding energy* referred to the ground state of the parent ion or the *transition energy* referred to the ground state of the neutral atom). What is usually computed in each one of these SCF calculations is the *average energy* of the configuration, i.e. excluding any multiplet structure (although this can also be included in the variational procedure – see below); (ii) the radial orbitals $u_{n\ell}(r)$ for all the occupied subshells in the system; (iii) values of the following integrals, which are computed from the $u_{n\ell}(r)$:

$$F^k_{n\ell,n'\ell'} = \int_0^\infty u^2_{n\ell}(r) \left(\frac{1}{r}\right) Y^k_{n'\ell',n'\ell'}(r)dr$$

(called the *direct* integrals) and

$$G^k_{n\ell,n'\ell'} = \int_0^\infty u_{n\ell}(r)u_{n'\ell'}(r) \left(\frac{1}{r}\right) Y^k_{n\ell,n'\ell'}(r)dr$$

(called the *exchange* integrals). These together determine the electrostatic multiplet intervals and

$$\eta_{n\ell} = \frac{1}{2m^2c^2} \int_0^\infty u^2_{n\ell}(r)\frac{1}{r^3}\frac{dV}{dr}dr \qquad (1.10)$$

where V is the central field potential, which determines the spin–orbit splittings. These integrals are known generically as the *Slater–Condon integrals* and, when substituted into the multiplet structure formulae for a given configuration (see, e.g., [17]), yield individual energy levels.

With some codes (e.g. [16]), it is also possible to include electrostatic integrals in the variational calculation. Since these are diagonal in LS coupling, a calculation of this type is called an LS-dependent Hartree–Fock calculation, and may become necessary when the configuration average approximation breaks down, and the orbitals become term-dependent.

1.10   Uniqueness of the solution

In principle, it may seem that there is no guarantee of finding a unique solution to such a complex problem by the SCF procedure. Fortunately, the Hartree–Fock method is based on a variational principle, which is reliable for ground states.[2]   Thus, the lowest energy state is uniquely defined. For excited states, the situation is less satisfactory: there may indeed be several solutions, a situation discussed by Koopmans [11], who reached the conclusion that, in such cases, one should choose the solution for which the diagonal $\varepsilon_{n\ell,n\ell}$ yields the closest ionisation energies for the electrons $n\ell$ to the observed values, and for which the off-diagonal energy parameters are zero. This is sometimes referred to as a 'theorem' but is more in the nature of a practical prescription.

There are cases for which more than one solution is found, and it is possible that both may possess physical reality under certain conditions [12] (this will arise again in chapter 11). Furthermore, the Hartree–Fock method can be made multiconfigurational, i.e. several configurations can be mixed or superposed. An electron is then 'shared' between different states, which goes beyond the independent particle approximation. The self-consistent method allows the mixing coefficients to be determined, but the configurations to be included must be specified at the outset, and there is no simple prescription as to which ones should be chosen or left out.

1.11   Relativistic central fields

For heavy atoms, it is often necessary to go beyond the Hartree–Fock approach, although the decision whether relativistic corrections are important depends on the problem in hand. For example, if it is desired to study atomic effects down a column of the periodic table, consistency may require that all the calculations are performed with the same code,

---

[2] The special significance of the ground state emerges as follows: let ψ be any wavefunction (not necessarily normalised) in the space of the exact solutions u_n of eigenenergies ϵ_n. Expand $\psi = \sum u_n c_n$. Then the expectation value

$$E = \frac{<\psi \mid \mathcal{H} \mid \psi>}{<\psi \mid \psi>} = \frac{\sum \mid c_n \mid^2 \epsilon_n}{\sum \mid c_n \mid^2} = \epsilon_0 + \frac{\sum \mid c_n \mid^2 (\epsilon_n - \epsilon_0)}{\sum \mid c_n \mid^2} \geq \epsilon_0$$

if and only if ϵ_0 is the ground state. This results is called the Rayleigh–Ritz variational principle.

i.e. that a relativistic approach is used from the outset. The relativistic Dirac–Fock method is available for this purpose. It is usually used in its multiconfigurational form, referred to as the MDF method for short. The relativistic Hamiltonian is defined as:

$$H = H_{DC} + \sum_{i>j}^{N} B(i,j) \qquad (1.11)$$

where H_{DC} is the N-particle relativistic Hamiltonian:

$$H_{DC} = \sum_{i} h_{DC}(i) + \sum_{i>j} \frac{1}{r_{ij}} \qquad (1.12)$$

The correction $B(i,j)$ to the Coulomb potential is treated as a perturbation of the zero-order Hamiltonian, and may include relaxation effects, correlations, quantum electrodynamic corrections and the relativistic retardation of the two-electron potential.

In the low energy limit, it has the form [13]:

$$B(i,j) = \frac{1}{2\,r_{ij}} \left[\alpha_i \alpha_j + \frac{(\alpha_i r_{ij})(\alpha_j r_{ij})}{r_{ij}^2} \right] \qquad (1.13)$$

where α_i are the Dirac matrices. This is the largest correction to the Dirac Hamiltonian. The most significant QED effects are the self-energy correction and the polarisation of the vacuum by the nucleus.

For open inner shells, the Breit correction is important and can be included perturbatively. QED effects can be allowed for by interpolation of tabulated data [14].

In relativistic calculations, L and S cease to be good quantum numbers, because the spin must be included within the single particle Hamiltonians $h_D(i)$, and so the results must be expressed in jj coupling. The constants of the motion become n, j and κ, where

$$\left. \begin{array}{ll} \kappa = l & \text{for } j = l - 1/2 \\ \kappa = -(l+1) & \text{for } j = l + 1/2 \end{array} \right\} \qquad (1.14)$$

Thus, each relativistic orbital $n\ell$ splits into two different relativistic orbitals for $\ell > 0$. In relativistic notation, the $\kappa = \ell$ functions are described as $\bar{\ell}$ and the $\kappa = \ell - 1$ are called ℓ functions. They are handled as separate orbitals in the SCF procedure, and the concept of configuration mixing in the MDF context must therefore be widened.

The eigenfunctions of the Dirac Hamiltonian have the form:

$$U_{njk}(r) = \left(\begin{array}{cc} P(r) & \Omega_{\kappa n}(\theta, \phi) \\ iQ(r) & \Omega_{-\kappa n}(\theta, \phi) \end{array} \right) \qquad (1.15)$$

where the radial functions resemble Hartree–Fock functions, except that there are now *two* radial functions $P(r)$ and $Q(r)$ which are, repectively, the large and small components. The atomic state function (ASF) representing an eigenstate of the many-electron atom is constructed from a linear combination of configuration state functions (CSFs). The CSFs are uniquely defined from the configuration specification, i.e. by the occupation numbers of the individual orbitals and by the coupling of the open shells to produce a given total J for the state.

Within the multiconfigurational scheme, one has the option of:(i) optimizing the total wavefunction for each individual state (optimised level approach) so that the generalised occupation numbers are obtained self-consistently; (ii) performing an average of configuration calculation for the CSFs and then, using coefficients determined from statistical occupation numbers, mixing the CSFs to determine the final wavefunctions, all of which have the same parity and total J (average level); and (iii) performing the average calculation as in (ii) but including appropriate CSFs for all relevant states of different parity and J-values (extended average level or EAL method [15]).

Approach (iii) is most useful for all calculations involving not only energies, but also the calculation of full transition arrays: it is the most convenient because orthogonality problems between initial and final state functions are thereby circumvented. In cases where ground state and excited state energies only are required, one can use approach (ii).

1.12   Computer codes

There exist several SCF codes for the solution of radial equations: the Hartree–Fock [16] equations are only one example, and the case described above is that of the single configuration approximation, in which each electron has well-defined values of n and ℓ. There exist several other possibilities: as stressed above, in Hartree's original method, the exchange term was left out; in the Hartree–Slater method [17], an approximate expression is used for the form of the exchange term. The Cowan code [20] is a *pseudorelativistic* SCF method, which avoids the complete four-component wavefunctions by simulating relativistic effects.

Full Dirac–Fock calculations are often performed using the GRASP code [18]. Exact criteria for the optimisation procedure depend somewhat on the accuracy required (for further details, see [15].

In the g-Hartree method [19] the Dirac equation is also used as the starting point, but the Lagrangian of quantum field theory is made stationary by altering the balance of direct and exchange terms in a very specific way. Like Hartree's original theory without exchange, this method is consistent with the fundamental principles of quantum field theory (the Hartree–Fock method is not), and allows the central field to be further improved by subsuming into it all the spherical part of the correlations, thereby producing the optimum central field.

This is by no means an exhaustive list of methods.

Configurations which arise by creating vacancies in the inner shells of atoms (chapter 7) are examples of highly excited states readily calculated by SCF methods. These techniques provide a very useful tool for an experimenter who is exploring a new spectrum and who wishes to know, ahead of making an observation, which energy range should be searched: the accuracy of the predictions is reasonably high when the transition is in the soft X-ray range. It is also possible to compute relative oscillator strengths which are often quite reliable, the Cowan code [20] being especially convenient for this purpose. However, one should bear in mind that the continuum is not included in any of these numerical methods, so that effects due to broadening of the transitions (chapter 6) are not taken into account. This can be a severe limitation for highly excited states.

1.13 Heisenberg's concept of holes

When an electron is removed from a closed shell or subshell deep inside the atom, we say that a *vacancy* is created. If the excited electron is removed far away, we call the vacancy a core *hole*. A hole looks very much like an electron, except that it has positive charge, and moves in the opposite direction. Insofar as it survives as an entity, it behaves like a particle and is referred to as a *quasiparticle*.

The quality of this approximation, which was introduced by Heisenberg to describe the deep core-hole excitations of X-ray spectroscopy, varies according to the element and also according to the excitation energy. This aspect will be discussed in chapter 7 (see in particular fig. 7.1 and the related discussion in section 7.2). The fact that holes are stable results in some extremely useful simplifications of atomic spectra: for example, if a vacancy is created in the f^{14} subshell, i.e. if we remove one electron to create the f^{13} hole, then this hole can behave like a single particle, i.e. a closed subshell minus one electron, despite the fact that it is made up from thirteen electrons.

The concept of a core hole is made more useful still by theorems relating to multiplet structure in the independent particle model. In general, we shall not be much concerned by multiplet structure in the present book: it is discussed in detail in many standard references. However, the theorems are interesting and relevant, and we therefore summarise them here:

(i) For a configuration of the type ℓ^{N-i}, where ℓ^N is a closed shell and i is the number of vacancies, the electrostatic multiplet structure is the same as for the configuration ℓ^i, but the spin–orbit splittings are inverted.

(ii) For two configurations $\ell^{N-1}\tilde{\ell}$ and $\tilde{\ell}^{\tilde{N}-1}\ell$, correlated to a configuration $\ell\tilde{\ell}$, the corresponding electrostatic multiplet structure is the same.

These theorems are further discussed in [21]. They show that, in the independent electron approximation, we can think, not only of independent electrons, but also of independent vacancies, which behave like electrons and even give rise to very nearly the same multiplet structure. This simplifies X-ray spectroscopy enormously, and further emphasises the significance of closed shells.

Of course, it is an approximation to regard quasiparticles as particles, and this approximation can be expected to break down in several ways. First, core holes always have a *finite lifetime*, i.e. they are broadened, and disperse on a short timescale. The effects of core-hole broadening will be discussed in chapters 8 and 11. Secondly, the very concept of a core hole may become inapplicable, i.e. it may prove impossible to identify a single structure in the spectrum as the result of exciting a quasiparticle. This form of breakdown is discussed in chapter 7. Experience shows that well-characterised holes tend to be the deepest ones, which are fully screened, while vacancies in the subvalence shells cannot always be described in this way. Thus, the concept of core holes is most useful in X-ray spectroscopy, but can sometimes break down quite severely at lower excitation energies.

An example where the hole has physical validity even for outer-shell electrons is the rare gas with one electron excited, leaving a nearly-closed shell as the core. This does not arise for He, but for the rare gases Ne, Ar, Kr, Xe, Rd, the resulting p^5 core has a ^2P structure, with a $J = 1/2$ and a $J = 3/2$ level. According to the usual ordering of levels within a multiplet, the $J = 1/2$ level should be the most tightly bound. However, this rule is violated for a hole, because the spin–orbit interaction is opposite in sign to that for a particle, and so the doublet is inverted. This is a special case of rule (i) above, which implies that the ordering of levels within a multiplet is regular for a less than half-filled shell and inverted for a more than half-filled shell. The energies of the nearly-closed shell states which are the ionisation thresholds of the rare gases are given in table 1.1. The spin–orbit interaction depends on the gradient of the central field potential (see equation (1.10)), and, since this is large inside the core, spin–orbit splittings tend to be much larger for holes than they are for excited particle states.

Table 1.1. Energies of the nearly-closed shell states of the rare gases.

Ar	Kr	Xe	Rn
15.76	14.00	12.13	10.75
15.93	14.66	13.43	

1.14   Periodic tables for ions

Just as there exists a periodic table for neutral atoms, one can in principle construct a table for any ionisation stage, by making use of binding energies rather than chemical properties. Such tables are similar, but not identical to those for neutral atoms. In particular, for reasons which will become clearer in chapter 5, as one increases the nuclear field with respect to the interactions between electrons, the filling of the long periods no longer occurs in the same way.

Another interesting exercise is to consider the systematics of double-ionisation thresholds, rather than merely the single-ionisation thresholds used to discuss the periodic table: the periodicities are then different (see fig. 1.3). For example, Li has the highest double-ionisation threshold in the periodic table, a fact which is important for the insertion of Li^+ ions in solids (see section 11.8) and applications in batteries. Apart from this exceptional case, which arises because Li^+ has an extremely high ionisation potential, the rare gases usually have the highest double-ionisation thresholds relative to neighbouring elements, while the lowest ones occur for Be, Mg and the alkaline earths (Ca, Sr and Ba), which have rare-gas configurations for the double ions, and for La, which lies at the onset of filling of the $5d$ subshell. Other neutral atoms with a closed s^2 subshell (Zn, Cd, Hg) lie near a local maximum in the double-ionisation potential, while atoms whose doubly charged ions possess an outer s^2 subshell (Si, Ge, Sn, Pb) lie at or near a local minimum in the double-ionisation threshold (an exception being C).

The systematics of the double-ionisation thresholds turn out to be very important in determining the properties of doubly-excited spectra. These are most prominent for elements lying close to local minima, which is why alkaline-earth elements play a special role (see chapter 7). Another important issue is the existence of crossing points between the curves for double ionisation and for ionisation from an inner shell. This is further discussed in section 7.14.

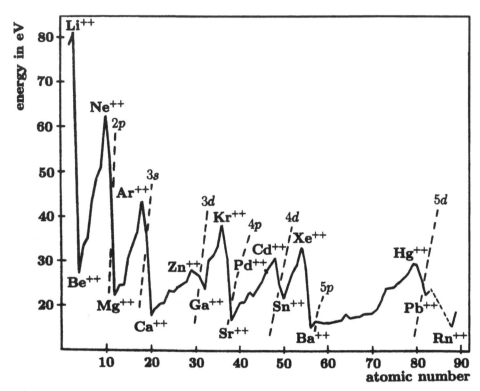

Fig. 1.3. Double-ionisation potentials as a function of atomic number. Some inner-shell thresholds are shown as dashed lines. Note the degeneracies or crossing points between inner-shell and double ionisation, which give rise to a variety of effects discussed in chapter 7.

1.15 Closed shells and negative ions

Under suitable conditions (more details will be given in section 2.24), it is possible to attach an extra electron to a neutral atom, and the resulting, very loosely bound object is called a *negative ion*. It is then an obvious step to search for stability of negative ions with closed shells, whose outermost configuration is that of a rare gas. One might expect that they would mirror the behaviour of the periodic table for neutral atoms, but this turns out not to be the case.

The H$^-$ ion possesses a single bound state and has an ionisation poten-

Table 1.2. Detachment thresholds for the negative ions

F⁻	Cℓ⁻	Br⁻	I⁻
3.40	3.61	3.36	3.06
3.45	3.72	3.82	4.00

tial of $0.75\,$eV, which turns out to be rather difficult to calculate *ab initio* (this is regarded as a severe test of atomic theory). According to the best models, it can be thought of as an electron loosely bound to a polarised H atom (we return to binding of an electron by a dipole potential in section 2.26). On the other hand, if we consider the sequence of halogens, then the negative ions have rare-gas configurations and one can draw up the table which corresponds to table 1.1. This yields the values listed in table 1.2. Note that these are *higher* than for H⁻. This arises because the halogens are more readily polarised than H, since their electrons occupy wider orbits, and are more susceptible to external perurbations. Thus, the resulting dipoles can be more binding. Note also that the thresholds increase in energy with increasing atomic number, which is the opposite of the behaviour for atoms, and again shows that larger systems are more readily polarised.

The fact that the stability rules of negative ions differ so much from those for neutral atoms is, again, a consequence of their radial properties. Binding by a polarisation potential is completely different from binding by a Coulomb well, even if the angular equations are identical.

1.16 Closed shells: summary

In summary, the existence of closed shells of electrons (magic numbers) has many observable consequences:

 (i) The existence and properties of the periodic table, many aspects of chemistry, the periodic behaviour of ionisation potentials, etc.

 (ii) A compact, spherical core in many-electron systems.

 (iii) The separation between the core, the valence, and the conduction bands in solids.

 (iv) The existence of X-ray and inner-shell spectra resulting from core-hole excitation.

 (v) The possible occurrence of cooperative effects involving the collective motion of all the electrons in a closed subshell.

Theoretical implications are:

(vi) The applicability of central fields for heavy atoms.

(vii) The possibility of introducing a fictitious radius r_0 to separate the spherical atomic core of filled shells from the outer reaches of the atom, where the valence electrons mainly are.

We note that the radial equation must be considered when discussing the stability of shells. It is worth adding that the radial Schrödinger equation contains a term $\ell(\ell + 1)\hbar^2/2mr^2$ (the centrifugal term) which has an important effect on the order of shell filling (see chapter 5). In chapter 12, we discuss examples of shell filling and magic numbers for systems made from atoms, in which the ordering is different from that of the periodic table, because the radial potential is not the same. This provides a tangible example of the importance of the radial equation.

List of elements with closed outer shells

He $(1s^2)$ Ne $(2p^6)$ Ar $(3p^6)$ Kr $(4p^6)$ Xe $(5p^6)$ Rn $(6p^6)$

List of elements with closed outer subshells

Be $(2s^2)$ Mg $(3s^2)$ Ca $(3p^64s^2)$ Zn $(3d^{10}4s^2)$ Pd $(4p^64d^{10})$ Sr $(4p^65s^2)$
Cd $(4d^{10}5s^2)$ Ba $(5p^66s^2)$ Yb $(4f^{14}6s^2)$ Hg $(5d^{10}6s^2)$ No $(5f^{14}7s^2)$

2

Rydberg states

2.1 Rydberg series in H and H-like systems

To an excellent approximation,[1] the energy levels E_n of H and hydrogenic
atoms or ions obey the formula:

$$E_n = E_\infty - \frac{R_M Z^2}{n^2} \qquad (2.1)$$

where n is the principal quantum number, E_∞ is the series limit or ionization potential, R_M is the Rydberg constant for the species of mass M
and Z is the nuclear charge (i.e. $Z = 1$ for a neutral species).

The energy formula (2.1) is obtained by solving the radial Schrödinger
equation, which can be written as:

$$-\frac{\hbar^2}{2m}\frac{d^2 u_{n\ell}}{dr^2} + \left(\frac{\ell(\ell+1)\hbar^2}{2mr^2} - \frac{Ze^2}{4\pi\varepsilon_0 r}\right)u_{n\ell} = E_n u_{n\ell} \qquad (2.2)$$

where the complete eigenfunction of the electron

$$\Psi_{n\ell}(r,\theta,\phi) = R_{n\ell}(r)Y_{\ell m}(\theta,\phi)$$

is separable in radial and angular coordinates[2] and $u_{n\ell}(r) \equiv r R_{n\ell}(r)$.

The energy levels E_n obtained by solving (2.2) are $(2\ell+1)$-fold degenerate in the angular momentum ℓ and n^2-fold degenerate in magnetic
sublevels, labelled by m, the projection of ℓ along an axis. The m degeneracy is called an *essential degeneracy*, and is simply due to the fact
that equation (2.2) is independent of m, whereas the ℓ degeneracy is an

---

[1] We neglect here the fine structure of H, hyperfine structure effects and QED corrections
which are not relevant to the argument of the present chapter. Readers interested in these
matters are referred to the excellent monograph on H by Series [22].

[2] Separability in radial and angular coordinates is exact for H, but not for all atomic systems.
For further discussion of the consequences when separability breaks down see chapter 10.

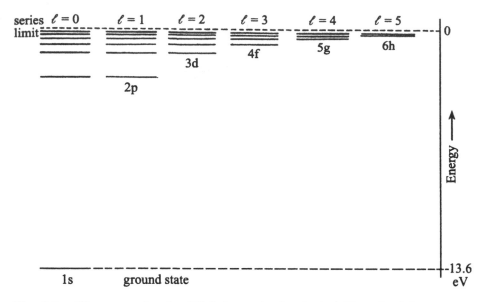

Fig. 2.1.   The energy levels of H (schematic, for clarity) Note the ℓ degeneracy, and absence of high ℓ states for low n.

accidental degeneracy: it is more surprising, because the centrifugal barrier term in (2.2) depends explicitly on ℓ. The m degeneracy survives *in any central field*, whereas the ℓ degeneracy is characteristic of the pure Coulomb field.

The distribution of energy levels and their ℓ degeneracies for H are illustrated schematically in fig. 2.1. We assume that the reader is familiar with the Schrödinger equation, and with its solution for the two-body Coulomb problem. One of the tasks of atomic spectroscopy is to retrieve diagrams such as fig. 2.1 from observed spectra and, having obtained them, to explain their properties. In other words, the experimental spectroscopist works backwards from the observations in contrast to the atomic theorist, who starts out from the Hamiltonian of the system. From this point of view, it is worth noting that the presence of just a few ℓ degeneracies alone does not suffice to characterise the Coulomb problem: for example, the $4s$, $4d$ and $4g$ levels of the three-dimensional harmonic oscillator are also energy degenerate. However, either the complete $(2\ell + 1)$-fold degeneracies of fig. 2.1 or the exact energy level spacings given by (2.1) are sufficient to characterise the spectrum. They are the *signature* of a purely Coulombic potential. In particular, we note that no $n < \ell + 1$ states exist in the pure Coulombic potential. The reason for this will be explained in section 5.2.

2.2 The series limit and the binding energy

The constant R_∞, for infinite atomic mass, is related to the fundamental constants m (the mass of the electron), e (the charge of the electron), h (Planck's constant) and c (the velocity of light) as follows:

$$R_\infty = \frac{me^4}{8\varepsilon_0^2 h^2} \tag{2.3}$$

where ε_0 is the permittivity of free space. Accurate measurements of R_∞ are therefore useful: it is known to better than 1 part in 10^{12}, and is presently the most accurately know constant in physics. Its value is $109737.31569 \pm 0.00006\,\mathrm{cm}^{-1}$.

For H-like systems, there is only one Rydberg series of energy levels converging on a unique limit E_∞. Since one cannot measure energy levels in experiments but only the transition energies between pairs of levels, the question of what value to use in equation (2.1) for E_∞ is a matter of convention: often, the value $E_\infty = 0$ is chosen, in which case E_1 is negative (binding energy) and a free electron has positive energy. As long as equation (2.1) applies, electrons of positive energy lie in a continuum which is completely segregated from the *infinite* manifold of the bound state spectrum and never mixes with it. This is a specific property of the long range Coulomb potential: if its binding strength is altered, bound states may rise or fall but may *never* cross over the associated ionisation threshold into the *adjoining* continuum. For short range potentials (as will be discussed in chapter 5) the same is not true: only a *finite* number of bound states exist, which may turn into resonances in the continuum if the binding strength is decreased.

Sometimes, one chooses $E_1 = 0$, whereupon for H in the approximation of infinite nuclear mass,

$$E_\infty = \frac{me^4}{8\varepsilon_0^2 h^2} = R_\infty \tag{2.4}$$

i.e. the Rydberg constant (which determines the energy interval between excited states) is equal to the binding energy (the energy interval between the ground state and the series limit), and all are determined by fundamental constants.[3]

The simple relationship between the series spacings $E_{n+1} - E_n$ and the binding energy, from which n can be recovered is *only* valid for two-particle systems. For a general many-electron atom exhibiting a Rydberg series, the value of n *cannot* be recovered in this way.

---

[3] This choice is useful for tables, e.g. those of Moore [23].

The Rydberg constant in equation (2.1) carries a subscript M for the mass of the positively charged particle. The reasons are that the nucleus is not infinitely massive, and that the formula also applies to any system containing two particles of opposite charge, i.e. it is valid for H, deuterium, exotic two particle atoms and H-like ions.

Strictly, the reduced mass:

$$\mu_M = \frac{mM}{m+M} \tag{2.5}$$

where m and M are the masses of the negatively and positively charged particles, should be used in place of m in equations (2.2) and (2.3), whereupon R_∞ is replaced by R_M. For exotic atoms, made up of charged particles other than just the proton and the electron, this can be a large correction but, for normal atoms, the nucleus is very much heavier than the electron, and so the mass shift is very small.

2.3   Units

There are several conventions for the energy scale, according to context. The quantity R_∞ (the Rydberg) sometimes serves in theoretical work as a unit of energy, but, more usually, the Hartree (equal to 2 × the Rydberg) is preferred and is termed the *atomic unit* of energy.[4] In atomic units, $\hbar = e = m = 4\pi\varepsilon_0 = 1$, and equation (2.2) assumes the particularly simple form

$$\frac{1}{2}\frac{d^2 u_{n\ell}}{dr^2} + \left(\frac{\ell(\ell+1)}{2r^2} - \frac{Z}{r}\right) = E_n u_{n\ell} \tag{2.6}$$

The unit of length is then a_0 (the Bohr radius of the ground state footnoteThe Å unit, equal to 10^{-10} m or 0.1 nm, finds favour with spectroscopists for the simple reason that it is roughly equal to the diameter of a H atom. of H = 5.29177×10^{-11} m), while the unit of time is set by the time taken for an electron in the first Bohr orbit to travel one Bohr radius, which turns out to be 2.41889×10^{-17} s. This last number is particularly interesting: even the fastest pulsed lasers (femtoseconds) are still slow on the fundamental atomic timescale. The fact that the atomic unit of time is so short is one reason why the static mean field approximation works so well for many-electron atoms. It gives some idea of the timescale on which the approximation might be expected to break down.

Electron spectroscopists use the eV (electron volt) as the unit of energy, in which case $R_\infty = 13.6058$ eV, while spectroscopists who measure the

---

[4] The velocity of light in atomic units is $\alpha^{-1} \sim 137$ where α is the fine structure constant.

wavelength of light express energy in reciprocal wavelength units (the wavenumber or Kayser) in which case

$$R_\infty = \frac{me^4}{8\epsilon_0^2 h^3 c} \qquad (2.7)$$

In these units, $R_\infty = 109737$ cm^{-1} and (for H) $R_H = 109678$ cm^{-1} which shows how small the mass correction is, even for the lightest atom.

Finally, true frequency units can be used directly, an option usually preferred by laser spectroscopists, if the value just given is multiplied by c, yielding $R_\infty = 3.2898274 \times 10^{15}$ Hz: the MHz and GHz are then popular measures for the bandwidth of high resolution lasers, while the kHz can be used to express the bandwidth of a truly excellent narrow band laser.

2.4   Rydberg states of many-electron systems

It is a remarkable fact that many-electron atoms, molecules and, indeed, *any* nearly spherical compact assembly of charged particles can, even in the presence of internal excitations, exhibit Rydberg series of energy levels (an example is shown in fig. 2.2, i.e. apparently infinite manifolds of levels E_n obeying the formula:

$$E_{n\ell} = E_\infty - \frac{R_M Z^2}{(n-\mu_\ell)^2} = E_\infty - \frac{R_M Z^2}{n^{\star 2}} \qquad (2.8)$$

where μ_ℓ is an approximate constant, known as the *quantum defect*, for a given series of angular momentum ℓ and $n^\star = n - \mu$.

This formula is very close to (2.1), apart from the introduction of μ_ℓ and $n^\star$. The similarity is in some ways deceptive because it conceals some essential differences, which are worth spelling out.

One important difference can be stated at once: the lowest level of the series may depend on the size of the compact assembly of charges. Clearly, if the radius of the charged core is greater than that of the first few Rydberg states of H, then some of the lowest energy series members will be missing. More generally, the nonhydrogenic core may have such a large effect on the lowest members that they may not seem to belong to the same Rydberg series.

2.5   Conservation of angular momentum and ℓ degeneracy

For a many-electron system, even if one is concerned with only the lowest ionisation threshold and even if one neglects the spin–orbit interaction, it is no longer true that a single Rydberg series occurs. As noted in section 2.1, it is a remarkable feature of the two-body problem that, although

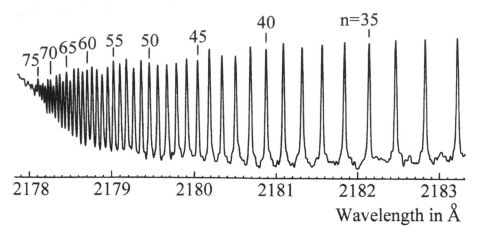

Fig. 2.2. Example of a long Rydberg series for a many-electron atom: the principal series of Sr (after M.A. Baig and J.-P. Connerade [24]).

the radial Schrödinger equation contains a centrifugal barrier term which depends explicitly on the angular momentum ℓ, the energy formula derived from it should be independent of ℓ. This accidental degeneracy is a property of the Coulomb potential, and is related to the fact that the classical two-body problem possesses closed orbits for an inverse square law of force:[5] the radial quantum number n_r (the number of nodes in the radial wavefunction) is related to both n and ℓ by:

$$n = n_r + \ell + 1 \tag{2.9}$$

Whereas, for non-Coulombic potentials, one can define n_r and ℓ, n is then no longer related simply to the binding energy. Indeed, for a complex, many-electron atom, it is not at all obvious how one should set about quantising the system, since there is no guarantee that the orbits of individual electrons will close.[6] In fact, conservation of the angular momentum ℓ for individual electrons is, at best, only an approximation. It would hold exactly for central fields. Even then, the same simple, precise relationship between n and ℓ as for H is not to be expected for many-electron atoms. As we shall see, the very meaning of n (the 'principal' or most important quantum number) becomes less clear-cut for many-electron systems. In a nutshell: the n and ℓ quantum numbers of

---

[5] More exactly, the introduction of relativistic dynamics, by spoiling strict orbit closure of the two-body problem, introduces a precession of the classical Runge–Lenz vector which, in the quantum case, lifts strict ℓ degeneracy: this is the origin of the Sommerfeld formula for H.

[6] We return to this question in chapter 10.

individual electrons both cease to be 'exact.' We actually have to define n from the number of nodes in the wavefunction, i.e. by reference to a calculation. Various forms of breakdown of the simple Rydberg formula also occur, to which we will often return in the course of this book.

If an approximate radial equation is constructed for a many-electron system, it must involve a non-Coulombic effective potential, because of the effects of electron–electron repulsion. Since the radial Schrödinger equation depends explicitly on ℓ, it is no surprise that the binding energy then depends on *both* n and ℓ, and so both must be specified for each electron when the configuration of the atom is written down.

2.6 Further limits: optical and inner-shell excitations

Even if only the outermost or *optical* electron is excited, several Rydberg series of energy levels, distinguished by their ℓ value, can occur. Indeed, once the spin-orbit interaction is introduced, not only for the optical electron, but also (and, as it turns out, more importantly) for the core electrons, a large number of possible series and, indeed, more than one possible series limit can occur even if only the optical electron is excited. An example is shown in fig. 2.3.

Of course, there is no reason why only the optical electron should be excited: the possibility exists to excite core electrons or *inner shells* by using radiation or collisions of greater energy. Since these electrons are more tightly bound, their removal to Rydberg states yields series converging on limits of higher energies.

Thus, not only does a many-electron system exhibit many series of energy levels, but it also possesses many distinct series limits or ionisation potentials. These higher thresholds may correspond to the removal of one electron, but leave the singly charged parent ion in some excited state. Indeed, it is even possible to remove more than one electron by direct excitation, but this further complication will be excluded for the time being (see chapter 7).

The first ionisation limit of a many-electron atom corresponds to the ground state of the corresponding or *parent* ion. Higher thresholds correspond to excited states of the parent ion. Apart from the special case of He, which has a hydrogenic parent ion, they are not simply related to fundamental constants. The many-electron Schrödinger equation must also be solved for the parent ion in order to determine the energies of the thresholds.

We have therefore generalised the Rydberg formula to many-electron atoms at one important cost: neither the intervals between series members nor the energy of the series limit bear any simple relation to fundamental

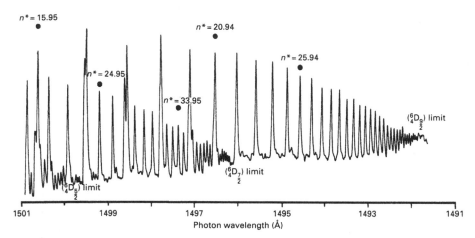

Fig. 2.3. Example of a many-electron system in which Rydberg series to several different limits occur. The series arise by excitation of a $3d$ electron in the Cr atom. The sextet term of the parent ion gives rise to five accessible limits, three of which are apparent in the present figure (after M.A. Baig *et al.* [25]).

constants.

Thus the beautiful simplicity of the hydrogenic Rydberg series, in which the spacings and the series limit are also related to one another, disappears, and n cannot be fully determined experimentally. An excitation is no longer a particle state but a *quasiparticle* state, whose parameters are modified from values expected for free particles, and can only be determined by solving a many-body equation.

In the present instance, experiment shows: (i) that high Rydberg states of complex atoms possess level spacings which are not too different from those of H, i.e. that the value of R_M for a complex atom is close to the value for H; but (ii) that the energies of the series limits are in general unrelated to R_M, so that *the n value cannot be uniquely determined merely from observations of a Rydberg series.* Indeed, it becomes necessary to separate the n value of the electron (deduced by counting the number of electrons in the system and applying the rules of quantum mechanics, i.e. by establishing how many nodes are present in the wavefunction, etc) from the *effective quantum number* $n^* = n - \mu$ which is the only quantity determined unambiguously by experiment. Finding the correct n value for an excited electron now involves an interplay between theory and experiment. In principle, indeed, the procedure could fail, since the theory for many body systems is only approximate. However, as we shall see, the ambiguities in n labelling, though real, can be handled, i.e. n remains a reasonably good label in a many-electron atom. The same is not always true for ℓ, and the ℓ characterisation more readily breaks down, as indeed one might anticipate from the fact that a $2\ell+1$ degeneracy occurs in H.

2.7 Scaling with n^*

Many properties in a Rydberg series scale in different ways. For example, the level spacing scales as $n^{*-1/3}$, which turns out to be an essential property when we come to discuss **K**-matrix theory in chapter 8. The same is true for core penetration, and all the properties which depend on the overlap between the core and excited electron wavefunctions (see chapters 4 and 6). The size of Rydberg states (discussed in section 2.14) scales as n^{*2}, while transitions between adjacent levels in the Rydberg manifold, which depend on the overlap between adjacent excited states, scale as n^{*4}. Yet more scaling rules for Rydberg series in external fields will emerge in chapter 10.

2.8 Relativistic dilation of level spacings

Sommerfeld has given a relativistic generalisation of the Rydberg formula. This has the form:

$$E_\infty - E_n = mc^2 \left\{ \frac{1}{\left(1 + [\alpha Z/n^*]^2\right)^{\frac{1}{2}}} - 1 \right\} \tag{2.10}$$

where the energies are in atomic units. This can be simplified by the substitution $\xi = \alpha Z/n^*$, leading to

$$\frac{E_\infty - E_n}{mc^2} = \frac{(1+\xi^2)^{\frac{1}{2}} - (1+\xi^2)}{(1+\xi^2)} \tag{2.11}$$

Note that ξ is always a small number. As $n^* \to \infty$, we have

$$\frac{E_\infty - E_n}{mc^2} \to \frac{1 + \frac{1}{2}\xi^2 + \ldots - 1 - \xi^2}{1+\xi^2} \to -\frac{1}{2}\xi^2 \tag{2.12}$$

which, bearing in mind that the energies are defined in atomic units, for which $m = 1$ and $c = 1/\alpha$, yields the nonrelativistic Rydberg formula.

Thus, we see that there is no relativistic dilation in the limit $n \longrightarrow \infty$. Any variation in Rydberg spacings due to relativistic effects in heavy

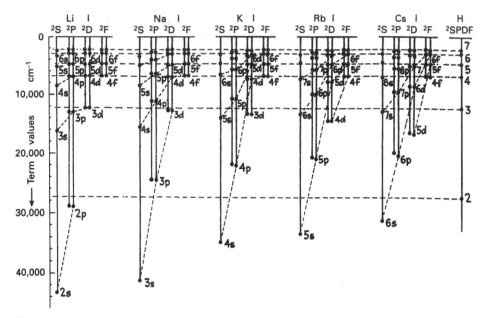

Fig. 2.4.   Energy level diagram for the alkali atoms, compared with the energy levels of the excited states of H. The hydrogenic energies are indicated by horizontal dashed lines across the figure. The alkalis are arranged from left to right in order of increasing atomic number. All the levels are doublets, but most of them are so close that they do not show on the present energy scale (after H.E. White [26]).

atoms is a very small effect indeed, and confined to low values of n, where core interactions dominate and (as pointed out above) Rydberg regularities tend to disappear anyway.

2.9   The alkali model

One of the most instructive energy level diagrams in the whole of atomic physics is plotted in fig. 2.4, which shows the comparison between the energy levels of the ground and first excited states of the alkali atoms and the excited states of H.

The salient points are as follows:

(i) The energy levels for different ℓ values occur at different energies, because the radial potential for alkali atoms is not truly Coulombic.

(ii) Moving across the diagram from left to right, i.e. from the light to the heavy alkalis, the ns ground states increase in n value by one each time, but remain below $n = 2$ of H. This is direct evidence for the existence of a compact electronic core, which screens the nuclear charge: the outermost electron in Cs I is so effectively screened from the nuclear charge that it behaves like a $2s$ excited electron in H. Indeed, as remarked above, the principal quantum number n must be defined either by counting the electrons in the core and filling shells according to the *aufbau* principle (see chapter 5) or by counting the nodes in the wavefunction: clearly, the binding energy in fig. 2.4 can no longer be used to determine n.

(iii) Moving across the diagram as before, the s states become more 'hydrogenic' (in the limited sense that they approach the horizontal dashed lines) towards the heavier elements (bear in mind, however, that the n values are *not* those of H).

(iv) Moving across as before, the $4f$ and higher nf lines are closely hydrogenic in the real sense for all the elements i.e. *they actually lie on the $n = 4$ line for H*. The f states become progressively more out of step with the n-numbering of the s states as one crosses over to the right of fig. 2.4.

(v) Moving across as before, corresponding positions are occupied by $3d$, $3d$, $3d$, $4d$ and $5d$ in Li, Na, K, Rb and Cs, and that these positions, as atomic number increases, actually *depart* from the $n = 3$ hydrogenic line. Indeed, the $3d$ orbital is occupied in Rb and lies deep inside the core, with a binding energy which lies off the scale of the diagram, so that the curve joining all $3d$ levels would suddenly plunge downwards between K and Rb. The situation is similar for $4d$ and $5d$ and, of course, another element with a single outer electron occurs beyond Cs: La has the outer configuration $6s^2 5d$, i.e. $5d$ is actually its ground state, with a binding energy somewhat greater than that of Li. The anomalous behaviour of the $\ell = 2$ and 3 levels when compared with the others is very apparent in table 2.1, which has been arranged according the n values: note how the starting entries differ from one alakali element to the next up the sequence.

(vi) If one tabulates the quantum defects μ obtained from the levels in fig. 2.4, then one finds that μ is fairly constant for a given ℓ up a Rydberg series, as demonstrated by the values in table 2.1.

All of these facts need to be explained, but the alkali model does not attempt to account for all of them. In particular, (v) requires principles beyond the simple alkali model which will be dealt with in section 5.5. Also, the alkali model does not seek to provide absolute magnitudes of μ_ℓ. Rather, it should be regarded as a conceptual framework within which

Table 2.1.   Effective quantum numbers $n^*_{n\ell}$ for alkali atoms.

Lithium

ℓ	$n=2$	$n=3$	$n=4$	$n=5$	$n=6$	$n=7$
0	1.589	2.596	3.598	4.599	5.599	6.579
1	1.960	2.956	3.954	4.954	5.955	6.954
2		2.999	3.999	5.000	6.001	7.000
3			4.000	5.004		

Sodium

ℓ	$n=3$	$n=4$	$n=5$	$n=6$	$n=7$	$n=8$
0	1.627	2.643	3.648	4.651	5.652	6.649
1	2.117	3.133	4.138	5.141	6.142	7.143
2	2.990	3.989	4.987	5.989	6.991	7.987
3		4.000	5.001	6.008	7.012	8.015

Potassium

ℓ	$n=3$	$n=4$	$n=5$	$n=6$	$n=7$	$n=8$
0		1.770	2.801	3.810	4.814	5.815
1		2.234	3.264	4.274	5.279	6.282
2	2.854	3.797	4.769	5.754	6.746	7.741
3		3.993	4.992	5.992	6.991	7.990

Rubidium

ℓ	$n=4$	$n=5$	$n=6$	$n=7$	$n=8$	$n=9$
0		1.805	2.845	3.856	4.861	5.863
1		2.288	3.323	4.335	5.341	6.344
2	2.767	3.706	4.684	5.673	6.667	7.663
3	3.988	4.986	5.989	6.984	7.984	...

Caesium

ℓ	$n=4$	$n=5$	$n=6$	$n=7$	$n=8$	$n=9$
0			1.869	2.920	3.934	4.940
1				3.390	4.405	5.412
2		2.551	3.532	4.530	5.529	6.529
3	3.978	4.974	5.972	6.971	7.970	8.969

experimentally or numerically determined properties of alkali atoms may be understood.

The fundamental assumption of the alkali model (which is vindicated by its success in interpreting fig. 2.4) is the following: *the space in which the electrons move can be partitioned into two regions: (a) a (small) inner region called the electronic* core, *within which the wavefunction of an electron is subjected to a complicated potential due to the presence of all the other electrons and (b) a (large) outer region accessible only to the one* valence *or optical electron involved in Rydberg excitation.*

We here anticipate that one consequence of the alkali model is the introduction of a minimum radius[7] r_0 at which solutions of the Schrödinger equation for the inner and outer reaches of the alkali atom can be joined (this idea will be useful in chapter 3).

2.10   Variation of wavefunctions with n and ℓ

Given these assumptions, it is clear that the optical electron spends most of its time in the large outer region, where the effective charge is equal to the charge on the nucleus minus the charge of the core, i.e. one atomic unit of charge. Consequently, we may obtain some idea of the solution by solving the hydrogenic radial equation with an external boundary condition (at large r) identical to that of H and an internal boundary condition determined by the properties of the core at r_0. A nice consequence of this approach is that, provided the core remains compact, the problem reduces to an *effective one-electron problem* which can be solved as a generalisation of the H atom. Wavefunctions resulting from such calculations are depicted in figs. 2.5 and 2.6. Their properties underpin the success of the alkali model.

The next point to note is the existence of so-called *penetrating orbits*,[8] i.e. the occurrence of wavefunctions of the type shown in fig. 2.5. These are wavefunctions whose innermost loop penetrates the core, but which are otherwise quite hydrogenic, as expected in the outer reaches of the atom. For large enough n, such wavefunctions *recapitulate* as indeed do the radial wavefunctions of H: this means that, apart from a normalisation factor, the inner part changes very slowly with increasing n beyond the first few values, as can be seen in fig. 2.5, and that the radial positions of all except the outermost nodes are stable.

Recapitulation will be discussed in more detail in section 3.4. As n increases, the number of nodes also increases and, as $n \to \infty$, the inner nodes coincide with those of a continuum functions. In fact, the positions of the nodes determine the phase of the continuum function (which is oscillatory) at threshold. There is a simple relation between the phase shift above threshold and the quantum defect of the bound states, which will be explained in chapter 3. If the eigenfunctions recapitulate, i.e. the positions of the nodes are nearly constant, then it follows that the

---

[7] Note that r_0 denotes the radius of the core, and a_0, the Bohr radius of H.
[8] This terminology is derived from the Bohr model of the atom, which was used in early descriptions of the alkali model, most notably by White [26].

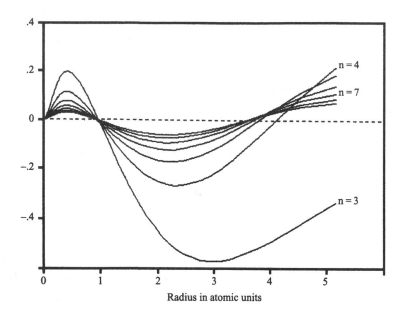

Fig. 2.5. The n dependence of atomic radial wavefunctions for Na. Note the *recapitulation* of atomic orbitals, a tendency of inner nodes to reappear at the same radius as n increases.

quantum defect is also nearly constant. This is the underlying principle of quantum defect theory.

   The behaviour as a function of the angular momentum ℓ is very different. Electrons of different ℓ experience a different centrifugal force through the $\ell(\ell+1)\hbar^2/2mr^2$ term in the Schrödinger equation. This force has a very large effect at small values of r. Consequently, for a given n, the degree of penetration (i.e. the amplitude of the $n\ell$ wavefunction within the core) varies greatly with ℓ. This is shown in fig. 2.6. It follows that μ_ℓ is not the same for different ℓ and must therefore carry ℓ as a subscript in the theory. A more thorough discussion of the properties of the quantum defect will be given in chapter 3, devoted to *quantum defect theory*, which is an important tool in the analysis of spectra.

2.11   Pseudo one-electron spectra

It is an observational fact, as stressed in the introduction, that many systems exhibit Rydberg series in some range of their spectrum and, indeed, that far more atoms possess Rydberg series than do not. We may understand this by noting that all atoms are of finite size, so that eventually, if an electron is promoted to sufficiently high n, it can find enough

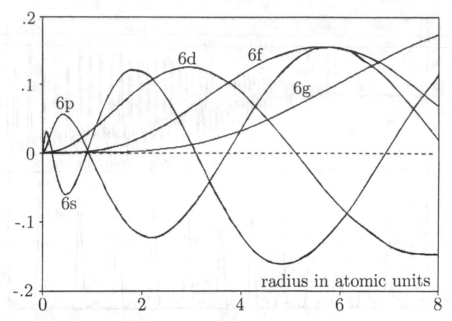

Fig. 2.6. The ℓ dependence of atomic radial wavefunctions for a given n in a typical alkali atom. Note the varying degree of penetration, and that the amplitude within the core decreases with increasing ℓ.

empty space around it to establish a quasihydrogenic wavefunction. As an example of this, we show, in fig. 2.7 very high Rydberg states in the principal series of Ba recorded by laser spectroscopy. Another instance was shown in fig. 2.2.

The interpretation just outlined is broadly correct in many cases, but must be qualified. The core is not always a passive *spectator*, and may develop a spectrum of its own, which intrudes upon Rydberg excitation. The question is then: how severe is the intrusion, and what will its effects be on Rydberg excitation? We will begin to answer this question by providing two examples of the manifestations of core effects in alkali-like spectra, one relating to the lower and one to the upper Rydberg spectrum.

It will not have escaped the reader that the alkalis are *not* the only elements with one external electron. If the alkali model had general validity, then it should be possible in principle to draw a similar diagram for the elements Cu, Ag and Au.[9] An attempt to do so is shown in fig. 2.8.

Note that fig. 2.8 reveals some new anomalies. The ordering of the elements does not, as one might expect, follow atomic number, and excited

---

[9] Indeed, one might even hope to bring all such elements onto a single diagram. We return to this issue in section 5.11.

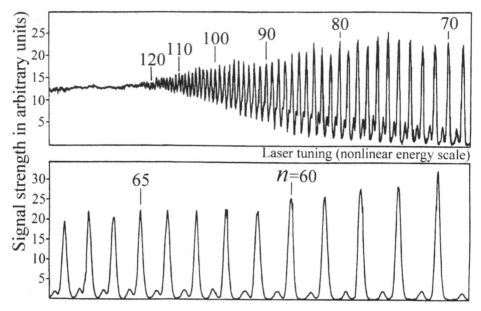

Fig. 2.7. Very high Rydberg states of the Ba atom recorded by laser spectroscopy using thermionic diode detection (after J.-P. Connerade *et al.* [27]).

configurations with one vacancy in the core suddenly appear in the midst of the optical spectrum. It is not obvious at first sight how the information in figs. 2.4 and 2.8 can be related. We might take the view that, when core excitation intrudes, the alkali model breaks down, at least for low-lying states. In fact, the situation is not so severe, as will be described in chapter 5 and it is possible to rationalise the behaviour shown in fig. 2.8.

As a provisional conclusion, let us merely note that the alkali model works best for optical electrons *when the Rydberg spectrum occurs far away in energy from any spectrum involving excitation out of the core.*

2.12   The extended alkali model

There is, however, one situation in which the alkali model can very readily be extended [28], namely when the atom contains a closed shell as the outermost configuration and the Rydberg electron itself is excited out of the core. We speak of subvalence or inner-shell excitation in this case, depending on how deeply the core hole is formed. In this case, since the atom in its ground state has *no* optical electron of its own it remains compact (see chapter 1). Once an inner electron is ejected from this compact core, it finds itself in the same situation as the valence electron of the alkalis, and begins to mimic their behaviour. Long Rydberg series

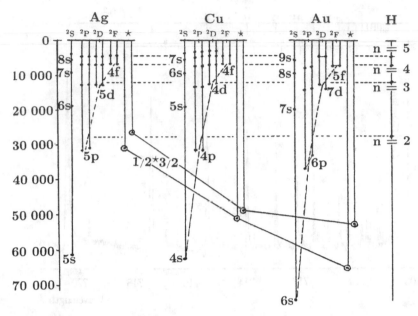

Fig. 2.8. Energy level diagrams for Cu, Ag and Au, all of which have one outer electron, drawn so as to emulate the trend for the alkalis. Note that the ordering of the atoms does not follow atomic numbers, and that core-excited configurations (plotted as circles and denoted with a star) appear amongst the low-lying excited Rydberg states (based on data from C.E. Moore [23]).

with quite constant quantum defects are then observed. An example is given in fig. 2.9 for the atom Zn.

However, there is one important difference in this new situation: the ionisation limits now correspond to *highly excited states of the core.* If these are fairly stable (as in fig. 2.9) then long series can be built on them. If, on the other hand, the *parent ion* states decay very rapidly, then Rydberg excitation can be quenched, especially for the highest Rydberg states which would have long natural lifetimes in the absence of core de-excitation.

To demonstrate how useful the alkali model is for d subshell excitation in Zn, Cd and Hg, consider the effective quantum numbers n^* listed in table 2.2 for np and nf orbitals. The behaviour of these numbers is closely similar to that for alkali spectra, with the most hydrogenic states (near integral n^*) being those of greatest ℓ, while the n values indicate that the outermost np electron, in its lowest available state, occurs at a 'binding energy' (referred to a core-excited threshold) intermediate between the values for $n = 1$ and $n = 2$ of H.

The 'lowest available state' in this case is obtained by counting which

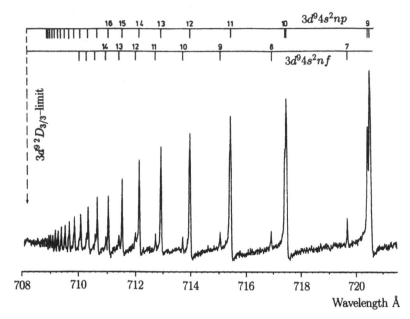

Fig. 2.9. Inner shell excitation series in the spectrum of Zn (after K. Sommer [29]).

Table 2.2.   Effective quantum numbers for Zn, Cd and Hg.

	Zn				Cd				Hg		
n	np	n	nf	n	np	n	nf	n	np	n	nf
4	1.51			5	1.58			6	1.60		
5	2.70			6	2.77			7	2.82		
6	3.74	4	3.98	7	3.80	4	3.96	8	3.84	5	3.96
7	4.75	5	4.97	8	4.81	5	4.96	9	4.86	6	4.96
8	5.75	6	5.97	9	5.81	6	5.96	10	5.86	7	5.96

orbitals are already occupied inside the core, which is so compact that the external Rydberg orbitals are only slightly displaced from those of H. Note also that the n values increase by one across the table from one element to the next in a given row, for all *except nf* orbitals. We postpone a discussion of why these differences in behaviour occur until chapter 5. Similarly, we postpone the question of how such highly-excited states, which lie in the continuum above the first ionisation potential, can remain stable, and how long Rydberg series can be built on excited states of the parent ion (see chapter 6).

2.13 The determination of ionisation potentials

The effective quantum numbers of table 2.2 were obtained from inner-shell excited series of Zn, Cd and Hg, which converge on excited states of the parent ion. Thus, the series limits bear no simple relation to any spacings in the Rydberg series. The Rydberg constant (neglecting finite mass effects, etc) depends only on fundamental constants. As already noted in section 2.2, there is an important difference between H and many-electron atoms which is obscured by the deceptively simple form of equation (2.8), as it appears to be similar for all atoms. The difference is that, for H, both the energy spacings between series members and the ionisation potential (i.e. the difference in energy between the series limit and the ground state) depend on R. For many-electron atoms, this is not the case: the spacings between Rydberg members are controlled by R and μ *but the series limit is independent of both R and μ*.

Thus, the observation of Rydberg series allows ionisation potentials to be determined. This is done by fitting the Rydberg formula to the observed series members, and hence extracting the series limit. Since the assumption that μ is a constant breaks down for the lower members, the simplest approach is to exclude them from the fit, and only to include series members for which μ has 'settled down' to a fairly constant value. This, however, is not the best approach, because even the lower members contain information on the nature of the variation of μ with energy, which needs to be included if the most accurate value of the series limit is to be obtained. Note that μ then ceases to be constant, and must be allowed to vary as a smooth function of energy.

Thus, for example, the principal Rydberg series of He, i.e. the manifold of the strongest allowed transitions from the ground state:

$$1s^2\,{}^1S_0 \longrightarrow 1snp\,{}^1P_1$$

has been observed up to $n = 35$ [30]. It has been found by Seaton [31] that the variation of μ with the energy E for this series of p states is well represented by the following procedure. First, one writes

$$\tan\left(\pi\mu(E)\right) = A(E)Y(E)$$

where, for the p states, $A(E) = 1 + E$. Then $Y(E)$ turns out to be a slowly varying function of E, and a good fit is given by

$$Y(E) = \frac{\alpha_0}{1 + \beta_1 E + \beta_2 E^2}$$

A least squares fit to the available experimental values determines $\alpha_0 = -0.03822716$, $\beta_1 = 0.402839$ and $\beta_2 = 1.27154$ and allows one to extract the series limit of helium as 504.2591 ± 0.0004 Å.

The fitting of Rydberg series to high resolution data, when performed with sufficient accuracy, can reveal very fine details of atomic spectra. Thus, by carefully fitting the principal series of $T\ell$, the influence of hyperfine structure on the determination of the ionisation potential has been revealed [32].

2.14   The size of Rydberg states

One of the most important aspects of Rydberg states is their rapid increase in size as the principal quantum number n is raised.

The mean radius of a Rydberg state is approximately $a_0 n^{*2}$, where a_0 is the Bohr radius. Although the atom in its ground state is a quantum object, it eventually ceases to be so as n^* increases: for $n^* \sim 100$, the diameter of the atom is $10^4 \times 2a_0$ which is roughly 1 μm, at which point the atom is definitely reaching macroscopic size. For $n^* \sim 1000$, the diameter of the atom is 1/10th of a millimetre.

Of course, such large states are rather fragile: as stressed above, Rydberg states can only develop if there is enough free space around the atom for the wavefunction of the electron to extend well outside the atomic core. This aspect seems to have been appreciated most clearly in the early days of quantum theory by Sommerfeld and Welker [33], who considered a H atom enclosed in a hollow sphere,[10] and showed that, if the radius of the sphere is less than $1.835\,a_0$, then the energy of the system becomes positive, i.e. the electron attempts to escape by exerting a pressure on the inner surface of the sphere.

This might seem to be an academic problem, with little chance of practical verification. In fact, individual atoms can be trapped in a rare-gas matrix, and the size of the lattice site in which they are confined is large enough that the first Rydberg states can be observed, albeit weakly, whilst higher members are quenched.

A related situation occurs in cluster physics: an aggregate of atoms may possess a hollow spherical shape (an example is buckmasterfullerene, made up of 60 C atoms in a symmetrical 'football pattern' arrangement). It is possible for a single atom to be trapped inside the spherical cage, and the situation is then rather similar to the one considered by Sommerfeld and Welker [33]. It is now possible to capture atoms inside hollow fullerene structures, and this has become a subject of study (see chapter 12.4). It does not seem that any Rydberg excitations of such atoms have yet been observed.

---

[10] It is perhaps interesting to note that this problem is exactly the opposite of the alkali model, presented above, in which the electron is free to move outside a radius r_0.

Atoms can also be trapped in a solid host of rare-gas atoms by freezing them onto a surface. The technique consists in mixing an effusive atomic beam, usually of metal atoms with a jet of rare-gas atoms which can then be deposited on a cold substrate and probed by using a beam of synchrotron radiation. This technique is known as *matrix isolation spectroscopy* because the individual atoms are isolated within the matrix of rare-gas atoms which constitute the host. There are many interesting interactions between the host and the metal atoms, which manifest themselves as shifts and splittings in the spectral lines of the metal atoms. In addition, it has been found that the lowest Rydberg states survive, which corresponds to the fact that the atom is trapped inside a cavity whose dimensions are only large enough to accommodate the first orbitals of a Rydberg series.

A specific example is that of Ag atoms trapped inside Ar, Kr and Xe matrices, and probed by synchrotron radiation, which is capable of ejecting an electron from the $4d^{10}$ subshell of the Ag atoms straight into Rydberg orbitals. The lowest orbital ($5p$) is seen in all cases, and the full multiplet structure has been analysed. In the case of an Ar matrix, in addition to $5p$, the strongest member of the $6p$ series is also seen, and can be identified by comparison with the spectrum of the free atom. This has been explained on the basis of the average radius of the orbitals, and is taken as evidence that higher Rydberg members are all quenched by a size effect [34].

Finally, it is possible to study assemblies in which a single metal ion is trapped inside a group of rare-gas atoms stuck together to form a cluster. The properties of such an ion are similar to those of the confined atom of Sommerfeld and Welker.

The confined atom can be regarded as a first step towards modelling solids, and the problem is of current interest now that numerical methods allow more complex atoms to be studied [35]. It is also a first step towards studying the compressibility of atoms, and their ability to partake in 'soft chemistry' (see section 11.8) [36].

2.15   A macroscopic 'sieve' for Rydberg atoms

An atomic site in a rare-gas matrix is, of course, of microscopic size. However, atoms in a beam can be excited by laser spectroscopy to extremely high n values: if such a beam is pointed at a metallic grating made up of micrometre-size slits, then a cut-off will be observed in the maximum n value of the atoms which can fly through the apertures unimpeded. Such an experiment has been reported by Fabre [37] and coworkers in Paris. While it is experimentally very challenging to realise, the principle of this

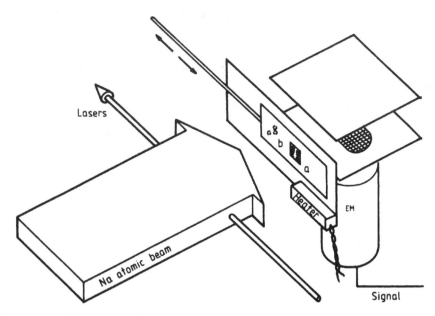

Fig. 2.10. Experimental set-up used by Fabre *et al.* for the observation of the transmission of Rydberg atoms through a mechanical grating. Rydberg atoms impinge on a gold foil (a) which contains the array of micrometre-size slits. A set (b) of large holes is used for calibration, and a movable shutter uncovers either (a) or (b). The transmitted atoms are detected by field ionisation, using a pulsed voltage $V(t)$ applied to a pair of parallel plates, which ionises the atoms and sweeps them into an electron multiplier (EM) (after C. Fabre *et al.* [37]).

measurement is very simple. It is probably the first direct determination of the size of a Rydberg atom using what one can regard as a 'classical' or macroscopic ruler.

The experiment was performed using the apparatus shown in fig. 2.10. The average size of the slits ($2\,\mu$m $\times$ $20\,\mu$m) was determined by recording their image on a scanning electron microscope. It was very important in the experiment (as in all experiments with very high Rydberg states) to minimise all external electric or magnetic stray fields, including, in this case those due to oxidised or dirty surfaces, contact potentials etc. After passing through the grating, the laser-excited Na atoms were detected by field ionisation, using a pair of parallel plates and a pulsed electric field which ionises all Rydberg states above $n = 22$. The transmission function of the foil for Rydberg atoms was thus determined as a function of n^2. This seems to be the first direct observation of the size of Rydberg atoms using apparatus of macroscopic dimensions.

Atomic size is important in the reversible intercalation of ions in solids (see section 11.8) which has many important applications. The problem of atomic size is also related to the question of endohedral capture of a small atom in a 'bucky ball', discussed in section 12.4: once the buckyball is formed, the atom must be able to penetrate the cluster through one of the hexagonal holes on the surface in order to become trapped inside.

2.16 The 'magic Rydberg states' of ZEKE spectroscopy

Another good example of the field ionisation technique is the so-called ZEKE method or zero kinetic energy photoelectron spectroscopy, pioneered by Müller-Dethlefs *et al.* [38] and now widely used, especially in laser spectroscopy of molecules. The basic principle is a very simple one. In conventional photoelectron spectroscopy, one detects photoelectrons which are energy analysed by an electron spectrometer, and the experimental resolution is limited by the apparatus to 5–10 meV. In the ZEKE method, instead of a spectrometer, one uses a pair of plates on either side of a molecular beam, which is excited by one or more pulsed lasers. The technique consists in detecting the ionisation current only a finite time (usually about 5 μs) *after* the laser pulse, by applying an electric field (the ZEKE pulse), which sweeps all the electrons which have remained in the interaction volume into the detector. Thus, all the electrons of moderate to high kinetic energy (the so-called 'hot' electrons) have had time to escape from the interaction region, and the only ones which are detected are those of very low or zero kinetic energy. By tuning the laser, one thus records a spectrum, whose resolution turns out to be limited only by the laser bandwidth. The ZEKE technique was first applied successfully to the molecule NO.

More recently, it has been established that the success of the method is related to the properties of some unusual Rydberg states (the 'magic' states for ZEKE) with $n \sim 150$, which are remarkably stable even for molecules, and sufficiently long-lived to survive the time delay between photoexcitation and the application of the ZEKE pulse. These states, because of their high n values, are extremely fragile, and are readily ionised by an external electric field. Electrons are thus 'parked' in these high Rydberg states until they are detected. By using a specially shaped ZEKE extraction pulse, which is ramped up slowly from zero, one can successively ionise the different Rydberg members, starting from the highest one and extending downward towards lower n. By selecting specific Rydberg states, the resolution has been further improved, and has allowed rotationally resolved spectroscopy of the benzene molecule to be performed.

The extraordinary stability of the 'magic' ZEKE states has been demonstrated in experiments in which a molecule is dissociated [39, 40]. For example, one can start by exciting HBr into the magic states, then pulsed-field ionise the molecule when dissociation has occurred and the H atom has been ejected: despite this dissociation, the signature of HBr is still found in the ZEKE spectrum. This may seem surprising, but can be understood in terms of Fermi's essentially free electron theory of collisions involving high Rydberg states (see section 2.17).

The precise nature of the 'magic' Rydberg states is still a matter of controversy. It seems likely that they are states of high angular momentum, since otherwise their lifetimes (which scale as n^5 rather than n^3) would not be high enough. If so, then photoexcitation alone would not allow them to be reached, and it is likely that collisions with atoms of the residual vacuum play some role in populating them.

2.17   Collisions: Fermi's model

By virtue of their enormous size and comparatively high stability against radiative decay, high Rydberg states are very susceptible to collisions. Indeed, some of the detectors which are used to observe them, such as the thermionic diode (see section 8.16) depend for their operation on the presence of collisions. Collisions between excited species affect ionisation and recombination rates in such diverse environments as gaseous nebulae [41], laboratory plasmas [42] and flames [43]; their study is therefore of some considerable intrinsic interest.

The fundamental model for collisions between a high Rydberg state and an extraneous atomic or molecular target is due to Fermi [44]. The basic hypothesis he makes is that the excited electron is so loosely bound to the ionic core that the target interacts *separately* with each of them. In fact, the Rydberg electron is treated essentially as though it were free, except that its momentum distribution is determined by its quantum state. As remarked by Fermi, in this approximation referred to as the *essentially free electron* (EFE) model, there is no real distinction between different atoms once Rydberg states of $n \sim 30$ or so have been excited. [11] There are a number of important situations in which this basic assumption of the EFE model actually breaks down (see section 4.7 for an example), and it is not always safe to neglect the dynamical properties of the core. Nonetheless, it is a very useful simplification and allows some general applications of the principle of detailed balance to be deduced for collisions.

---

[11] In fact, there is a slight distinction, readily allowed for within Fermi's EFE hypothesis, which arises from the fact that states of the same n but different ℓ are distributed differently in energy for different atoms (for reasons discussed more fully in chapter 3), and that, at high n, ℓ-mixing is induced by collisions. The subject of ℓ-mixing by collisions is discussed in section 4.7. Experimentally, another clear manifestation of the difference in the energy distribution of ℓ states occurs in the ℓ-mixing range of the diamagnetic problem, discussed in chapter 10.

Thus, Matsuzawa [45] has considered various kinds of collisions and predicts from the EFE model that, for those in which the Rydberg electron is transferred and attaches to the target, the rate constant for the transfer should equal the rate constant for attachment of free electrons of the same velocity.

2.18 Cross sections for attachment: Wigner threshold law

Experiments on attachment are performed at very low kinetic energy, since the resulting negative ions are fragile, which is why Rydberg electrons are very appropriate. The aim of the experiments is to measure the attachment cross section, or the attachment rate (which is essentially the product of the velocity and the cross section – see below) as a function of the incident energy.

For a two-particle capture process, a general rule emerging from the work of Bethe and Wigner in nuclear physics, usually known as *Wigner's threshold law* describes the variation of the cross section close to the threshold, i.e. at very low kinetic energies, and leads to a cusp in the cross section. This law states that

$$\sigma(E, \ell) \sim E^{\ell - \frac{1}{2}} \text{ for } E \to 0 \qquad (2.13)$$

In particular, for $\ell = 0$ or s-wave scattering, the cross section varies as $1/E^{1/2}$ or $1/v$, which is at first sight a surprising result, since it differs from the $\pi \lambda_D^2$ value. where λ_D is the de Broglie wavelength. This arises because of the attractive nature of the binding force: a polarisation potential yields a $1/E^{1/2}$ variation, while a Coulomb potential yields a $1/E$ variation, so that most situations lie in between.

The attachment rate at finite temperature is given by a thermal average:

$$< k_e(T_e) > = \int_0^\infty \sigma_e(v) v f(v, T_e) dv \qquad (2.14)$$

in the EFE model, the same approach is used to describe the attachment of Rydberg electrons $n\ell$ except that these electrons are now 'quasifree', whence

$$< k_{n\ell} > = \int_0^\infty \sigma_e(v) v f_{n\ell}(v) dv \qquad (2.15)$$

where $f_{n\ell}$ is now the distribution appropriate for the $n\ell$ electrons. In principle, the two should be the same as $E \to 0$. There are appreciable deviations, due to the so-called 'post-attachment interactions', when the binding energy is significant, and so the rule is only valid for $n > 50$ or so.

The important question of the energy range over which Wigner's threshold law is valid has been considered by Vogt and Wannier [46]. These authors derive a more specific law:

$$\sigma_e = 4\pi a_0^2 \left(\frac{\alpha}{2E}\right)^{\frac{1}{2}} \tag{2.16}$$

with a definite proportionality constant, and a range of validity limited to $(8\alpha E)^{1/4} < 0.5$.

2.19   Measurements of electron attachment

The EFE model (see previous section) describes the Rydberg electron as a nearly-independent particle moving in the field of a very distant, compact ionic core. During a collision, this core acts as a spectator, and binary electron–target interactions dominate. The time-averaged kinetic energy of a Rydberg electron is the same as its binding energy, which is in the meV range. This suggests that Rydberg atoms are suitable projectiles to probe low energy attachment processes, in which a neutral target with an affinity for electrons (there are many such targets, extending from atoms to complex molecules or clusters) collides with Rydberg atoms A⋆ (the ⋆ conventionally denotes excitation), and the loosely bound electron transfers from the projectile A to the target T as follows:

$$A^\star + T \rightarrow A^+ + T^-$$

The electron affinity is the binding energy with which a neutral target can capture an electron. It is not sufficient for a target to have a high electron affinity in order for it to capture electrons efficiently [601]. For example, SF_6 does not have a very high affinity, but it does possess a very high cross section for the capture of low energy electrons, and therefore forms negative ions very efficiently. Molecules of this type have practical applications. Quite often, they contain halogen atoms. Because they easily take up electrons (a process often referred to as scavenging), they act as flame retardants.

For this process to be regarded as a binary encounter involving only the Rydberg electron and the target, as suggested by the EFE model, it is necessary for the range of the interaction between target and electron to be smaller than the dimensions of A⋆ and that the interaction between A^+ and T^- after the collision (the *post-attachment* interaction) should also be very small. This is indeed the case [49] provided A⋆ is in a high n state (in practice, $n \geq 40$) because, on average the ions A^+ and T^- are formed at large separations, so that their trajectories are unaffected.

The study of collisions between Rydberg atoms and e^- attaching targets therefore serves as an excellent probe of interactions in an energy range below that accessible by other methods, and free electron attachment cross sections for the formation of negative ions in the 1–100 meV range have been measured by this technique (negative ions are discussed in section 2.24).

Attachment to a molecular target from an atom in a Rydberg state is described by:

$$\mathrm{A}(nl) + \mathrm{CD} \xrightarrow{\kappa_i} \mathrm{A}^+ + \mathrm{C} + \mathrm{D}^- \qquad (2.17)$$

where CD is a molecular target and κ_i is the ionisation rate constant for the Rydberg state A(nl) due to collisions. The observed lifetime T_{obs} of the Rydberg state is given by:

$$T_{obs} = \frac{1}{\frac{1}{\tau_{nl}} + N_T \kappa_i} \qquad (2.18)$$

where N_T is the number density of targets and τ_{nl} is the natural lifetime of the Rydberg state.

In this situation, the number $N(t)$ of high Rydberg states which survive at a time t after a short exciting laser pulse is:

$$N(t) = N_0 \exp -\frac{t}{T_{obs}} \qquad (2.19)$$

while the ion production rate:

$$R_i \equiv N(t) N_T \kappa_i = N_0 N_T \kappa_i \exp -\frac{t}{T_{obs}} \qquad (2.20)$$

where N_0 is the number of Rydberg atoms initially excited by the pulse. The ions may be collected by applying a small electric field, which is not sufficiently large to ionize the Rydberg atoms present in the interaction volume. By applying a sufficiently large electric field to ionize the remaining Rydberg atoms almost instantaneously a short time after the laser pulse, the information required to deduce the rate constant κ_i may be extracted experimentally, and the value can be compared with rate constants for the attachment of free thermal electrons to the target.

Such experiments have been reported by Stebbings [47] and collaborators. They show that, provided collision induced mixing is not significant, Matsuzawa's deduction[45] from the EFE is verified.

The electron attachment rate is basically the cross section times the velocity (see equation (2.14)). For attachment rates to a neutral target, since the cross section varies as $1/E^{1/2} = 1/v$, this means that the attachment rate tends asymptotically to a constant value. Thus, for SF$_6$, which is important as a flame retardant, the rate at around $n \sim 40$ is

$$k_n = (4 \pm 1)\, 10^{-7} \text{ cm}^3 \text{ s}^{-1}$$

which is not too far from the Wannier prediction of 5.2×10^{-7} cm^3 s^{-1}.

2.20   Collisions and high Rydberg states

Collisions can complicate attachment from high Rydberg states: in practice, three different processes arise. In the first, the Rydberg electron is directly removed from the excited atom and becomes attached to the target. In the second, collision-induced mixing between different excited levels of the atom occurs, so that the Rydberg state can no longer be regarded as pure. In the third, some kind of excimer state is formed[12] for which the selection rules (i.e. rules governing which states can be excited – see section 4.2) are rather different from those of the atom. The presence of collisions is an important consideration in many experiments involving high Rydberg states. For example, when photoabsorption is used as the means of observation, densities of about 10^{15} atoms per cm^3 are normally used, implying an interatomic spacing of roughly $1/10^5$ cm or 10^3 Å . This is 10 times smaller than the size of an $n = 100$ Rydberg state, despite which high Rydberg states are successfully observed in such experiments. It follows that, in real experiments, there are many perturbations due to collisions which must be considered, a subject to which we return in section 4.7: in fact, an electron need not describe the full semiclassical orbit for a Rydberg excitation to be observed. Even a portion of the trajectory suffices for the curvature of the orbit to be determined, and it is this curvature which determines the energy. One can thus understand how high Rydberg states survive even under what seem to be quite unfavourable circumstances.

Further remarks on relative intensities of Rydberg members affected by collisions will be found in section 4.7.

2.21   A gas of Rydberg atoms

With the development of far infrared and millimetre wave amplifiers and detectors, in which Rydberg atoms are used as the active medium, researchers have been led to study the possibility of making a dense gas of highly excited atoms. Unfortunately, there are fundamental constraints

---

[12] An excimer is defined as a molecule which exists only in a highly excited state and possesses no stable ground state (see section 2.27).

to the pressure which can be achieved, and these are set by the interactions between Rydberg atoms, which cause them to break up or ionise at a very fast rate above a certain density, so that the gas evolves towards a plasma as the density is raised. The ion–electron production is found to exhibit a marked threshold behaviour, with a sharp onset for a well-defined atom density, depending on the principal quantum number n. The existence of a threshold of this kind can be ascribed to an avalanche effect due to Rydberg atom–electron collisions. This problem has been studied by Vitrant *et al.* [50], who have shown that there is a limit to the achievable density of Rydberg atom samples at room temperature: the smallest average interatomic distance to Rydberg atom size ratio is of the order of 15. The evolution from a gas of Rydberg atoms, with each atom containing one loosely bound electron to a plasma containing singly charged ions and electrons (Rydberg atom–plasma transition) competes with the order–disorder or Mott transition (see chapter 11 for a discussion of Mott transitions): it turns out that the orders of magnitude of the two effects are somewhat different and that the Rydberg atom–plasma transition is actually an obstacle to the observation of the more fundamental Mott transition for a gas of excited atoms. Various suggestions have been made to overcome plasma formation: in practice, very low temperatures (a fraction of a kelvin) would be required, or the experiments would have to be done very quickly, so that the excited atoms do not have time to move, which would require picosecond excitation and detection times.

2.22   Excitation of Rydberg states by very short pulses

Because of the energy broadening associated with very short timescales (Fourier's theorem), there is a similarity between the effect of a collision and that of a pulsed laser field: one loses the wavelength selectivity normally associated with photoexcitation when the pulses are short, but the extent to which normal selection rules are violated depends on the intensity and duration of the radiation pulse. Very high power pulses will be discussed in chapter 9. Here, we address the question of the influence pulsed excitation has on the observation of Rydberg states when the pulses are just short enough for wavepackets to be formed from several Rydberg orbitals. The finite duration of a picosecond laser pulse corresponds to a frequency interval of about 20 cm^{-1}. Thus, if atoms are excited to high Rydberg states (principal quantum number $n > 30$) by such a pulse, several levels of the Rydberg series can be simultaneously and coherently excited.

We can calculate their number simply by differentiating equation (2.8),

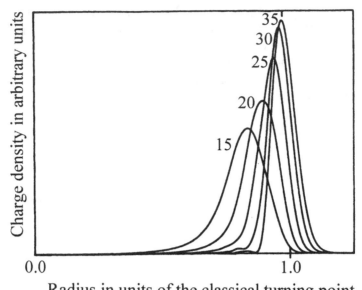

Fig. 2.11.   Rydberg wavepackets after excitation by a 10 ps laser pulse tuned
to a frequency around principal quantum number 85 in the Rydberg spectrum.
The times chosen correspond to the first half of the orbit. They are indicated in
ps above each wavepacket (after J.A. Yeazell and C.R. Stroud [52]).

which yields

$$\Delta n = \frac{n^{\star 3}}{R_M Z^2} \Delta E \tag{2.21}$$

where ΔE is the laser bandwidth in energy units. Since $R \sim 109737\,\text{cm}^{-1}$,
we find that $\Delta n \sim 5$ for a neutral atom. With the proper choice of pulse
length and laser frequency, it has been shown that a wavepacket is created
which oscillates and radiates with many of the characteristics of a classical
electron in a Kepler orbit [51].

   The wavepacket can be written as follows:

$$\Psi_R(r,t) = \sum_n a_n(t) u_n(r) \exp -i\omega_n t \tag{2.22}$$

where $u_n(r)$ are the hydrogenic radial wavefunctions, $a_n(t)$ are time-
dependent amplitudes of the excited Rydberg states n, which also depend
on the envelope of the laser pulse and the frequency of the laser (see section 9.9) and the sum is taken over all the Rydberg states. If the time t is
fixed, then the spatial distribution of the charge density $r^2 \mid \Psi(r,t) \mid^2$ can
be plotted after excitation by the laser pulse. This is shown in figs. 2.11
and 2.12 for the first and second halves of the orbit [51].

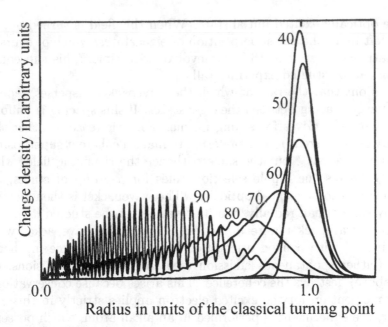

Fig. 2.12. Rydberg wavepackets after excitation by a 10 ps laser pulse tuned to a frequency around principal quantum number 85 in the Rydberg spectrum. The times chosen correspond to the second half of the orbit. They are indicated in ps above each wavepacket (after J.A. Yeazell and C.R. Stroud [52]).

After formation, the wavepacket travels outwards towards the classical turning point, where it becomes very narrow. Close to the turning point, such a wavepacket has an uncertainty product $\Delta r \Delta p$ very close to the minimum allowed by Heisenberg's uncertainty principle ($\hbar/2$). It then reverses direction and accelerates towards the core (or the nucleus, in the case of H), where it is dispersed. For a pure Coulomb potential, the classical period in a Kepler orbit with $n \sim 85$ is about 93 ps. After a few periods, the wavepacket appears to disperse completely, although it can revive at later times.

In the above example, the spatial localisation produced was in the radial coordinate. It is also possible, by using Rydberg states, to produce localisation in the angular coordinates, although the method of preparation of the wavepacket is then different.

The observation of angularly localised wavepackets made up from a coherent superposition of Rydberg states is achieved by short pulse optical excitation in the presence of a strong rf field. Such excitations are possible in practice because a very short optical pulse has a large coherent bandwidth, which may extend over several Rydberg transitions. The role of the rf field is to mix high angular momentum eigenstates together with the optically excited initial state. When this field is turned off, the atom is left in a coherent superposition of eigenstates, which of course is not a stationary state and therefore evolves with time. This subsequent evolution can be studied experimentally.

It turns out that the rate at which the wavepacket disperses depends on the energy spacing between the eigenstates. If this spacing is uniform, the wavepacket survives for a longer time. For this reason, rather than excite states of different n, it is preferred to make a coherent superposition of states of different ℓ but the same n (hence the rf mixing field, which effectively breaks the dipole selection rules for a group of states, and allows them all to be excited optically). The wavepacket is then localised angularly rather than radially, and its rotation can be studied [52].

If such a wavepacket were formed in H, then the wavepacket would remain intact, with a fixed orientation in space, until some incoherent process (either spontaneous emission (see chapter 4) or collisions, discussed above) destroys the coherence. This arises because conservation of angular momentum for the excited electron applies strictly in this case. However, the experiment is performed in an alkali atom, which possesses a core, and there is a back reaction of the excited electron on this core (core polarisation), which depends on the degree of penetration of the excited electron into the core, i.e. on the quantum defect, which itself is a function of the angular momentum. Thus, the wavepacket precesses under the influence of a small potential due to the quantum defect of the alkali. It is found to follow a classical trajectory determined by the core polarisation potential.

While the wavepackets just described possess many of the features of classical Kepler orbits, they are not strictly speaking quantum analogues of the classical orbits. The nature of the true quantum Kepler state is an interesting problem, first raised by Lorentz [53] and Schrödinger [54]. They conjectured that it should be possible to define genuine quantum states which would mimick the classical behaviour of an electron on a Bohr–Sommerfeld–Kepler ellipse, with minimum quantum fluctuations. This issue was extensively studied by Gay [55], who pointed out that such semiclassical states are known for the harmonic oscillator, but not for atoms. He described how to solve a related, more restricted problem: how to build states localised with minimum fluctuations on a Kepler ellipse. This new class of elliptic states is a coherent superposition of spherical states which can be realised in practice, and this step represents an important advance in resolving the Lorentz–Schrödinger conjecture. By using such states, it may prove possible to define Coulomb wavepackets of minimum spread which mimic the classical behaviour of an electron on a Kepler orbit: this would finally provide the analogue of coherent states for the Coulomb field of the atom, and would address the unresolved question of how classical a Rydberg electron can become, which is central to the study of the correspondence principle in the limit where the corresponding classical problem becomes chaotic (see chapter 10).

2.23   The ionisation of Rydberg atoms by very short pulses

Very short and intense pulses of radiation can be applied, not only to excite Rydberg states as described in the previous section, but also to ionise a Rydberg atom which is already in a Rydberg state. Ionisation can of course be used as a simple method of detecting Rydberg atoms (an example of ionisation used for detection is the experiment of section 2.15). However, it can also arise that one wishes to study the influence of a strong field on a Rydberg atom, in which case ionisation becomes an impediment, and one may seek to minimise it. One way to achieve this is to make use of *ultrashort pulses*, i.e. such that the duration of the pulse is shorter than the classical Kepler orbit period of the atomic electrons. Under such conditions, the ionisation probability can be drastically reduced, a phenomenon sometimes referred to as stabilisation. The suppression of ionisation by strong ultrashort laser pulses has been widely discussed: it has been shown that, if only the electric field is considered, there is an upper bound on the ionisation probability [56], and that, under quite general conditions, this bound tends to zero as the laser pulse duration itself tends to zero. Thus, there is a clear advantage in making very short laser pulses in order to study high electric field effects. We return to this theme in chapter 9.

2.24   Negative ions

Negative ions are formed by the attachment of a supplementary electron to a neutral species. First, it may seem strange that this is possible, and the mechanism by which it can happen must be explained. Second, we may ask whether it is always possible, and how many electrons $N(\mathcal{Z})$ can in fact be bound to a nucleus of charge $\mathcal{Z}$.

Taking the second point first, there are some rigorous results in this area. For example, it has been proved that, in nonrelativistic quantum theory, $N(\mathcal{Z})/Z \rightarrow 1$ as $\mathcal{Z} \rightarrow \infty$, a result which is only true for fermions.[13] The Pauli principle plays a crucial role in its derivation [57]. Upper bounds on $N(\mathcal{Z})$ have also been obtained. For an atomic nucleus, $N(\mathcal{Z}) \leq \mathcal{Z}$. It follows that H^{2-} is not stable. For a molecule,

---

[13] For bosons, it turns out that $N(\mathcal{Z})/Z > 1.2$ for large $\mathcal{Z}$.

$N(\mathcal{Z}) \leq 2\mathcal{Z} + K - 1$, where K is the number of atoms in the molecule, and $\mathcal{Z}$ is now the total nuclear charge [58].

Concerning mechanisms, the force which holds a negative ion together is usually the field due to a dipole, but the origin of this dipole can vary, and different examples will be given. In section 2.19, we saw how negative ions can be formed by collisions between electron attaching targets and atoms in high Rydberg states. More generally, they can be formed either by a three-body process or by the dissociation of a parent molecule.[14]

Negative ions of atoms are exceptional in that they do contain confined electrons but do *not* exhibit Rydberg series, the reason being that, on detachment of the electron, there is no long range Coulombic force. A negative ion can be formed from any neutral species (atom, molecule or cluster) which binds an extra electron to itself through a weak, non-Coulombic potential. This interaction is of short range. The electron is bound to the system by a polarisation potential which falls off very fast with radius. For example, in H, the potential binding a further electron to the system falls off as the inverse fourth power of the radius. It is a common property of short range potentials (see chapter 5) that they can hold only a *finite* number of bound states, this number being a direct measure of their binding strength. Similarly, negative ions do not give rise to Rydberg series under single electron excitation. Mostly, we expect a small and finite number of bound states (although in principle the number can be infinite under conditions described in section 2.26). In fact, excited bound states of atomic negative ions are infrequent. They are more likely for molecules, because the dipole holding the system together is larger (see section 2.26)

Negative ions were first studied by astrophysicists [59] who recognized that absorption by H$^-$ is a major source of opacity in the infrared spectrum of the sun. Not all the elements may be able to form negative ions, although their nonexistence is hard to prove. Initially, it was believed (for plausible theoretical reasons) that atoms with closed outer subshells like the alkaline earths could not form negative ions. This has now been shown to be untrue, both experimentally [60] and theoretically [61].

The binding energy of negative ions is very small (usually of the order of 1 eV or less), which means that the potential is usually too weak to support an excited state. However, there are exceptions, such as Li$^-$ [62] and Be$^-$ [63] and molecules such as NaCℓ^-, NaBr$^-$ and NaI$^-$ [64], and theory suggests there are several others [65], for reasons discussed in section 2.26.

The fact that few negative ions with excited bound states exist does

---

[14] See also the discussion of polarisation radiation in section 12.15.

not mean that the spectra contain little structure: again, as expected for a short range potential, there are many resonances. These are excited states which lie in the continuum and are therefore broadened (cf chapter 6). Since they are broad, they live only for very short times. They are therefore described as *autodetaching*, which means that they fall apart spontaneously, on timescales of 10^{-12}–10^{-15} s, giving a neutral atom and an electron. Atomic negative ions thus acquire a rich spectroscopy despite the paucity of bound states.

Because of the very weak binding field and this complexity of detail, they pose a formidable challenge to theorists: accurate predictions of their properties require an excellent understanding of electron–electron correlations, and computational methods extending beyond the customary basis: normally, negative ions do not exist within the Hartree–Fock approximation, precisely because polarisation of the core is not allowed for.

Our purpose in mentioning negative ions here is to contrast their properties with those of systems which do exhibit Rydberg states. In particular, for negative ions, short pulse excitation is unable to produce narrow wavepackets of the kind described above, which arise by superposition of several Rydberg states. Thus, the response of negative ions to short pulses is markedly different from that of neutral atoms or positive ions, a matter which will be taken up again in chapter 9.

Readers interested in negative ions may refer to the excellent review by Andersen [66].

2.25 The Langevin interaction in molecules

Most molecules and molecular ions are held together by forces in which electron–electron correlations and exchange interactions play a decisive role. An exception is the hydrogen positive ion H_2^+, which contains only one electron but is a very stable system. It provides us with an interesting example of how a shallow well can occur near the dissociative limit under conditions reminiscent of those involved in the formation of negative ions.

In the elementary theory of H_2^+, it is considered as a simple system in which vibrational, electronic and rotational motions can be separated (the Born–Oppenheimer principle) and fully analytic solutions exist (uniquely for a molecule) which show that the molecule is stable. This, however, is not the complete story. In fact, as H_2^+ is separated into H and H^+, one encounters an additional shallow minimum near the dissociation limit, at much larger internuclear distances than its equilibrium separation. This second minimum, which arises from a dipole in the neutral fragment induced by the presence of the charged fragment, is capable of supporting three bound vibration-rotation levels in the ground state of H_2^+. They appear as highly excited states, very close to the dissociation threshold [68, 67].

This charge-induced dipole force is the most fundamental interaction between a charged particle and a neutral species with no permanent dipole moment of its own. It is known as the Langevin interaction . It is a $1/r^4$-dependent force through which neutral and charged species can interact. There is, however, another way in which an interaction of this kind can occur. In the example just given, the polarisation of the neutral species can only be induced, because H_2^+ has no permanent dipole moment. More complex molecules can, however, possess permanent dipole moments, which leads to a different situation.

2.26   Negative ions of molecules

Negative ions of molecules are usually different from their atomic counterparts, because a molecule can possess a permanent dipole moment capable of binding an extra electron to the system. The problem of an electron bound by a dipole field was first addressed in the context of a polarised H atom interacting with negative mesons [69]. It has been established theoretically [70, 71, 72, 73], for a dipole $+e$ and $-e$ of moment eR that the minimum dipole moment to bind an electron is $Re = 0.639a_0$. It has further been shown [74] that a rigid dipole of moment greater than this in fact supports an *infinite* number of bound states. Experimentally, the binding of an electron by the field of a molecular dipole was first observed for $LiC\ell^-$ [75].

The simple rigid dipole theory is not truly applicable to a molecule, because the polarisation is altered by the presence of the additional electron, i.e. we have a combination of the two mechanisms. Jordan and Luken [76] have set up a more elaborate *ab initio* model using a Hartree–Fock scheme, which accounts satisfactorily for molecular relaxation.

In the spirit of section 1.15, since $LiC\ell$ is a closed shell molecule, the bonding of an extra electron to it might seem surprising at first sight. The fact that the mechanism involved is due to the presence of a dipole moment shows how cautious one must be in applying 'united atom' models too literally to molecules.

2.27   Excimers and external field effects

In section 2.12 which dealt with the extended alkali model, it was pointed out that inner-shell excitation of an atom with a closed outer shell results in a compact core with an outer electron occupying wavefunctions reminiscent of alkali orbitals. The same is also true when the valence electron of a rare-gas is excited: a vacancy is produced in the compact closed shell and an extended Rydberg state is formed around it. An atom with a closed outer shell is of course chemically inert (see chapter 1). Since the high chemical activity of an alkali atom is due to the outer electron, it follows that a rare-gas, although it is inert in its ground state, becomes very active in the excited Rydberg state and can then participate in chemical reactions. Excitation to a Rydberg orbital can be regarded as a form of optical catalysis for a chemical reaction which, otherwise, would not take place. In the well-known chemical reaction:

$$2\,Na + C\ell_2 \rightleftharpoons 2\,NaC\ell \qquad (2.23)$$

which proceeds by 'borrowing' the valence electron of an alkali metal and adding it to the seven outer electrons of $C\ell$ to complete a closed shell, one can replace Na by a Rydberg-excited Xe* inert gas atom, thereby forming a new type of molecule Xe*Cl which exists only in the excited state. Such a molecule is known as an *excimer*.[15]

Excimers are very important to laser physicists: since they possess no stable ground state, they may emit radiation but are incapable of reabsorbing it. They are therefore fully transparent to the radiation which they can produce and, since they also are naturally formed in a condition of population inversion, provide an excellent laser medium.

2.28   Very high Rydberg states and external fields

Just as Rydberg states become active and interact with their surroundings (examples occur in collisions and in the formation of excimers) they are also very fragile and sensitive to externally applied fields. Indeed, as already noted above, the observation of high Rydberg states has the implication that stray fields must be kept well under control, since high Rydberg states are very easily perturbed. Rydberg states of very high principal quantum numbers were detected by Rinneberg *et al.* [77, 78] by excitation with CW lasers: these authors have followed a Rydberg series in Ba, first up to $n \sim 290$, i.e. only 40 GHz in energy from the series limit [77] and, in a second experiment, up to $n = 520$ [78]. External fields (including the earth's magnetic field) were compensated by using three mutually perpendicular pairs of Helmholtz coils and a parallel plate capacitor; nine concentric rings within the capacitor plates served to correct

---

[15] Strictly speaking, the word is inappropriate, since it comes from 'excited dimer' and the molecules described are heteronuclear. Various other names have been proposed, but the common usage is followed here.

for fringe fields, and great care was taken to direct the atomic beam along the symmetry axis of the capacitor, in order to minimize perpendicular components of any spurious electric fields. Even so, the experiment was limited by stray fields of mV per cm or less, which are sufficient to induce Stark mixing at the high values of n involved. At the time, it seemed that larger Rydberg states had only been detected by rf astronomers, who had observed transitions to Rydberg states with $n \sim 733$ [79].

However, in experiments on electron attachment from very high Rydberg states, in which a specially designed cubic cage was used, Stark components were minimised in order to adjust the voltages on the box and thereby compensate external electric fields to a precision of $5\,\mu V$ per mm. Thus, Rydberg states of $n \sim 1000$ have been observed [80, 81], which is higher than any astrophysical value yet reported: thus, it is very important to achieve precise cancellation of external electric fields.

If we turn this requirement around, it is clear that high Rydberg states themselves may be used as excellent probes to reveal the influence of external fields on quantum systems. In fact, Rydberg atoms are widely used to probe the influence of either static electric and magnetic or AC fields, or combinations of fields. Examples of how the influence of external fields can be studied will be presented in chapters 9 and 10.

2.29 Rydberg states in molecular spectra

As remarked in section 2.4, Rydberg states are by no means exclusive to atomic systems. Any compact, neutral or positively charged assembly containing electrons can in principle exhibit Rydberg states. The criteria are: (i) that a large, undisturbed and empty volume is available for excited wavfunctions to spread into; and (ii) that the singly ionised assembly should be stable. On the other hand, negative ions (section 2.24) do not exhibit Rydberg states, because an excited electron does not experience a Coulomb force. The emergence of Rydberg series therefore depends on the stability of the core, and whether it is a passive spectator or participates in the excitation, as explained in section 2.12. Related factors may intervene such as the presence of excitations to other Rydberg states of similar energy (known as *intruders*) discussed in chapters 3 and 8. The size and shape of the core (i.e. how close it is to a spherical, compact, ionic core) must also be considered, especially as regards the lowest series members and, last but not least, one must examine whether the ionic core is associated with a specific atom or *chromophore* or whether it is distributed over several atoms.

In molecular spectroscopy, one distinguishes between two classes of orbitals. For some states, the occupied orbitals are all comparable in size to the internuclear separation. The molecular states are then built up from a superposition of atomic orbitals whose principal quantum number is equal to or smaller than the occupied orbitals of the ground states of the constituent atoms. These are called *valence states*, because they are involved in bonding, and may correspond either to ground or to excited states. An excited valence state is often formed by promoting one electron of the ground state configuration from a bonding to an antibonding valence orbital.

However, it is also possible to construct an excited state of a molecule by promoting one electron to an orbital made up from a superposition of atomic orbitals whose principal quantum numbers are much larger than those of occupied orbitals from the ground states of the constituent atoms. Such states are called *molecular Rydberg states*. For a diatomic molecule AB , they are generally well represented as $(AB^+)n\ell\lambda$, where n and ℓ are the usual quantum numbers of the Rydberg electron, and λ is the projection of angular momentum along the internuclear axis. The selection rules on λ are as follows: $\Delta\lambda = 0$ occurs, for example for $\sigma \longrightarrow \sigma$ transitions, when the polarisation vector is parallel to the internuclear axis, and $\Delta\lambda = \pm 1$ when the axis is perpendicular to the electric field vector. However, one must also bear in mind that the molecular core (AB^+) has a nonspherical potential, so that conservation of angular momentum does not really apply. Thus, ℓ is only an approximate quantum number.

Molecular Rydberg states are of importance, because different perturbations appear which can be analysed in detail and constitute various forms of breakdown of the Born–Oppenheimer approximation [82].

2.30 Rydbergisation

A question which arises immediately when considering molecular orbitals is whether their character involves fairly equal contributions from different constituent atoms, or whether they are primarily associated with one of the atoms. For a diatomic molecule AB, the answer to this question may depend on the internuclear separation R_{ab}, i.e. on the degree of vibrational excitation.

When the Rydberg states belong to series which converge to the *same* state of the ion but to different vibrational levels, as occurs for identical symmetry states of H_2 converging to the H_2^+ ground state in different vibrational states [86], then the properties of the series and in particular the electronic couplings or perturbations between them are nearly independent of R_{ab}. However, it can also happen that an orbital localised on A at small R_{ab} turns into an antibonding orbital at intermediate R_{ab} and then becomes localised on B at large R_{ab}, eventually turning into a molecular Rydberg state at infinite R_{ab}. When the character of the orbital changes smoothly as a function of R_{ab}, this phenomenon is called the *Rydbergisation* of a valence orbital [83]. It is used to explain why the lowest members of the $ns\sigma$ Rydberg series are absent in the hydrides CH, NH and OH [84].

In absence of Rydbergisation, there are still complexities which can hinder the full development of series. A molecule is a compact assembly of electrons, but (a) is larger than an individual atom and (b) is usually not spherical. Provided the threshold corresponds to a sufficiently stable excited state of the molecular ion, neither (a) nor (b) need prevent the formation of Rydberg states: if sufficiently external, the orbitals of the excited electron will only be slightly perturbed by the electrons of the molecular core. Thus, while Rydberg states are often less regular in molecules than in atoms, they can nevertheless extend to very high n.

2.31   Molecular H

A well-known example occurs in H_2, which is very compact, because it contains two spin-paired electrons which are not too dissimilar from the paired electrons of helium. The series limit of H_2 therefore occurs somewhat higher in energy (at more than 15.4 eV), as compared to 13.6 eV for H (but lower than the 24.6 eV for He, because H_2 is much larger).

Although H_2 is the simplest molecule, its Rydberg manifold is actually very complicated. Photoabsorption from the ground state results in two interacting Rydberg series, converging to each of two vibrational levels of the ion core. The lower members of the series are described as $^1\Sigma^+$ and $^1\Pi$ series (Σ and Π stand for projections 0 and 1 of the total angular momentum along the internuclear axis) arising from the excitation of $np\sigma$ and $np\pi$ electrons (i.e. np electrons with projections 0 and 1 of the individual angular momenta along the internuclear axis). As n increases, the coupling changes as a result of ℓ uncoupling, and the two series to the same limit perturb each other strongly, resulting in large energy shifts and intensity variations. Strong perturbations, due to lower n states of series converging on higher vibrational states of the ion, also occur, and so the detailed situation is actually quite involved. Despite these perturbations, the two long Rydberg series have been observed up to $n > 40$ (see fig. 2.13), by using para-H_2 cooled down to liquid N_2 temperatures, so that every vibrationalIelectronic transition results in a single line (vibrational structure is inhibited) and the Rydberg pattern emerges more clearly. Thus, the perturbations have been analysed in detail [85].

In effect, the single ionisation threshold of the atom is split into a large number of closely spaced thresholds which correspond to the vibrational

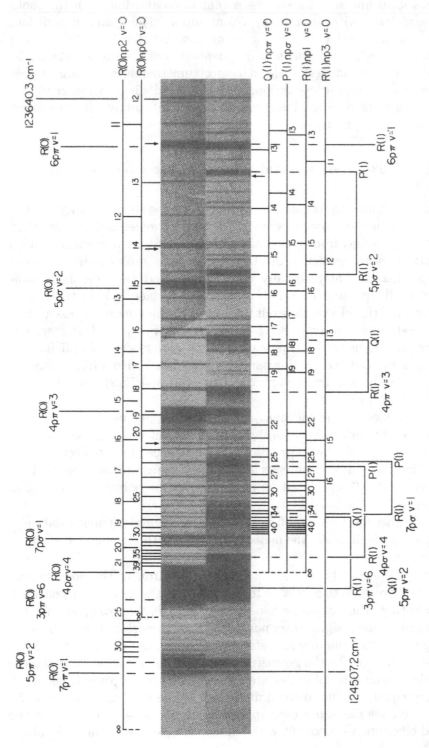

Fig. 2.13. Photoabsorption of H_2 at low temperature and high resolution, showing two long Rydberg series, and complex patterns due to vibrational structure of the molecular ion (after G. Herzberg and Ch. Jungen [86]).

and rotational fine structure of the parent molecular ion, each threshold being associated with an ionisation continuum. There are, in addition, vibrational progressions, each one being associated with a dissociation continuum, which corresponds to fragmentation of the excited states. In general, even the simplest molecular spectrum therefore contains a vast number of overlapping excitation, ionisation and dissociation channels, which render its detailed interpretation very dificult as the ionisation thresholds are approached.

2.32 Rydberg states in triatomic molecules

Although diatomic molecules may appear to be the simplest, they are not in fact as similar to atoms as some of the heavier molecules, because they depart very strongly from spherical symmetry. Thus, diatomics often have the least atom-like molecular orbitals. Under favourable conditions, more complex molecules may actually yield simpler and more regular 'quasi-atomic' Rydberg series than light molecules. In fact, it is instructive to consider how the electronic excitation of molecules more complex than simple diatomics can give rise to Rydberg spectra. As a first step, tri-atomics provide us with a simple example of how spectral simplicity (i.e. Rydberg structure and regular bands) is associated with a high degree of symmetry, while spectral complexity (irregular spacings) results from a lack of symmetry.

For molecules, just as for many-electron atoms, one uses equation (2.8) to describe the series. As before, the quantum defect μ characterises the whole Rydberg series but now absorbs the influence of the molecular core. In practice, its value still depends largely upon the atomic symmetry of the Rydberg orbital (more detail on its properties for polyatomic molecules will be given in section 3.11).

Two further features are often observed in vacuum ultraviolet absorption spectra which reflect the molecular core structure associated with the Rydberg transitions.

First, the proximity of a low n Rydberg electron to the nonspherical molecular core leads to partial or total removal of the atomic ℓ-degeneracy. The resulting group of states is described as a *Rydberg complex*. Within this complex, one assigns states possessing Rydberg orbitals with the same principal quantum number and atomic symmetry, but different molecular symmetry dictated by the geometry of the core. This yields more allowed transitions, whose energies deviate somewhat from the predictions of the Rydberg equation, due to the finite extent of the nuclear framework. As n increases, the molecular core appears more like a central charge to the excited electron. Consequently, the multiplet spacings within a Rydberg complex decrease rapidly with increasing n, and a return to an atomic degenerate state (greater 'atomicity') results.

The second feature that may become apparent in vacuum ultraviolet spectra is rotational and/or vibrational structure. This can be more or less obtrusive, depending on the molecule concerned and, in some cases, may mask the Rydberg series unless steps (e.g. cooling of the molecule) are taken to simplify the observed spectra.

With regard to the vibrational structure, it is essential, for a Rydberg series to be well developed (or sharp), that the oscillator strength for a particular electronic transition be concentrated predominantly in the 0–0 transition. This means that Rydberg excitations will be most atomic-like if the excited electron originates from a lone-pair or weakly bonding molecular orbital, which is an orbital located on a specific atomic site, rather than one in which the electron is distributed over a bond. A dense set of Rydberg states will then be readily accessible for an electron making a transition from its initial (nonbonding) orbital to Rydberg orbitals of high principal quantum numbers. A high resolution recording of the Rydberg spectrum should then permit absorption features of high n to be resolved. From such observations, an extremely accurate value of the first ionisation potential of the molecule can be determined. A high resolution recording of the low n Rydberg transitions is also of great interest if the rotational fine structure can be resolved (low n transitions are specified here, because, for the high n transitions, ℓ uncoupling occurs, resulting in rotational collapse). An analysis of the rotational lines not only gives information on the geometry of the excited state, but often allows the molecular symmetry (the irreducible representation) of the accepting Rydberg orbital to be determined. It may also provide details on the spontaneous fragmentation of the molecule when excited to a high Rydberg state, an effect known as *predissociation*.

The group VI dihydride molecules (H_2O, H_2S, H_2Se and H_2Te) provide a specific illustration of these rather general statements. Early studies [87] at low resolution showed prominent Rydberg series and the tantalising hint that rotational structure might easily be resolved. More recently, studies at high spectral resolution [88, 89, 90] using the radiation continuum of the electron synchrotron as a 'background' source to probe photoabsorption over a wide range of wavelengths have revealed a lot of interesting detail which exemplifies the behaviour of Rydberg series in polyatomic species.

The most important molecule of this sequence is of course H_2O [88]. In terms of analysis, this is actually the most complex case. This is because: (i) H_2O is a strongly asymmetric rotor, which complicates the rotational spectra; the properties of a symmetric rotor are shown in fig. 2.14; (2) a large number of vibronic transitions are observed; and (3) the elec-

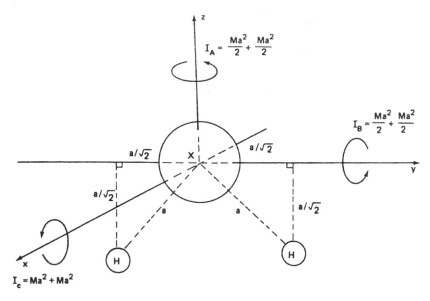

Fig. 2.14. The geometry of the group VI dihydride molecules: when the moments of inertia about the two rotation axes are equal (as illustrated), the rotor is described as a symmetric top and has simpler rotational structure than when they are not (asymmetric top). The condition for the moments of inertia about both axes to be equal is that the bond angle should be $\pi/2$. M is the mass of an H atom. (after J.-P. Connerade and J. Hormes [91]).

trostatic splitting of the Rydberg orbitals intrinsic to H_2O is not large, and significantly less than for the other molecules in the sequence, causing rotational overlap for the low n transitions within the same dominant Rydberg complex. Nevertheless, for high n transitions, as mentioned above, the rotational fine structure collapses. The resulting Rydberg series for both H_2O and D_2O are shown in fig. 2.15. By using the requirement that the quantum defect for a Rydberg series approaches a constant value at high n, the ionisation potentials of H_2O and D_2O have been determined as 101740 ± 10 cm^{-1} and 101910 ± 8 cm^{-1}, respectively, which are in good agreement with high resolution photoelectron measurements [93].

The dominant Rydberg series of H_2O and D_2O shown in fig. 2.15 can be assigned as a 'd' series because, by $n = 4$, the atomic ℓ-degeneracy is beginning to be quickly restored. Only for the first member of this Rydberg series is the electrostatic splitting sufficient to give a reasonable separation between the Rydberg orbitals with different molecular symmetry, and even then severe overlap occurs. This is illustrated in fig. 2.16, which shows the '$3d$' Rydberg complexes of H_2O and D_2O. Note the remarkably irregular nature of the structure, despite the fact that the corresponding

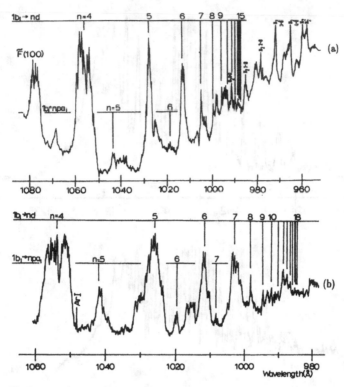

Fig. 2.15. High members of the Rydberg series in both (a) water and (b) heavy water (after J.-P. Connerade *et al.* [92]).

bands in other molecules of the sequence are simple in appearance.

This provides a simple example of how a spectrum becomes complex in structure and exhibits irregularity of energy spacings as a result of symmetry being broken. The dramatic effects of broken symmetry on the complexity of spectra and the emergence of erratic level spacings will be discussed in more detail in chapter 10.

The other molecules of the series, H_2S, H_2Se and H_2Te, approximate very well to being symmetric (oblate) rotors. Consequently, the rotational fine structure of the low n members of the Rydberg series is much less complicated [94, 95] than that of the H_2O molecule. In fig. 2.17, we show a survey of all the leading members of the molecular sequence to illustrate this point. The rotational fine structure is even more simplified in appearance owing to predissociation which broadens the individual lines in each branch. Rotational analysis provides details of the predissociation mechanisms and allows the correct assignment of the Rydberg transitions to be established.

As for the H_2O spectrum, ℓ-uncoupling processes causes the breadth

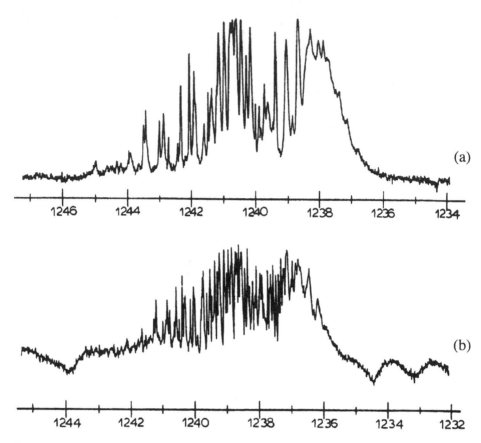

Fig. 2.16. The $3d$ members of the Rydberg series of (a) H_2O and (b) D_2O, showing the enormous complexity of rovibronic detail and the effect of overlap with orbitals of different molecular symmetry (after J.-P. Connerade *et al.* [92]).

of the individual bands to decrease and become almost atom-like as n increases. Together with the nonbonding nature of the $1b_1$ orbital, this satisfies the requirement for sharp, well-separated Rydberg members at high n, as demonstrated in fig. 2.18 for H_2S, H_2Se and H_2Te. From these spectra the first ionisation potentials of H_2S, H_2Se, and H_2Te, have been determined as $84428\pm3\,\mathrm{cm}^{-1}$, $79824\pm2\,\mathrm{cm}^{-1}$, and $73749\pm2\,\mathrm{cm}^{-1}$ respectively [96, 97]. Similar spectra have been recorded for D_2S, D_2Se and D_2Te [96, 97], from which their first ionisations potentials are obtained as $84466\pm2\,\mathrm{cm}^{-1}$, $79854\pm2\,\mathrm{cm}^{-1}$ and $73760\pm6\,\mathrm{cm}^{-1}$ respectively. The observation of Rydberg series in molecules provides useful information on their ionisation thresholds.

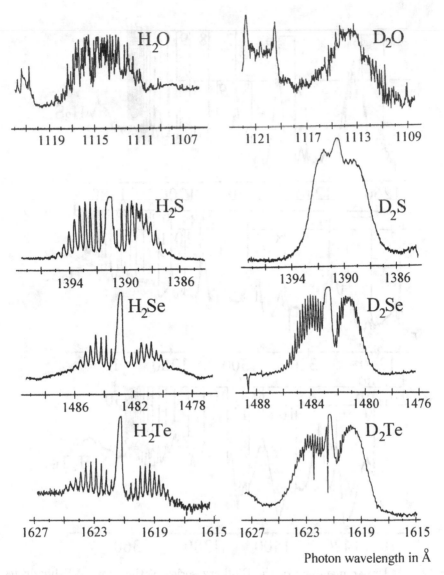

Fig. 2.17. Leading members of the Rydberg series of the group VI dihydrides and deuterides. Note the complex structure for asymmetric top molecules, as compared with the much more ordered and regular patterns for symmetric tops (after J.-P. Connerade *et al.* [92]).

2.33 Polyatomic molecules

We turn now to more complex molecules and give examples of Rydberg series in polyatomic species. The interesting point is that, because rovibronic structure may intrude less upon the spectrum, heavier and more

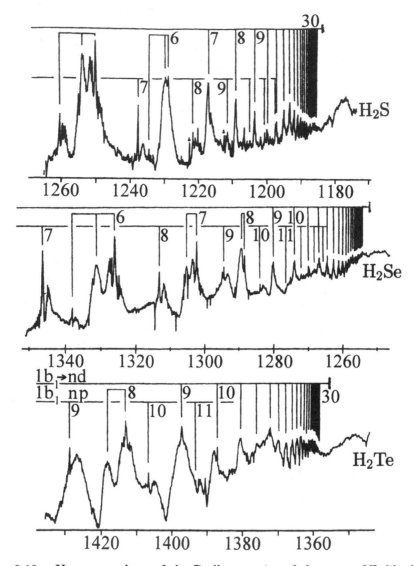

Fig. 2.18. Upper members of the Rydberg series of the group VI dihydrides (after J.-P. Connerade *et al.* [92]).

complex species can occasionally give rise to longer and more distinct series. We begin by an example of this type, drawn from studies of the methyl halides.

The following case may clarify this point, and also provides an example of how the 'corresponding atom' is chosen. Consider the molecules HI and HBr. The presence of the light H atom has the consequence that the rotational structure is very open, so that, at high principal quantum numbers, the overlap between the structures built on successive Rydberg members is severe. Thus, clear Rydberg series to very high n are not readily observed, and the spectra cannot be sorted out without rotational cooling.

On the other hand, if the H atom is replaced by the heavy group CH_3, then the rotational structure disappears from view, and beautiful Rydberg series result, which can be traced back without difficulty from the series limits. Examples of the spectra of CH_3I [98] and CH_3Br [99] are shown in figs. 2.19 and 2.20. The Rydberg series in these molecules converge on series limits which are associated with one specific atom or *chromophore*, whose environment is probed selectively by the excitation. Because the energy of the threshold is very close to that of one atom in the molecule, and also because it remains fairly constant when the CH_3 group is substituted by any other member of the sequence C_nH_{2n+1}, we know that the chromophore in these examples is the halogen atom.

The substituents radicals act like an H atom, donating one electron to complete the closed shell around the halogen chromophore. Thus, the spectra of CH_3I and CH_3Br are very similar to those of the rare-gases which follow the chromophore in the periodic table, as will be further described in section 3.11. For example, this particular excitation spectrum of CH_3I can be compared with the $5p$ spectrum of the corresponding atom Xe. This spectrum possesses a prominent doublet structure of the parent ions, due to the spin-orbit splitting of the nearly-filled shell, as described in section 1.13.

The case of fig. 2.20 is also interesting because it demonstrates the effect of vibrational excitation of the core: in the figure, one notes, in addition to the long series, that further structure converges to higher energy than the limits. This structure is attributed to vibrational excitation of the molecular core, which produces a slight disturbance or *perturbation* of the Rydberg manifold, without severely affecting its properties.

The Rydberg series of another member of the sequence, C_2H_5I, have also been studied [100]. They are more diffuse due the opening of new ionisation and dissociation channels and, as a result, cannot be followed to such high members.

There are many other examples of long Rydberg series in still heavier molecules: for instance, very high Rydberg states of C_6H_6 have been reported [101].

2.34 Impurities and excitons in solids

The previous three sections have demonstrated that Rydberg series do not merely arise for atoms. We now take the matter further, and explore

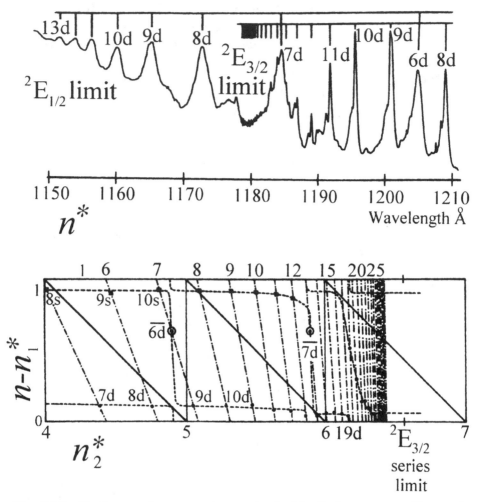

Fig. 2.19. Rydberg series converging on the doublet limits corresponding to the halogen parent ion in CH_3Br. The series are long and well developed, with little intrusion of rovibronic structure. A quantum defect plot is shown alongside. For the meaning of this kind of plot, see chapter 3 (after M. A. Baig *et al.* [99]).

a system which is not even related to any equivalent atom.

At first sight, it might seem, from the example outlined briefly in section 2.14, that Rydberg excitations would arise mainly from impurity atoms in a solid, and would only contain a few low members, small enough to fit within the interstices of the lattice. Indeed. it might seem impossible to associate any other form of Rydberg excitation with condensed matter.

There are, however, Rydberg excitations in solids closely related to

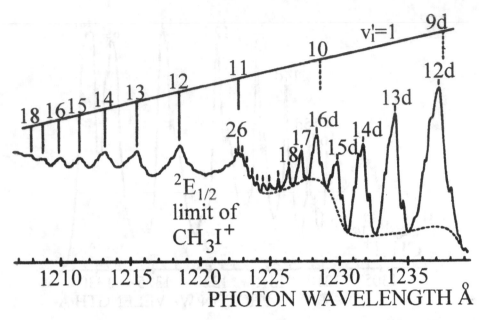

Fig. 2.20. Rydberg series converging on the $J = 1/2$ doublet limit corresponding to the halogen parent ion in CH_3Br. The series are long and well developed; a vibrational progression due to excitation of the molecular core occurs towards higher energies (after M.A. Baig *et al* [98]).

those we have discussed, except that the value of the Rydberg constant (and of the effective mass of the electron) are totally different from those for free atoms.

If an electron is ejected from the valence band directly into the conduction band of a solid, it leaves behind it a hole which behaves exactly like a particle, except that its charge and momentum are both equal and opposite to that of the electron, so that overall neutrality is preserved and the total momentum is zero. Thus, the electron and hole move apart from one another, usually rather rapidly.

If, however, the process takes place in excited states near the band edge, then the momentum of the particle and of the hole are almost zero, and they experience a mutual Coulomb attraction. The problem is essentially the same as that of the H atom or that of the energy levels of positronium: the electron and hole possess discrete energy levels given by the equation:

$$E_n = -\frac{m_r}{m}\frac{R}{n^2} \qquad (2.24)$$

where n is the principal quantum number, m is the mass of a free electron,

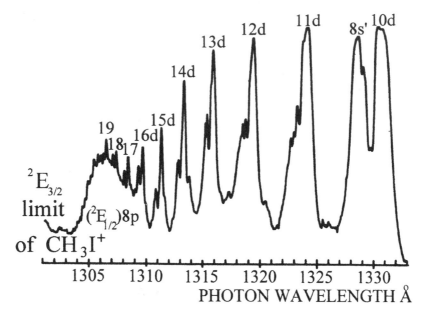

Fig. 2.21. High members of the Rydberg series to the $J = 3/2$ doublet limit corresponding to the halogen parent ion in CH_3I. The series limit occurs close to the $n = 8$ member of the series in the previous figure (after M.A. Baig *et al.* [98]).

m_r is the reduced mass, given by the equation:

$$m_r = \frac{m_e^\star m_h^\star}{m_e^\star + m_h^\star} \qquad (2.25)$$

with the quantities $m_e^\star$ and $m_h^\star$, the effective masses of the electron and hole respectively. In practice, these are rather different from those for free particles, and so the *apparent* Rydberg constant $m_r R/m$ is very different from that for atoms (instead of an energy scale in eV as for atoms, one finds energy scales in tens of meV). Such aggregates, involving two or more charged particles which are effective mass particles (e.g. electrons in the conduction band or holes in the valence band) are referred to generically as 'effective-mass-particle complexes.'

The bound electron–hole system is known as an *exciton*. It can be thought of as a 'quasiatom', formed from the particle and the hole, with the two objects rotating about a common centre of mass, the angular momentum being quantised. Just as, in atomic physics, one generalises by considering the electron in a many-electron atom as a quasiparticle, we can now replace the positive centre of charge by a hole. This system is often

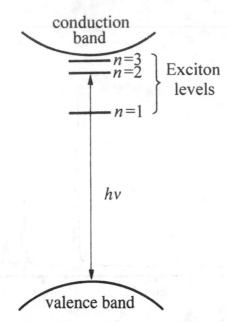

Fig. 2.22. Bound excitonic levels below the base of the conduction band: the photon energy required to promote valence electrons into excitonic states is slightly less than the absorption edge produced by exciting electrons into the base of the conduction band.

regarded as the solid state analogue of positronium;[16] 'ionisation' of this electron–hole 'pseudoatom' or 'pseudohydrogen' can correspond to such a large spatial separation that the Coulomb attraction is negligible, which is consistent with excitation to the base of the conduction band, as shown in the energy-level diagram of fig. 2.22 (see, e.g. [103] for discussion).

Such states are created when an impurity atom acts as a donor, i.e. when an atom with a low ionisation potential is substituted for one of the atoms in a semiconductor host. The ionisation energy of an electron bound to the donor impurity can be much smaller than the energy gap of the semiconductor. One then refers to a *shallow donor impurity* [104]. An example of the observed 'Rydberg' series is shown in fig. 2.23 Excitonic states are identified by the presence of a Rydberg-like spectrum in some semiconductors with large enough energy gaps.

It is interesting that, because of the much smaller value of the effective Rydberg constant in the solid, the highest n states are actually much

[16] More generally, there are two kinds of exciton: when the electron and hole are completely delocalised from any specific atomic site and form bound states, one has a Wannier–Mott exciton. When both the electron and hole are localised on or near a specific atomic site in the solid, so that the exciton is formed from atomic or molecular states perturbed by the crystalline environment, one has a Frenkel–Peierls exciton [102].

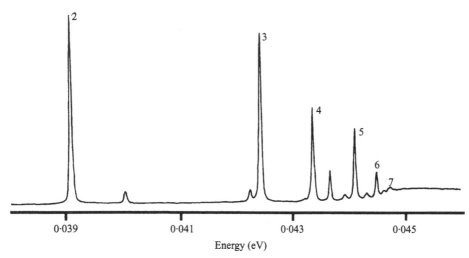

Fig. 2.23. Example of a Rydberg series whose energy splittings are determined by solid state effects: the absorption spectrum of P as a donor impurity in Si (courtesy of R.A. Stradling).

larger than are atomic states of the same n. Thus, the highest states in fig. 2.23 possess radii of 100 nm or more. Such giant Rydberg states are very important to explore an otherwise inaccessible regime of atomic physics, as will be explained in section 11.16. Another interesting point is the presence of displaced lines in the spectrum, which do not fit the main Rydberg series. These can be attributed to a departure from sphericity: in the solid, the effective mass is not a scalar but a tensor quantity, and the symmetry is ellipsoidal. Thus, the resonances are split into two modes along the axes of symmetry. A similar situation will arise again in chapter 12, when the properties of giant resonances in clusters are discussed.

Excitons may also pair to form a quasimolecule. This possibility, for the analoguous situation of the free electron–positron or 'positronium molecule' was considered by Wheeler [105], who concluded that such quasi-molecules would not be stable. Subsequently, Hylleraas and Ore [106] showed that, in fact, they should be stable against dissociation. Lampert [107] pointed out that Wheeler's molecular entities should also exist in nonmetallic solids, with positive holes playing the role of positrons, and introduced the notion of an 'excitonic molecule'. Sharma [108] then computed its binding energy as a function of $\sigma = m_e/m_h$ (m_e being the mass of the electron and m_h the mass of the hole) by using a variational method, and found stability for some values of σ. It was later shown that, in fact, the excitonic molecule is stable over the whole range of σ [109]. The first obervation of the excitonic molecule was reported by Haynes [110].

2.35   Conclusion: the Rydberg signature and its implications

The alkali model and its extensions are fundamental to the development of atomic physics, and will be referred to several times in the course of the present monograph.

In view of the fact that the Coulomb interaction between the electrons is long range, it may seem surprising that the atomic core should be of finite extent and, moreover, possess a fairly well-defined radius. This aspect is further discussed in chapter 5.

The alkali model demonstrates that there are two distinct ways in which n, the principal quantum number, can increase in value. The first is through excitation for a given atom. In this case, as the first ionisation threshold is approached, one moves towards semiclassical solutions in the sense of the correspondence principle. The second is through the filling of shells: as we move down a column of the periodic table (across fig. 2.4) the transition energy generally decreases, but n must nevertheless be increased to take account of the occupation of orbitals. In the latter case, n ceases to be a purely Coulombic label, and this manner of increasing it does not produce any shift towards semiclassical solutions in the sense described above. The distinction between these two different ways of raising n will be important when we come to consider double excitations in chapter 7 and in the discussion of centrifugal barrier phenomena in chapter 5.

From the examples given in the present chapter, we can draw one general conclusion: Rydberg series are common, even in complex many-electron systems, provided conditions exist which allow one electron to be promoted into a region of space where it can become *dynamically independent* of the others. A Rydberg series is therefore the spectral signature of independent electrons, all the many-electron interactions being subsumed into: (a) the value of the threshold energy and (b) the magnitude of the quantum defect, both of which are determined inside the core.

While Rydberg series are usually prominent, on occasions they may also be *absent*, i.e. they may not be found at excitation energies at which they would have been expected. Such an observation can also be important, and may provide a valuable clue as to the nature of the excitation channel.

The breakdown of exact one-electron quantisation may be more or less pronounced: for a highly correlated system, i.e. one in which the electrons are not independent, it may even be sensible to search for new quantum numbers, since the breakdown is likely to be complete. This topic will be discussed in section 7.11.

This chapter, though containing many different examples, by no means exhausts all possible kinds of Rydberg series. Other cases, with different peculiarities, will arise throughout this book: there is a truly amazing variety in the nature of Rydberg excitations – see in particular chapters 7 and 8. Finally, we reemphasize that the Schrödinger equation for many-electron atoms is solved with good accuracy in two basic situations: one is for the spherical closed shell, and the other for one electron outside a closed shell. The first, which describes rare-gases, gives rise to the concept of the atomic core, and the second, which describes alkalis, to the notion of the Rydberg atom.

3

Quantum defect theory
for bound states

3.1   Introduction

Quantum defect theory (QDT) was developed by Seaton [111] and his collaborators, from ideas which can be traced to the origins of quantum mechanics, through the work of Hartree and others. They relate to early attempts to extend the Bohr theory to many-electron systems (see e.g. [114]).

In chapter 2, we saw how the quantum defect is defined from a slight modification of the Rydberg formula for H. It is found experimentally to be nearly constant for different series members, especially for unperturbed series in atoms with a compact core. The first task of QDT is to 'explain' this fact, and to extract from this empirical observation an appropriate wavefunction, consistent with an effective one-electron Schrödinger equation, such that the quantum defect would turn out to be be nearly constant as the principal quantum number n is changed.

QDT is not an *ab initio* theory, i.e. it is not an attempt to solve the many-body problem from first principles. Rather, it is a theoretically-based parametrisation. One seeks a form for the wavefunctions and for their dependence on n; this in turn leads to precise rules for the variation of many other quantities with n because, in quantum mechanics, once the wavefunctions are known, many observable properties of the system become calculable.

The benefits of QDT do not stop there. Amongst the calculable properties is the variation of transition strengths, and it is possible to show that the behaviour of df/dE plots as a function of energy, together with their smooth connection to the continuum cross sections described in chapter 4 are natural consequences of QDT. Indeed, in QDT, the Rydberg series as a whole, together with the adjoining continuum, are regarded as a single entity, known as an *excitation channel*. The simple theory described so

79

far, in which one electron is excited independently of the others, leads to an unperturbed Rydberg series with a constant quantum defect and is called *single channel QDT*. However, it is also possible for series to be coupled to each other, in which case they are perturbed, i.e. the quantum defects are no longer constant: this is described by *multichannel QDT* (MQDT), which is one of the most successful representations of interactions lying beyond the independent particle model. In other words, MQDT provides a description, through its representation of interchannel coupling, of an important class of many-body interactions.

This coupling exists not only amongst the bound states, but can be extrapolated into the adjoining continuum. If interchannel couplings involving the bound states are well understood, one can predict the breadths and line profiles of the associated autoionising resonances (cf chapter 6). Thus, MQDT is not only a convenient parametrisation of empirical information (or information derived from *ab initio* calculations) but is also a unifying conceptual framework, bringing together into one logical scheme properties of bound and of continuum states. For this reason, it has been described as a unification of spectroscopy and scattering.

The organisation of the present chapter is as follows. First, we give the basic principles of single channel QDT. Next, we move on to series perturbations, and to simple graphical methods which can be used to analyse them by MQDT. Finally, we give examples of how such methods can be extended to more complex situations, and we deal with extrapolations beyond the series limit which allow autoionisation profiles to be predicted from the data for interacting bound states.

3.2   Analytic approach to the quantum defect

A simple way to introduce the quantum defect is to suppose that the potential of the atom is nearly Coulombic, but differs at short range by an α/r^2-dependent term. We thus replace the effective potential of equation (2.2) by:

$$V_{eff}(r) = -\frac{Ze^2}{4\pi\varepsilon_0 r} + \frac{\hbar^2}{2m}\frac{\alpha}{r^2} \qquad (3.1)$$

where α is merely an adjustable parameter and the factor $\hbar^2/2m$ is introduced for convenience. Hence

$$-\frac{\hbar^2}{2m}\frac{d^2 u_{n\ell'}}{dr^2} + \left(\frac{\ell'(\ell'+1)\hbar^2}{2mr^2} - \frac{Ze^2}{4\pi\varepsilon_0 r}\right)u_{n\ell'} = E_n u_{n\ell'} \qquad (3.2)$$

where $\ell'(\ell'+1) = \ell(\ell+1) + \alpha$. If we let $\ell' = \ell - \mu_\ell$, then

$$\mu_\ell = \ell + \frac{1}{2} \pm \left\{ \left(\ell + \frac{1}{2} \right)^2 + \alpha \right\}^{\frac{1}{2}}$$

where the $+$ sign is unphysical, as $\lim_{\alpha \to 0} \mu_\ell = 0$. It follows that

$$E_{n\ell} = -\frac{R}{n - \mu_\ell}$$

This approach shows that the quantum defect results from the short range part of the potential. It provides parametrised wavefunctions associated with the μ_ℓ, and suggests a more versatile method of handling the atomic core, described below.

3.3 Single channel QDT

As already explained, the basic task of QDT is to construct from the empirical observation that the quantum defect μ is constant a simple, i.e. one-electron wavefunction for a the Rydberg electron of a many-electron atom. The relationship to alkali spectra (cf section 2.9) is useful: alkali atoms possess a compact electronic core with one external or valence electron; this immediately suggests that the space of the atom can be divided into two distinct regions, an inner region, in which an exact treatment would be difficult, because all the electrons are present together and interact strongly (this is the many-body region) and an outer region, away from the core, where essentially only one electron is present, so that the problem reduces to solving a one-electron equation. We further suppose that the many-body region is small, and confined within a radius r_0, which may actually be any radius larger than the core, so that the results of QDT should not depend on the precise value chosen for r_0. Since this region is small, we will not actually be concerned about the solution within r_0, but rather with its form outside r_0, and the purpose of the theory will be to show that, because of the presence of the core, this differs from the wavefunction for H by an amount which depends entirely upon μ. Thus μ will appear as a parameter which subsumes the many-body effects of the core and allows them to be included in an effective one-electron wavefunction.

At a general energy ε in atomic units, measured relative to the ionisation threshold, or in terms of the reduced energy variable ν of QDT, defined by $\varepsilon \equiv E_\infty - 1/2\nu^2$ (note that ε is negative for bound states), the one-electron Schrödinger equation outside r_0 is just the same as for H. Thus, the solution involves two functions, $f(\nu, r)$ and $g(\nu, r)$, whose behaviour at the origin is different. We have

$$f(\nu, r) \sim r^{\ell+1} \quad g(\nu, r) \sim r^{-\ell} \quad r \to 0 \tag{3.3}$$

the first of these tends to zero at the origin, which satisfies the boundary conditions for H and is therefore called the *regular function*, while the other diverges at the origin, which does not satisfy the boundary conditions for H, and is therefore usually discarded. It is called the *irregular function*. In our case, the origin is excluded from the problem, because the internal boundary is at $r = r_0$. Consequently, we must retain *both* $f(\nu, r)$ and $g(\nu, r)$ in the general solution outside r_0, with numerical coefficients c_f and c_g to be determined by the boundary conditions of the problem, i.e. at $r = r_0$ and $r \to \infty$. For reasons which will emerge, it is more convenient to write these two numerical coefficients in terms of two new quantities N_ν and μ as $c_f = N_\nu \cos(\pi\mu)$ and $c_g = N_\nu \sin(\pi\mu)$, where N_ν is a normalisation constant which depends on the energy and μ is an additional phase shift between the regular and irregular functions. We thus have:

$$\psi(\nu, r) = N_\nu \left\{ f(\nu, r) \cos(\pi\nu) - g(\nu, r) \sin(\pi\mu) \right\} \quad \text{for } r \geq r_0 \qquad (3.4)$$

So far, we have not specified the energy, and so ν is not yet fixed at any specific value, but is a running variable. In order to quantise the energy, we must as usual impose boundary conditions appropriate for a bound state. These will turn out to quantise ν, as we now demonstrate.

From the inner boundary condition, μ is now determined such that ψ joins smoothly at $r = r_0$ onto the wavefunction inside the core region. There are two ways of treating this condition. We might, for example, possess some *ab initio* theory (not contained within QDT) which allows us to solve the many-body equations at $r < r_0$ in some approximate way, and thereby determine μ from first principles. The other possibility is to fix μ semiempirically, simply by using the experimental value. *Thus, QDT as such has nothing to say about the value of μ except that it is assumed to remain constant as n is changed for a channel free of perturbations*

At large r, on the other hand, we need the asymptotic boundary condition, and this requires us to use the asymptotic forms of $f(\nu, r)$ and $g(\nu, r)$, the regular and irregular Coulomb functions, as $r \to \infty$. These are

$$f(\nu, r) \to u(\nu, r) \sin(\pi\nu) - v(\nu, r) \exp(i\pi\nu)$$
$$g(\nu, r) - \to u(\nu, r) \cos(\pi\nu) - v(\nu, r) \exp(i\pi\nu + \tfrac{1}{2})$$

In these expressions, $u(\nu, r)$ increases exponentially with increasing r, and its coefficient in the final expression for ψ must therefore vanish at large r, whereas $v(\nu, r)$ tends to zero at large r and may therefore have a nonzero coefficient. Substituting into equation (3.4), we have

$$\psi(\nu, r) \to N_\nu \left\{ u(\nu, r) \sin\left[\pi(\nu + \mu)\right] - v(\nu, r) \exp\left[i\pi(\nu + \mu)\right] \right\} \quad \text{for } r \to \infty$$
$$(3.5)$$

from which, by applying the boundary condition that ψ must tend to zero at infinity for a bound state, we have the condition

$$\sin\left[\pi(\nu + \mu)\right] = 0 \qquad (3.6)$$

As will be shown later, the tangent function is a more fundamental quantity, and so this equation is also written in the equivalent form

$$\tan\left[\pi(\nu + \mu)\right] = 0 \qquad (3.7)$$

which is satisfied only when $\nu + \mu = n$, where n is an integer.[1]

At this point, ν must now satisfy the quantisation condition

$$\nu = n - \mu \qquad (3.8)$$

from which we recover the expected Rydberg formula:

$$\epsilon = E_\infty - \frac{1}{2\nu^2} = E_\infty - \frac{1}{(n - \mu)^2} \qquad (3.9)$$

in atomic units, so that μ is now recognized as the same quantity as defined empirically in section 2.9. We are now able, given a quantum defect μ, to deduce the form of the wavefunction $\psi(\nu, r)$ outside the radius r_0 (taken here as the effective radius of the core).

3.4 Comparison with *ab initio* theory: recapitulation

The form of the wavefunction deduced by QDT is rather interesting. It differs from that of H, as we have seen, by the fact that the irregular Coulomb function $g(\nu, r)$ is necessarily present in the solution. Mathematically, this means that the zeros or nodes in ψ occur at values which no longer depend solely on n as in the case of H, but on both n and μ. At large enough n, the wavefunctions of H have the interesting property that their inner nodes *recapitulate*, i.e. that the radii at which all except the outermost nodes occur for successive values of n become stable as the series limit is approached.

This property is actually not surprising: as the series limit is approached, the bound state wavefunction acquires more and more nodes, and tends to the oscillatory function of the continuum. The position of the nodes is related to the phase in the continuum, and we may expect that the two are connected, since the wavefunction at very high n must change smoothly into the free electron's wavefunction just above the series limit. In QDT, as for H, the wavefunction for $r > r_0$ preserves this

---

[1] Actually, QDT alone does not determine whether n should be positive or not. Indeed it says nothing at all about the correct value of n, a point which recurs repeatedly in applications of QDT to real data.

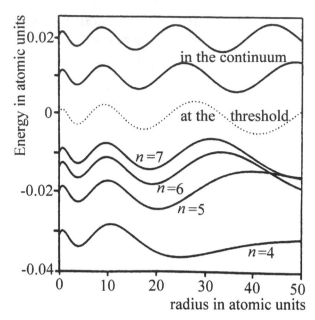

Fig. 3.1. Recapitulation of the inner nodes for radial Coulomb wavefunctions: also shown are the near-threshold continuum function with delta function normalisation (dotted curve) and two continnum functions above the threshold (adapted from H. Friedrich [112]).

useful property, i.e. for successive values of n the inner nodes in the wavefunction recapitulate provided only that μ is a constant.

The shift between the position of the nodes for H and for a general QDT wavefunction is controlled entirely by μ and may be considered as a *phase shift* at high enough n.

This is actually the most fundamental property of QDT, and if one examines numerical calculations of wavefunctions by, say, the Hartree–Fock method, the question whether QDT can be used to summarise the results depends on how accurate this recapitulation of the nodes turns out to be. The simplest check of the applicability of QDT to a theoretical problem is therefore to plot the wavefunctions and observe whether the radii of the inner nodes behave as in fig. 3.1.

Recapitulation is an important property because it provides us with an immediate interpretation of the physical meaning of the quantum defect μ: for large enough n, the bound state wavefunctions possess an oscillatory inner part which defines a *phase*, and is nearly independent of energy if μ is nearly constant in energy. A change in the value of μ corresponds to a shift in the radial position of all the nodes. As one tends to the series limit, the oscillatory part grows. Continuum functions, of course, become completely oscillatory. Clearly, their properties are very similar to those of the bound states of very high n. If we define the phase δ at the base of the continuum, then the relationship

$$\delta = \pi\mu \qquad (3.10)$$

expresses this continuity across the threshold (see section 3.8). It also explains why μ, as an essentially constant quantity, is only really defined for high enough n, since one can only infer a phase when there are sufficient nodes in the wavefunction, and recapitulation breaks down at very low n.

There are instances in atomic physics where the recapitulation of nodes actually breaks down even at fairly high n. While such situations are rare, they are extremely interesting, because they provide insight into the real meaning of the quantum defect, and allow one to understand how it may cease to be constant even in the absence of any perturbation (see section 5.18).

3.5   Transition to MQDT

A simple analogy may help to understand how and why the transition to a multichannel theory can be performed. It is a matter of common observation that, as one approaches a bridge with equally spaced railings on each side, viewing it laterally, a beat frequency or *moiré fringes* are seen. A similar phenomenon occurs with two noninteracting Rydberg series: if the series limits I_1 and I_2 are shifted with respect to each other by an amount $\Delta_\infty = I_2 - I_1$, then different spacings occur at a given energy, so that moiré fringes also occur. The condition for levels from each series to coincide is

$$E = I_2 - \frac{R}{n_2^2} = I_1 - \frac{R}{n_1^2} \qquad (3.11)$$

leading to

$$\Delta E_\infty n_1^2 n_2^2 = R(n_2^2 - n_1^2)$$

and to

$$n_2^2 = \frac{n_1^2}{1 - \dfrac{n_1^2 \Delta E_\infty}{R}} \qquad (3.12)$$

Clearly, if $\Delta E_\infty = 0$, all the levels of the two series coincide and there are no Moiré fringes. Also, if

$$\frac{\Delta E_\infty n_1^2}{R} \gg 1$$

then $n_2^2 = R/\Delta E_\infty$ and $n_1 \gg n_2$, which means that the series member n_2 is nearly coincident with the series limit $E_{1\infty}$ and many near-degeneracies occur. Between these extremes, fringe patterns occur. These fringes are due to a periodicity in the recurrence of regions containing a high density of levels.

Now suppose that these two series are coupled, i.e. that they interact with each other. Since the coupling between two levels (spectral repulsion) is greatest when the levels lie closest together, there will clearly be a Rydberg periodicity in the strongly perturbed levels, while the intermediate levels (in between the moiré fringes) see much smaller perturbations. This cyclic nature of perturbations can be exploited to draw a much more complete picture of the interchannel interaction: by an appropriate choice of coordinates, the recurrences can be mapped into a single, compact diagram. Also, given information within one Rydberg cycle, one can predict what happens in all the others, provided they repeat or recapitulate accurately. This technique of parametrisation and extrapolation is known as multichannel QDT (MQDT) because it can be extended well beyond the situation of just two interacting Rydberg series considered first for simplicity.

3.6   Inclusion of two channels: the Lu–Fano graph

The theory described in section 3.3 is called *single channel QDT*. It reduces an unperturbed Rydberg series to a small number of constants, which include μ. Were this the only achievement of QDT, it would not have assumed much importance in atomic physics, because most series in the spectra of many-electron atoms are in fact perturbed, so that a constant value of μ is a rarity.

Consider now two Rydberg series, one converging to a series limit I_1, and the other to a limit I_2, in such a way that the bound states of the series to the lower limit (I_1, say) are perturbed by lower members of the series to I_2. Let us further suppose that the coupling between the two series is adjustable, so that we may begin by setting it to zero, in which case both series are unperturbed. For brevity, each one of these series, including all its bound states and the adjoining ionisation continuum, will be described as one excitation *channel*. Thus, in the first instance we are setting the interchannel coupling to zero.

When a series is unperturbed, μ is constant. Suppose the unsorted levels have been observed in an experiment. In order to assign them to one or other series, the procedure is to calculate quantum defects referring them to both possible limits, and then to choose the 'best fit' to two series, one of constant μ_1 and the other of constant μ_2, converging to the limits

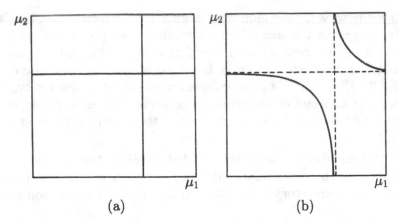

Fig. 3.2. Plot of quantum defects for the bound states of two Rydberg series converging to different limits: (a) for no interchannel coupling; (b) for nonzero interchannel coupling.

I_1 and I_2 respectively. As a matter of convenience, each of the Rydberg members i would have two numbers μ_1^i and μ_2^i associated with it, and we can plot all the points i on a plane using these two coordinates.

The result is the two-dimensional graph pictured in fig. 3.2(a): since the principal quantum number n is not relevant to the determination of the fractional part of μ, the graph may be plotted modulo 1 on both axes. For the case we have chosen (constant quantum defects), the points will lie on two intersecting straight lines, and we can use such a plot to read off directly which lines belong to which series.

It is clear that this procedure only works if both μ_1 and μ_2 are constant. If a point lies far off the lines, then it becomes difficult to decide to which of the two series it might belong. Let us now suppose that a weak interaction between the two series is turned on. In this case, the quantum defects can no longer remain constant because both series will be perturbed, but the extent of the perturbation will depend on the energy interval between the series members. In a case of zero interchannel coupling, there is no relation between values of μ_1 and μ_2, so that a member of series 1 might possess *any* value of μ_2. If, however, the coupling is turned on, then, for values of μ_2 close to the original (unperturbed) values, the series member will be perturbed by members of the other series, of which there is necessarily at least one close in energy since all values of n are present in the graph. Consequently, its value of μ_1 will be altered from the unperturbed value. Since (by standard second-order perturbation theory) the strength of the perturbation falls off as the inverse of the energy interval, we see that the branches of the graph must affect each other most at the point where they cross. There is in fact a fundamental rule (Wigner's no-crossing theorem) due to von Neumann and Wigner [115] which states that two curves representing the energies of interacting staes (i.e. states with the same quantum numbers J) as a function of a continuously adjustable perturbation cannot actually cross, but must exhibit an *avoided crossing* (see fig. 3.2(b)), which only becomes exact in the unattainable limit of zero coupling. For an exact or true crossing to occur, the series must possess different quantum numbers, in which case there can be no interaction between them.

For series members close to the avoided crossing, the distance to the two branches in the graph may easily become equal. It is then impossible to decide which series they belong to, and we say that they are completely mixed.

The advantage of such plots is that all the members of a series (excepting, perhaps, the first few, for which μ in the unperturbed condition would not have settled down to a constant value anyway) can be included. The plot is a means of converting information which, at first sight, differs from one member to the next because the energy intervals are ever decreasing, into strictly comparable data in which the doubly-periodic (i.e. periodic in both ν_1 and ν_2) nature of Rydberg states is properly displayed.

We have earlier commented that ℓ characterisation may readily break down for many-electron atoms, because it is dependent on the validity of the central field approximation. Here, we encounter a new form of breakdown of the n characterisation: it is no longer clear which series certain members belong to when they are strongly coupled. There is no particular meaning in attributing n values to individual series members under these conditions. Thus independent particle labelling of the states becomes a matter of convention or compromise: in fact, the magnitude of the avoided crossing in the diagram is a measure of many-body effects, or of the breakdown of the independent electron approximation.

We have introduced the diagrams of fig. 3.2 as plots of μ_1 against μ_2. More usually, since theory yields a continuous function $\mu_1(\nu_2)$ representing a continuous variation in μ_1 as a function of the continuous energy variable ν_2 (see below), it is conventional to plot μ_1 against ν_2 modulo 1 (i.e. taking account only of the fractional part), which (neglecting a sign change) is the same diagram. Such graphs are called Lu–Fano graphs [111, 116] and are commonly used to analyse interactions between series when they are simple enough to involve only two distinct limits. Of course, there is no reason in principle why more than two series limits may not be involved, and the interactions may easily become too complex to be represented in one plane. In its most general form, the interaction would be represented in an N-dimensional cube, where N is the number of limits involved (see section 3.9).

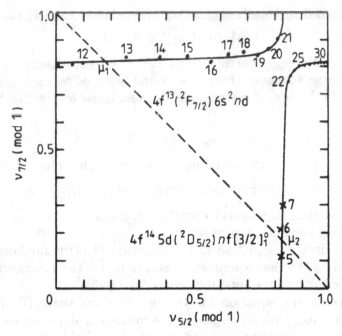

Fig. 3.3. Experimental example of a Lu–Fano graph showing the avoided crossing between just two interacting series in the spectrum of Yb. The series involved are an inner-shell and a doubly-excited series: this kind of excitation is discussed in chapter 7 (from M.A. Baig and J.-P. Connerade [113]).

3.7   Properties of Lu–Fano graphs

The most obvious property of a Lu–Fano graph is its branch structure which, in the case of two series, involves only one avoided crossing. An experimental example with two series is shown in fig. 3.3.

An interesting rule (due to the periodic or cyclic nature of the perturbation in overlapping Rydberg series discussed above) is that, as long as the parameters have no strong energy dependence, the curves enter and leave the diagram on opposite sides in the same position and with the same slope, i.e. they form a multiperiodic continuous function. The experimental data correspond to fixed points, defined by effective quantum numbers n_1^* and n_2^*. In order to obtain smooth curves, we introduce energy variables ν_1 and ν_2. The defining relationship between them is

$$I_1 - \frac{R}{\nu_1^2} = I_2 - \frac{R}{\nu_2^2} \qquad (3.13)$$

and is independent of the coupling strength. The bound states of series

1 are determined by solutions of an equation similar to (3.7) *viz.*

$$\tan\left[\pi(\nu_1 + \mu_1)\right] = F_1(\nu_2)$$

where $F_1(\nu_2)$ represents perturbations due to the second series, and must clearly be large wherever there is a bound state of series 2, but small otherwise. Both series must be treated on an equal footing, so we must also have

$$\tan\left[\pi(\nu_2 + \mu_2)\right] = F_2(\nu_1)$$

where F_1 and F_2 have the same functional form. The solution is

$$\tan\left[\pi(\nu_1 + \mu_1)\right]\tan\left[\pi(\nu_2 + \mu_2)\right] = R_{12}^2 \qquad (3.14)$$

where R_{12}^2 is the interchannel coupling strength. When $R_{12}^2 = 0$, we clearly recover two independent series, each one with a constant quantum defect, as defined by equation (3.7). Equation (3.14) is fundamental to two-channel QDT: the experimental points lie on the intersections between this curve and the function defined by (3.13).

By differentiating equation (3.14), one can show that: (i) the curve $\mu_1(\nu_2)$ always has positive slope; (ii) the maximum slope occurs at $\nu_2 = n_2 - \mu_2$ or $\nu_2 = -\mu_2$ when plotted modulo 1; (iii) this maximum slope is $1/R_{12}^2$; (iv) the minimum slope occurs at $\nu_2 = (n_2 + \frac{1}{2}) - \mu_2$ or at $\nu_2 = \frac{1}{2} - \mu_2$ when plotted modulo 1; (v) the minimum slope is R_{12}^2.

There are three limiting situations to consider: we may have $R_{12} \gg 1$ (strong mixing), $R_{12} \ll 1$ (weak mixing) and $R_{12} = 1$ (equal mixing). The physical distinction between these three situations will appear more clearly when we come to consider the associated resonances in the continuum (section 6.10).

We can also write equation (3.14) in its determinantal form:

$$\begin{vmatrix} \tan\left[\pi(\nu_1 + \mu_1)\right] & R_{12} \\ R_{21} & \tan\left[\pi(\nu_2 + \mu_2)\right] \end{vmatrix} = 0 \qquad (3.15)$$

which has the advantage that it can readily be generalised to many channels as

$$\mid \tan\left[\pi(\nu_i + \mu_i)\right]\delta_{ij} + (1 - \delta_{ij})R_{ij}\mid = 0 \qquad (3.16)$$

for situations in which many series interact. Further generalisations of equation (3.14) are discussed in section 8.36.

3.8   Continuum states

As was emphasised above, continuum states are included in the description of the complete excitation channel. The part lying below threshold (bound states) is called *closed*, while the continuum states are the *open* part of the channel. The energy range where channels begin to open is actually rather complex, as it contains both a structure of discrete origin and continua.

In fig. 3.1, wavefunctions for bound states and for the continuum state lying near threshold ($\epsilon \to 0$) are shown. In order to compare them, one has to normalise the continuum functions, which are not square-integrable. This difficulty is resolved by requiring

$$< \epsilon \mid \epsilon' > = \delta(\epsilon - \epsilon') \tag{3.17}$$

which is the delta function normalisation condition. With this choice, one obtains the comparison shown in the figure, which emphasises that the radii of the inner nodes of the bound states are continuous with those of the continuum states across the threshold. The radii of the nodes define the phase shift δ in the open channel. For an increase of one unit in the quantum number n, the bound state solutions gain one extra outer node, i.e. are shifted by half a wavelength, which means that the phase is shifted by π. The phase shift δ in the open channel is thus related to the quantum defect by

$$\delta = \pi\mu \tag{3.18}$$

(sometimes called Seaton's theorem) which expresses continuity across the threshold.

For two-channel QDT, between the thresholds I_1 and I_2, there are further series members belonging to series 2 which lie in the continuum of series 1 and interact with it. We postpone a discussion of these series members, which give rise to broadened resonances, until section 6.10. Above I_2, two continua are present, and possess phases δ_1 and δ_2 which are related as above to μ_1 and μ_2.

More generally, in any problem involving coupled series, equation (3.16) determines the behaviour of the observed spectrum in the full energy range. Below the thresholds, in the energy range where all channels are closed, the solutions determine the bound state energies. In the region between the thresholds, some channels are open and others closed, and this equation then determines the phase shifts due to resonances, as will be described later. There is also a region above all the thresholds, where only continua occur, in which case this equation determines the connections between all the smooth phases, or the compatibility conditions connecting the continua.

3.9   Extension to more channels and to more limits

Of course, there is no reason to limit the problem to two series. We could in principle have any number of series converging to two limits. As long

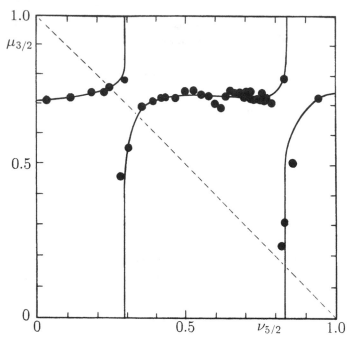

Fig. 3.4. Example of a Lu–Fano graph involving more than two interacting series, and consequently more than one avoided crossing. This particular graph occurs in the spectrum of Yb, and involves doubly-excited configurations, discussed in chapter 7. The number of series in the graph is equal to the number of intersections of the curves with the diagonal, i.e. 3 in this case (Kaenders and Connerade – unpublished).

as only two series limits are present, there is no problem in including within the graph as many interacting series as are actually present, and so diagrams of far greater complexity are created, with a greater number of avoided crossings, according to the number of series converging to each limit. Reasoning as before, if the interchannel couplings all tend to zero, then the number of lines parallel to one axis is equal to the number of series associated with the other. As a general rule for interacting series, if we draw a diagonal line across the graph, the maximum number of intersections with the quantum defect curves tells us how many series are involved. Hence, one deduces the number of avoided crossings and the general structure of the graphs, and the associated equations are readily generalised from those given above.

An example of a two-dimensional Lu–Fano graph involving more than two interacting series is shown in fig. 3.4. A more difficult generalisation (because it cannot be represented graphically in two dimensions) is the case in which more than two series limits are present. For three limits, the

Lu–Fano plot is contained in a unit cube, and more generally, it will give rise to a curve in an N-dimensional cube, at which point the simplicity of visualisation is lost, although a mathematical representation is still completely feasible.

One should remember, also, that near an avoided crossing it is often the case that one interaction dominates the others, so that a projection onto one face of the cube contains the most relevant information. The generalisation of MQDT to many limits has been discussed by Armstrong *et al.* [117].

3.10   Limitations of empirical QDT

So far, we have discussed what is called empirical QDT, namely the fitting of experimental data using the finite number of adjustable parameters required to represent interchannel coupling. Unfortunately, while the procedure works well for simple cases, one eventually runs into two problems: (1) for complex cases involving many parameters, it can turn out that more than one set is found which will represent the same experimental data equally well; and (2) it may prove necessary to introduce energy-dependent parameters in order to represent observations.

There is, of course, no objection in principle to the occurrence of energy-dependent parameters, provided this dependence is weak:[2] after all, the constancy of μ was the approximation in the first place, as the numbers in table 2.1, in the absence of perturbations, show quite clearly. However, the consequence of an energy dependence is that the variations in μ_1 and μ_2 are no longer perfectly periodic, and it may then become unclear which part of the variation with energy is intrinsic (i.e. existed before interchannel coupling was turned on) and which part is produced by coupling between channels.

The first difficulty is more fundamental, and suggests that, in complex cases, more information is needed to determine quantum defect parameters completely than is available in a set of experimental transition energies.

One therefore seeks a more fundamentally based theory. In fact, the theory was originally developed to supplement *ab initio* Hartree–Fock calculations, in which case uncertainties such as the indeterminacy of n are automatically removed. With a similar purpose in mind, it has been necessary to develop a relativistic version of MQDT. This was done by Johnson and Cheng [120].

---

[2] The case where the variation is strong in the absence of perturbations is also of interest: we return to this issue in section 5.20.

Greene [118] and Aymar [119] have described an approach in which the
Schrödinger equation is solved for a parametric potential, and phase shifts
are determined unambiguously from the solution. The potential can be
fixed by using experimental ionisation thresholds, and this turns out to
be sufficient information for a very accurate representation of even quite
complex spectra (see chapter 7 for some examples).

3.11 Lu–Fano graphs for polyatomic molecules

Rydberg series in polyatomic species were discussed in section 2.33. Because of ℓ-uncoupling, increased 'atomicity' at high n and the availability
of well-resolved experimental spectra of polyatomic molecules at high n,
one can reexamine the assumption of nearly constant, quasiatomic quantum defects used in early studies of molecular spectra [87].

The idea is to incorporate some of the very important developments of
MQDT made in atomic physics. An ambitious programme (not addressed
here) is to develop and extend MQDT to molecular species by including
full rovibronic structures for simple molecules within the framework of an
extended theory by using frame transformations. Instead, we describe a
far more restricted phenomenological approach, akin to MQDT for atoms,
and applicable only to high n states of polyatomic species in which the
rotational and most of the vibrational structure has collapsed. We can
then compare the Lu–Fano graphs directly with those of 'corresponding'
atoms, and discuss both similarities and differences.

The reason for choosing fairly complex molecules is that this approach
makes no pretence to include rotational or vibrational structure. Heavier
molecules have the advantage of suppressing rotational detail and making
Rydberg regularities more apparent.

The next example illustrates how the 'corresponding atom' is chosen.
Consider the molecules HI and HBr. The presence of the light H atom
results in a very open rotational structure, so that, at high principal
quantum numbers, successive Rydberg manifolds overlap. Since separated
Rydberg members do not occur at very high n, they can only be sorted
out by rotational cooling.

On the other hand, if the H atom is replaced by the heavier group CH_3,
the rotational structure disappears from view, and beautiful Rydberg series result, which can be followed back without difficulty from their series
limits. Examples from the spectra of CH_3I and CH_3Br were shown in
figs. 2.21 and 2.19. The chromophore in these molecules is the halogen
atom, and the equivalent or united atom is the one following in the periodic table, *viz.* a rare gas, the extra electron being provided by the CH_3
group. Like the spectra of the corresponding rare gases, those of CH_3I

and CH_3Br exhibit a prominent doublet structure, due to the spin–orbit interaction in the excited core. However, the splittings are different: since I is heavy and the C–I bond is relatively weak, the molecule resembles a diatomic molecule with symmetry $C_{\infty v}$ [122], which removes the $\frac{3}{2}$ factor from the atomic spin–orbit formula for p^5.

The history of these spectra is quite interesting: pioneering work on the absorption spectra of the methyl halides was performed by Price [121] who found the Rydberg series converging to the two ^2E limits, the spin–orbit splitting of which had earlier been predicted by Mulliken [123]. Following the work of Price, Mulliken [124] identified the excited configurations. A more complete bibliography can be found in the paper by Baig *et al.* [125].

Of course, in addition to Rydberg series, features specific to molecules such as a vibrational progression in CH_3I can appear. Nonetheless, the electronic structure can be sorted into two groups of series closely reminiscent to those of the atom.

We now plot the usual Lu–Fano graph, using, as before:

$$\left. \begin{aligned} E_n &= I_1 - \frac{R}{(n_1 - \mu_1)} \\[2mm] E_n &= I_2 - \frac{R}{(n_2 - \mu_2)} \end{aligned} \right\} \tag{3.19}$$

on the understanding that μ_1 and μ_2 now have a more complicated meaning, because they may absorb some molecular distortions. We obtain the plot of fig. 3.5(a). Note that; (i) in contrast to the procedure of Price we make no assumption as to the constancy of quantum defects; and (ii) the procedure is at this stage completely empirical. For comparison, we also show in fig. 3.5(b) the Lu–Fano graph for the corresponding atom. It emerges that the quantum defects for the molecule are not the same as for the corresponding atom, but the topological structure of the Lu–Fano plot (i.e. the number of branches, and the number of times they enter and leave the diagram) is the same for corresponding series. In other words, the analogy between atomic and molecular Rydberg spectra can be sought not in the constancy or numerical values of the parameters, but in the general structure of the plot and the nature of the perturbations or couplings between the series.

This method can be put on a somewhat more formal footing [122]. There there are some interesting details, such as the presence of weak, supernumerary series in the molecular plots which do not occur for the atom. However, exactly as for coupled channels in atoms, the MQDT interpretation gives no information on the precise nature or assignment labels to be attached to the states. It seems natural, in view of the

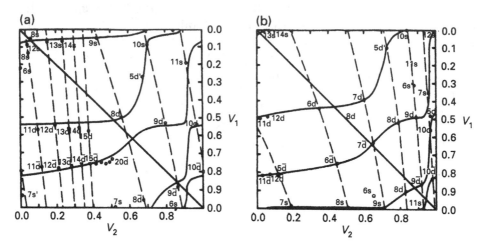

Fig. 3.5.   Two-dimensional Lu–Fano graphs: (a) for CH₃I and (b) for the corresponding series in the united atom Xe. Note the structural similarity of the graphs despite the differences in magnitude of the quantum defects (from J.A. Dagata *et al.* [122]).

similarities between the atomic and molecular spectra, to describe series as '*ns*' or '*nd*' at high enough values of *n* the principal quantum number, but in reality this is no more than an analogy, valid only at high values of *n*, which cannot be expected to apply strictly in a molecular field.

3.12   Conclusion

MQDT is an evolving subject with a vast literature, covering both atomic and molecular physics. It finds extensive applications. Thus, the summary in the present chapter is of necessity incomplete. The reader should be warned that several different formulations of the theory exist, which can be transformed into each other but differ somewhat in apparent structure. It is unwise to use equations borrowed from different papers on MQDT without checking quite carefully which version the author uses and whether the formulations are consistent with each other. The full apparatus of MQDT is quite complex. We have emphasised only the basic principles: for a complete discussion of the mathematical detail of MQDT, the reader is referred to a review paper by Seaton [126].

4

Atomic f values

4.1 Line strengths of discrete transitions

In the present chapter, we consider line strengths of transitions between bound states, i.e. of lines whose only form of natural broadening is radiative, and which lie below the first ionisation potential. The simplest situation is encountered in the *photoabsorption* or *photoexcitation* of an atom, initially in its ground state $| i >$, in which case one transition is observed to each excited final state $| f >$. The price one pays for this simplicity is that all excited states cannot be reached in this way because of selection rules.

The distribution of intensities is an essential property of a spectrum. We may consider: (a) the distribution in energy over the whole spectrum; or (b) the distribution within an individual spectral line.

It was noted in the previous chapter that spectral lines of interacting channels can differ greatly in intensity within a narrow energy range. However, with some significant exceptions, if one can approach the series limit closely enough to be clear of perturbers, the intensities of successive members decrease monotonically with increasing principal quantum number n. This is described as *the normal course of intensity* for a Rydberg series.[1].

In the presence of perturbations, the course of intensities becomes far less regular than that of transition energies, and it is generally more likely that intensities will exhibit fluctuations or departures from the expected

---

[1] It is perhaps worth noting here that the 'normal' course of intensities does not necessarily apply even in absence of perturbers. One such situation is illustrated, in fig. 4.4 Another possible cause exists, which is sufficiently rare to be regarded as an interesting anomaly: centrifugal barrier effects (see chapter 5.) can be responsible for remarkable variations in the course of intensities, a good example being the $5d \longrightarrow nf$ spectrum of Ba^+. Both effects are features of the independent electron model of the atom, although they are strongly altered in the presence of correlations

behaviour. Intensity variations are more sensitive indicators of the break-down of simple approximations than transition energies, and their study is therefore particularly important.

At a yet finer level of understanding, one is concerned, not just about intensities of lines, but also about how the intensity is distributed within the width of a spectral line. The lineshape, or line profile, carries within it much more information about the condition of excited atoms and, pos-sibly, their environment than could be deduced from the gross structure of spectra.

We shall concentrate our attention on homogeneous broadening effects, i.e. those which are the same for all the atoms in an ensemble, and exclude from our discussion effects due temperature and density.[2]

4.2   Selection rules

The time-dependent Schrödinger equation for an atom in an electromag-netic field reads

$$i\hbar\frac{\partial}{\partial t}\Psi(\mathbf{r},t) = \left\{-\frac{\hbar^2}{2m}\nabla^2 + V(r) - \frac{i\hbar e}{m}\mathbf{A}\cdot\nabla + \frac{e^2}{2m}\mathbf{A}^2\right\}\Psi(\mathbf{r},t) \quad (4.1)$$

where $V(r)$ is the central field of the atom, and the vector potential is introduced by replacing $\mathbf{p}$ in $\mathcal{H} = p^2/2m$ by $\mathbf{p}+e\mathbf{A}$ and using the relation $\nabla\cdot\mathbf{A} = 0$.

The vector potential of a plane wave is periodic and, if terms in $\mathbf{A}^2$ are neglected, the perturbation due to the field corresponds to a matrix element

$$M_{ij} = <i\,|\,(\exp i\,\mathbf{k}\cdot\mathbf{r})\hat{\mathbf{v}}\cdot\nabla\,|\,j> \quad (4.2)$$

where $\mathbf{k}$ is the propagation vector, and $\hat{\mathbf{v}}$, the polarisation vector of the wave.

The exponential can be expanded as:

$$\exp i\,\mathbf{k}\cdot\mathbf{r} = 1 + i\,\mathbf{k}\cdot\mathbf{r} + \frac{1}{2!}(i\,\mathbf{k}\cdot\mathbf{r})^2 + ... \quad (4.3)$$

For small $\mathbf{k}\cdot\mathbf{r}$, which is satisfactory at long wavelengths, but breaks down towards the X-ray range, the exponential can be replaced by unity and higher order terms are neglected. In this case

$$M_{ij} = \hat{\mathbf{v}}\cdot<i\,|\,\nabla\,|\,j> = \frac{im}{\hbar}\hat{\mathbf{v}}\cdot<i\,|\,\dot{\mathbf{r}}\,|\,f>$$

---

[2] For a discussion of these, we refer the reader to standard texts, e.g. [149].

because $\mathbf{p} = m\dot{\mathbf{r}} = -i\hbar\nabla$. Next, we use Heisenberg's equation to find the time evolution of $\mathbf{r}$:

$$\dot{\mathbf{r}} = \frac{1}{i\hbar}[\mathbf{r}, \mathcal{H}]$$

whence

$$<i\,|\,\dot{\mathbf{r}}\,|\,f> = \frac{1}{i\hbar}<i\,|\,\mathbf{r}\mathcal{H} - \mathcal{H}\mathbf{r}\,|\,f> = \frac{1}{i\hbar}(E_i - E_f)<i\,|\,\mathbf{r}\,|\,f>$$
$$= i\omega_{fi}<i\,|\,\mathbf{r}\,|\,f>$$

from which we deduce that

$$M_{ij} = -\frac{m\omega_{ij}}{\hbar}\hat{\mathbf{v}}\cdot<i\,|\,\mathbf{r}\,|\,j> \qquad (4.4)$$

Since $-e\hat{\mathbf{v}}\cdot<i\,|\,\mathbf{r}\,|\,j>$ is the component of the electric dipole moment connecting states i and j, this is called the electric dipole approximation.

The strength of a transition in this approximation depends on the value of $<i\,|\,\mathbf{r}\,|\,j>$: if it is zero, the transition is *electric dipole forbidden*. Even if it is dipole forbidden, the transition can still occur through higher terms of the multipolar expansion (4.3) which, in order, are the magnetic dipole, electric quadrupole, etc. If the whole series vanishes, then the transition is said to be *strongly forbidden*. However, even in this case, the transition can sometimes occur if the intensity of the radiation is high enough, through the influence of the $\mathbf{A}^2$ term in the Schrödinger equation, which was neglected at the outset. Thus, the selection rules for two-photon or multiphoton transitions induced by a strong laser field are not the same (see section 9.3) as for excitation by a weak radiation field.

For reasons of symmetry, as described in elementary texts, the dipole matrix element may vanish for all but a few values of the quantum numbers characterising the levels. The general rules which govern whether it vanishes or not are called *selection rules*.

In the present book, we assume that the reader is familiar with the elementary theory of selection rules and with the properties of the spherical harmonics which are used to describe the angular parts of atomic eigenfunctions. However, a few further remarks may be useful, since selection rules have such important consequences when one seeks to recover the structure of energy level diagrams from experimental observations.

Selection rules are the real key to spectroscopy. Without them, transitions between so many levels would be possible that reconstructing energy level diagrams from observational data would become an almost insuperable task. However, since selection rules may hinder accessing certain excited states, it is also necessary to know experimental means of violating them, so that one can, where necessary, reach otherwise inaccessible states. Thus, in a spectrum in which only singlet states can be photoexcited, it may prove useful that electron impact excitation can violate the

normal selection rules for photoexcitation and allow some triplet states
to be detected.

Selection rules may be rigorous (strong) or approximate (weak), depending on whether the quantum numbers involved are well-characterised
constants of the motion for the complete many-electron system or whether
they are only approximate constants, or describe only one electron or a
subgroup of electrons. Thus, for electric dipole transitions, in the absence
of external fields, the parity selection rule, which allows only transitions
from even to odd or odd to even states to occur is a strong rule. Likewise,
if ΔJ is the difference in total angular momentum quantum number between the initial and final states, then $\Delta J = 0, \pm 1$ and $J = 0 \longrightarrow J = 0$
is also a strong rule. On the other hand, the rules $\Delta L = \pm 1$ and $\Delta S = 0$
are valid only for LS coupling, and will give way to different rules (or,
perhaps, no strict rules) when the coupling conditions change. These
are called *weak* selection rules. Likewise, although the parity rule is a
strong one, the individual ℓ_i angular momentum of electron i is not an
exact quantum number for a many-electron system. Thus, while the parity, which is defined as $(-1)^{\sum_i \ell_i}$ remains a good quantum number, the
individual values of ℓ_i which are involved in defining it are not unique.
This is how configurations involving more than one excited electron can
be accessed by photoexcitation.

One can also regard the selection rules for radiative excitation as an
expression of the conservation of angular momentum: since the angular
momentum of the photon is ± 1 atomic units, the angular momentum
of the atom must also change by ± 1 upon absorption or emission of one
photon, which may be coupled more or less strictly to the change $\Delta \ell$ in the
angular momentum of an individual electron via the independent particle
approximation. This way of envisaging selection rules will be useful in
chapter 9, when excitation by intense beams of light (laser radiation) will
be considered.

For excitation by a weak beam of radiation, such that a single photon is
involved per absorption event, one is mainly concerned with electric dipole
transitions, because they are usually the strongest. Other selection rules
will apply if the transitions are not due to an induced electric dipole.

When the radiation is intense, more than one photon may become involved in a single transition (this is called multiphoton excitation), and
the selection rules are, again, no longer the same (chapter 9).

In the presence of external DC fields, selection rules break down, and
one may even, in extreme cases, need to introduce new quantum numbers. All states of different projections of total angular momentum are
degenerate in energy for a spherical atom in the absence of external fields,
but the degeneracy is lifted when a magnetic field is applied, and differ-

ent polarisations of the light beam then give rise to different transitions between the magnetic sublevels (see section 4.9) below. In *strong* electric and magnetic fields, the breakdown is more pronounced. Selection rules for orbital angular momentum are broken in strong electric fields, because the full rotational symmetry of the atom no longer exists, so that its eigenfunctions no longer belong to a representation of the three-dimensional rotation group [150]. The same is true in strong magnetic fields, as explained in detail in chapter 10.

Dipole selection rules apply for excitation by single photons in the perturbative regime. Selection rules for multiphoton excitation are different (see chapter 9). For excitation by collisions with charged particles or by light beams of high intensity, turned on so fast that the normal conditions of perturbation theory do not apply, then there are no strict selection rules, although various *propensity* rules may still apply.

Unlike a selection rule, which acts by forbidding a large number of transitions, a propensity rule is interpreted positively as encouraging some specific transition. It can be violated (i.e. is not strict), but nevertheless gives a fair idea of which transition is likely to be strong in a spectrum. Allowed transitions between states of similar mean radius will tend to have a large matrix element, and this may be regarded as the basis for a propensity rule. Although such transitions tend to be strong, other transitions are also of significant strength, which is why it is only a propensity rule.

Thus, a propensity rule is not strict. One example is Bethe's rule, which states that, for dipole transitions, the orbital angular momentum ℓ changes with overhelming probability in the same direction as n. This rule is actually violated for small ℓ, but as ℓ increases its validity grows. Bethe's rule has interesting connections with quasiclassical limits [127].

A selection rule is stricter, but only applies in a negative sense: the rule does not guarantee that a transition which satisfies it will be strong: it may still happen that a transition is very weak or disappears for some other reason (usually connected with the radial equation). What *is* implied is that any transition which *violates* the rule is so weak as to be barely observable. While selection rules involving angular momentum quantum numbers have quite general validity, it may still happen, because of the radial wavefunctions, that a particular allowed transition between states with specific principal quantum numbers vanishes. This is due to a phase cancellation effect.

As an example of a propensity rule, transitions between two states of similar n are usually more intense than transitions between states of very different n. While the radial Schrödinger equation does not give rise to genuine selection rules in the same manner as those derived for spherical harmonics, radial matrix elements govern the intensities of allowed transitions, and are responsible for the more subtle propensity rules.

4.3   Einstein rate coefficients

Transition strengths can be given in terms of Einstein rate coefficients. For a pair of states $\mid \jmath >$ and $\mid k >$ it is shown in elementary texts that these are related in a simple way. If one assumes that, for any pair of microstates i and j, the rate from i to j is equal to the rate from j to i one has the principle of detailed balance). Then, the relation between the coefficients is consistent with thermodynamics (Planck's black-body radiation law).

We assume the reader is familiar with this theory and merely restate the relations to define units and notation. If $E_\jmath < E_k$ we have:

$$\frac{dN_k}{dt} = -g_k N_k B_{k\jmath} I(\nu_{k\jmath}) + g_\jmath N_\jmath B_{\jmath k} I(\nu_{k\jmath}) - g_k N_k A_{k\jmath} \qquad (4.5)$$

and

$$\frac{dN_\jmath}{dt} = g_k N_k B_{k\jmath} I(\nu_{k\jmath}) - g_\jmath N_\jmath B_{\jmath k} I(\nu_{k\jmath}) \qquad (4.6)$$

and the Einstein–Milne relations state that

$$g_\jmath B_{\jmath k} = g_k B_{k\jmath} \qquad (4.7)$$

and

$$A_{k\jmath} = \frac{2h\nu^3}{c^2} B_{k\jmath} \qquad (4.8)$$

This last relation is actually a purely microscopic equation, and its derivation via thermodynamics in the elementary texts is somewhat circuitous: in reality, quantum electrodynamics is required for a proper treatment (see, e.g., [128]).

In the absence of any radiation field, $I(\nu_{k\jmath}) = 0$ The population N_k of level k simply decays exponentially with a lifetime $A_{k\jmath}^{-1}$ to level $\jmath$. This provides one way of defining the intensity of a spectral line.

It is also shown in elementary texts that the spontaneous transition rate for emission of a photon by a single atom in the dipole approximation is given by:

$$S_{k\jmath} = \frac{4\alpha}{3c^2} \omega_{k\jmath}^3 \mid \mathbf{r}_{k\jmath} \mid^2 \qquad (4.9)$$

where $-e\mathbf{r}_{k\jmath} = -e < k \mid \mathbf{r} \mid \jmath >$ is the electric dipole moment coupling states $\jmath$ and k. This expression allows calculation of transition strengths directly from the eigenfunctions of the atom if they are known and allows selection rules to be derived.

4.4   Oscillator strengths

4.4.1   Classical dispersion theory of the driven oscillator

Intensity variations reveal a variety of effects, both in the bound spectrum and above threshold. In order to study them, one needs to introduce the concept of *oscillator strength*, as a preliminary to discussing various factors which influence the intensities of spectral lines.

The equation of motion of the electron, driven as an oscillator by the electric field vector $\vec{\mathcal{E}} = \vec{\mathcal{E}}_0 \exp i\omega t$ of an electromagnetic wave,[3] is:

$$\ddot{\mathbf{r}} + \gamma \dot{\mathbf{r}} + \omega_0^2 \mathbf{r} = -\frac{e}{m} \vec{\mathcal{E}}_0 \exp i\omega t \tag{4.10}$$

for which the steady state solution is:

$$\begin{aligned}
\mathbf{r} &= \frac{-e/m}{\omega_0^2 - \omega^2 + i\omega\gamma} \vec{\mathcal{E}}_0 \exp i\omega t \\
&= \frac{-e/m}{\omega_0^2 - \omega^2 + i\omega} \vec{\mathcal{E}} \gamma
\end{aligned} \tag{4.11}$$

leading to a current density in the medium

$$\mathbf{J} = -Ne\frac{d\mathbf{r}}{dt} = \frac{Ne^2}{m} \frac{1}{\omega_0^2 - \omega^2 + i\omega\gamma} \frac{d}{dt} \tag{4.12}$$

where N is the number density of oscillators. Using Maxwell's equations:

$$\begin{aligned}
\nabla \times \mathbf{B} &= \frac{\mu_0}{\kappa} \left(\epsilon_0 \frac{d}{dt} \vec{\mathcal{E}} + \mathbf{j} \right) \\
&= \frac{\mu_0}{\kappa} \left(1 + \frac{Ne^2}{m\epsilon_0} \frac{1}{\omega_0^2 - \omega^2 + i\omega\gamma} \right) \frac{d}{dt} \vec{\mathcal{E}} \\
&= \frac{\mu_0}{\kappa} \epsilon_0 \epsilon \frac{d}{dt} \vec{\mathcal{E}}
\end{aligned} \tag{4.13}$$

where

$$\epsilon \equiv 1 + \frac{Ne^2}{m\epsilon_0} \frac{1}{\omega_0^2 - \omega^2 + i\omega\gamma} \tag{4.14}$$

is the complex dielectric constant relative to the vacuum. We define a complex refractive index $\tilde{n} = n - ik$ such that

$$\epsilon = (n - ik)^2 = n^2 - 2ink - k^2 \tag{4.15}$$

---

[3] We use $\mathcal{E}$ for electric field with an arrow to denote the vector in order to avoid confusion with E the total energy.

Then, provided $|\epsilon| \sim 1$, we can write

$$
\left.
\begin{array}{rl}
n & \simeq \; 1 + \dfrac{Ne^2}{2m\epsilon_0} \dfrac{\omega_0^2 - \omega^2}{(\omega_0^2 - \omega^2)^2 + \gamma^2 \omega^2} \\[3mm]
k & \simeq \; \dfrac{Ne^2}{2m\epsilon_0} \dfrac{\omega\gamma}{(\omega_0^2 - \omega^2)^2 + \gamma^2 \omega^2}
\end{array}
\right\}
\tag{4.16}
$$

and the absorption coefficient a defined by $a \equiv (2\omega/c)\,k$ is given by

$$
a \simeq \frac{Ne^2}{m\epsilon c} \frac{\omega^2 \gamma}{(\omega_0^2 - \omega^2)^2 + \omega^2 \gamma^2}
\tag{4.17}
$$

4.4.2   Approximation for narrow profiles

For narrow profiles at high frequencies, $\gamma << \omega_0$ and $\omega - \omega_0$ the detuning is small, so we can write $\Delta\omega \equiv \omega - \omega_0$ and $\omega + \omega_0 \simeq 2\omega$. It follows that

$$
\begin{aligned}
(\omega_0^2 - \omega^2)^2 + \gamma^2\omega^2 & \simeq \; (2\omega\Delta\omega)^2 + \gamma^2\omega^2 \\
& \simeq \; 4\omega^2 \left\{ \Delta\omega^2 + (\gamma^2/2) \right\}
\end{aligned}
\tag{4.18}
$$

and

$$
\omega_0^2 - \omega^2 \simeq 2\omega\Delta\omega
\tag{4.19}
$$

whence

$$
a(\omega) \simeq \frac{Ne^2}{m\varepsilon_0 c} \frac{\omega^2 \gamma}{4\omega^2 \left[(\omega_0 - \omega)^2 + (\gamma/2)^2 \right]}
$$

$$
\simeq \frac{Ne^2}{2m\epsilon c} \frac{(\gamma/2)}{(\omega_0 - \omega)^2 + (\gamma/2)^2}
\tag{4.20}
$$

and

$$
n(\omega) - 1 = \frac{Ne^2}{4m\epsilon_0\omega_0} \frac{\omega_0 - \omega}{(\omega_0 - \omega)^2 + (\gamma/2)^2}
\tag{4.21}
$$

Usually, spectroscopists work not with the angular frequency ω but with real frequency $\nu = \omega/2\pi$, in which case

$$
a(\nu) = \frac{Ne^2}{4\pi m\epsilon_0 c} \frac{\gamma/4\pi}{(\nu_0 - \nu)^2 + (\gamma/4\pi)^2}
\tag{4.22}
$$

and

$$
n(\nu) - 1 = \frac{Ne^2}{16\pi^2 m\epsilon_0\nu_0} \frac{\nu_0 - \nu}{(\nu_0 - \nu)^2 + (\gamma/4\pi)^2}
\tag{4.23}
$$

where the ratio of the constants for $a(\nu)$ and $n(\nu)$ is, as expected, $4\pi/c$. In units for which $\varepsilon_0 = 1/4\pi$, one has:

$$
a(\nu) = \frac{Ne^2}{mc} \frac{\gamma/4\pi}{(\nu_0 - \nu)^2 + (\gamma/4\pi)^2}
\tag{4.24}
$$

and

$$n(\nu) - 1 = \frac{Ne^2}{4\pi m\nu_0} \frac{\nu_0 - \nu}{(\nu_0 - \nu)^2 + (\gamma/4\pi)^2} \tag{4.25}$$

4.4.3 Absorption integrated over all frequencies

To form an estimate of the total spectral strength of the classical oscillating electron, we integrate the absorption over all frequencies thus:

$$S_{class} = \int_0^\infty a(\nu)d\nu \tag{4.26}$$

If we assume that the natural frequency ν_0 of the oscillating or dispersion electron occurs at a very high value so that the kernel of the integral is nearly zero at the lower limit, we can drop the lower limit to $-\infty$, in which case the integration can be performed exactly and

$$S_{class} \simeq \int_{-\infty}^{+\infty} a(\nu)d\nu$$

$$= \frac{Ne^2}{4\pi m\epsilon_0 c} \int_{-\infty}^{+\infty} \frac{(\gamma/4\pi)d\nu}{(\nu_0 - \nu)^2 + (\gamma/4\pi)^2}$$

$$= \frac{Ne^2}{4m\epsilon_0 c} \tag{4.27}$$

yielding the remarkable result that the strength is completely independent of the linewidth γ. This means that the oscillator strength of a spectral line can be defined quite independently of damping or broadening mechanisms.

There is also the implication in this procedure that the line must be discrete, i.e. this procedure cannot be applied to a continuum, since an integral over all frequencies would then diverge. The question of how to describe continua is considered in subsection 4.6.1.

4.4.4 Extension to the simple quantum theory

According to the Einstein theory, the rate of energy absorption from a field I_ν is given by $N_j B_{jk} I_\nu h\nu = N_j(c^2/2\nu^2) A_{jk} I_\nu$, where there are N_j atoms per unit volume in the ground state $| j >$ and N_k atoms per unit volume in the excited state k. The corresponding classical expression if all the absorption strength of one electron were concentrated into a single line would be the one in subsection 4.4.3, *viz.* $(Ne^2/4m\epsilon_0 c) I_\nu$.

However, in a real atom with one dispersion electron (for example, in the Rydberg series of an alkali atom, as considered in chapter 2), the absorption strength of one electron is not concentrated into a single line, but is split up into several frequencies. Various attempts were made, by Ladenburg [129] and others, to relate the classical and quantum expressions by considering that only a fraction f of the strength of the classical oscillator is available in each line, f being a dimensionless quantum mechanical quantity called the *oscillator strength* of a transition. The early attempts were fraught by the difficulty that induced emission processes had been left out.

Eventually, Kramers [130] gave the correct formulation, by considering $| \jmath >$ not as the ground, but as a general excited state. In this case, two processes must be included, *viz.* $E_k > E_\jmath$, which is absorption, and $E_\imath < E_\jmath$, which is emission. We thus arrive at:

$$a_\jmath(\nu) =$$
$$\frac{e^2}{4\pi m\epsilon_0 c} \left\{ \sum_k \frac{f_{k\jmath}(\gamma_{\jmath k}/4\pi)}{(\nu_{k\jmath} - \nu)^2 + (\gamma_{\jmath k}/4\pi)^2} - \sum_\imath \frac{f_{\jmath\imath}(\gamma_{\jmath\imath}/4\pi)}{(\nu_{\jmath\imath} - \nu)^2 + (\gamma_{\jmath\imath}/4\pi)^2} \right\}$$
$$(4.28)$$

and

$$n_\jmath(\nu) - 1 =$$
$$\frac{e^2}{16\pi^2 m\epsilon_0} \left\{ \sum_k \frac{f_{k\jmath}(\gamma_{\jmath k}/4\pi)(\nu_{k\jmath} - \nu)}{(\nu_{k\jmath} - \nu)^2 + (\gamma_{\jmath k}/4\pi)^2} - \sum_\imath \frac{f_{\jmath\imath}(\gamma_{\jmath\imath}/4\pi)(\nu_{\jmath\imath} - \nu)}{(\nu_{\jmath\imath} - \nu)^2 + (\gamma_{\jmath\imath}/4\pi)^2} \right\}$$
$$(4.29)$$

so that each transition from $| \jmath >$ to $| k >$ carries an oscillator strength $f_{\jmath k}$ related to the Einstein–Milne A coefficient by the expression:

$$f_{\jmath k} = \frac{2\pi m\epsilon_0 c^3}{e^2 \nu^2} A_{\jmath k} \qquad (4.30)$$

For transitions between bound states, the oscillator strength $f_{\jmath k}$ is a fundamental new quantity which quantum mechanics introduces in dispersion theory. We now have not only one characteristic frequency ν_0 but a whole series of transition frequencies $\nu_{\jmath k}$, and the presence, not only of absorption terms, but also of 'negative absorption' terms due to stimulated emission. The theory of Kramers [130] was checked experimentally by Ladenberg and Kopfermann [131] for a line in the spectrum of Ne.

In terms of the quantum-mechanical dipole matrix element $| \mathbf{r}_{\jmath k} |^2$ which determines the strength of an allowed spectral line from the wavefunctions of states $| \jmath >$ and $| k >$, the oscillator strength is defined as

$$f_{\jmath k} \equiv \frac{2m\omega_{\jmath k}}{3\hbar} | \mathbf{r}_{\jmath k} |^2 \qquad (4.31)$$

from which it is also clear that $f_{\jmath k}$ changes sign with $\hbar\omega_{\jmath k} = E_\jmath - E_k$ according to whether E_k is above or below $E_\jmath$.

4.5   The Thomas–Reiche–Kuhn sum rule

From the previous discussion, if an electron is in state j, the total oscillator strength available for transitions to all other levels must add up to the classical value for one oscillator, i.e.

$$\sum_k f_{jk} = 1 \qquad (4.32)$$

This is called the Thomas–Reiche–Kuhn sum rule for a one-electron spectrum. If $\mathcal{Z}$ electrons are involved, then the sum is simply $\mathcal{Z}$.

The sum rule can also be derived by elementary quantum mechanics from the definition of the dipole transition probability and its relation to the f value.

The simplest way is from Heisenberg's equation for the time evolution of an operator $\hat{A}$, viz:

$$\frac{d\hat{A}}{dt} = (i\hbar)^{-1}[\hat{A}, \hat{H}] \qquad (4.33)$$

If we apply it to the operator $\hat{r}$, we have

$$< j \mid \frac{d\hat{r}}{dt} \mid k > = \frac{1}{i\hbar} < j \mid \hat{r}\hat{H} - \hat{H}\hat{r} \mid k > = (i\hbar)^{-1}(E_k - E_j) < j \mid \hat{r} \mid k > \qquad (4.34)$$

of course, $m\,(d\hat{r}/dt)$ is none other than the momentum We can write this equation as

$$\mathbf{p}_{jk} = im\omega_{jk}\mathbf{r}_{jk} \qquad (4.35)$$

in vector notation for two states $\mid j >$ and $\mid k >$, with $\hbar\omega_{jk} = E_k - E_j$, and we have that the momentum $p_{jk} = m\,(dr/dt)_{jk}$. Thus, if we take the x component of $f_{jk} \equiv (2m\omega_{jk}/3\hbar)\mid r_{jk}\mid^2$, written out more explicitly as

$$(f_{jk})_x = \frac{2m\omega_{jk}}{3\hbar} \mid x_{jk}\mid^2$$

$$= \frac{2m\omega_{ba}}{3\hbar} < j \mid x \mid k >< k \mid x \mid j > \qquad (4.36)$$

we can now use this last expression together with (4.35) which relates $< j \mid x \mid k >$ with $< j \mid p_x \mid k >$ to contruct the commutator of p_x and x by substituting once for the left factor and once for the right factor of $\mid x_{jk}\mid^2$ in (4.5) and subtracting, then summing over all the intermediate states $\mid k >$ and using the closure relation $\sum_k \mid k >< k \mid \equiv 1$. The result is

$$\sum_k (f_{jk})_x = \sum_k \frac{i}{3\hbar} |<j \mid p_x \mid k><k \mid x \mid j>$$

$$-\sum_k <j \mid x \mid k><k \mid p_x \mid j>|$$

$$= \frac{i}{3\hbar} <j \mid p_x x - x p_x \mid k>$$

$$= \frac{i}{3\hbar} [x, p_x] = \frac{1}{3} \qquad (4.37)$$

whence, by adding the x, y and z components, one has

$$\sum_k f_{jk} = 1 \qquad (4.38)$$

Note that, for the closure relation to be applied in this way, the summation must extend over all bound states and adjoining continua. The implicit assumption is that one electron is involved. Within the independent electron approximation, each electron has its own complete set of associated wavefunctions. Thus for N electrons, the f value summation over the complete spectrum yields N. If they all come from the same subshell, and if their excitation spectrum is in a separate energy range from the spectrum of the other electrons in the same atom, then their integrated oscillator strength can be evaluated separately and should yield this value. In the fully interacting atom, oscillator strength is exchanged between channels as a result of coupling, and so the sum rule may turn out to apply only over the whole spectrum, in which case it will simply yield the total number of electrons in the atom. This last rule is not altered by correlations, and will apply in any case.

A good example occurs in the alkali spectra: the integrated oscillator strength evaluated over the optical range is approximately 1 for all alkalis. This proves that the other electrons (i.e. those in the core) hardly participate in the visible spectrum. If, however, observations are extended to very high excitation energies, one eventually breaks into deeper electronic shells (see chapter 7), and so a separate sum rule will apply for inner-shell excitation. For instance, we may expect an oscillator strength of 10 for excitation from a d subshell.

Sum rules can be obtained, not only for absorption or emission, but for many other optical functions, e.g. dielectric permeability, refractive index, rotatory power and ellipticity or circular dichroism. In fact, any quantities in physics which are related by a Kramers–Kronig transformation[4] provide

---

[4] For another example of the use of Kramers–Kronig relations, see section 6.14.

a path for the derivation of further sum-rules [132].

4.5.1 Different ways of describing the strength of a line

There are several different ways in which we may describe the strength of a line, all of which are useful in different contexts.

(i) First, we may define a *transition rate* W_{if} from a rate equation of the type

$$\frac{dN_i}{dt} = -W_{if}N_i \qquad (4.39)$$

where N_i is the number of atoms in the state $\mid i >$, and the rate W_{if} can then be calculated for an individual atom, by using an appropriate model for the excitation mechanism. For example, it is shown in elementary textbooks that the *photoexcitation rate* by unpolarised radiation of intensity $I(\omega_{if})$ is given by

$$W_{if} = \left(\frac{\pi e^2}{3c\hbar^2\epsilon_0}\right) I(\omega_{if}) \mid< i\mid r\mid f >\mid^2 \qquad (4.40)$$

This form is useful when discussing rate equations.

In order to specify a microscopic quantity, one usually tabulates the spontaneous transition rate A_{21}, which does not depend on the external field $I(\omega)$ at least for small field strengths. One may also give its inverse, which is the lifetime.

(ii) To introduce a cross section, simply note that the rate of absorption of energy per atom from a beam $I(\omega_{if})$ is $\hbar\omega_{if}W_{if}$. The simplest way is to think in terms of the flux of photons in the beam, which is $I(\omega)/\hbar\omega$. Thus, the rate absorbed per second is

$$W_{12} = \int_{line} \frac{I(\omega)}{\hbar\omega}\sigma_{12}d\omega \qquad (4.41)$$

Since the rate and the angular frequency are per second, σ_{12} has the dimensions of area. It is related to the f value by integrating over the line:

$$\int_{line} \sigma_{12}d\omega = \frac{\pi e^2}{2m\varepsilon_0 c}f_{12} = \frac{\pi c r_e}{2\varepsilon_0} = 0.167(\text{cm}^2 \text{ s}^{-1})f_{12} \qquad (4.42)$$

where r_e is called the classical radius of the electron. The magnitude of the constant connecting σ to f of course depends on the variable chosen to display the spectrum. The connection is found from conservation of energy, by noting that

$$I(\omega)d\omega = I(\nu)d\nu = I(E)dE = I(\lambda)d\lambda = ...$$

and that

$$\frac{d\omega}{\omega} = \frac{d\nu}{\nu} = \frac{dE}{E} = \frac{d\lambda}{\lambda}$$

since $c = \nu\lambda$ etc.

(iii) One may also use the absorption coefficient $a(\omega)$, which was introduced together with the refractive index when considering a system of N oscillators driven by the electromagnetic wave. More generally, if we have N_i atoms in the initial state, then the integrated absorption coefficient

$$\int_{line} a(\omega)d\omega = \frac{2\pi e^2}{4\varepsilon mc}N_i f_{ij} \tag{4.43}$$

(iv) Finally, one may specify the oscillator strength or f value of the transition, f_{ij}, or its gf-value $g_i f_{ij}$, where g_i is the statistical weight of the initial state. Note that the f value for absorption is conventionally defined as positive, whereas for emission it is negative. In the presence of a nonresonant absorption background, one should subtract the slowly-varying part before integrating over the feature to extract an oscillator strength. This procedure arises in section 6.13.

All these quantities are equivalent since they all depend on the same transition matrix element, although their units are not the same. The f value has the advantage of being a dimensionless quantity. With broad band illumination, the appropriate quantities are those which are integrated over the spectral feature, such as the f value or the Einstein coefficient. With narrow band illumination (i.e. a monochromatic source narrower than the spectral feature), it is appropriate to use a quantity which is defined point by point within the line profile, such as the absorption coefficient, the cross section, or the differential oscillator strength df/dE.

4.6   The course of f values in a Rydberg series

For an unperturbed Rydberg series accessed from the ground state, the normal course of intensity is for the first member to be the strongest, with a rapid drop in intensity as the excitation energy increases. One can show that for many systems f_n, the oscillator strength for photoexcitation to the nth level falls off as n^{-3} at large n. This is also the rate at which the energy spacings in a Rydberg series fall off with increasing n, which suggests that a good way of plotting information on oscillator strengths is to involve the spacings between Rydberg members, so that the magnitudes do not fall off too rapidly with n. The way in which this is done is

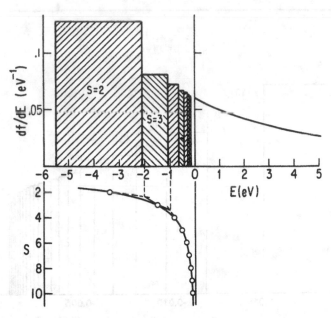

Fig. 4.1. The geometrical construction by which df/dE is plotted as a function of energy, so that the discrete and continuous spectra join at the threshold: tangents are drawn to the curve of linestrength s against energy at the positions of the lines, and their crossing points determine the bases of 'boxes', whose surface area is made equal to the line strength for each transition. (after A.R.P. Rao [133]).

illustrated in fig. 4.1 for the case of H. Since f values in Rydberg series generally fall off rather quickly with increasing n, this histogram representation is very useful. The 'boxes' are each given the area of an f value and the bases are drawn so that they join smoothly without overlap. At large enough n one can draw the base from an energy corresponding to $n-1/2$ to an energy corresponding to $n+1/2$ and the result is that midpoints at the top of the boxes lie on a smooth curve, which may locally be close to a straight line. For atoms other than H and a fairly unperturbed seris, the base of the boxes can be defined using n^* rather than n, and plotting from $n^* - 1/2$ to $n^* + 1/2$, which is satisfactory as long as the energy separation between successive members varies approximately as $1/n^{*3}$.

In fig. 4.2, we show a typical df/dE plot for an unperturbed Rydberg series in a many-electron atom. Such plots are very convenient for the Rydberg series of any atom, because they immediately show up irregularities in the course of intensities as a function of energy while putting f values onto comparable scales despite wide variations in magnitude. The method works well because (as noted above) irregularities in the intensities are generally more pronounced than those in the energy differences.

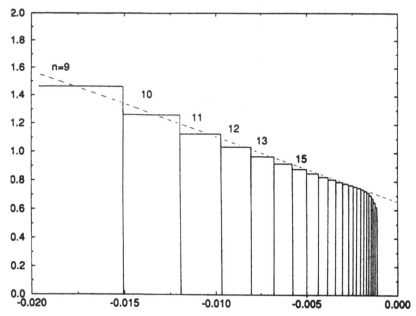

Fig. 4.2. Plot of the differential oscillator strength df/dE as a function of energy for the principal series of Na (experimental data): each 'box' represents one transition, arranged (see the previous figure) so that the df/dE graph has no gaps. A smooth curve then connects the points shown, and should, in this example, tend to a straight line. Since the measurements were made in the presence of a magnetic field, the *very* high Rydberg members appear to lose intensity because satellite lines emerge, as explained in chapter 10 (after M. Nawaz *et al* [166]).

There may be such strong series perturbations that the simple rule for drawing the edges of the 'boxes' no longer works, and overlaps or spaces between adjacent boxes then appear in the diagram. In this case, the geometrical construction should be used. This commonly arises when the lowest member of a Rydberg series is included in the plot.

4.6.1   Intensities in the continuum

The concept of f value is appropriate for discrete or isolated features, over which an integration can readily be performed, but not for a continuum extending over a wide energy range. In this case, one can measure the *differential oscillator strength* or f value per unit energy interval, which clearly depends on the units which are chosen to represent energy. The quantity which represents absorption per unit energy, frequency or wavelength interval is the total cross section σ, written as $\sigma(E)$, $\sigma(\nu)$ or $\sigma(\lambda)$ respectively, and the connection to the df/dE curve is given by the

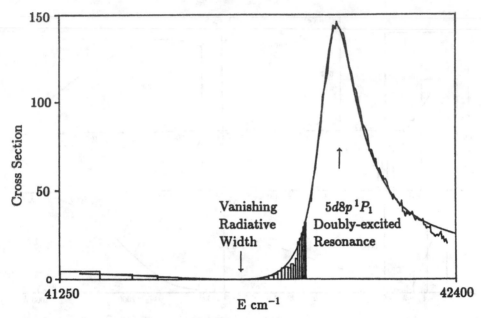

Fig. 4.3. The principal series of Ba I in the vicinity of the first ionisation threshold, showing how the df/dE curve of the high series members joins smoothly onto that of the continuum, even in the presence of a strong perturber, in this case a doubly excited resonance straddling the threshold, which has a pronounced effect on the course of intensities (see fig. 8.22 – after J.-P. Connerade *et al* [136]).

relation:

$$\left.\frac{df}{dE}\right|_{n\to\infty} = K\sigma(E)|_{E=I} \qquad (4.44)$$

where I is the appropriate ionisation threshold and the constant K is chosen so that the Thomas–Fermi–Kuhn sum rule is satisfied. Once df/dE curves for the discrete transitions are plotted in the manner of fig. 4.1, continuity of the absorption strength as one crosses over the threshold into the continuum expresses itself as a smooth join between the curve defined by the 'boxes' which represent the discrete states and df/dE in the continuum, obtained from measurements of $\sigma(E)$. This continuity across the threshold is a fundamental principle of QDT. It was obtained from first principles by Sugiura [134] for H. Gailitis [135] then suggested an averaging method, taking into account maxima and minima of Rydberg members as the series limit is approached, leading to a smooth join with the continuum, and Seaton [137] established quantum defect plots in the manner described here, so that the smooth behaviour across the threshold is sometimes referred to as Seaton's theorem.

Continuity across the threshold is a fundamental property, which de-

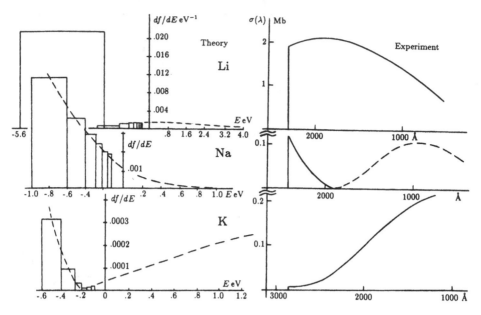

Fig. 4.4. Seaton–Cooper minima for a selection of alkali spectra (schematic), showing their evolution as a function of the atomic species. Theoretical curves are shown on the left and corresponding plots of the experimental photoionisation cross sections on the right. The figure demonstrates the large difference between the alkalis (after M. Nawaz *et al* [166]).

rives from the continuity in the wavefunctions across the threshold illustrated in fig. 3.1: quite clearly, if the wavefunctions are continuous across the threshold, then all the physical observables derived from them will also be continuous in the same way.

Since continuity across the threshold is so fundamental, it is worth testing the principle under conditions where f values can be measured to very high principal quantum numbers n, and where a strong intruding resonance located in the photoionisation continuum just above the threshold perturbs the course of intensities in the Rydberg series. An example of this kind occurs in the spectrum of Ba and is shown in fig. 4.3. The fact that, even in such a case, the df/dE curve joins completely smoothly shows that perturbations do not upset this principle, i.e. that it is of very general validity.

4.6.2   Minima in the continuum

The course of df/dE in the continuum is normally such that it decreases monotonically with increasing energy roughly as $1/E$, a trend which is strictly correct for H, but applies to many other cases in photoabsorption from the ground state. If the excited state wavefunction possesses a node, then the matrix element for photoionisation, which involves an integration over the radius, splits into two contributions of opposite sign. As the kinetic energy of the emerging photoelectron changes, its degree of penetration into the atomic core is also changed. Consequently, the balance between the two contributions changes as a function of the kinetic energy. There may thus exist an energy E at which there is cancellation, and the matrix element vanishes exactly. There will then be zero absorption (a minimum in the observed cross section) at this energy. Although this effect exists within the independent particle model, the positions of such minima depend on cancellations, and are thus sensitive to correlations, which modify the wavefunctions and may move them considerably in energy.

The possibility of such minima in the observed cross section, and the explanation of how they arise were first discussed by Rudkjøbing [138], in a pioneering paper which is unfortunately hard to find and has therefore been rarely quoted. Accurate calculations of cross sections near such minima in the alkalis were first reported by Seaton [139]. Similar calculations were also performed by Cooper [140]. In line with current usage, we shall refer to such features as Seaton–Cooper minima. They are not to be confused with another type of minimum of the photoionisation cross section (the Combet–Farnoux minimum), which occurs in the presence of a centrifugal barrier (see section 5.4).

For a more recent example of how Seaton–Cooper minima arise in laser spectroscopy and a more up-to-date comparison between theory and experiment, see [141].

The Seaton–Cooper minima of alkali spectra are the best known. Because of the continuity between discrete and continuous spectra noted in the previous subsection, if a Seaton–Cooper minimum drops below the threshold, it will turn into a minimum in the course of intensities of the corresponding Rydberg series, in such a manner that continuity of the df/dE plot is preserved (see above). In the discrete part of the spectrum, one may find a minimum rather than a zero, because it is not necessary that a transition should exist at precisely the energy where cancellation occurs, i.e. the Seaton–Cooper minimum may very well fall between two members of a Rydberg series. However, the anomaly in the course of intensities in the series will be apparent.

An instance of this type is found in the spectrum of Li I, as shown in fig. 4.4: note in particular how the df/dE curve for Li is distorted from the expected shape by the minimum below threshold, so that the curve rises rather than falls towards the threshold. The effect is *not* a perturbation: alkali spectra have double excitations and inner-shell spectra very far in energy from the optical spectrum, so there are no intruders in this range.

In an interesting theoretical study, Msezane and Manson [142] have considered the opposite situation where, instead of becoming more compact, as when the Seaton–Cooper minimum drops into the bound states, the wavefunctions become more diffuse, and the minima tend outwards. They have taken as their initial state an excited rather than a ground state. For photoexcitation from ground state wavefunctions, at most a single Cooper minimum is found, but this is not necessarily true for photoexcitation from excited states. In this case, even if the initial state wavefunction has no nodes, a Cooper minimum may occur. When the initial state wavefunction has nodes, then in addition to the usual Cooper minimum, further minima also appear. Some of these may also drop into the bound state spectrum, and an example of this kind is discussed in section 5.18.

4.6.3   The Fano effect

It was noted by Seaton [139] that, because of the slight term dependence of wavefunctions which is caused by the spin–orbit interaction, Seaton–Cooper minima occur at slightly different energies for excitation to continua of $^2P_{1/2}$ and $^2P_{3/2}$ character, respectively, which differ only in the orientation of the spin of the escaping electron. By choosing the energy of the incident photon correctly, it is possible to excite selectively *only* into one of the two continua. Suppose we also choose to excite the atom using circularly polarised light, of angular momentum $j = 1$. Conservation of angular momentum then requires that the spin of the emerging photoelectron be either parallel or antiparallel to the angular momentum of the incident photon. Thus, merely by irradiating an unpolarised ensemble of atoms at the frequency of the Seaton–Cooper minimum in one of the two available photoionisation channels, using circularly polarised light of the appropriate frequency, spin-polarised photoelectrons are produced.

This phenomenon, first pointed out by Fano [143] and called the *Fano effect*, has been used to produce spin-polarised electron sources [144]. Sources based on irradiating atoms are now mainly of academic importance, because solid state sources based on GaAs devices yield a higher current more simply.

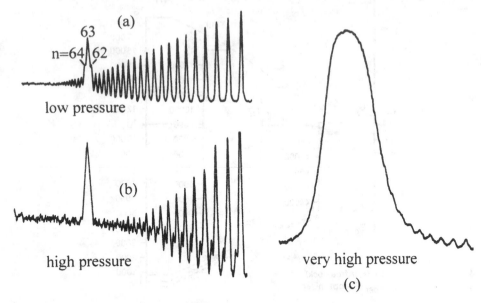

Fig. 4.5. The uppermost members of the two-photon excitation series of Ca in the presence of a small pressure of rare gas, showing the prominent series perturbation which arises by collision-induced mixing with a doubly excited intruder level. Cases (a) and (b) show the effect of increasing the pressure of the gas, whilst (c) shows the effect of a very high pressure and a perturber of high atomic weight (after K.S. Bhatia *et al* [145]).

4.7 The influence of collisions on high Rydberg members

High Rydberg states rapidly become very large (their radius grows as n^2). Thus, although transitions coupling them to the ground state become weaker and weaker with increasing n, they also become more and more fragile, and susceptible to the influence of external fields, because the Rydberg electron is further and further away from the nucleus. The influence of high external magnetic fields on Rydberg states is discussed in chapter 10.

This is particularly true with regard to collisions. Care must be taken when performing experiments on high series members to investigate whether collisions between atoms (especially if other atomic species, e.g. a noble gas used as a 'buffer gas' to impede diffusion of the sample, is present) can influence the result.

The most dramatic effect of collisions is to induce mixing between nearby excited states of the atom, whose presence would not normally manifest itself. Consider for example a spectrum in which a regular Ryd-

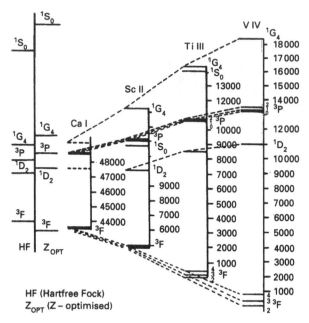

Fig. 4.6. The energy of the intruder level in relation to the ionisation threshold for the Ca-like isoelectronic sequence, showing how the doubly excited intruder state can be tracked down to the approximate position where the series perturbation is observed by using a combination of theory and experiment (after K.S. Bhatia *et al.* [145]).

berg series is present, and in which an intruder is inaccessible because of selection rules, in the absence of foreign gas. At zero buffer gas pressure, the spectrum will consist of a series with the normal course of intensities. As the pressure is raised, an anomaly will appear near the energy of the intruding state, because the collisions are then able to break the normal dipole selection rules by inducing mixing between the Rydberg states and the intruder level.

Such an example occurs in the Ca I spectrum, and is shown in fig. 4.5, taken from the observations of Bhatia *et al.* [145]. In this experiment, an even parity Rydberg spectrum was excited by two-photon absorption from the ground state, using a tunable laser source. At very low buffer gas pressures, a normal course of intensities was observed. As the pressure was raised, however, a pronounced series perturbation was found at very high n values. It turns out that, in Ca, the doubly excited $3d^2\,{}^1G_4$ state occurs within the high members of the Rydberg series (see fig. 4.6). Because it has $J = 4$, it is not accessible by one-, two- or even three-photon absorption from the ground state. However, collisions can mix it with neighbouring Rydberg members, and thereby break the selection rules, leading to a pronounced intensity anomaly which could readily be confused with an ordinary series perturbation if the influence of the buffer gas were not considered.

The theory of collisions involving high Rydberg states was first studied by Fermi [146], who took as a guiding principle (cf. section 2.17) that the excited electron is very far from the core so that all Rydberg atoms would resemble each other for $n > 30$, i.e. that the excited electron responds as a single essentially free electron. The example of fig. 4.5 shows that, in fact, this is not always the case: the situation considered by Fermi only applies for atoms whose doubly excited states lie rather far away from the Rydberg spectrum. Atoms such as the alkaline-earths, which have low lying, doubly excited states, may be expected to differ significantly from Fermi's theory. Other, related aspects involving laser-excited atoms are discussed in section 5.15.

A more usual situation is that several different Rydberg series converging on the same threshold have high members lying very close to each other in energy. The series may differ from each other in their angular momentum quantum numbers ℓ so that, say, only two are accessible under dipole selection rules from the ground state, while others are not. In the presence of collisions, each Rydberg member becomes mixed with its nearest neighbours in energy, so that new series emerge which appear to violate the selection rules. An example of this type is shown in fig. 4.7 in high lying members of the spectrum of Sr.

As a result of complexities introduced by the presence of doubly-excited states, in the energy range of Rydberg excitation, the theory of collisions with highly excited atoms is a good deal more complex than predicted by simple theory. Although attempts have been made to include doubly-excited states in calculations of broadenings and shifts [147], there does not seem as yet to be a complete theory for such effects.

4.8   Measurements of *f* values for high Rydberg members

There exist well-established methods for the measurement of f values. These are discussed at length in standard references [149]. More recent experimental developments have been quite comprehensively reviewed [151]. The present section is not intended as a substitute for this general literature, but concentrates specifically on rather sensitive techniques based on differential refractive index determinations, which have been successfully applied to observations of high Rydberg members in the laboratory.

In general, the most sensitive techniques are those based on measurements of the refractive index rather than of the absorption coefficient. Since the absorption coefficient and the refractive index are connected

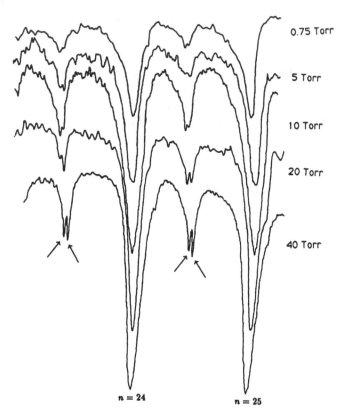

Fig. 4.7. The spectrum of Sr, showing high Rydberg members and the emergence of supernumerary series members (denoted by arrows), due to the influence of collisions. Note their dependence on buffer gas pressure (after M. Nawaz *et al.* [148]).

through the Kramers–Kronig relation, measurement of one is equivalent to measurement of the other, so that the same information is in principle available from either. Measurements of the refractive index have the intrinsic advantage that they are less affected by opacity than measurements of absorption. They are also very suitable because they can be combined with either interferometry or magneto-optics, both of which possess high accuracy. Of these, the Faraday rotation method is capable of the highest sensitivity. Interferometric techniques (the Hook method) have proved less capable of reaching the highest Rydberg members. They are reviewed in detail in [151].

4.9   Faraday rotation

The use of Faraday rotation spectroscopy to study high Rydberg members and to measure their f values is a recent development. Little early work was done on magneto-optical effects in the vacuum ultraviolet, which can only be explained in terms of technical difficulties, since the subject is not intrinsically new.

Magneto-optical rotation (MOR) has a long history, stretching back as early as the work of Faraday [152] and Macaluso and Corbino [153], who related MOR to the Zeeman effect [154] well before the advent of quantum mechanics. The Zeeman effect is much used in classical spectroscopy for the determination of J values and g factors, and this is discussed in many standard texts.

The new developments concern the application of MOR to the measurement of f values of transitions to high Rydberg states, using both synchrotron radiation and laser sources. Since both these sources are normally polarised (the polarisation properties of accelerated electrons were first studied by Schott [155]), the method is naturally suited to them. Also, one can make use of either DC superconducting magnets with synchrotron radiation or of long pulse magnets (typically, milliseconds) [156] synchronised with pulsed lasers (typically, tens of nanoseconds) to achieve high magnetic fields.

The use of a high field is an important feature of the method. A high field, in the context of MOR (as opposed to the definition used in chapter 10) is a field strong enough to produce rotation angles of several π in the vicinity of a spectral line. Experiments can then be performed with a fixed (crossed) polariser and analyser, in the classic arrangement for observing the dispersed Faraday effect, thereby greatly simplifying the optical arrangement – a significant advantage in the vacuum ultraviolet range.

Although this method is currently one of the most sensitive, and has allowed f value measurements for neutral atoms to be extended to very high n values, the need for a high field is its ultimate limitation. As will be seen, there are other high field effects (the quadratic effects, discussed in chapter 10) which, at very high n, intrude by modifying the atomic f value (see fig. 4.2 for an example). If the effect of the very high field can be understood and allowed for, then the method is capable of some further extension, but does lose in simplicity of application. The exact point where quadratic effects become important varies from atom to atom. It depends not only on the diamagnetic polarisability of the atom (which is fairly constant from atom to atom in high Rydberg states), but also on the magnitude of f values at high n, which is much more variable.

Table 4.1. Classification of electro- and magneto-optical effects

Field-induced birefringence (electronic effects)			
Magnetic field		Electric field	
Voigt or Cotton– Mouton effect	Zeeman effect (transmission)	Lo-Surdo– Stark effect	Electric double refraction
Magneto-optic rotation		Electro-optic effects	
Kerr effect (reflection)	Faraday effect (transmission)	Kerr effect (quadratic)	Pöckels effect (linear)

Thus, in alkali spectra, it turns out that most of the available f value of the optical spectrum (~ 1) is concentrated in the first resonance line, leaving very little available for high Rydberg states, whose intensity falls off rapidly and is actually very sensitive to correlations [157]. In alkaline-earth elements, on the other hand, with twice as much f value available for the optical spectrum and similar magnitudes, much more oscillator strength is available for high members (apart from local perturbations) and the series can therefore be followed to much higher values of n both by photoabsorption and by MOR. We now outline the principles of Faraday spectroscopy for high Rydberg states.

4.10   Electro- and magneto-optical effects

A good introduction to electro- and magneto-optical effects can be found in the book by Harvey on *Coherent Light* [158]. The main effects and the relationship between them are indicated in table 4.1. Many atoms are readily produced as vapour columns, using standard laboratory methods [159]. The natural mode in which to conduct experiments on unperturbed free atoms is therefore in transmission. As table 4.1 emphasises (the reason is given below), the Faraday effect contains equivalent information to the Zeeman effect in transmission. Actually, what Harvey calls the *Zeeman effect in transmission* is usually referred to as the *inverse Zeeman effect* [160], to distinguish it from the Zeeman effect observed in emission.[5]

---

[5] There is another use of the expression *inverse Zeeman Effect* which can be found in [162], namely to describe the magnetic moment induced in a sample by the passage of a polarised light beam through a transparent medium. However, the more common usage is the one adopted in the text.

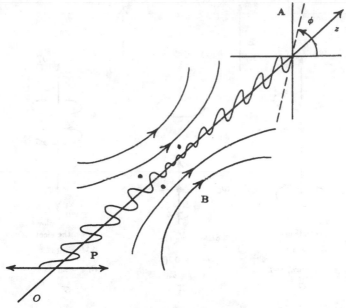

Fig. 4.8. Basic geometry for the observation of the dispersed Faraday effect: the polariser P is crossed with an analyser A, and propagation along an axis Oz is parallel to the field lines **B** in the region of the atomic absorption cell. Rotation of the plane of polarisation through an angle φ occurs as a result of magneto-optical birefringence (after J.-P. Connerade [161]).

An important point is that the inverse Zeeman effect depends on atomic absorption coefficients, whereas the Faraday effect depends on refractive indices. Since the two are connected by the Kramers–Kronig transformation, it follows that the same atomic physics is, in principle, contained in both types of measurement. However, the advantage of studying optical rotation rather than line intensities is the fact that the latter are distorted by opacity in photoabsorption, whereas angles of magnetic rotation are not similarly affected. In practice, this means that experiments in which f values are determined can be performed at high absorption densities, which is a huge advantage when dealing with high Rydberg members.

4.11   Basis of the magneto-optical method

The experimental arrangement is sketched in fig. 4.8. Plane polarised radiation is incident on an absorption cell placed between two crossed polarisers in a magnetic field. The direction of the magnetic field is parallel to the direction of propagation. It is helpful to consider the three classical geometries associated with σ^+, π and σ^- polarisations. These are

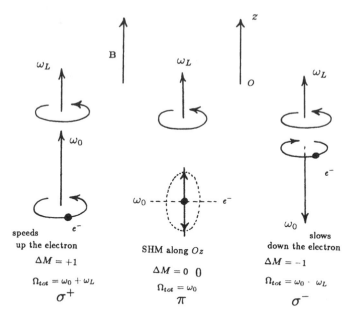

Fig. 4.9.   Classical geometries associated with the three polarisations σ^+, π and σ^-. The orbital angular momentum of the electron is ω_0, the Larmor precession is ω_L and the total angular momentum is Ω_{tot}. Any plane wave can be resolved as the coherent sum of two opposite circularly polarised waves of equal amplitudes. Because, according to electromagnetism, a dipole cannot radiate along its length, the π motion is not observed in the geometry of the previous figure. (after J.-P. Connerade [161]).

sketched in fig. 4.9. Let ω_0 be the orbital angular momentum of the electron and ω_L the Larmor precession. The resulting total angular momenta are then formed as shown. From the figure, one can see that, because a dipole cannot radiate along its length (electromagnetism), the geometry of the experiment excludes the π components. Fortunately, we can represent *any* plane wave travelling along Oz as the coherent sum of two circularly polarised waves. Thus, the problem is reduced to combining the effects of the two refractive indices and the two absorption coefficients for circularly polarised light on the propagation of the beam.

Magneto-optical activity results from the asymmetry in the indices of refraction for the left ($-$) and right ($+$) hand polarised circular components of a light wave induced in a medium by a longitudinal **B** field. It subdivides as shown in table 4.2.

The plane of polarisation of the wave is rotated as a function of detuning from the line centres, because the two circular components suffer relative retardation as they traverse a length ℓ of the absorbing column, owing to the differences of refractive index for σ^+ and σ^- light. This effect is

Table 4.2. Magneto-optical activity
(see [168] for details of notation).

Magnetic circular dichroism (MCD)	Magnetic circular birefringence (MCB)
Due to the difference between the absorption coefficients $a_+(\nu)$ and $a_-(\nu)$, we have an ellipticity angle for initially plane-polarised radiation:	*Due to the difference between the refractive indices $n_+(\nu)$ and $n_-(\nu)$, we have a rotation angle of plane-polarised radiation:*
$\Theta(\nu) = (\omega\ell/2c)\left\{k_+(\nu) - k_-(\nu)\right\}$	$\Phi(\nu) = (\omega\ell/2c)\left\{n_+(\nu) - n_-(\nu)\right\}$

referred to as *magnetic circular birefringence* (MCB), and is the cause of MOR.

A beam of plane-polarised light tuned exactly to the wavelength of one of the members of the Zeeman doublet (either σ^+ or σ^-) will be completely absorbed at high enough opacity. The emerging beam is then circularly polarised, and no rotation can occur. This effect is called *magnetic circular dichroism* or MCD for short.

Pure MCD with no rotation occurs if the sample is optically thick and completely absorbing in one circular polarisation, but not in the other. Pure MOR will occur when both circular polarisations are equally absorbed, i.e. the absorption coefficients $\alpha_+(\nu)$ and $\alpha_-(\nu)$ are equal, but the refractive indices $n_+(\nu)$ and $n_-(\nu)$ are not equal. The latter condition is satisfied at the centre of symmetry of the rotation pattern, *viz.* the field-free resonance frequency ν_0.[6] In principle, the situation seems simpler when either pure MOR or pure MCD occurs, which is why most of the effort has traditionally been expended in separating one from the other, leading to MOR and MCD spectroscopies.

The most widely used is MCD spectroscopy, in which MOR is removed by using circularly polarised light: it finds application in the study of molecules [169], where one recognises different types of patterns, according to whether the magneto-optical effect arises from a temperature-independent ground state, from a state whose degeneracy is resolved in the presence of the field, so that populations become temperature-sensitive and different dispersive curves are combined, or again when there is field-induced mixing of wavefunctions in a molecule of low symmetry with no degenerate states.

The observation of pure MOR by tuning a laser to the field-free transi-

---

[6] This is only strictly true if one neglects diamagnetic effects, which are quadratic in the magnetic field strength, a consideration which can become important for high Rydberg members (see chapter 10).

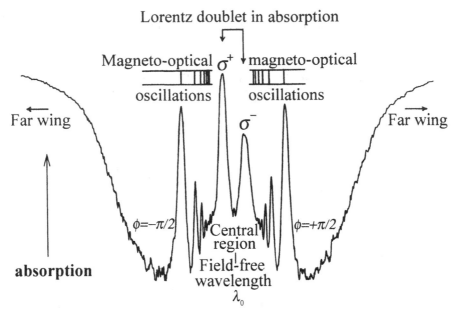

Fig. 4.10.   General appearance of MOV patterns for a Zeeman doublet. In the example shown here, the polarising optics were not perfect, which results in a Lorentz doublet superimposed on the centre of the rotation pattern (after J.-P. Connerade *et al.* [163]).

tion frequency ν_0 and measuring the rotation as a function of the magnetic field strength belongs to *forward scattering* spectroscopy. It has been applied to very low Rydberg states [170], for which complications due to quadratic shift of MOR patterns cannot occur.

Neither approach is really suitable for high Rydberg members. Instead, it is better to let *both* MOR and MCD occur together scanning through all available frequencies for a fixed magnetic field strength, and accounting correctly *for both the birefringence and the dichroism terms at each point of the line profile*. This method of determining relative f values is called the magneto-optical vernier (MOV) technique to distinguish it from MOR and MCD. The appearance of an MOV pattern for a wide range of field strengths, ranging from overlapping to fully resolved patterns of a Zeeman doublet, is illustrated in fig. 4.10 Oscillations of intensity occur on either side of the well-resolved Zeeman doublet, extending far beyond the absorption wings. If $\Phi(\nu)$ is the rotation angle, then these oscillations simply correspond to the successive maxima and minima of the function $\sin^2 \Phi$ as a function of frequency as the absorption line is approached. If a Zeeman component were completely isolated from all the others, then the resulting pattern would be symmetrical about its centre. The problem is to compute such patterns taking account: (a) of the presence of other Zeeman components; and (b) of the effects of absorption. The calculations can then be extended to all nearby frequencies ν.

From equation (4.28), considering only absorption from a single (e.g. $^1S_0$) ground state to a Zeeman doublet (e.g. $^1P_1$) and allowing for the presence of N absorbers, we have

$$
\left.
\begin{aligned}
a_\pm(\nu) &= \frac{e^2 N f}{m \epsilon_0 c} \frac{(\gamma/4\pi)}{(\nu - \nu_0 \pm \alpha)^2 + (\gamma/4\pi)^2} \\[2ex]
n_\pm(\nu) - 1 &= \frac{e^2 N f}{4\pi \epsilon_0 m} \frac{(\nu - \nu_0 \pm \alpha)}{(\nu - \nu_0 \pm \alpha)^2 + (\gamma/4\pi)^2}
\end{aligned}
\right\} \quad (4.45)
$$

We can substitute these formulae in the expressions for $\Theta(\nu)$ and $\Phi(\nu)$ of table 4.2. We then know the variation of the rotation and ellipticity angles as a function of wavelength. Notice that they depend on the product $Nf\ell B$, where N is the number density of absorbers and ℓ is the length of the absorbing column, and that the Zeeman components are located at $\pm\alpha$ where $\alpha = eB/4\pi mc$, which involves the magnetic field strength B. Thus, if optical rotation patterns of different transitions originating, for example, from the ground state are obtained under identical experimental conditions (same N, ℓ and B), the relative f values can be determined. By working with high rotation angles (many π radians) one can observe intensity oscillations, which provide an absolute determination of the angle as a function of frequency without the need for moving parts in the optical system.

We compute the intensity I_t transmitted through crossed polarisers according to the expression:

$$
I_t = I_0 \left\{ \sin^2 \Phi + \sinh^2 \frac{\Theta}{2} \right\} \exp \left\{ -\frac{k_+ + k_-}{2} \ell \right\}
$$

which involves *both* $\Theta(\nu)$ and $\Phi(\nu)$, i.e. combines the effects of rotation and dichroism.

There are some further steps, which should also be included, and are quite straightforward in practice. The true profiles of the absorption lines are not Lorentzian, as assumed in the simple theory above, but are broadened by the Maxwellian velocity distribution:

$$
P(v)dv = \frac{1}{\pi^{1/2} v_0} \exp - \left(\frac{v}{v_0} \right)^2 dv
$$

where $v_0 = (2kT/m)^{1/2}$ is the most probable velocity in the absorption cell, the temperature T of which we can measure. When the Lorentzian form corresponding to the natural linewidth is convolved with the Gaussian velocity distribution, bearing in mind that the moving atom absorbs at a frequency $\nu\,(1 - v/c)$, the result is a Voigt profile, and we have:

$$a_{\pm}^{Voigt}(\nu) = \int_{-\infty}^{+\infty} a_{\pm}^{Lorentz}\left\{\nu\left(1 - \frac{v}{c}\right)\right\} P(v)dv$$

$$n_{\pm}^{Voigt}(\nu) = \int_{-\infty}^{+\infty} n_{\pm}^{Lorentz}\left\{\nu\left(1 - \frac{v}{c}\right)\right\} P(v)dv$$

Hence, we obtain the expressions:

$$a_{\pm}(\nu) = \frac{e^2 Nf}{\Delta\nu_D m\epsilon_0 c}\frac{\nu}{\pi}\int_{-\infty}^{+\infty}\frac{e^{-y^2}dy}{(\bar{\nu}\mp\bar{\alpha}) - y)^2 + \gamma^2}$$

and

$$n_{\pm}^{Voigt}(\nu) - 1 = \frac{e^2 Nf}{4m\epsilon_0(\nu_0 \pm \alpha)\Delta\nu_D}\frac{1}{\pi^{\frac{3}{2}}}\int_{-\infty}^{+\infty}\frac{e^{-y^2}\left\{(\bar{\nu}\mp\bar{\alpha})\right\}dy}{\left\{(\bar{\nu}\mp\bar{\alpha}) - y\right\}^2 + \gamma^2}$$

where some new quantities have been introduced, *viz.* the Doppler width $\Delta\nu_D = \nu_0\,(v/c)$ and the widths, shifts and detunings expressed in Doppler widths: $\gamma = \Gamma/4\pi\Delta\nu_D$, $y = (\nu' - \nu_0)/\Delta\nu_D$ (ν' is the integration variable), $\bar{\alpha} = \alpha/\Delta\nu_D$ and $\bar{\nu} = (\nu - \nu_0)/\Delta\nu_D$

When calculating simulated profiles, it is actually more convenient to compute a complex error function point by point across the profile, using a fast algorithm [171]: the real part yields the absorption coefficient as a Voigt profile, and the imaginary part, the refractive index. Finally, the intensity calculated from the theoretical expressions above is convolved with an apparatus function, to determine the observed intensity. The apparatus function must take account of the finite linewidth effects, the stability of the furnace and magnets, etc. Various different forms for the apparatus function (top hat function, triangular, Gaussian, Lorentzian, etc.) have been tried, and the results for the f values are remarkably insensitive to the choice (less than 0.1% change for 10% change in the apparatus function).

A very accurate technique has been developed to reduce the data under the most favourable conditions of observation. It relies on the different sensitivities in different regions of the magneto-optical pattern when the structures due to different Zeeman components overlap in energy but are resolved. One finds that the centre of the patterns changes fast as a function of $Nf\ell B$, while the outer reaches of the pattern change much more slowly. By balancing one against the other, full advantage is taken of the complete theory, which is the origin of the MOV principle. This is treated in detail in [172]. Magneto-optical methods are very powerful: they have been used for the detailed study of the vanishing radiative width effect (section 8.30). They can also be applied to the study of autoionising resonances (section 6.15). Rotation of the plane of polarisation of light

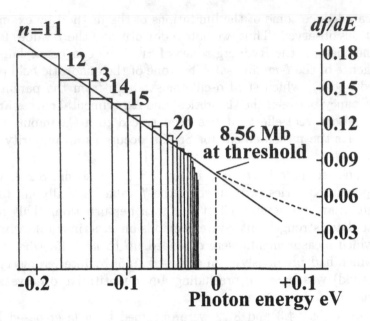

Fig. 4.11. Differential oscillator strength plot for Sr, showing measurements for high Rydberg members up to $n = 28$ (note: this is the same Rydberg series as shown in fig. 2.2 – after J.-P. Connerade *et al.* [163]).

(optical activity) has also been used for the detection of laser-induced continuum structure (sections 8.20 and 8.21). Alternative methods exist which are also based on the determination of refractive indices, either by interferometric measurements (the so-called 'hook method' [151]) or by exploiting nonlinear optics (section 9.7).

4.12   Results of measurements

Typical results of f value determinations are illustrated here by two examples. The first relates to the principal series of Sr in a synchrotron radiation experiment, while the second relates to f value determinations for the principal series of Ba obtained in a laser-based experiment. Full descriptions of the experimental details for both synchrotron [163] and laser[164]-based MOV experiments have been published. The data are summarised as df/dE curves, which were introduced in section 4.6.

The first example relates to a fairly unperturbed series and is shown in fig. 4.11. The data were obtained by synchrotron radiation spectroscopy, using a conventional grating spectrograph to provide dispersion. Above $n = 28$, effects due to the strong field did not allow reliable data to be obtained. Since the actual variation is closely approximated by a straight line, we can deduce some of the limitations of the method by examining the variations observed. Thus, we note a downturn of the f values for the highest members of the Rydberg series of Sr. This can be explained as the emergence of the ℓ-mixing satellites (one of the quadratic field effects mentioned above), which steal oscillator strength from the parent line. The discrepancy between the theoretical and experimental curves around $n = 18$ is also a real effect: it has been traced to a Ca impurity line, which distorts the measurement for Sr (Ca occurs as an impurity in Sr samples).

The second example is the one in fig. 4.3 (see also fig. 8.22), which shows a perturbed series, within which the f values actually fall to zero as a result of an interference effect due to a perturbation. This reduction in oscillator strength put severe demands on experimental technique, despite which measurements were extended up to $n = 45$, whereas the highest which had previously been achieved [165] by interferometry (the Hook method) was $n \sim 15$, precluding any quantitative observation of the minimum.

The data of figs. 4.3 and 8.22 were obtained by a laser-based MOV technique using a pulsed magnet: the high spectral resolution of the laser allows very high Rydberg members to be reached.

Although all the examples chosen involve singlet states, for which the theory is especially simple, there is no problem in extending the method to more complex Zeeman patterns, or indeed in including the effect of Paschen–Back uncoupling on the MOV spectrum [166]. The influence of ℓ-mixing on MOV patterns has also been studied, and is in principle well understood [167]. If the experiment is performed with lasers, the influence of laser power on Faraday rotation arises both by population transfer and by the Autler–Townes splitting (section 9.10)[173].

4.13   A more general formula for the rotation angle

The MOV method involves computing the *full* line profiles, without resorting to the so-called *far-wing approximations*. In general, it is always better to include the full expressions in the calculations of simulated spectra, since they can easily be built into a computer programme, and each one of the terms has a physical meaning which is readily identified. If, however, one seeks an analytic expression for the rotation angle, then one may use the following explicit formula:

$$\Phi(\nu) = -\frac{Nf\ell Be^3}{8\pi m^2 c^2} \frac{\nu(\nu - 2\nu_0)}{(\nu_0^2 - \alpha^2)\left[(\nu - \nu_0)^2 - \alpha^2\right]}$$

obtained from the full expressions; it is vastly superior to the far-wing approximation described by Zemansky [168].

It is worth noting that collisions in high Rydberg states and their influence on both linewidths and f values can also be studied by MOR spectroscopy [174].

4.14 Nonlinear effects

As already noted in section 4.12, nonlinear effects due to laser power can occur. In experiments with laser light in which f values are measured, it is important to ensure that the intensity of the radiation is not high enough to alter the quantities being measured. This can happen in several ways. The first is a straightforward effect: atoms are easily pumped into an excited state, and this will reduce the ground state population, thereby also reducing the observed rotation. In addition, there may also be nonlinear effects due to intense laser light [175]. Thus, it is known that, for intense radiation, there exists a self-induced Faraday rotation produced by the light beam itself (without any external **B** field), which rotates the plane of polarisation of elliptically polarised light. Similarly, it is possible to use an intense beam of circularly-polarised light to rotate the plane of a colinear, linearly polarised, weaker probe beam, without any external **B** field, or to confer ellipticity on a colinear, weaker beam (the Buckingham–Kerr effect). There are many other effects of higher order.

To test for freedom from these effects, the laser intensity must be varied at constant B and MOV patterns recorded. Of course, the nonlinear effects are also of great intrinsinc interest. Their detection by Faraday spectroscopy has been considered by Karagodova and coworkers [176].

4.15 Conclusion

The main theme of this chapter has been the distribution of oscillator strengths amongst transitions between bound states in a Rydberg spectrum. This subject is closely related to the properties of autoionising resonances, which will be discussed in chapters 6 and 8. As regards measuring f values, the MOV technique is used to determine the refractive index of an autoionising resonance (section 6.15).

5

Centrifugal barrier effects

5.1   Introduction

Centrifugal barrier effects have their origin in the balance between the repulsive term in the radial Schrödinger equation, which varies as $1/r^2$, and the attractive electrostatic potential experienced by an electron in a many-electron atom, whose variation with radius differs from atom to atom because of screening effects. In order to understand them properly, it is necessary to appreciate the different properties of short and of long range potential wells in quantum mechanics.

As the energy of the incident photon is increased above the ionisation threshold, centrifugal barrier effects often come to dominate the response of many-electron atoms, which totally alters the spectral distribution of oscillator strength in the continuum from what might be anticipated by comparison with H. This applies not only to free atoms, but also to the same atoms in molecules and in solids: many of the changes due to centrifugal effects occur within a small enough radius that they are able to survive changes in the environment of the atom.

Since the centrifugal term is present in the radial Schrödinger equation for *all* atoms, we must explain why centrifugal effects only dominate the inner valence spectra of fairly heavy atoms. Centrifugal barrier effects are present even in H. However, they act differently in transition elements or lanthanides.

One reason is straightforward: the ground state of H has $\ell = 0$, and therefore only p states are accessible directly by a dipole transition from the ground state. For $\ell = 1$, the centrifugal barrier term is small at the mean radius of the np electrons, so centrifugal effects do not intrude. More profound reasons are the emergence of an electronic core, and the interplay between the noncoulombic potential and the repulsive barrier.

Centrifugal effects in many-electron systems explain many features in the periodic table of the elements. A deeper understanding of the table requires one to reformulate the building-up or *aufbau principle* of Bohr [177] and Stoner [178]. One must include centrifugal barrier effects in the rules for 'building-up of inner-shells' [179]. Since the explanation of the periodic table is one of the major achievements of quantum mechanics, centrifugal effects are extremely important, and their neglect in any discussion of shell filling is unjustified.

In the early days of atomic theory, it was often assumed that only an empirical understanding could be achieved beyond the first few rows of the periodic table, where the simple formulation of the aufbau principle breaks down. The modern view is that shell and subshell filling can be accounted for within the central field approximation, provided the centrifugal barrier effects are included.

Of particular significance are the *long periods* which arise when filling the d and f subshells. These rows of the table contain the *transition elements* for the d subshells and the *lanthanides* and *actinides* for the $4f$ and $5f$ subshells respectively.

The plan of the present chapter is as follows. First, we describe some observations of centrifugal barrier effects. Second, we give an account of the Göppert-Mayer–Fermi theory for the filling of the long periods and, third, we bring the two subjects together within the conceptual framework of the double-valley potential model. Finally, we discuss the inadequacies of the independent electron approximation in relation to centrifugal barrier effects, and present more modern alternatives which call upon many-body theory.

5.2   Minimum principal quantum numbers for H

The minimum n value for a given ℓ is equal to $\ell + 1$, and this is a consequence of the existence of a centrifugal barrier, as follows readily from the form of the effective radial potential

$$V_{eff}(r) = -\frac{1}{r} + \frac{\ell(\ell+1)}{2r^2} \tag{5.1}$$

where we have chosen atomic units with $m = \hbar = e = 1$ for simplicity.

The second term is readily identified: $\ell(\ell+1)\hbar^2/2mr^2 = M^2/2mr^2$, and thus, for circular motion, there is an apparent force:

$$-\frac{\partial}{\partial r}\left\{\frac{M^2}{2mr^2}\right\} = \frac{M^2}{r^3} = \frac{mv^2}{r}$$

which we recognise as the centrifugal term.

V_{eff} describes a potential which rises to infinity at $r = 0$, and crosses the $V = 0$ axis at $r = \ell(\ell+1)/2$. Outside this radius, it has a minimum, readily found by differentiation to lie at $r = \ell(\ell+1)$, at which point $V_{min} = -1/2\ell(\ell+1)$.

It is clear that, for nonzero ℓ, no bound state can lie below this minimum. Since $E_n = -1/2n^2$, this means that

$$-\frac{1}{2n^2} > -\frac{1}{2\ell(\ell+1)}$$

whence $n^2 > \ell(\ell+1)$, which we immediately recognize as the rule for the values of ℓ in the H problem, viz. $\ell = 1$ for $n = 2$; $\ell = 1,2$ for $n = 3$, $\ell = 1,2,3$ for $n = 4$, etc.

Thus, the origin of this rule is immediately apparent as a centrifugal barrier effect: the repulsive centrifugal term is responsible for excluding radial wavefunctions from the centre for $\ell > 0$. Also note that there is only one radius at which the electrostatic and centrifugal potentials balance exactly.

As will emerge, these properties are somewhat altered in heavy atoms: there may be more than one radius for which exact balance occurs. Also, the centrifugal barrier may not rise high enough in a many-electron atom to exclude the radial wavefunction from within the core, although of course it will always exclude the wavefunction from the geometrical centre.

5.3   The delayed onset effect

The delayed onset effect was discovered experimentally by Ederer [180] in 1964 in the photoionisation spectrum of Xe gas. A more recent compilation of results for the same spectrum is presented in fig. 5.1. If one considers the ionisation continuum of H, then just above the ionisation threshold, there is a maximum in the cross section, which is followed by a monotonic decline with increasing energy. Such situations are illustrated in fig. 4.1: until the observations of Ederer, it had been believed that this would be the most general behaviour of continuum cross sections and that, apart from the Seaton–Cooper minima discussed in section 4.4, or the local influence of interchannel perturbations, revealed in spectra such as that of fig. 4.3, unperturbed continuum cross sections would usually follow a monotonic course of declining intensity comparable to that of H.

The case of Xe demonstrated that this is definitely not the case. That this is a gross effect is apparent from the scale of energies involved: the maximum in the cross section is *delayed* (in energy) by many eV above the associated threshold, which completely distorts the appearance of the photoionisation continuum.

The delayed onset effect was soon explained in terms of a centrifugal

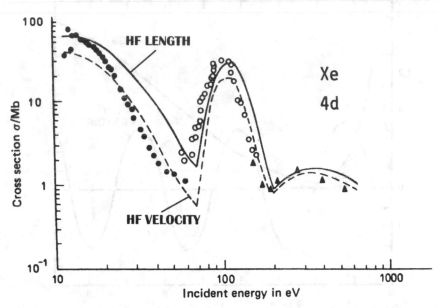

Fig. 5.1. The spectrum of Xe above the $4d$ thresholds, showing the delayed onset effect. The dots are experimental results from [181], the open circles are from [182] and the triangles from [183], while the theoretical curves (Hartree–Fock length and velocity approximations – see equation (5.31)) are due to [184]. Note the logarithmic scales.

barrier by Cooper [185]. It is readily understood on the assumption that the centrifugal barrier rises high enough to exclude the nf and low energy f waves from within the core. The principle is as follows. Under excitation from the $4d$ subshell, the final continuum state will be either ϵp or ϵf, in accordance with the selection rules. (The symbols ϵ or ε are conventionally used to denote an unquantised energy variable, just as n is used for the principal quantum number in the energy range of the bound states.) The latter, with $\ell = 3$, experiences a large centrifugal barrier, and will exhibit the largest effect. As a result of the barrier, which is positive and therefore repulsive, electrons of low kinetic energy cannot penetrate into the electronic core of the atom. Consequently, the spatial overlap between the core states and states of $\epsilon \sim 0$, i.e. states near the threshold, is very small for continua of high angular momentum, and such transitions are weak (this is an example of a *propensity rule* due to the radial equation, as discussed in section 4.2). On the other hand, as shown in fig. 5.2, in the case of $3d \rightarrow \epsilon f$ transitions, as the kinetic energy ϵ of the escaping electron rises, it begins to penetrate or overcome the centrifugal barrier, and the spatial overlap between initial and final states

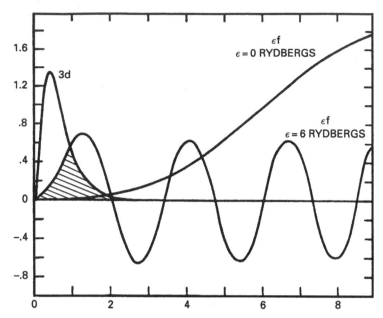

Fig. 5.2. Explanation for the delayed onset of photoionisation and subsequent maximum in terms of spatial overlap between initial and final state wavefunctions and phase cancellation effects (see text). The spatial overlap between the $3d$ orbital and the εf orbital with $\varepsilon = 6$ is shown as a hatched area, and corresponds to the maximum value of the cross section, just before phase cancellation sets in. (after S.T. Manson and J.W. Cooper [186]).

increases. Consequently the dipole matrix element, which determines the photoexcitation cross section, begins to *increase* with energy. This process continues until, as a result of the shortening in de Broglie wavelength of the wavefunction of the continuum electron, negative loops of the εf wave begin to penetrate the core and overlap with the core wavefunctions. This produces a change of sign or *phase cancellation* in the dipole matrix element, and so the cross section thenceforward *decreases* with increasing energy (see fig. 5.2).

This elegant and simple explanation was backed up by independent particle model calculations for a number of cases [186, 187], which demonstrated the effect semiquantitatively, and confirmed the observed dependences on atomic number for most elements.

5.4   The Combet-Farnoux minimum

It will not have escaped the reader that the dipole selection rules allow *two* continua to be accessed from the ground state in the situation just described, i.e. that the cross sections for $d \to \epsilon p$ and $d \to \epsilon f$ excitation must be added together in order to determine the total cross section. Since the ϵp continuum experiences little or no delayed onset effect, it exhibits a monotonic decline in cross section above threshold, whereas a strong delayed onset occurs in the $d \to f$ channel. Consequently, a shallow (nonzero) minimum appears just above the ionisation threshold in the total cross section. It is called a Combet-Farnoux minimum.

5.5   Relationship with the alkali and extended alkali models

It is worth noting the connection between the delayed onset effect and the alkali and extended alkali models, discussed in sections 2.9 and 2.12. In particular, it is clear that the bound states of high angular momentum possess little kinetic energy, and are therefore unable to overcome a strong centrifugal barrier. The p, d and f bound electrons therefore extend further and further into the outer reaches of the atomic potential, as the repulsive centrifugal force increases rapidly with increasing ℓ. Since the region they explore becomes more and more Coulombic as penetration of the core decreases, the centrifugal term is responsible for the greater and greater hydrogenicity of the quantum defects with increasing ℓ noted in connection with table 2.1.

The centrifugal force is also responsible for another, more subtle change, which is also apparent in table 2.1: in addition to the increase in hydrogenicity which, according to QDT is associated entirely with the fractional part of the effective quantum number $n^\star$, we note that the magnitudes of the effective quantum number and the principal quantum number become increasingly out of step as the atomic number of the element increases. For $\ell = 0$, this is a simple reflection of the filling of subshells: as commented in section 2.9, it becomes necessary to count the number of electrons in the core in order to determine n, because the Rydberg formula no longer yields this number directly in the presence of the parameter μ, and all $\ell = 0$ Rydberg series are observed to begin with an effective quantum number between 1 and 2. The situation is not so simple with d and f electrons, because, as a result, the labelling of, say, the nf series eventually becomes out of step with the labelling of the ns series in the same spectrum. Thus, for example, in Li, the first members of the s and f series are labelled $2s$ and $4f$ with $n^\star = 1.589$ and $n^\star = 4.000$, respectively, but, for Cs, the leading series members are $7s$ and $4f$ with $n^\star = 1.869$ and $n^\star = 3.978$, respectively. This is a consequence of the filling of subshells of high angular momentum, which affects the behaviour of the associated Rydberg series. It results from a breakdown in the elementary aufbau principle of Bohr and Stoner which cannot be resolved without a detailed

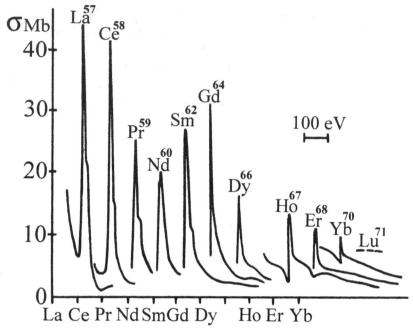

Fig. 5.3.   A summary of data from the experiments of Zimkina and Gribovskii, showing the systematic trend in the giant resonances of solids through the lanthanide sequence (after T.M. Zimkina [188]).

study of solutions of the radial equation (see below).

5.6   Giant resonances in solids

In an apparently unrelated series of experiments on the soft X-ray transmission properties of thin films of solids, Zimkina and co-workers (see, e.g., [188]) observed very intense absorption peaks, varying systematically both in energy and in intensity as a function of the atomic number. An overal summary of their observations is shown in fig. 5.3. The systematic trend with atomic number was unexpected: prior to Zimkina *et al.*'s observations, it was believed that soft X-ray spectra of solids would exhibit mainly solid state properties, and that broad resonances were indicative of band structure. Zimkina and her colleagues had the insight to resist this view, and recognised the fundamentally atomic nature of the resonances she and her colleagues had discovered. Henceforth, we shall use the word *quasiatomic* to denote a feature or spectrum which is intrinsically of atomic nature but has been observed in another phase (i.e. in a cluster or a solid) and is therefore slightly modified by solid state effects.

The experimental discovery of *giant resonances* as quasiatomic resonances can thus be attributed to Zimkina and her coworkers in the Soviet Union, although the designation of these features and the recognition that they are indeed atomic in origin took some time to become generally accepted.

It seemed at first surprising that isolated and rather intense atomic transitions should be present in the solid. In order to establish the correctness of this interpretation, it was necessary to investigate the spectra of the corresponding free atoms which, largely for technical reasons, had escaped observation until then. They were not in fact uncovered until 1974 [189, 190], when the first observations of a giant resonance in a free atom were reported independently.

Detailed observations and assignments of the giant resonances in many free atoms then followed rapidly, guided by the earlier observations for the solid. It was soon established experimentally that the quasiatomic giant resonances occur for nearly all solids for which the corresponding atomic transition exists, and that they may persist both in molecules and in ions. A detailed discussion of the properties of giant resonances in atoms, molecules and solids may be found in the book *Giant Resonances in Atoms, Molecules and Solids* [191], which summarises much of the research in this area prior to 1986.

A significant observation, which confirmed the earlier findings of Zimkina and her colleagues, is that giant resonances only occur for elements belonging to or immediately preceding the long periods of Mendeleev's table. A proper understanding of this requires us to reconsider the Bohr–Stoner building-up principle, and the manner of d and f subshell filling, which holds the key to the properties of the resonances in fig. 5.3

5.7 The Göppert-Mayer–Fermi theory of orbital contraction

As explained in elementary texts on atomic physics, e.g. [192] or [193], the Bohr–Stoner principle describes how the periodic table is built up in terms of angular momentum and the filling of subshells. The order in which subshells are filled is basically that of the energy levels of a fictitious one-electron atom, whose energy level spectrum would resemble that of an alkali, and it is assumed that the ordering of the levels remains unchanged while filling takes place, so that it never becomes necessary to solve the radial equation for a many-electron system: a few simple rules suffice. Oddly enough, although this 'explanation' of the periodic table was regarded as a triumph of quantum mechanics, the lack of consistency in the manner of treating the radial part of the problem and the failure of the Bohr–Stoner principle beyond the first three rows did not cause too much concern. In those texts which discussed the long periods, the standard way of patching up the argument was to invoke the notion of 'screening' together with a semiempirical description [193]. The first person to cast doubt on the correctness of this explanation was Fermi [194], who noted that the filling of the d and f subshells takes place *inside* the outermost shells of the transition elements, i.e. beneath the outer surface of the atom, and that electrostatic screening alone cannot explain this without invoking a new force. He therefore pointed to the centrifugal term $\ell(\ell+1)\hbar^2/2mr^2$ in the radial equation, and noted that it was the only term which might differ for the states involved, and thereby, perhaps, account for the peculiar behaviour of subshells of $\ell = 2$ and 3.

This idea was followed up by Göppert-Mayer [195] in 1941, who solved the radial equation for a number of elements in the d and f sequence. Unfortunately, she used the Thomas–Fermi model, which gives a fairly poor description of the radial potential and, as is now appreciated, does not account properly for the shell structure of the atom. Thus, although she found a certain number of interesting properties of $4f$ elements (lanthanides), she was unable to account for the filling of the d subshells, and her paper did not therefore have the impact it might otherwise have achieved.[1]

She had, through numerical calculations, uncovered that the effective radial potential experienced by $4f$ electrons, namely

$$V_{\text{eff}}(r) = V(r) + \frac{\ell(\ell+1)\hbar^2}{2mr^2} \tag{5.2}$$

where $V(r)$ is the purely electrostatic term due to the central mean field, exhibits a double well and (correctly) speculated that the filling of the $4f$ subshell out of sequence and deep inside the atom might be associated with a sudden transfer (orbital contraction or collapse[2]) of the wavefunction from the outer to the inner well. For this reason, in the present book, we refer to this explanation as the *Göppert-Mayer–Fermi theory*. In fig. 5.8, we show the form of the double well potential investigated by Göppert-Mayer [195].

---

[1] It has since been discovered that a more accurate equation, which takes account of radial correlations: the g Thomas–Fermi equation, does allow d-orbital collapse to occur. Thus Mayer was actually not far from her objective

[2] In early literature [208], the term wavefunction collapse was frequently used to describe this property. We avoid it here: it causes confusion with wavefunction collapse in the quantum theory of measurement. For this reason, we prefer the term orbital collapse, following [196].

5.8 Wu's semiclassical theory

Double-well problems are common in physics. The most celebrated example is the symmetric double well, whose semiclassical or WKB solution was obtained by Dennison and Uhlenbeck [197] and accounts for the doubling of the energy levels in the ammonia molecule.

A less familiar situation is the asymmetric double well, into which semiclassical models such as the WKB should also provide some insight. The full WKB solution to the double-well problem was discussed in detail by Fröman [198]. Simplified solutions had, however, been given earlier by Wu [200] in a much neglected paper which considerably predates the work of Göppert-Mayer. Wu concentrated his efforts on explaining the change $\Delta n \sim 1$ of one unit in the effective quantum number which accompanies contraction of the lowest energy orbital as all Rydberg solutions are essentially 'stepped down' from n to $n-1$, etc. This property was rediscovered numerically at a much later date [208] (see fig. 5.7), and its term dependence was analysed [201] by researchers who seem to have been unaware of Wu's much earlier work.

The radial Schrödinger equation can be written in its eikonal form as

$$\frac{d^2\psi}{dr^2} + P(r)^2\psi = 0 \qquad (5.3)$$

Fröman writes the solution in terms of two phase shifts, which we call Δ and δ, for the inner and outer wells respectively, and a penetration factor γ representing the barrier between them. The inner well extends from r_1 to r_2, the barrier from r_2 to r_3, and the outer well from r_3 to r_4. Thus

$$\gamma^2 = \exp\left\{\frac{1}{\hbar}\int_{r_2}^{r_3} P(r)dr\right\} = \exp\left\{-K\right\}$$

The WKB joining condition has the form:

$$\tan\delta\tan\Delta = \left\{\frac{\exp-K}{1+(1+\exp-2K)^{1/2}}\right\}^2$$

with

$$\Delta = \left(n+\frac{1}{2}\right)\pi - \frac{1}{\hbar}\int_{r_1}^{r_2} P(r)dr$$

and

$$\delta = \left(n+\frac{1}{2}\right)\pi - \frac{1}{\hbar}\int_{r_3}^{r_4} P(r)dr$$

For a symmetric well with $\ell = 0$ and $\exp-2K \ll 1$, $\Delta = \delta$ and

$$\tan^2\left\{\left(n+\frac{1}{2}\right)\pi - \frac{1}{\hbar}\int_{r_1}^{r_2} P(r)dr\right\} = \frac{1}{4}\exp-2K$$

leading to

$$\frac{1}{\hbar} \int_{r_1}^{r_2} P(r)dr = \left(n + \frac{1}{2}\right)\pi \pm \frac{\exp -K}{2}$$

which is the well-known doubling of levels referred to above.

For an asymmetric well, again with $\gamma \ll 1$, one has

$$\tan \Delta \tan \delta = \frac{1}{4}\gamma^2$$

From which, after some algebra

$$\int_{r_3}^{r_4} Q(r)dr = \left(n - \ell - \frac{1}{2}\right)\pi - \tan^{-1}\left[\frac{\gamma^4}{4}\tan \int_{r_1}^{r_2} P(r)dr\right] \qquad (5.4)$$

in atomic units,[3] where

$$Q(r) = \left\{\left|E - V - \left(\ell + \frac{1}{2}\right)^2 r^{-2}\right|\right\}^{-1/2}$$

(The reason for using the term $(\ell + \frac{1}{2})$ rather than $\ell(\ell+1)$ is not a feature of the present problem, but is a property of the WKB method [199].) Equation (5.4) is formally solved by noting that, in the external region r_3–r_4, the outer well is supposed to be Coulombic (this is exactly the same assumption as in MQDT, see chapter 3) and the integral is exact. Thus, one obtains

$$\int_{r_3}^{r_4} Q(r)dr = \pi(-E)^{-1/2} - \pi\left(\ell + \frac{1}{2}\right)$$

whence, writing

$$\pi\mu = \tan^{-1}\left\{\frac{\gamma^4}{4}\tan \int_{r_1}^{r_2} P(r)dr\right\}$$

one has $E = -1/(n - \mu)^2$, which relates the properties of the quantum defect to the accumulated phase in the inner well and the tunnelling rate through the barrier. Wu [200] describes various numerical approximations to estimate the change in μ which results from changes in the properties of the inner well. It is clear from the structure of the solution that μ will change by one unit when the argument of the inverse tangent changes smoothly from $-\infty$ to $+\infty$, i.e. that the effective quantum number n^* changes by 1. Note that *all* the radial solutions change when μ changes,

---

[3] It may be helpful to note here that, in the case of only one classically allowed region of motion, the analogue of this equation is simply $\oint Pdq = (n + \frac{1}{2})h$, i.e. the Bohr–Sommerfeld quantisation condition.

i.e. the complete Rydberg manifold is displaced together in energy, which is precisely what is observed (see fig. 5.7).

This theory gives a very useful picture, with the following shortcomings: (i) We assumed that $\gamma \ll 1$, whereas, as will emerge, the barriers can also be weak. This is not a serious problem. Referring to the above, one readily obtains a more correct expression:

$$\pi\mu = \tan^{-1}\left\{\left[\frac{\exp -K}{1+(1+\exp -2K)^{1/2}}\right]^2 \tan \int_{r_1}^{r_2} Q(r)dr\right\}$$

which is not confined to $\gamma \ll 1$ and may indeed be used above the potential barrier. Approximate forms for K can be derived, e.g. for parabolic barriers. (ii) The treatment assumes a semiclassical or WKB approach. This is probably the most serious limitation, since the effect of the centrifugal barrier is largest for the lowest energy solution, whose de Broglie wavelength becomes comparable to the scale size of the inner well. Despite this limitation, this model captures the spirit of the problem and the general behaviour of the solutions remarkably well. Note that orbital collapse is a quantum effect: as $n \to \infty$, the energy difference due to transfer of the innermost lobe of the solution from the outer to the inner well tends to zero.

5.9 Numerical approaches

The understanding of orbital collapse remained in a rather unsatisfactory state until around 1969 when, as a result of considerable progress in numerical methods and the advent of fast computers, it became a relatively simple matter to solve the radial Schrödinger equation in the Hartree–Fock and related schemes, which take better account of the shell structure of the atom.

Researchers then began to plot effective radial potentials [204] and to compute in detail the behaviour of radial wavefunctions [208]. It thus came to light that the behaviour described by Mayer applies not only to f electrons, but also, in a slightly different way, to d subshells, so that the same *ab initio* theory could be used to account in a quantitative way for the order of filling without any recourse to empirical information.

In fig. 5.4, we show the double-well radial potential as computed by Griffin *et al.* [208] for elements with $Z \sim 56$. Note that the radial scale used in plotting this figure is highly nonlinear: on a linear scale, the double-well or double-valley potential looks rather similar to that of fig. 5.8. Fig. 5.4 is plotted in this way to show how very small changes in the wells can precipitate a very large change in the radius of the $4f$ wavefunction: from a radius of about 13 atomic units in Ba I, it collapses

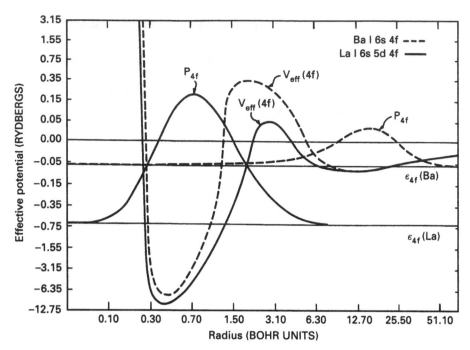

Fig. 5.4. The double-well radial potential for $4f$ elements close to the point at which orbital collapse occurs. Note the highly nonlinear scales on both axes, designed to show as much detail as possible for both wells (after D.C. Griffin *et al.* [208]).

to a radius of about 0.3 atomic units in La – a dramatic and very sudden contraction, since it occurs for a change in atomic number of just one unit.

5.10   The d transition elements

For the d electrons, the situation is similar, but not as dramatic: the centrifugal term in this case is smaller, and the potential, instead of a positive barrier between two wells, exhibits only a knee, as shown in fig. 5.5. In spite of this, orbital collapse occurs, but over intervals of more than one atomic unit.

The exact point at which collapse occurs depends rather critically on the presence and strength of the knee at the edge of the inner well. Being less dramatic, the collapse of d orbitals is also more easily disturbed. Thus, it was found [209] that the radius of the $4s$ electron in the $3d$ transition elements is very close to the position of the knee. When the $4s$ orbital is occupied, the potential experienced by the $3d$ electrons is

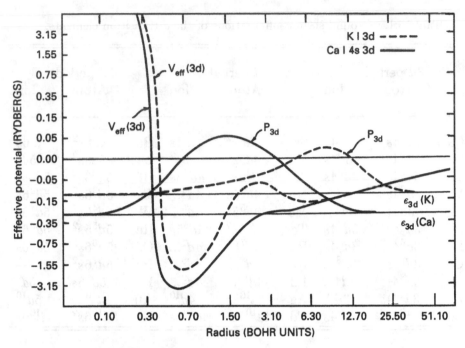

Fig. 5.5.   The effective radial potential for $3d$ electrons, showing the knee at the edge of the inner well (after D.C. Griffin *et al* [209]).

changed, and there occurs a subtle interplay between $3d$ and $4s$ filling which is a process described as *competition* between the filling of the d and s subshells. Since this process depends on the radial equation and on the balance between the two wells, it does not necessarily repeat in the same way in each period, and indeed there are so-called anomalies in the order of filling which derive from this fact. For example, there is only one element in the whole periodic table (Pd) which has a closed outermost $4d^{10}$ subshell.

A good way to see the breakdown of the aufbau principle is to draw up a table of configurations not only for the ground states of the atoms, but also for the corresponding ions. From table 5.1, one can see that, usually, it is easier to ionise the outermost s than the outermost d electron. The exceptions are:

(i) Pd: as just noted, the only element with a closed outer subshell which is neither s^2 nor p^6.

(ii) Y, for which it is easier to ionise $4d$ than $5s$: this is not unexpected from the order of filling, since $4d$ was also the 'last electron' in the building-up process.

Table 5.1.   Ground state configurations of the d transition elements.

3d period			4d period			5d period		
Atom	Ion		Atom	Ion		Atom	Ion	
Ca	$...4s^2$	$...4s$	Sr	$...5s^2$	$...5s$	Ba	$...6s^2$	$...6s$
Sc	$3d4s^2$	$3d4s$	Y	$4d5s^2$	$...5s^2$	La	$5d6s^2$	$...5d^2$
Ti	$3d^24s^2$	$3d^24s$	Zr	$4d^25s^2$	$4d^25s$	Hf	$5d^26s^2$	$5d6s^2$
V	$3d^34s^2$	$3d^4...$	Nb	$4d^45s$	$4d^4...$	Ta	$5d^36s^2$	$5d^36s$
Cr	$3d^54s$	$3d^5...$	Mo	$4d^55s$	$4d^5...$	W	$5d^46s^2$	$5d^46s$
Mn	$3d^54s^2$	$3d^54s$	Tc	$4d^55s^2$	$4d^55s$	Re	$5d^55s^2$	$5d^56s$
Fe	$3d^64s^2$	$3d^64s$	Ru	$4d^75s$	$4d^7...$	Os	$5d^66s^2$	$5d^66s$
Co	$3d^74s^2$	$3d^8...$	Rh	$4d^85s$	$4d^8...$	Ir	$5d^76s^2$	—
Ni	$3d^84s^2$	$3d^9...$	Pd	$4d^{10}...$	$4d^9...$	Pt	$5d^96s$	$5d^9$
Cu	$3d^{10}4s$	$3d^{10}$	Ag	$4d^{10}5s$	$4d^{10}$	Au	$5d^{10}6s$	$5d^{10}$
Zn	$3d^{10}4s^2$	$3d^{10}4s$	Cd	$4d^{10}5s^2$	$4d^{10}5s$	Hg	$5d^{10}6s^2$	$5d^{10}6s$

(iii) V, Co, Ni and La, *for which it is impossible to remove one electron from one occupied orbital and leave the ion in its ground configuration.*

Case (iii) is actually the most interesting, because it demonstrates most clearly the connection between breakdown of the aufbau principle and the appearance of many-body effects: excitation of one electron to an extended orbital in V, Co and Ni *must* be accompanied by a rearrangement of the parent ion core (cf section 7.3). In other words, for these three elements, one cannot specify which electron ($3d$ or $4s$) should be removed 'first.' Whichever is chosen, the result would be to leave the ion in an excited state.

The only possible explanation for this behaviour is that the $3d$ and $4s$ labels do not describe an individual electron correctly. Thus, *either* the configuration labels, *or* the notion that individual electrons can be excited, have broken down, which means that single particle orbitals cannot really be defined in this region of the periodic table. In other words, it is no longer possible to identify individual particles with specific orbitals. Although this is an extreme example of this type of breakdown, there are many other situations where a confusion between the nature of two one-electron labels can arise, especially involving s and d electrons. In subsection 7.14.6, we return to this question and to its implications in terms of the simultaneous excitation of two electrons.

It is worth noting, however, that there will be no such thing as a regular Rydberg series converging to the ground state of the ion in V, Co or Ni, and this is in line with the notion that a regular Rydberg series is associated with electrons behaving independently of each other (see section 2.35).

5.11   Homologous orbital collapse

So far, we have discussed orbital contraction as an effect occurring across the periodic table. However, centrifugal barrier effects grow as one moves to heavier elements, and this suggests that one should also study orbital collapse down columns of the table. The difficulty in doing this is that we can assume small changes in the electrostatic potential occur in moving from one fairly heavy element in the periodic table to the next one, but we can no longer make such an assumption when moving from one row to the next. We therefore need some index or measure of the strength of the central field.

This problem has been addressed by Connerade [205] who argues that nature provides this information for each element through the binding energies of the ns and, to a lesser degree, the np states, which undergo no dramatic changes as a result of orbital contraction and, indeed, in the case of ns states, experience no centrifugal force whatsoever. One can therefore define a dimensionless number ξ which is simply the ratio of the binding energies of such states to the binding energies of the corresponding states in H. The value of ξ is not very dependent on which of these states is chosen but, in order to have just one value for each element and to be sure it is a comparable measure, one should define it the same way in all cases, for example, by using the binding energy for elements or ions with just one outermost ns electron in the ground state.

Once this has been done, a plot of bound state energies against ξ yields: (i) in the case of ns and even np states, a set of nearly straight lines, showing that these energies depend only on electrostatic screening; and (ii) in the case of nd states, curves which may suddenly cross over the grid of ns and np lines whenever orbital contraction occurs. Such a graph is shown in fig. 5.6. The advantage of having a measure ξ of the electrostatic binding strength is that we are no longer confined to a single homologous sequence. Consider, for example, the sequences of singly-charged ions of Zn, Cd and Hg. Quite clearly, we may consider them alongside the singly-charged ions of Ca, Ba and Sr, since all of them involve an s electron in the ground state, and refer their binding energies to the singly-charged ion of He. Such a plot is called an *extended homologous sequence* [205], and has the advantage that much more information may be included in the

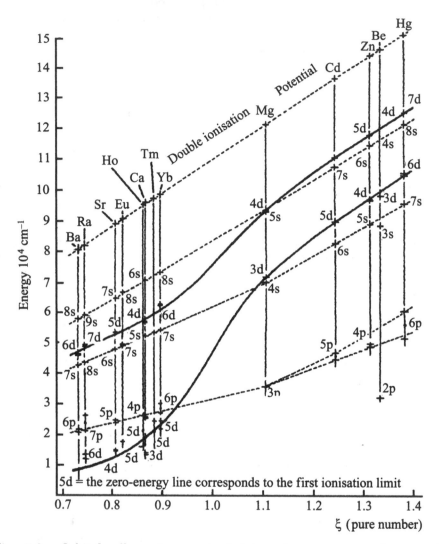

Fig. 5.6. Orbital collapse in an extended homologous sequence (see text for details) (after J.-P. Connerade [205]).

same graph, thereby allowing one to track orbital collapse systematically over a wide range of electrostatic binding strengths.

Such graphs provide a wealth of information. First, they explain the difference in character and chemistry of different sequences of elements which are otherwise very similar. Second, they allow us to understand how orbital contraction is controlled from within the atom. Externally, all the atoms in the plot are similar: an electron in the outer reaches of

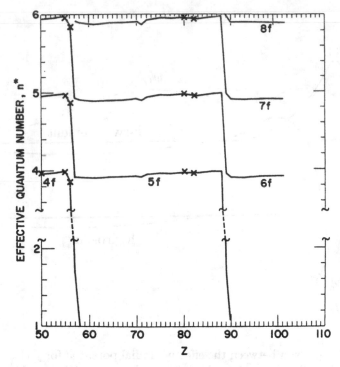

Fig. 5.7. The trend of effective quantum numbers for a Rydberg manifold of high angular momentum, showing the discontinuity in all the quantum numbers as a result of orbital collapse (after D.C. Griffin *et al* [208]).

the atom always sees more or less the same effective nuclear charge. The differences between the atoms are due entirely to differences in effective screening as a function of atomic number.

5.12   Orbital collapse and Rydberg excitation

As will be clear from Wu's model (section 5.8), when orbital collapse occurs, it is not only the lowest state of high angular momentum which is altered. All the excited states belonging to the same channel also experience a sudden contraction. This expresses itself as a discontinuity in all the effective quantum numbers, as shown in fig. 5.7. In summary, it is fair to say that the Göppert-Mayer–Fermi theory allows a complete understanding of the filling of the periodic table to be achieved for both the d and the f subshells of the elements. It even allows tentative estimates of how filling of the $5g$ subshell might occur in the transuranide or superheavy elements [208], although this calculation includes only first-order relativistic corrections.

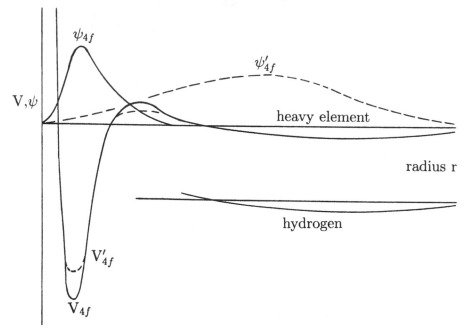

Fig. 5.8. Comparison between the effective radial potential for f electrons in H and the double-well radial potential in a heavy element by Göppert-Mayer–Fermi theory (not to scale).

5.13   The origin of the double well

Numerical procedures provide accurate solutions of the SCF problem, but are perhaps not enough to understand why a phenomenon appears. The double well occurs for rather simple reasons which we now endeavour to make clear. Consider first the effective radial potential of H for $\ell = 3$, which is shown in the lower inset of fig. 5.8. This potential has a 'hard centre', due to the repulsive centrifugal force, which varies as $1/r^2$, and therefore dominates at small r. On the other hand, it also has an attractive well at large r, which is due to the Coulomb term. At one point of intermediate radius r_c, the two terms cancel each other exactly.

It is a basic principle of QDT (cf chapter 3) that all atoms resemble H in the region outside the core, where the effective charge is one atomic unit. Thus, we can infer the potential for the outer well of a heavy atom, whose core is smaller than r_c, as shown in the fig. 5.8.

Now consider what happens inside the radius r_c. As the electron penetrates into the core, the attractive part of the potential dominates, and increases fast with decreasing radius, because screening by the core be-

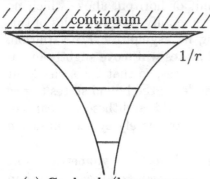

(a) Coulomb (long range
The number of bound states
is always infinite

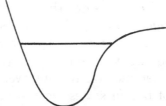

1. There is always at
least one bound state
2. The number of bound
states is always finite

(b) symmetric short range

(c) asymmetric short range

For (b) and (c), resonances
can appear in the continuum,
drop below threshold and turn
into bound states as the potential becomes more attractive

1. It is possible to have
no bound state at all
2. The number of bound
states is always finite

Fig. 5.9. Different kinds of potentials in quantum mechanics and the resulting
spectra (schematic).

comes less and less effective. Eventually, however, once the core has been
penetrated, the centrifugal, repulsive term must take over again, because
it varies as $1/r^2$, and therefore continues to grow rapidly as r decreases.

Thus, the double-well potential originates: (i) from the presence of the
centrifugal barrier; and (ii) from the existence of an atomic core. These
are sufficiently general for the double-well potential to develop slowly and
deepens systematically with increasing atomic number [204].

5.14   The mechanism of orbital collapse

We turn now to the mechanism of orbital collapse, which is not immediately obvious from the numerical calculations of the various authors
quoted above. It has been explained by Connerade [210, 211] using analytic potentials and elementary quantum theory. The key feature to note
is the difference in nature between different kinds of potential in quantum mechanics. This is illustrated in fig. 5.9(a).   First, we have the
familiar Coulomb well or *long range* potential. This, as we have seen in
chapter 2, gives rise to Rydberg series containing an *infinite* number of

bound states converging on a definite limit or horizon, above which no associated bound state can ever rise.[4] If we modify the properties of this potential at short range, while preserving its asymptotic $1/r$ variation, QDT tells us that each one of the bound states will move slightly in energy, but by a decreasing amount as n increases, so that the series limit remains the same. Indeed, it is never possible for the bound states to rise into the continuum above the ionisation threshold, and these two parts of the spectrum are separated from each other completely by the ionisation threshold.

Now consider the potential in fig. 5.9(b). This is a symmetric, *short range potential*, by which is meant simply that its edges fall off faster than $1/r$. There are many potentials in this class; for example, the symmetric square well, from which it is readily shown that:(i) there always exists at least one bound state, even if the well is infinitely shallow; (ii) the number of bound states increases as the well is made more binding, but always remains *finite*. The binding strength of the well depends on its depth times the square of its width, with a numerical coefficient which depends on the shape of the well; and (iii) the bound states, as the binding strength of the well is decreased, rise up to the threshold and then disappear one by one, turning into short lived resonances or modulations of the continuum which broaden quickly as they rise yet further in energy and disappear from view. Such potentials are of frequent occurrence in nuclear physics, and Ramsauer minima arise from the resonances near threshold. There is a theorem (Levinson's theorem) which connects the number of observable resonances with the number of bound states inside the well, and states that *the number of phase changes $\sim \pi$ in the continuum is equal to the number of bound states in the well*. A good example of this kind of potential is the Woods–Saxon well in atomic clusters (see chapter 12).

The third kind of well we need to consider is the *asymmetric* short range well, shown in fig. 5.9(c): this well shares most of the properties of the symmetric short range well, except for (i); it can, in fact, be completely empty of states, a fact which is of crucial importance in the present context. Asymmetric short range wells include the asymmetric square well with one hard wall, (whose ground state is made up from the first excited state of the symmetric well), the Morse potential, the Lennard–Jones potential, etc, several of which will be of interest to us.

The inner well of the double-valley potential of fig. 5.8 has the characteristics of a short range potential. It was speculated by Connerade [210]

---

[4] It is worth noting also that a dipole field of more than a certain strength, despite falling off faster than the Coulomb potential, can also hold an infinite number of bound states, as pointed out in section 2.24.

that orbital contraction is due to the following cause: there is a quantum-mechanical condition which an asymmetric short range potential must satisfy before it can support one bound state; its binding strength must exceed a certain number, which depends on the shape of the well. He suggested, using the example of an asymmetric square well, that orbital contraction occurs as soon as the inner well satisfies the quantum-mechanical condition for the inner well to support a bound state. This was confirmed quantitatively in a later paper [211], from which the following argument is extracted.

Consider the Morse potential:

$$V_{Morse}(r) = D \left\{ \exp[-a(r - r_0)] - 1 \right\}^2 - D \qquad (5.5)$$

This is an asymmetric short range potential, with analytic bound state eigenfunctions (confluent hypergeometric functions[5] see [206]). We can fit equation (5.5) to the inner well of the double-well potential in the numerical calculations of [208] to a high degree of accuracy by adjusting the two parameters a and D (a^{-1} is proportional to the width and D is the depth of the well). The quantity $\xi = (2mD)^{1/2}/a\hbar$ is a measure of the binding strength of the well.

In common with all other short range wells, the Morse potential is only capable of supporting a *finite* number of bound states. If $\xi < \frac{1}{2}$, there are no bound states at all. If $\frac{1}{2} < \xi < \frac{3}{2}$, there is one bound state. The condition for the existence of the nth bound state is $n + \frac{1}{2} < \xi < n + \frac{3}{2}$, so that bound states are uncovered one by one as ξ increases. This behaviour holds the key to the phenomenon of orbital collapse: when the inner-well binding strength is less than $1/2$, the orbitals are constrained to exist in the outer reaches of the atomic potential. When the strength is greater than $1/2$, the quantum-mechanical condition for an inner-well bound state is satisfied, and the orbital collapses as soon as the binding energy falls below the binding energy of the outer well.

The SCF potential is well represented by a Morse potential in the inner well (see fig. 5.10); we obtain an analytic form for the wavefunction of the bound state, and we can compare the analytic wavefunction with the numerical solution. It is apparent from the figure that they agree excellently. Since the Morse potential has no outer well, the properties of the collapsed orbital are entirely determined by the inner well and, since $\xi \sim 0.6$, it is clear that the binding condition is satisfied in the case of La.

When the case of Ba is examined in the same way, it turns out that $\xi \sim 0.53$, so that in fact it is *only just* possible to hold a bound state in

---

[5] In reality, these solutions are not completely exact, but are accurate to a very high degree of approximation. The error involved is of the order of the overshoot at the origin towards negative values of r in fig. 5.10.

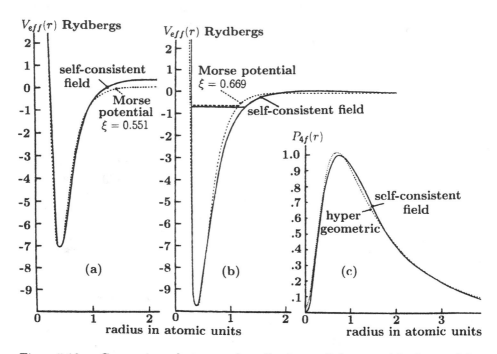

Fig. 5.10. Comparison between the effective radial potential obtained by Hartree–Fock calculations and a Morse potential adjusted to it. The cases shown are (a) for Ba and (b) for La, while (c) shows the hypergeometric and SCF wavefunctions for case (b). In all the graphs, the full curve is the numerical Hartree–Fock result, while the dotted curves are the Morse fits. The case of Ba is just at the critical binding condition (after J.-P. Connerade [211]).

the inner well, but at an energy slightly higher than the outer well. Thus, the ground state falls in the outer valley (interesting anomalies in the Ba$^+$ spectrum related to the critical condition are discussed in section 5.18).

In this way, we may picture orbital contraction as a purely quantum-mechanical effect, which arises from the existence of a short range well within the atom. The binding strength of this well increases with atomic number and, as a result, the critical condition for the appearance of the first bound state is satisfied around $Z = 56$. The condition for two bound states to occur inside the inner well is satisfied in a similar way at the onset of the $5f$ period, giving rise to the actinide sequence.

The mean radius of the collapsed orbital is so small that it can be described as *localised* deep inside the atom. The localisation of $5f$ states is less complete than that of $4f$ states because of the extra node in the wavefunction.

5.15 The influence of orbital collapse on collisions

An interesting consequence of orbital collapse is the fact that some excited states can be of very small dimensions. This makes them invulnerable to collisions. In fact, this is the reason the intruder state in fig. 4.5 survives in the presence of collisions with a foreign gas, whereas the high Rydberg states in which it is embedded are wiped out: the $3d^2\,{}^1G_4$ state has two $3d$ electrons which are quite well localised, and therefore the excited state has very small dimensions. This kind of behaviour has been studied in laser-excited Ca atoms in the presence of collisions: it has been demonstrated by studying the fluorescence from highly-excited states that all the strong lines originate from compact excited states, while the Rydberg excitations are quenched [207].

5.16 Giant resonances in free atoms

Note that the terms giant resonance and shape resonance are both prevalent in the literature. Usually, the term giant resonance is preferred by theorists involved in calculating many-body contributions to the cross section, while the term shape resonance is used by those who represent the features by using an effective independent electron potential. The atomic giant resonances share many of the characteristics of plasmons (see chapter 12), however they are not usually considered to be atomic plasmons, because they are too strongly damped. Atomic giant resonances vary systematically with atomic number, as demonstrated by the experimental data of fig. 5.3. This systematic variation is due to the progressive deepening of a short range inner well in the double-valley effective potential of the atom as atomic number increases (see [204, 208, 209, 210, 211] for more detailed discussions). The variation is well represented as a mean-field effect and, although its detailed calculation may involve many-body theory, it is qualitatively accounted for within the independent electron approximation. This is demonstrated in fig. 5.11 for a sequence of atoms just prior to the emergence of a giant resonance.

The physical origin of the double well is also readily understood. In H (as opposed to many-electron atoms), if the centrifugal repulsive term is included within an effective potential then, since it grows as $\ell(\ell + 1)\hbar^2/2mr^2$, there is a repulsive potential at small radius ($r < r_0$, say) can only expel wavefunctions of high angular momentum into the outer Coulombic reaches. Note that, at all radii, the effective potential is the net

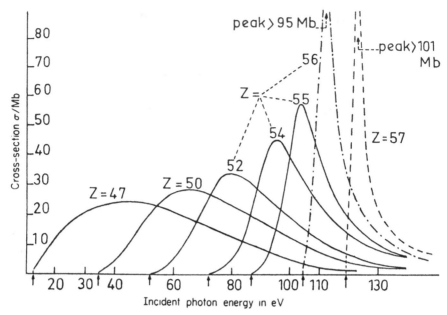

Fig. 5.11. Emergence of giant resonances in the 4d spectra of the elements preceding the lanthanide sequence (after F. Combet-Farnoux [224]).

result of a centrifugal repulsive term, varying as $1/r^2$, which dominates at small r, and an attractive Coulombic term, varying as $1/r$, which dominates at large r. In heavier atoms, a core develops within a radius r_0. Coming in from large radii, the potential starts off as in H. Thus, on the outside, it is a shallow Coulombic well with a repulsive shoulder around r_0. However, within r_0 the potential becomes strongly attractive again, because the electron penetrates the core and sees an incompletely screened nuclear charge, leading (for heavy atoms) to a deep inner well. This inner well has with a fast-rising repulsive wall near the centre, because the repulsive term varies as $1/r^2$ and must eventually dominate over all electrostatic forces at small enough r. near the centre. The inner valley is a short range well and (like all short range potentials in physics) can support only a finite number of bound states. If the well is 'squeezed' these states rise into the continuum and become virtual states or short lived resonances. In this respect, its behaviour is quite different from that of the long range outer or Coulombic outer well, which supports an infinite number of bound states. These can never cross above the threshold and rise into the adjoining continuum.

5.17   Sum rules and the disappearance of Rydberg series

One consequence of the appearance of a giant resonance in one excitation channel and the existence of sum rules for the oscillator strength
(section 4.5) is the weakening or disappearance of oscillator strength for
Rydberg series of the same channel. This can be explained in several
different ways.

First, we may say that, since a giant resonance exhausts the sum rule
within its width, all the oscillator strength available within an excitation
channel is consumed within the width of the giant resonance. Thus, for excitation from $4d^{10}$ to $4d^9 f$, the available oscillator strength, corresponding
to a subshell of ten electrons, will be concentrated within a single feature,
leaving none for any Rydberg states, which occur at a different energy.

Another description stresses the spatial overlap between the initial and
final state wavefunctions: the radius of the $4d$ initial state orbital is such
as to have a large overlap with the inner well of the $\ell = 3$ double-well
potential. Consequently, final state orbitals which are collapsed have a
large spatial overlap with the initial state, giving rise to a large dipole
matrix element and transition probability. On the other hand, final state
orbitals which are eigenfunctions of the outer well (Rydberg orbitals) have
very little spatial overlap with the $4d$ orbital, and give rise to very low
transition probabilities.

Both descriptions are, of course, equivalent. They suggest that the
disappearance of Rydberg series is associated with the emergence of a
giant resonance. The fact that the transitions are absent does not mean
that the states do not exist: they must be present, since the outer well
is Coulombic. It is only the transition probabilities which tend to zero,
i.e. the states are stabilised against radiative decay directly to the ground
state. Were the initial state to lie *outside* rather than inside the potential
barrier, it would be possible to reverse this situation and to enhance
to observation of Rydberg states despite the centrifugal barrier. This
situation occurs in the example of the next section.

5.18   The Ba$^+$ problem

Thus far, it may have surprised the reader that the double well was always
considered as a bistable situation, leading either to orbital collapse into
the inner well or to its expulsion into the outer reaches of the atom. The
possibility has not been considered that the wavefunction might be split
evenly between both wells, with comparable probabilities of the particle
being on either side of the centrifugal barrier. In fact, although such a cir-
cumstance is rare, there is nothing in principle to prevent its occurrence.

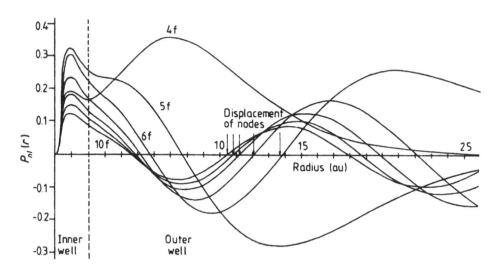

Fig. 5.12. Nonrelativistic Hartree–Fock calculations for Ba^+, showing the bi-modal behaviour of the nf orbitals resulting from centrifugal barrier effects (after J.-P. Connerade and M.W.D. Mansfield [212]).

Only one example is known to occur naturally in which an ion possesses excited nf orbitals which are neither diffuse nor collapsed, but remain poised in the 'knife-edge' situation: this case arises in the spectrum of Ba^+.

The *existence* of an anomaly in the spectrum of Ba^+ had been known for many years. It was commented on by Saunders *et al.* [213], who remarked: 'the perturbation ... appears to be of a novel type' but could advance no explanation for it. The realisation that this is due to a near-critical potential barrier effect came much later [212]. Hartree–Fock calculations were used to establish that the nf wavefunctions of Ba^+ become resonantly localised in the inner well of a double-well potential around $n = 5$. As a consequence, all the low n orbitals become distinctly bimodal or hybrid in character (as can be seen in fig. 5.12) with peaks in both the inner and the outer potential wells. This results in: (i) an unusual course in the quantum defects for the nf series: they are not constant, but depend strongly on the energy, despite the absence of any local perturbation; (ii) a pronounced intensity anomaly in the $5d \to nf$ excitation series, which was first noted by Roig and Tondello [214]; and (iii) an anomaly in the course of the spin–orbit splittings for the first members of the series, both of which can be explained on the basis of the bimodal character of the orbitals. In particular, as regards intensities, the overlap between the $5d$ orbital, which is external to the centrifugal barrier, and excited nf orbitals in the outer well allows a fairly long Rydberg series to be observed, so that the intensity goes through a minimum and then a maximum before dropping off at very high n.

By fitting a Morse potential to the numerical data for the Hartree–Fock potential of Ba^+ in its inner well, one can show [211] that it will support a bound state between the computed energies of the $n = 4$ and $n = 5$ solutions. In essence, we can regard the occurrence of the aforementioned anomalies as the signature of a giant resonance which has fallen in energy below the ionisation threshold and into the bound states, as a result of a deepening of the inner well due to the increased nuclear charge in Ba^+ as compared with the neutral Ba atom.

5.19   Levinson's theorem and the nature of the resonance

The hybrid nature of the solutions and the form of the wavefunctions in fig. 5.12 allow us to be more specific about the nature of the resonance involved. Indeed, it is a general feature of scattering problems (applicable also, for example, to photodetachment from negative ions or to molecular scattering) that orbitting resonances arise by trapping of an electron within a centrifugal barrier, while Feshbach resonances arise by coupling of two different channels, or transitions from one adiabatic curve to another [215]. In discussions of continuum structure, one also distinguishes between two kinds of resonance: the shape (or accordion) resonance, which arises when there is a sudden change in the phase shift or a 'breathing' motion of the orbitals and the Feshbach (or rope) resonance, which arises when the wavefunction actually changes sign at the origin. This difference is illustrated in fig. 5.13, from which it is clear that the resonances discussed in the present chapter, when considered within an independent particle model as resulting from centrifugal barrier effects, are shape resonances, while autoionising states, which involve the notion of a 'compound state' and its interference with the continuum, are Feshbach resonances. This difference is of fundamental importance in understanding many properties of giant resonances, as stressed by Temkin and Bhatia [216].

Since the behaviour of all short range wells is rather similar, and since the gross features of giant resonances are determined by the inner well, we may take over many well-known results of quantum scattering theory as applied to nuclear physics in our discussion of their properties.

One of these is *Levinson's theorem* (see section 5.14), which tells us that, for *any* short range well corresponding to an angular momentum ℓ

$$\Psi = \frac{U(r)}{r}\, \varphi_0$$

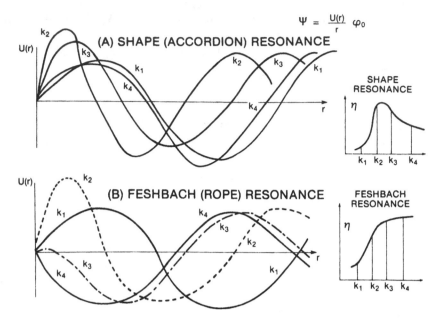

Fig. 5.13. Typical behaviour of the radial wavefunctions in the case of: (a) a shape or accordion resonance; and (b) a Feshbach or rope resonance (after A. Temkin and A.K. Bhatia [216]).

and capable of supporting m bound states, the phase shift δ_ℓ at the base of the continuum is given by

$$\lim_{\epsilon \to 0} \delta_\ell(\epsilon) = m\pi \tag{5.6}$$

5.20 QDT in a double well

In the usual applications of QDT (chapter 3), μ and δ are assumed to be weakly varying functions of energy. Indeed, it is often argued that QDT, to be applicable at all, requires a weak variation of μ. Slow variations of μ and δ were parametrised by Seaton [31], but situations where μ varies rapidly as a function of energy in the absence of any interchannel coupling require an extension of the theory [217].

The discussion proceeds by analogy with effective range theory for scattering by a short range force. It is pointed out by Schiff [219] p 144–5 that the asymptotic behaviours of the waves scattered by a pure short range force and by one involving a long range Coulombic part are not the same, so that the phase shifts derived will be different. Thus [223] $\tan\delta$ behaves as $\epsilon^{-1/2}$ in the limit of $\epsilon \to \infty$ for potentials of finite range which remain finite as $r \to 0$, while for r^{-1} potentials, $\tan\delta$ behaves as

$\ln \epsilon/\epsilon^{1/2}$. However, in standard QDT, Seaton [31] obtains useful results by neglecting the logarithmic factor: we therefore do the same and neglect the asymptotic differences.

The phase shifts at two energies E_1 and E_2 corresponding to wavevectors $\mathbf{k}_1$ and $\mathbf{k}_2$ in a short range well are related in the following way:

$$\left.\begin{array}{c} \dfrac{d^2u_1}{dr^2} + \mathbf{k}_1^2 u_1 - Vu_1 = 0 \\[2mm] \dfrac{d^2u_2}{dr^2} + \mathbf{k}_2^2 u_1 - Vu_2 = 0 \end{array}\right\} \qquad (5.7)$$

Multiply the first equation by u_2 and the second by u_1 and integrate the difference between them over r from $r = 0$ to a distance R somewhat larger than the effective range of the potential. This yields

$$\left(u_2\frac{du_1}{dr} - u_1\frac{du_2}{dr}\right)\Big|_0^R = (\mathbf{k}_2^2 - \mathbf{k}_1^2)\int_0^R u_1u_2 dr \qquad (5.8)$$

The same equation holds for the free particle wavefunctions ψ_1 and ψ_2 which are defined by $V(r) = 0$. Thus:

$$\left(\psi_2\frac{d\psi_1}{dr} - \psi_1\frac{d\psi_2}{dr}\right)\Big|_0^R = (\mathbf{k}_2^2 - \mathbf{k}_1^2)\int_0^R \psi_1\psi_2 dr \qquad (5.9)$$

Subtract this last equation from the previous one, noting that, as $r \to R$, the terms are the same because $u(R) = \psi(R)$. Also, write k for the magnitude of the vector $\mathbf{k}$. It then follows that

$$\left(\psi_2\frac{d\psi_1}{dr} - \psi_1\frac{d\psi_2}{dr}\right)_{r=0} - \left(u_2\frac{du_1}{dr} - u_1\frac{du_2}{dr}\right)_{r=0} = (k_2^2-k_1^2)\int_0^R (\psi_1\psi_2-u_1u_2)dr$$
$$(5.10)$$

The function $u(r)$ is the product of r and the radial wavefunction, and its normalisation is chosen so that

$$\psi(r) \equiv \frac{\sin(kr + \delta)}{\sin\delta} \qquad (5.11)$$

in order that its asymptotic form $\lim_{r\to\infty} u(r) \to \psi(r)$. The quantity δ is called the *phase shift* [219]. Thus

$$k_2\cot\delta_2 - k_1\cot\delta_1 = \frac{1}{2}(k_2^2 - k_1^2)\rho(E_1, E_2) \qquad (5.12)$$

where

$$\rho \equiv \rho(E_1, E_2) \equiv 2\int_0^\infty (\psi_1\psi_2 - u_1u_2)d\tau \qquad (5.13)$$

We pick E_1 such that $\delta_1 = \pi/2$ and $\cot\delta_1 = 0$. Then, noting that $k_1^2 = \tilde{k}$

is a constant and dropping the subscript 2

$$k \cot \delta = \frac{1}{2}(k^2 - \tilde{k})\rho(E) \qquad (5.14)$$

gives the phaseshifts above threshold, i.e. for $E > V$. For $E < V$, $k = iK$ say, becomes pure imaginary, and k^2 changes sign. Thus we obtain

$$iK = -\frac{1}{2}(K^2 - \tilde{k})\rho(E) \qquad (5.15)$$

where K, by definition, is real. This corresponds to the fact that δ cannot have the same definition below threshold, where bound state wavefunctions do not oscillate as $\sin(kr + \delta)$, but decay exponentially at large r. Nevertheless, it is useful to continue δ below threshold. One could define Δ such that $\delta = i\delta$ and

$$\cot \delta = \cot(i\Delta) = -i \coth \Delta \qquad (5.16)$$

However, we must take care that $\cot \delta$ is cyclic, whereas $\coth \Delta$ is not. Thus, the more correct choice is $\delta + m\pi = i\Delta$, when equation (5.16) is satisfied and the value of m must be chosen to satisfy Levinson's theorem and to allow a smooth continuation.

As shown by Berry [220], the phase shift increases by π at each shape resonance for positive energies above threshold. When the $4f$ function drops into the inner well, $5f$ moves to occupy the position of $4f$ before collapse, $6f$ moves to play the role of $5f$, etc [201]. As far as QDT is concerned, μ relates to the Rydberg energy formula applied in the outer well. Thus, if $\mu = n - \nu$ is written with n obtained by node counting, there will be an error of 1, because in fact one should use $\mu = n - 1 - \nu$ when the resonance drops below the bound states. Consequently, Seaton's theorem becomes

$$\delta = (\mu - 1)\pi \qquad (5.17)$$

and we have

$$K \coth(1 - \mu)\pi = \frac{1}{2}(K^2 + \tilde{k})\rho(E) \qquad (5.18)$$

Near a shape resonance [221]

$$\tan \delta = \frac{\Gamma}{2(E_0 - E)}$$

where

$$\Gamma = \frac{\exp -2\gamma E_0}{2\left(\dfrac{\partial \psi}{\partial E_{E0}}\right)}$$

and γ are slowly varying quantities. Thus

$$C(E) = \coth(1 - \mu(E))\pi = \frac{2}{\Gamma}(E_0 - E) \tag{5.19}$$

Plots of quantum defects and of $C(E)$ against energy are shown in fig. 5.14 (a) and (b). They demonstrate how this equation 'linearises' QDT for a single channel containing a giant resonance.

$C(E)$ is a more appropriate function to plot than μ itself, as it leads to a straight line plot from which E_0 and Γ_0 for the shape resonance can be determined. In the case shown, $m = 1$ because the well is just able to hold one bound state, as explained above: the energy E_0 obtained in this way from experimental data is consistent with the determination from Hartree–Fock calculations using a Morse potential.

In essence, the approach outlined here [217] provides a QDT for the double-well problem. Equation (5.6) applies to any shape of short range well. There are other general results of this kind: for example, it was pointed out by Schwinger in connection with problems of nuclear physics that there is no possibility of determining the shape[6] of a short range potential whose scale size is of the order of the de Broglie wavelength of a scattered particle merely from observations of the low energy scattering spectrum and the bound states. In a sense, this is disappointing news: the implication is that one cannot, in the present instance, deduce the shape of the inner well from observations of the spectrum. However, there is one benefit in this indeterminacy. We can, in fact, find *many* short range potentials which are capable of mimicking the bound state spectrum and the density of states near threshold, and this includes especially simple ones, from which analytic expressions for the phase shift can be deduced. This possibility is put to good use in the next section where a simple formula for the lineshape of a giant resonance and a relationship between its energy and its breadth are worked out, and where the differences between shape and autoionising resonances are explained in more detail.

5.21 Shape and giant resonances

The isolated Beutler–Fano resonance appears, at first sight, to be the simplest situation which can give rise to a resonance in the continuum. In principle, the resonance can appear at any energy above the threshold, since resonance and continuum belong to distinct channels. Thus, there

---

[6] It is a general principle that effective range theory is independent of the shape of the potential. Strictly speaking, however, one should adapt effective range theory to the case of nonzero angular momentum. This has been done explicitly by Drukarev [218], who shows that the width Γ then depends on the effective range of the potential

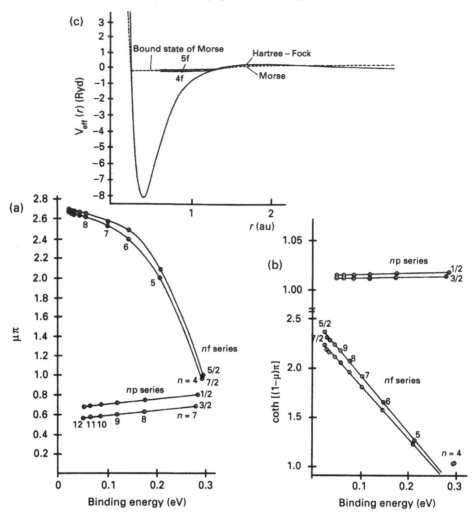

Fig. 5.14. Quantum defect plots for the centrifugally distorted nf series in Ba$^+$:
(a) shows the ordinary QDT plot, while (b) shows the plot obtained from the
same experimental data using the generalised theory in the text. Note that the
lowest point in the nf channel lies off the graph: this is normal, since it has no
node except at the origin and the corresponding wavefunction lies mostly in the
non-Coulombic part of the potential; (c) shows how the energy of the bound state
can also be obtained by fitting a Morse potential to the Hartree–Fock potential
of the inner well (after J.-P. Connerade [217]).

is no implied connection between the width of a Beutler–Fano resonance
and its energy above threshold.

   As explained above, resonances appear which do not owe their existence to the presence of two channels but could occur as broad features
within a single channel even if correlations were turned off. They are not

due to the occurrence of discrete structure of one channel embedded in the continuum of another, but to intrinsic properties of the continuum itself, i.e. to a rapid variation in the density of states. Such resonances can become huge, and dominate the cross section completely, dwarfing transitions to all other channels in the same energy range. We say that they exhaust the oscillator strength sum rule within their range. For this to be true, the overlap between initial and final states must be large, and this is also a condition favouring rearrangements of all kinds (many-body effects) within the atom.

A distinction has already been made in section 5.19 between giant and autoionising resonances. In the present section, we point out further differences between them, and establish a simple connection to quantum scattering theory.

The unique feature of giant resonances in atoms is that they occur in a potential which combines both a long range and a short range well in the same physical system, with a high probability that the electron will transfer between the two. It was pointed out in section 5.20 that the bound states and low energy scattering spectrum of a short range well are not dependent on its detailed shape but mainly on how binding it is. We can take advantage of this to 'mock up' a very simple potential which will give us an analytic formula for the phase shift namely a spherical square well with angular momentum, which possesses analytic solutions in the continuum [223]. The resulting expression for the resonant part of the phase shift (cf equation (8.2)) is

$$\tan \Delta_0 = \frac{z j_\ell(z\prime) j_{\ell-1}(z) - z\prime j_\ell(z) j_{\ell-1}(z')}{z j\ell(z\prime) j_{-\ell}(z) + z\prime j_{-\ell-1}(z) j_{\ell-1}(z\prime)} \qquad (5.20)$$

where ℓ is the angular momentum ($\ell > 1$), j are the spherical Bessel functions, $z = ka$, $z\prime = k\prime a$, where k and $k\prime$ are the wavevectors inside and outside the well and a is its radius.

The cross section curves in fig. 5.15 show how the giant resonance evolves as the binding strength of the short–range well increases. Note the similarity between fig. 5.15(a), obtained from detailed Hartree–Fock calculations [224], and fig. 5.15(b) which is the present very simple model based on the principles of effective range theory. In particular, the broken curve of fig. 5.15(b) provides an immediate explanation for the corresponding feature of fig. 5.15(a): in the expression for the cross section

$$\sigma(k) = \frac{4\pi}{k^2} (2\ell + 1) \sin^2 \Delta_0 \qquad (5.21)$$

for a giant resonance, the background due to other continua can be neglected, since the giant resonance structure dominates the cross section. The presence of a resonance is, as usual, signalled by Δ_0 passing through

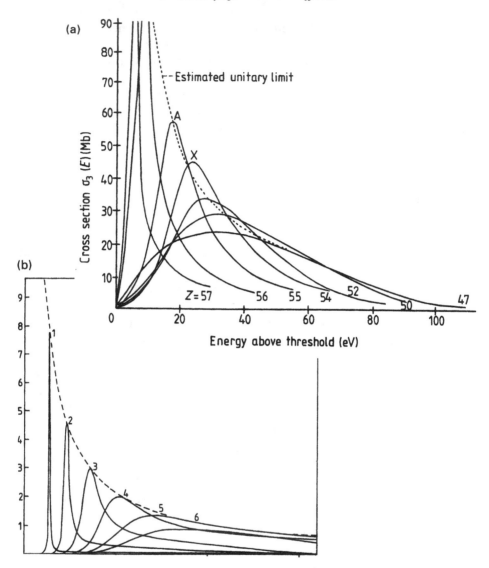

Fig. 5.15. Evolution of 'giant resonances' as the binding energy of the short-range inner well becomes progressively smaller: (a) according to Hartree–Fock calculations due to Combet-Farnoux; and (b) according to the spherical square well with angular momentum described in the text (after J.-P. Connerade [222]).

π as energy is increased. Thus, the cross section curves all lie below a $1/k^2$ envelope (the dashed curve in the figures), which is known in nuclear scattering theory as the *unitary limit*.

Now, we can see an observational difference between the Beutler–Fano resonance and the giant resonance, due to the fact that the former is an interchannel effect while the latter would be present even as a single-channel phenomenon; for a giant resonance, the width does depend on how high the resonance lies above the associated threshold, whereas for a Beutler–Fano resonance, it depends on a coupling strength, which is not related to the position of a threshold. If the giant resonance is far above threshold, the resonance is broad. As the well is made more binding and the resonance approaches the threshold energy, it becomes progressively sharper.

One can argue from the uncertainty principle [211, 225] that there should be a linear relation between energies and widths for giant resonances, and this deduction is obeyed quite well by experiment, as shown in fig. 5.16. This has been termed by Connerade a *universal curve* for giant resonances. In this respect, giant resonances and Beutler–Fano resonances behave rather differently from each other.

In fig. 5.17, we show two experimental examples of giant resonances, together with computed profiles using the simple phase-shift formula given above. The giant resonance has a characteristic asymmetric shape, with a low value of the cross section close to the threshold and a long 'tail' extending towards high energies. Note that the point of lowest cross section associated with a Beutler–Fano resonance is not linked to any threshold energy: in fact, the Beutler–Fano window is adjacent to the resonance energy E_0 of the Fano profile formula (6.20).

Too much significance should not be attached to the parameters describing the short range well. *It is only an effective potential.* Although the behaviour of the breadth of shape resonances as a function of energy has been studied theoretically for different short range wells [226] and attempts have been made to relate their observed behaviour to properties of atoms [227, 228] or to changes in the dimensions of molecules [229], Schwinger's theorem precludes a precise solution of the inverse scattering problem is such cases, just as it does for the equivalent problem in nuclear physics. It is mainly the binding strength of the well which is significant.

In the case of a giant resonance, the inner, short range well is almost, but not quite sufficiently binding to trap a bound state, and so the resonance appears in the continuum, just above the mouth of the well. However, it is sufficient for a very slight increase to occur, either in the depth of the well or in the height of the barrier, for the resonance to become trapped in the inner well.

The two terms *giant resonance* and *shape resonance* are sometimes used as synonyms in the literature. However, a distinction between the two should be made: a shape resonance is a resonance whose properties and profile are well described by a simple theory involving a realistic potential. A giant resonance, on the other hand, involves a large spatial overlap between the orbital of the initial state and that of the final, continuum

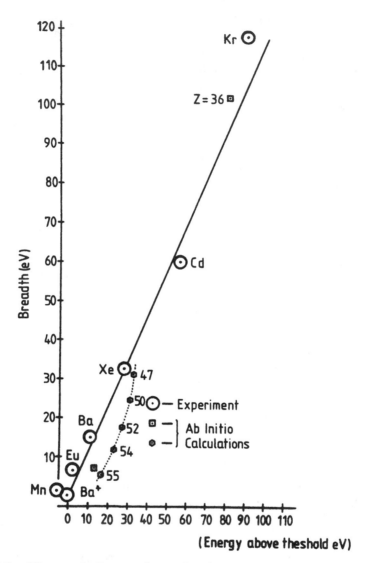

Fig. 5.16. The nearly linear relationship between the breadth of a giant resonance and its energy above the threshold illustrated using both experimental
data (large circles) and theoretical, *ab initio* calculations (small circles and boxes)
(after J.-P. Connerade [225]).

state. It follows that the escaping electron is influenced by the internal
rearrangements inside the atom. Under these conditions, a simple theory
with a realistic potential fails, and many-body theory is required for a
proper description (see section 5.26). The term giant resonance, borrowed
from an analogous situation in nuclear physics, emphasises the collective

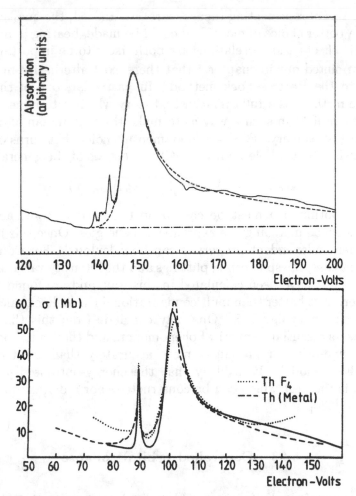

Fig. 5.17. Experimental examples of giant resonances, and their representation using the simple profile formula in the text: (a) for the $4d$ resonance of Gd atomic vapour (full curve); and (b) for the $5d$ resonance of Th, both in the condensed phase (dashed curve) and the vapour (dotted curve); the other curves are plotted from the formulae in the text (after J.-P. Connerade [222]).

nature of the excitation.

5.22   The g-Hartree theory of orbital collapse

It is clear that, in the Ba$^+$ problem, with an nf wavefunction poised on the knife edge between the two potential wells, there is an extremely sensitive dependence on the accuracy of the effective mean field. Hartree–Fock theory provides a first-order description of the phenomena involved, but is not capable of representing them accurately. Indeed, the question arises whether a better choice of mean field could be made, bearing in mind that relativistic effects and correlations are both likely to be important.

It was pointed out in chapter 1 that there exist alternative mean-field theories to the Hartree–Fock method. In particular, one of these, the g-Hartree method, is a fully relativistic theory which determines the 'optimum' mean field in such a way as to make the Lagrangian of quantum field theory stationary. This is a fundamental choice, but turns out [230] to be satisfied by a whole family of SCF potentials of the general form

$$V_{gH}(r) = gV_{dir}(r) + (1 - g)V_{exch}(r) \qquad (5.22)$$

where the optimum g must be chosen on the basis of some additional criterion such as matching the experimental energies. Once this has been done, other, independent properties can be calculated. For example, in the present case, the spin–orbit splittings and the intensity variation in the Rydberg series have been calculated in this way, and are found to agree with experiment better than multiconfigurational Dirac–Fock calculations [231], as shown in fig. 5.18. One may conclude from this that orbital collapse is an essentially spherical phenomenon, and that radial potentials can be found which represent it more accurately than the traditional Hartree–Fock model. It is likely that the energy-optimised g Hartree potential is the best which can be constructed. For a discussion of these points, see [230].

5.23   Controlled orbital collapse

Consideration of the knife-edge problem led to the suggestion [232] that orbital collapse in certain atoms might be controlled externally by modifying the potential in the outer reaches through excitation of a valence electron. The validity of this suggestion was confirmed by Hartree–Fock calculations in which an atom was progressively excited to higher and higher Rydberg states and, for each one of these, inner-shell excited configurations involving the promotion of an inner d electron to an nf orbital were calculated. It emerged that the $4f$ orbitals could be made to exhibit a bimodal structure (see fig. 5.19), and to transfer from one well into the other as a function of n.

While the calculations shown in the figure illustrate the general mechanism involved in the controlled collapse problem, it is also necessary to take account of term dependence, which becomes critical in the knife-edge situation. For photoexcitation in free atoms, configuration average calculations are not the most appropriate: to observe controlled collapse in the

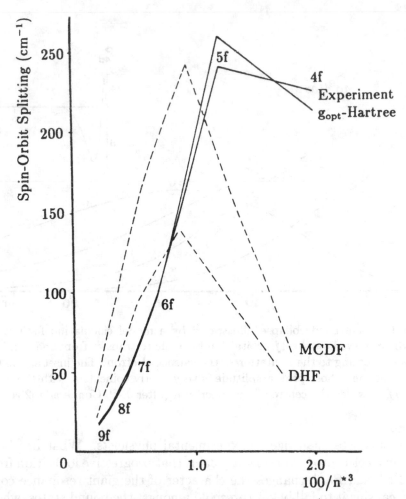

Fig. 5.18. Spin–orbit splittings for the nf series of Ba^+, showing that good agreement is obtained between g Hartree and experimental values. For comparison, Dirac–Fock (labelled DHF) and multiconfigurational Dirac–Fock (labelled MCDF) curves computed from the same code are also shown (after J.-P. Connerade and K. Dietz [230]).

d^9f singlet channel, Ba is a more suitable atom than Cs.[7] On the other hand, configuration average calculations are the best choice for comparison with similar effects in clusters and solids which will be discussed in chapters 11 and 12.

The observation of controlled collapse through the excitation of Rydberg states must undoubtedly be possible, but has never been achieved,

---

[7] I am grateful to Dr C. Clark of NIST Washinton for this remark.

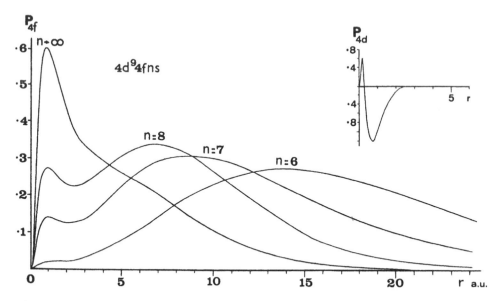

Fig. 5.19. Controlled collapse, illustrated by a model calculation for the Cs atom which shows how the $4f$ orbital can be made to transfer from one well into the other according to the excitation of the valence electron. The inset shows the $4d$ wavefunction, whose peak amplitude is very nearly coincident with the inner lobe of $4f$, towards the centre of the inner well (after J.-P. Connerade [234]).

and remains as a challenge to experimental physicists. What has been shown, in a celebrated experiment [233] is that progressive ionisation from Ba to Ba$^+$ and Ba^{2+} changes the character of the giant resonance completely, causing it to fall below threshold amongst the bound states, where it becomes a perturbation of a similar nature to the one discussed above. These dramatic spectra are shown in fig. 5.20.

It has also been shown [235] that similar effects occur for d electrons in certain compounds as the chemical environment is changed. Reinforcement of the centrifugal barrier by inserting the atom in a molecular F cage has also been demonstrated [236]. The effect of the cage may be studied by observing the relative strengths of Rydberg transitions and shape or giant resonances, which tend to gain or lose intensity at each other's expense, because of the Thomas–Reiche–Kuhn sum rule (section 4.5).

It is also worth noting that a different mechanism exists for producing double-well potentials and orbital segregation of a purely molecular kind in rather large, symmetrical molecules, without the need for atomic centrifugal barriers. This subject and its connection to the effects described here are discussed by Robin [237]. In connection with the corresponding

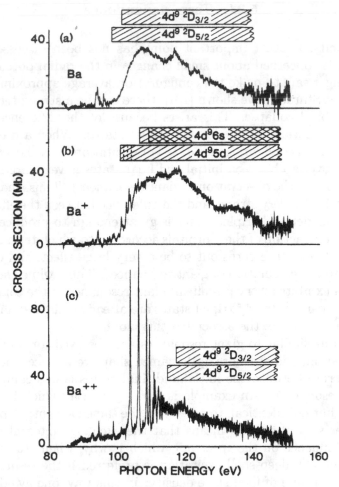

Fig. 5.20. Observation of controlled collapse through neutral, singly- and doubly-ionised Ba (after T.J. Lucatorto *et al.* [233]).

or united atom model for the spectra of molecules, it is interesting to study the 4*d* giant resonances in HI and the sequence $C_nH_{2n+1}I$, whose Rydberg excitations were discussed in section 2.33, because their electronic configuration relates them to Xe [238]. Such considerations also extend to the spectroscopy of solids. Thus, luminescence and photon-stimulated desorption from CsI, both of whose constituent ions are in configurations equivalent to Xe, serve as probes in a model system to study 4*d* relaxation in ionic crystals [239]. The relaxation of 4*d* vacancies, forming zones with high densities of excited states, plays a crucial role in determining the luminescence of the solid [240].

5.24   Term dependence

For simplicity, another important point has not been discussed so far. When one is concerned about small changes in the radial potential, then it is no longer safe to make the configuration average approximation. As Wendin and Starace have shown [201], there can be a strong term dependence of orbital collapse. This arises because of the difference between the radial potentials for singlet and triplet states. When a d electron is excited, the final configuration is $d^9 f$. For excitation from the $4d$ subshell, the spatial overlap between initial and final states is very large (which is the condition for the observation of giant resonances). Thus, the exchange interaction is also very strong, and the difference between the effective potential for singlet and triplet states is great enough to produce different degrees of contraction of the f orbitals according to which term is excited. This term dependence turns out to be a very large effect, and cannot be neglected in the calculation of giant resonances. Thus, while the $4d$ spectrum of La exhibits a very prominent giant resonance in the continuum of the singlet channel, the $4f$ triplet states are already collapsed, and appear as bound states *below* the associated threshold.

Excitations similar to giant resonances are observed for much deeper-shell excitation, and are more aptly named shape resonances, because the spatial overlap between the initial and final states is less complete than for giant resonances. An example is $3d \rightarrow f$ in Ba, which is similar in character, but not identical, to $4d \rightarrow f$ in the same element: in particular, the smaller spatial overlap means that the exchange integral has a less dramatic influence on the spectrum, which is why, when studying solids (see chapter 11), deep shell excitation is preferred if the resonance is to be used as a probe of final state density: in this way, one avoids most of the term dependence.

5.25   Shape resonances in molecules

A more interesting situation occurs even in a very simple molecule, as Dehmer and coworkers have demonstrated [202]: the $1s^2$ shell absorption spectrum of N_2 was found to exhibit unexpected behaviour for a diatomic molecule from the first row of the periodic table. Its Rydberg series is extremely weak, and an intense broad peak appears at $\sim 14\,\mathrm{eV}$ in the continuum above the ionisation potential. This anomalous behaviour is due to a centrifugal barrier effect acting on the $\ell = 3$ component of the excited state: although the electric dipole selection rule produces an escaping ϵp electron, it scatters off the anisotropy of the molecular field into the entire range of contributing angular momenta. Also, the spatial extent of the molecular field is such as to overcome its centrifugal barrier and penetrate more easily into the molecular core. This combination of circumstances produces a dramatic shape resonance of f-wave character, and is an example of how centrifugal effects may emerge in an unexpected way when combined with the molecular field.

This interpretation has been confirmed experimentally [203] by an elegant experiment in which the different angular distributions of the escaping photoelectrons were used to separate the partial waves and determine directly how much f-wave character is present in the photoionisation spectrum

5.26 Many-body perturbation theories and giant resonances

A simplification we have made throughout this chapter is to discuss the excitation of giant resonances using the language of effective radial potentials. While this provides a useful zero-order picture, the remarks made above about strong term dependence show that it has severe limitations. A potential is only useful if it can be at least approximately correct for a group of states. If the potential needs to be recalculated for initial and final states, there are already problems in using it to describe the excitation process. Moreover, if *each* excited state requires a different potential to represent it, then the theoretical model is unsuitable.

An alternative approach is to work in a frozen basis and to calculate excitations by systematic perturbation expansions, which are most elegantly written out using Feynman graphs. An excellent introduction to Feynman graphs in this context is given by Kelly [241].

The strategy just described defines what is called *many-body theory* by atomic physicists. It is usually performed by using a frozen core Hartree–Fock basis, which is regarded as the reference because it is an independent electron approximation with a single potential. The two main approaches to many-body perturbation theory are then: (i) to compute the Feynman graphs order by order, counting and summing all the diagrams for a given order, to make sure that none are left out, or at least picking the most significant ones so that one can be fairly confident that the largest interactions have been included; or (ii) to pick a class of diagrams with a convenient algebraic structure, forming a series which can be summed exactly to all orders of the perturbation expansion.

Stategy (i) is the one followed by the school of Kelly [242] and his coworkers. It is known as many-body perturbation theory (MBPT) and has the advantage of being very flexible, as it can be applied to all atoms, including those with open shells, but the disadvantage that the calculations can only be taken to finite (and in practice rather low) orders of perturbation, since the number of graphs grows as $N!$, where N is the order of perturbation. Strategy (ii) is the basis of the RPA (random phase approximation), used in nuclear and plasma physics (for an introduction to the RPA method, see, for example, the book by Bertsch and Broglia [243]). It has the advantage that the interactions which are calculated are treated to all orders of perturbation theory, but the disadvantage that only a specific class of diagrams can be treated, while the others are left out. In atomic physics, exchange is a very important interaction and must also be included. This was achieved by Amusia and his school, and the resulting theory (RPA with exchange) is called the RPAE. The RPAE has the limitation (as compared with the MBPT) that it is only well adapted to atoms with closed or half-filled subshells.

Although the MBPT and RPAE approaches differ, they suffer from the same basic difficulty in that neither method is complete. To some extent, the methods complement each other. In practice, each one has different advantages for different situations.

5.26.1   MBPT

The problem of N identical fermions interacting with each other through a two-body potential V_{ij} and with a centre of charge through a one-body potential T_i leads to the Hamiltonian

$$\mathcal{H} = \sum_i^N T_i + \sum_{i<j}^N V_{ij} \tag{5.23}$$

For a nuclear charge $\mathcal{Z}$,

$$T_i = -\frac{\nabla_i^2}{2} - \frac{\mathcal{Z}}{r_i}$$

in atomic units.

In order to define the basis states, the interaction potential V_{ij} is, as usual in atomic physics, approximated by a sum over individual potentials V_i, where V_i must be Hermitian, but can be chosen arbitrarily. We can thus define an 'unperturbed' or independent particle Hamiltonian

$$\mathcal{H}_0 = \sum_{i=1}^N (T_i + V_i) = T + V \tag{5.24}$$

The unperturbed wavefunction of the ground state is Φ_0. It is made up from a determinant containing the N single particle orbitals ϕ_n which are solutions of the equation

$$\mathcal{H}_0 \phi_n = \epsilon_n \phi_n \tag{5.25}$$

where the ϵ_n are single particle energies, while Φ_0 is itself a solution of the equation

$$\mathcal{H}_0 \Phi_0 = E_0 \Phi_0 \qquad (5.26)$$

and E_0 is the total energy of the unperturbed ground state.

Those of ϕ_n which are occupied in Φ_0 are called unexcited states, and those of ϕ_n which are empty are called excited states. Unoccupied unexcited states are referred to as 'holes' and occupied excited states are referred to as 'particles'; the words hole and particle, being part of the language of many-body theory, describe properties of the complete system rather than those of actual holes and particles.

Correlations are then included by the normal procedures of perturbation theory, which allow them to be treated systematically, order by order. The full Hamiltonian is

$$\mathcal{H} = \mathcal{H}_0 + \mathcal{H}'$$

where

$$\mathcal{H}' = \sum_{i<j}^N V_{ij} - \sum_{i=1}^N V_i$$

The method uses diagrams to keep track of the different interactions and represent their matrix elements. Some diagrams for $\mathcal{H}'$ are shown in fig. 5.21. A key result of the perturbation theory of Brueckner [244] and Goldstone [245] is that the perturbed ground state becomes

$$\Psi_0 = \sum_L \left(\frac{1}{E_0 - \mathcal{H}_0} \mathcal{H}' \right)^n \Phi_0 \qquad (5.27)$$

and the shift in the total energy

$$\Delta E = E - E_0 = < \Phi_0 \mid \mathcal{H}' \mid \Phi_0 > \qquad (5.28)$$

The subscripted sum $\sum_L$ is used to indicate that the only terms which are included are the so-called 'linked' diagrams [245], i.e. those Feynman graphs which contain no external lines or any part of the diagram completely disconnected from the rest.

The first-order corrections to Φ_0 are described by the Feynman graphs in fig. 5.22. These diagrams correspond to the algebraic term

$$(\epsilon_p + \epsilon_q - \epsilon_k - \epsilon_k')^{-1} < kk' \mid V \mid pq > n_k^+ n_{k'}^+ n_q n_p \, \Phi_0 \qquad (5.29)$$

where n_i^+ are creation and n_i anihilation operators which create an occupied (particle) state i within Φ_0 or destroy it.

It turns out that if V_i is chosen to be the Hartree–Fock potential, then diagrams (b), (c) and (d) in the figure cancel, so that only (a) needs to be evaluated in the first order.

(a)   H'   - - - - - - - ✗

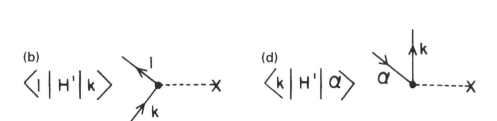

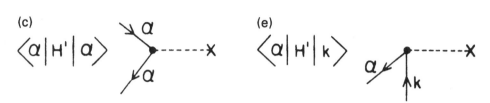

Fig. 5.21.   Feynman graphs for the first-order correction to the energy: (a) shows the basic interaction, while (b), (c), (d) and (e) show different matrix elements. Lines drawn upwards and labelled k and ℓ correspond to particle lines or occupied excited states, while lines drawn downwards represent a hole or the absence of an electron from the state α (after H.P. Kelly [241]).

In the procedure usually applied by Kelly [242], bound state wavefunctions were calculated up to about the tenth Rydberg member, and then the remainder obtained by extrapolation. In the continuum, a sufficiently large number of k values is chosen to represent all the interactions. Typically, this means that about 30 continuum states are required for each ℓ value.

An important feature is a certain flexibility in the choice of the Hermitian potential V. It was pointed out by Kelly that any term of the form $(1 - P)\Omega(1 - P)$, where $P = 1 - \sum_{i=1}^{N} | i ><i |$ is a projection operator and Ω is an arbitrary Hermitian function, can be added in to give V.[8] This has the advantage that Ω can be varied to give the best description of the physical situation.

The next step is to describe excitation by an external optical field. This is a time-dependent perturbation of the form $\mathcal{E}(t) = \mathcal{E}z \cos \omega t$ which induces a dipole moment $P = \alpha(\omega)\mathcal{E}(t)$, where $\alpha(\omega)$ is called the *polar-*

---

[8] This is similar in some ways to the freedom to choose g in the potential of the g Hartree method.

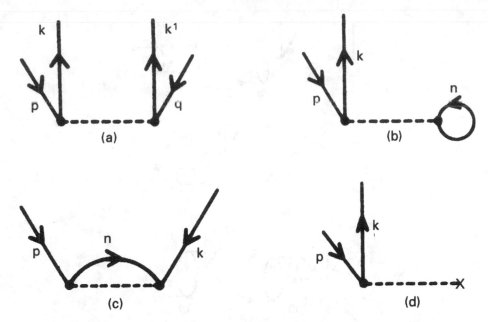

Fig. 5.22. First-order corrections to the wavefunction: (a) two–body correlation: p and q are excited to k and k';(b) p is excited to k by interacting with unexcited states n; (c) is the exchange interaction from p to k via n; (d) is the excitation from p to k through the perturbation. The dashed line is the Coulomb interaction while the cross indicates an interaction via the perturbation. For a Hartree–Fock potential, (b), (c) and (d) sum to zero (after H.P. Kelly [241]).

isability of the atom. A calculation of $\alpha(\omega)$ allows the cross section $\sigma(\omega)$ to be determined through the fundamental relation

$$\sigma(\omega) = \frac{4\pi\omega}{c}\mathcal{I}m\alpha(\omega) \qquad (5.30)$$

As an example, the lowest-order Feynman graphs contributing to the calculation of α corresponding to the excitation of an electron from the state p to the state k are shown in fig. 5.23. The heavy dot in the diagram indicates that the matrix elements of the dipole z are involved. Again, the graphs are time-ordered in the sense that they must be read from bottom to top. Thus, an upward vertical arrow represents a particle, which travels forwards in time, whereas a downward arrow represents a hole, which travels in the opposite sense to a particle, i.e. backwards in time. The Coulomb interactions (i.e. interactions due to $\mathcal{H}'$) are represented by horizontal dashed lines. Those which occur below the dot describe *initial state correlations*, while those which occur above the dot describe *final state correlations*. From each diagram, one can write the corresponding term in the perturbation expansion. For example, diagram (a) reads

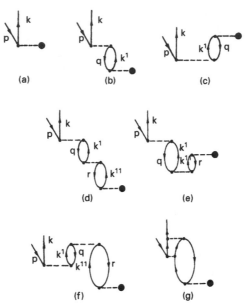

Fig. 5.23. Lowest-order diagrams contributing to the calculation of the dipole polarisability (after H.P. Kelly [241]).

$< k \mid z \mid p >$ and is a direct single particle excitation from $\mid p >$ to $\mid k >$. Diagram (b) reads

$$\sum_{k}' \frac{< kq \mid V \mid pk' >< k' \mid z \mid q >}{\epsilon_q - \epsilon_{k'} + \omega + i\eta}$$

where $i\eta$ is a vanishingly small imaginary part. Likewise, diagram (c) reads

$$\sum_{k'} \frac{< q \mid V \mid k' >< kk' \mid z \mid pq >}{\epsilon_p + \epsilon_q - \epsilon_k - \epsilon_{k'}}$$

where the ordering of the terms matches the time ordering of the diagrams: bottom to top in the figure corresponds to left to right in the mathematical expression. For each diagram, an exchange counterpart (not shown) must also be included. The number of dashed lines gives the order of the term in the perturbation expansion. Thus (a) is a first-order term, (b) and (c) are second-order terms and (d), (e), (f) and (g) are third-order terms. The number of open particle–hole pairs at the top of the diagram indicates the number of real transitions which have occurred. Thus, all the diagrams in the figure correspond to a single $p \to k$ transition. Diagrams for the simultaneous excitation of two electrons will contain two open pairs at the top. The 'bubbles', on the other hand, represent fluctuations, in which excitations are created and destroyed on a very short timescale within the system.

The notes above are not intended as a summary of the theory, but merely to give the reader some feeling for the use of Feynman graphs and their physical content. For a more complete account, see, e.g., Kelly [242].

MBPT has been used very successfully to calculate the photoionisation and autoionisation spectra of a number of different atoms, including complex structure involving many interchannel interactions. Its main feature is completeness up to a given order. A large number of diagrams representing different processes can be included, and there is no difficulty of principle in tackling systems with open shells, although the calculations rapidly become very complex. An example of MBPT applied to a real situation is given in fig. 5.24, which shows the double excitations[9] above the ionisation threshold of Ca from [246]. Note that all the main features of the spectrum, including the continuum strength, the autoionising lines, their strengths and profiles, are calculated point by point for each value of the energy ϵ on a grid which is adjusted to be fine enough for an accurate representation of experimental data. No assumptions are made about the shapes of the autoionising structure, and therefore the complex interactions between configurations and the effects they have in changing the form of the profiles are included, at least to the order of accuracy to which the perturbation expansion has been taken.

5.26.2 The RPAE method

The random phase approximation or RPA comes from nuclear physics. When applied to atomic problems, it must be extended to include exchange, which was achieved by Amusia and his school [247]. The method can be qualitatively described using the diagrams of fig. 5.23: diagrams (b) and (d) exemplify leading terms of a series (the bubble graphs) which, it turns out, defines a geometric progression which can be summed to infinity algebraically. This is the structure of the RPA: it contains the strings of bubbles up to infinite order. It turns out that their exchange counterparts can also be summed exactly, and inclusion of the two yields the RPA with exchange or RPAE method: the RPAE finds application in atomic physics, where exchange interactions become important.

The RPAE includes the bubble graphs and their exchange counterparts to all orders, but does not include any of the other graphs. This turns out to be sufficient if the atom has a closed shell. It is also possible to

---

[9] This same series will be discussed again several times in this book; for example, it is shown in fig. 6.4 It exhibits a particularly interesting perturbation, which is further discussed in section 8.29.

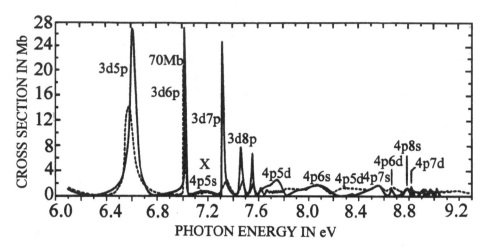

Fig. 5.24. MBPT calculation of the doubly-excited spectrum of calcium in the region above the first ionisation threshold. The curves are: — many-body calculation by Kelly and others; - - - experiments by Newsom and by Connerade and others. This example is of particular interest in connection with symmetry reversals and width fluctuations (see chapter 8) and concerns the same series as the data shown in fig. 8.21. Note the symmetry reversal of the lines about the point marked 'X'. (after Z. Altun *et al.* [246]).

treat half-filled shells and to allow for a spin dependence of the potential, which is done in the SRPAE method. However, the RPAE has difficulties if the atom has other than closed or half-closed subshells. The relativistic version of the RPAE is called the RRPAE method. There are also various methods which use the diagrammatic structure of the RPAE, but replace the basis functions by other than Hartree–Fock functions (see, e.g., [248]). This leads to various developments beyond the pure or exact RPAE.

Far from the ionisation threshold, where the escaping photoelectron has a large kinetic energy, the normal method of calculation for the continuum states is to compute the orbitals in a field determined by using the frozen orbitals of the neutral atom with one electron removed. However, if one is calculating a resonance which lies close to the threshold, this approach may fail. This happens because the escaping photoelectron moves slowly, so that the residual ion has time to relax as it escapes. It is then better to compute the orbitals in the relaxed field of the ion. This approximation is called the GRPAE or the RPAER, and is referred to as the RPAE with relaxation.

One can also extend the RPAE by including the self-energy part of the photoelectron's interaction (SEP). In lowest order, this correction is due to the polarisation potential induced by the escaping photoelectron. One can include both SEP and relaxation, which can be regarded as the second approximation. Finally, one can perform hybrid calculations by combining the RPAE with MBPT so as to include more diagrams and attempt to benefit from the advantages of each. The problem, in all such extensions, is that a desirable feature of the RPAE is then lost, as we now explain.

Referring back to the derivation leading to equation (4.2), the matrix element for photoionisation may be expressed in the equivalent forms

$$< \Psi_0 \mid \hat{\mathbf{v}} \cdot \sum_i \nabla_i \mid \Psi_f >= (E_f - E_0) < \Psi_0 \mid \hat{\mathbf{v}} \cdot \sum_i \mathbf{r}_i \mid \Psi_f > \qquad (5.31)$$

The left hand side is known as the dipole velocity form and the right hand side as the dipole length form. In principle, both should yield the same results, as indeed they do *exactly* if a single central field Hamiltonian and the independent electron approximation are used. If other approximations are made, they may, and generally will, yield different results. The accuracy with which the dipole length and velocity calculations agree then gives some idea of the validity of the approximate methods used in the calculation (although satisfying this criterion alone gives absolutely no guarantee that the results of the calculation are realistic).

Many-body theory starts out from the principle that all wavefunctions (for both ground and excited states) should be calculated in the *same* atomic field, i.e. from the same Hamiltonian. The perturbative expansion then allows the higher-order corrections to be calculated systematically. It can then be shown [250] that in the pure RPAE, the dipole length and dipole velocity forms of the cross section are precisely equal, by construction. For this reason, the pure RPAE is often referred to as 'exact,' which means simply that it satisfies equation (5.31) exactly, and not that one should necessarily expect it to agree exactly with experiment.

If relaxation is included, the initial state and the final state are eigenstates of different Hamiltonians. As already explained, if a many-body theory is constructed with differing initial and final state functions [249, 250], higher-order corrections can be included which normally lie outside the RPAE, hence the name *generalised* or GRPAE. Under these conditions, it is no longer the case that dipole length and velocity forms are made equal by construction. Further calculations must then be made to optimize the ground state functions in order to recover this equality. This further work usually takes the form of incorporating experimental ionisation potentials, etc.

The quality of a many-body calculation can, to some extent, be assessed by looking at the difference between the length and velocity forms. For example, in the MBPT, which does not sum diagrams to all orders, the two are *not* equal, and they tend to bracket the experimental curve. It has been found that taking ground state correlations into account brings them into agreement with each other and rather close to the geometric mean of the length and velocity forms.

This suggests a very simple way to include a substantial part of the correlations. One may simply perform a Hartree–Fock calculation for the ground and for the excited states, calculate the length and velocity forms of the cross section and take the geometric mean of the two. This is referred to as the HFU approximation if the continuum states are unrelaxed, and as the HFR if the Hartree–Fock geometric mean is calculated using relaxed final states.

In MBPT also, it is a goal to achieve equality of dipole length and velocity forms. This test is often applied to give some idea whether the approximations made in the calculation are sufficiently accurate.

It is perhaps useful here to give an idea of the precision expected from many-body theories in general and the RPAE in particular for the calculation of giant resonances. figs. 5.25 and 5.26 illustrate a particularly difficult case, *viz.* the $4d \rightarrow f$ giant resonance in Ba. The experimental data in the figures were actually obtained for the solid, which avoids some of the complex structure due to double excitations intruding in the spectrum.

The case of Ba is a rather difficult one. It is found that, if all second order diagrams are included in the calculation, the cross section peaks too sharply near the threshold. It is also found that the inclusion of relaxation improves the situation a lot. In principle, of course, the pure RPAE and the RPAER are opposite extremes. Since the escaping photoelectron is present, the ion cannot be fully relaxed, and on the other hand its orbitals cannot be frozen either. Thus, the situation really lies between the two. In the case of Ba, the fully relaxed approximation is best, but this is not true in all cases. Which approximation is best can be regarded as a peculiarity of the shell under study, and of its radial position within the atom.

Despite a very considerable amount of effort to improve the quality of such calculations, they represents a formidable challenge, and the accuracy is only of the order of 20–30%. Currently, the RPAE remains the most intellectually consistent and successful theory for the *ab initio* calculation of giant resonance cross sections.

From experience gathered with many calculations using both MBPT and RPAE, it is found that correlations can be divided into two classes, namely correlations involving electrons within the same shell (intrashell correlations) and correlations between electrons in different shells (intershell correlations). Their relative magnitudes vary from one atom to another, because the energy separation between shells itself varies markedly,

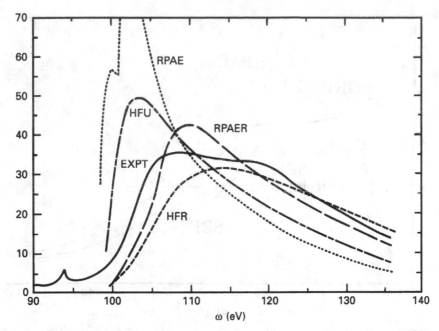

Fig. 5.25. Examples of calculations of the giant resonance in Ba using many
different theoretical approaches, and their comparison with experimental data
for the solid: note the wide scatter in the curves, which may be taken as an
indication of how challenging this particular problem is, since all the different
approximations differ significantly from experiment and between themselves (after H.P. Kelly *et al.* [251]).

as stressed in the present chapter.

Many-body theory has also been applied to calculate the response of
negative ions [253], where intershell interactions and core rearrangement
effects are even greater than for neutral atoms. It turns out that the
spectra of photodetachment in negative ions are dominated by correlation
effects to an even greater degree than photoionisation for neutral atoms.

5.27   Conclusion

Centrifugal barriers have a profound effect on the physics of many-electron
atoms, especially as regards subvalence and inner shell spectra. One aspect not discussed above is how energy degeneracies arising from orbital
collapse can lead to breakdown of the independent electron approximation and the appearance of multiply excited states. Similarly, we have not
discussed multiple ionisation (the ejection of several electrons by a single
photon) enhanced by a giant resonance. Both issues will be considered in
chapter 7.

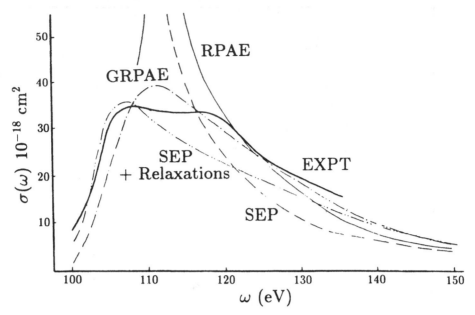

Fig. 5.26. As for the previous figure, but with a different selection of theoretical methods (after M. Ya. Amusia *et al.* [252]).

An important property of giant resonances is that, owing to localisation of the excited f orbitals within the atom, they survive in the solid. Examples of how localisation can be probed by inner-shell excitation in condensed matter will be given in chapter 11.

6

Autoionisation

6.1 Beutler–Fano resonances

The photoionisation continuum of H is clean and featureless. Its intensity declines monotonically with increasing energy. Many-electron systems, in general, always exhibit structure embedded in the continuum. Such features are neither purely discrete nor purely continuous, but of mixed character, and are referred to as *autoionising resonances*. They were discovered experimentally by Beutler [254], and the asymmetric lineshape which they can give rise to follows a simple analytic formula derived by Fano [256]. For this reason, they are often referred to as *Beutler–Fano resonances*. A typical autoionising resonance is shown in fig. 6.1

Autoionisation is a correlation effect. It occurs for all many-electron atoms in highly-excited configurations which lie above the first ionisation threshold. Many spectra used as illustrations in the present volume provide examples of autoionising lines (see in particular chapter 7).

The origin of autoionising structure can be either of the following mechanisms or a combination of both. First, it is possible to excite more than a single electron at a time. Although forbidden in the independent particle model of the atom, many-electron excitation is physically possible, and indeed likely. It provides tangible evidence that the independent particle model is only an approximation. The fact that double excitation can give rise to very intense resonances shows that the breakdown of the independent particle model is by no means a small or negligible effect. The magnitude of this breakdown depends on the proximity in energy between single and double excitations. Since the first are allowed in the independent particle model while the second are forbidden, one can think of the double excitations arising perturbatively, in which case the mixing terms coupling the two are inversely proportional to the energy difference between them. They are thus very prominent in alkaline-earth spectra

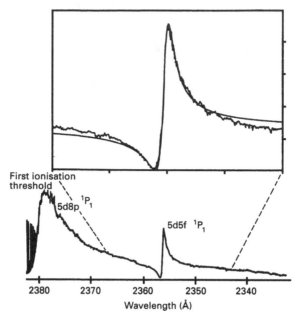

Fig. 6.1.   A typical autoionising resonance, as observed in the spectrum of Ba by laser spectroscopy. Note the broadening and asymmetry of the observed line profile. The inset shows the $5d5f$ resonance on an expanded scale, together with a theoretical fit (smooth curve) based on the Fano formula (see text). Note that the discrepancies between theory and experiment occur mainly in the wings of the line (after J.-P. Connerade [257]).

(which have low double-ionization potentials), but less intrusive in the spectrum of He (which has the highest double-ionization threshold).

The second mechanism (inner-shell excitation) occurs within the independent electron model, because there is nothing within this model to determine *which* electron in the atom should be excited: thus, provided photons of sufficient energy are available, it is always possible to excite electrons of higher energy than the most external or optical electron. This again gives rise to discrete excitations embedded in the continuum. Within the independent electron approximation, autoionisation does not occur, because the different channels for excitation of a single electron are uncoupled. Even in this case, autoionisation only appears once correlations, or the coupling between different channels, are turned on.

In chapter 7, spectra due to inner-shell and double excitations will be discussed. Here, we concentrate on the lineshapes which occur when resonances are embedded in the continuum.

Autoionisation is one of the most fundamental correlation phenomena. There are different ways of arriving at the Fano lineshape formula for an autoionising resonance. Since these are also alternative approaches to the theory of electron–electron correlations, it is quite useful to describe them. We then see the different theories in action, and recognise their respective advantages. In the present chapter, we tackle the problem in three different ways: (i) by degenerate, time-independent perturbation theory; (ii) by MQDT; (iii) by diagrammatic MBPT. In chapter 8, we return to the problem by yet another route using Wigner's scattering theory.

6.2 General background

As just noted, autoionisation provides a clear example of an electron–electron correlation effect, since coupling between the discrete state of one channel and the continuum of another is excluded in the independent particle model. One sees over what range interchannel coupling acts, because it is a broadening effect, involving a continuous band of energies. The observed interference yields important information on the nature of electron–electron correlations. Its study reveals the interplay between single- and many-electron interactions.

The widths of Beutler–Fano resonances can vary widely. Their appearance is usually asymmetric, although the degree of asymmetry decreases with increasing energy within a given spectrum, and eventually becomes hardly noticeable in the X-ray range, as other causes of broadening intrude.

For a pure autoionising resonance embedded in a single continuum, absorption by the underlying continuum is cancelled out at one energy close to the resonance energy, and the resulting transparency is referred to as a *transmission window*.

Autoionisation was discovered experimentally by Beutler [254], whose pioneering investigations as a student of Paschen [255] in Berlin in the early 1930s laid the foundations of vacuum ultraviolet spectroscopy. It was realised quite early that they correspond to the excitation of quasidiscrete states from which the atom can spontaneously ionise by ejecting an electron into the continuum. Another experimenter, Shenstone [258], subsequently introduced the name *autoionisation* to describe this phenomenon. The nature of the asymmetry remained somewhat mysterious until the theory of autoionisation was established by Fano [256] in 1961. Autoionising resonances are closely related to the Breit–Wigner resonances of nuclear physics. An alternative theoretical description of Beutler–Fano resonances in terms of Wigner's scattering theory will be discussed in chapter 8.

Another connection which deserves mention is with MQDT, discussed in chapter 3: MQDT is also a scattering theory, and is equivalent to the theory of Wigner. The connection between them emphasises that autoionisation is an *interchannel* coupling effect. An excitation channel contains both the complete manifold of Rydberg states converging on a threshold and the adjoining continuum with the same quantum numbers which is connected to it. Bound states and continua belonging to the same channel therefore share common features even when they are perturbed. Autoionisation is related to the Lu–Fano graphs described in section 3.6.

It is useful to consider it from several different theoretical perspectives: it provides a working example of the relation between apparently distinct theoretical methods.

6.3   The theory of autoionisation

We approach the theory of autoionisation from several different perspectives.

Historically the first theoretical approach to autoionisation makes use of the fact that, in the independent electron model, it would simply not occur. The perturbation theory of autoionisation was established by Fano [256], along lines which will be described below.

One can also obtain the Fano formula explicitly from a diagrammatic many-body expansion. There exists a wide variety of alternative theoretical approaches, some of which (e.g. coordinate rotation, projection operator methods, etc) will not even be described in the present monograph.

Since the subject of autoionisation is quite central, we shall return to it several times from different points of view, giving alternative derivations of the basic profile formula. We begin with Fano's formulation, which forms the most suitable introduction.

6.4   The Fano formula

The independent electron model serves as the reference basis. Fano's theory of autoionisation consists in describing the consequence of turning on an interaction between a sharp state and the underlying continuum, which are presumed initially to be devoid of correlations. Of course, the perturbation is a hypothetical one, since it cannot really be turned off. The independent electron atom, as such, does not exist. Hypothetical interactions are familiar in perturbation theory. They carry with them the implication that, if they could be removed, the zero-order Hamiltonian which would result can be solved exactly, providing the basis for a perturbative expansion. For a many-electron atom, this is clearly not so, but the idea is nevertheless convenient. It is a case of pretending that, somehow, we are able to solve the Schrödinger equation for the system in absence of autoionisation, this being the *only* effect which has not been included. This assumption goes by the name of *prediagonalisation*. Thus, the unperturbed basis consists of a discrete bound state $\mid \varphi >$ degenerate in energy with a continuum $\mid \varepsilon >$ and the two must in some way be combined in order to describe the observed resonances.

The origin of the perturbation is not of great importance in setting up the theory: if the system is an atom, the most usual form for the operator is the interelectronic repulsion e^2/r_{12}, although spin–orbit or spin–spin, etc. interactions are equally possible. For molecules, in addition to electronic and spin-dependent interactions, autoionisation can be driven by rotational and vibrational interactions, and so the possibilities are even wider than for atoms.

To keep the problem simple, we also assume that only one, isolated excited state $\mid \varphi >$ is present in the energy range of interest, and that the unperturbed continuum would be flat, i.e. would contain no structure at all within the range of the resonance.

Let H_0 be the Hamiltonian of the independent electron atom. We use the formalism of time-independent, degenerate perturbation theory to describe the problem, the variation being that, in the present case, the states which are degenerate in energy belong to the continuum on one hand and to the discrete spectrum on the other. This is a very interesting complication: it is fundamental to quantum mechanics that discrete energy levels appear in what would otherwise be a fully continuous spectrum. Autoionisation is a mechanism which couples bound states of one channel to continuous states of another.

In standard degenerate perturbation theory,[1] one forms a linear combination of degenerate states, and then selects the appropriate initial combination of coefficients. Likewise, in the present instance, we use the superposition theorem to write the wavefunction of a hybrid state – part discrete and part continuous – of energy E as:

$$\mid E >= a_\varphi \mid \varphi > + \int \mid \varepsilon > b\left(\varepsilon\right) d\varepsilon \qquad (6.1)$$

The unperturbed states φ and ε are eigenstates of the unperturbed Hamiltonian H_0 and this can be expressed in a generalised matrix form as follows:

$$\begin{pmatrix} H_{\varphi\varphi} & 0 \\ 0 & H_{\varepsilon\varepsilon'} \end{pmatrix} \begin{pmatrix} \mid \varphi > \\ \mid \varepsilon > \end{pmatrix} = \begin{pmatrix} E_\varphi \mid \varphi > \\ \varepsilon \mid \varepsilon > \end{pmatrix} \qquad (6.2)$$

---

[1] See, e.g., [259].

where the generalisation consists in allowing H_0 to have part continuous and part discrete subscripts.

In another form:

$$\left.\begin{array}{c} <\varphi\,|\,H_0\,|\,\varphi>= E_\varphi \\[2mm] <\varepsilon'\,|\,H_0\,|\,\varepsilon>= \varepsilon\delta(\varepsilon-\varepsilon') \end{array}\right\} \tag{6.3}$$

where the off-diagonal terms $<\varepsilon\prime\,|\,H_0\,|\,\varphi>$ are all zero, i.e. the unperturbed Hamiltonian is diagonal. For this reason, Fano calls this the *prediagonalised* state.

To couple the discrete and continuous channels, we introduce an off-diagonal perturbation V such that the total Hamiltonian is $H = H_0 + V$. The specific form of the operator which appears in V does not matter as far as the general theory is concerned: it can be any interaction capable of coupling the discrete and continuum channels to each other. Most usually, it is the interelectron electrostatic interaction e^2/r_{ij}, but many other operators will do.

The corresponding equation for the perturbed system is then:

$$\begin{pmatrix} H_{\varphi\varphi} & V_{\varphi\varepsilon} \\[2mm] V_{\varepsilon\varphi} & H_{\varepsilon\varepsilon} \end{pmatrix} \begin{pmatrix} a_\varphi\,|\,\varphi> \\[2mm] \int b_\varepsilon\,|\,\varepsilon> d\varepsilon \end{pmatrix} = E \begin{pmatrix} a_\varphi\,|\,\phi> \\[2mm] \int b_\varepsilon\,|\,\varepsilon> d\varepsilon \end{pmatrix} \tag{6.4}$$

From this equation, noting that $V_{\varphi\varepsilon} = V_{\varepsilon\varphi}^*$ because H is Hermitian, and dropping the subscript φ on V (for just one bound state), we obtain, by the usual device of premultiplication by the complementary ket, a pair of coupled equations:

$$\left.\begin{array}{c} E_\varphi a_\phi + \int V_\varepsilon^* b_\varepsilon d\varepsilon = E a_\varphi \\[2mm] V_\varepsilon a_\varphi + \varepsilon b_\varepsilon = E b_\varepsilon \end{array}\right\} \tag{6.5}$$

which are the equations from which Fano starts [256]. Solving the second of these for b_ε, we have at once:

$$b_\varepsilon = \frac{V_\varepsilon a_\phi}{E-\varepsilon} \tag{6.6}$$

which diverges if $E = \varepsilon$. This form is not unexpected, as such operations are well known in scattering theory. We proceed in the standard manner (cf Dirac [260], section 50) by defining a function $\mathcal{Z}(E)$ such that

$$(E-\varepsilon)^{-1} = \frac{1}{E-\varepsilon} + \mathcal{Z}(E)\delta(E-\varepsilon) \tag{6.7}$$

where one takes the principal part of any integral over $1/(E-\varepsilon)$ and the function $\mathcal{Z}(E)$ is determined by the details of the problem. Resubstituting for b_ε in the first of the two equations (6.5), we now find that:

$$E_\varphi a_\varphi + \mathcal{P}\left\{\int \frac{V_\varepsilon^* a_\varphi V_\varepsilon}{E - \varepsilon}d\varepsilon\right\} + \mathcal{Z}(E)\mid V_E\mid^2 = E \qquad (6.8)$$

Defining

$$F(E) \equiv \mathcal{P}\left\{\int \frac{\mid V_\varepsilon \mid^2}{E - \varepsilon}d\varepsilon\right\} \qquad (6.9)$$

where $\mathcal{P}$ denotes the principal part of the integral, we obtain:

$$\mathcal{Z}(E) = \frac{E - E_\varphi - F(E)}{\mid V_E \mid^2} \qquad (6.10)$$

In common with all problems of the same type in degenerate perturbation theory (see [259]), since the coefficient a_φ has factored out of the equations, it must be determined by normalising the perturbed function thus:

$$< \tilde{E} \mid E > = a_\varphi^*(\tilde{E})a_\varphi(E) + \int b_\varepsilon^*(\tilde{E})b_\varepsilon(E)d\varepsilon = \delta(\tilde{E} - E) \qquad (6.11)$$

Substituting for b_ε one has

$$a_\varphi^*(\tilde{E})\left[1 + \int d\varepsilon V_\varepsilon^*\left\{\frac{1}{\tilde{E} - \varepsilon} + \mathcal{Z}(\tilde{E})\delta(\tilde{E} - \varepsilon)\right\}V_\varepsilon\right.$$

$$\left. \times \left\{\frac{1}{E - \varepsilon} + \mathcal{Z}(E)\delta(E - \varepsilon)\right\}\right]a_\varphi(E) = \delta(\tilde{E} - E) \qquad (6.12)$$

which defines $\mid a_\varphi(E)\mid^2$. The product inside the integral has to be handled rather carefully, as discussed in [256], but if this is followed through, one finds:

$$\mid a_\varphi(E)\mid^2 = \frac{\mid V_E \mid^2}{[E - E_\varphi - F(E)]^2 + \pi^2 \mid V_E \mid^4} \qquad (6.13)$$

which shows that the discrete state E_φ is diluted into a continuous band of energy states with a half-width $\Gamma/2 \equiv \pi \mid V_E \mid^2$, about a mean position $E_0 = E_\varphi + F(E)$, provided we assume that $F(E)$ is approximately constant over the width of the resonance. Note that this dilution of the energy level to form a resonance is completely symmetrical and has a Lorentzian shape as a function of the energy E which we can express as

$$\mid a_\varphi \mid^2 = \frac{1}{\pi}\frac{\Gamma/2}{(E - E_0)^2 + (\Gamma/2)^2} \qquad (6.14)$$

6.5   The Fano parametrisation

So far, we have only considered the dilution by mixing of the originally sharp level $| \varphi >$ into the featureless continuum $| \varepsilon >$. In practice, one does not observe energy levels but transitions between them, so we must now introduce an initial state $| i >$, say, and consider transitions to the 'mixed' state $| E >$. In fact, the most sensible quantity to consider is the ratio of the transition probability ending in the mixed state to the transition probability ending in the featureless continuum, both being induced by the dipole operator $\tilde{D}$, i.e. the quantity

$$R = \frac{|< E \,|\, \tilde{D} \,|\, i >|^2}{|< \varepsilon \,|\, \tilde{D} \,|\, i >|^2} \tag{6.15}$$

which is a measure of the modulation produced by the resonance in an otherwise featureless continuum cross section $|< \varepsilon \,|\, \tilde{D} \,|\, i >|^2$.

One can obtain the result somewhat more directly than in [256]: we have

$$R = \frac{|\, a_\varphi < \varphi \,|\, \tilde{D} \,|\, i > + \int b_\varepsilon < \varepsilon \,|\, \tilde{D} \,|\, i > d\varepsilon \,|^2}{|< \varepsilon \,|\, \tilde{D} \,|\, i >|^2} \tag{6.16}$$

but we know that

$$b_\varepsilon = \left\{ \frac{1}{E - \varepsilon} + \mathcal{Z}(E)\delta(E - \varepsilon) \right\} V_\varepsilon a_\varphi$$

whence

$$R = \frac{|\, a_\varphi \,|^2}{|< \varepsilon \,|\, \tilde{D} \,|\, i >|^2} \, |< \varepsilon \,|\, \tilde{D}i > + P \left\{ \int V_\varepsilon \frac{< \varepsilon \,|\, \tilde{D} \,|\, i >}{E - \varepsilon} d\varepsilon \right\}$$

$$+ V_\varepsilon \mathcal{Z}(\varepsilon) < \varepsilon \,|\, \tilde{D} \,|\, i >|^2 \tag{6.17}$$

If we define the shifted state

$$| \Phi > = | \varphi > + P \left\{ \int V_\varepsilon \frac{< \varepsilon \,|\, \tilde{D} \,|\, i >}{E - \varepsilon} d\varepsilon \right\} \tag{6.18}$$

then

$$R = |\, a_\varphi \,|^2 \frac{< \Phi \,|\, \tilde{D} \,|\, i >}{< \varepsilon \,|\, \tilde{D} \,|\, i >} + V_\varepsilon \mathcal{Z}(\varepsilon) < \varepsilon \,|\, \tilde{D} \,|\, i >|^2 \tag{6.19}$$

From (6.10) we note that $|\, V\varepsilon \,|^2 \, \mathcal{Z}(\varepsilon) = \varepsilon - E_0$ is simply the detuning from the resonance energy. Defining a new energy variable

$$\epsilon \equiv \frac{2(\varepsilon - E_0)}{\Gamma} = \mathcal{Z}(\varepsilon)\pi$$

and a parameter

$$q \equiv \frac{< \Phi \mid \tilde{D} \mid i >}{\pi V_\varepsilon < \varepsilon \mid \tilde{D} \mid i >}$$

we find that

$$\mid a_\varphi \mid^2 = \frac{1}{\pi^2 V_\varepsilon^2 (\epsilon^2 + 1)}$$

and

$$R = \mid a_\varphi \mid^2 \mid \pi V_\epsilon q + \pi V_{\epsilon\epsilon} \mid^2$$

whence

$$R = \frac{(q + \epsilon)^2}{1 + \epsilon^2} \tag{6.20}$$

which is known as the *Fano lineshape formula*.

It is often convenient to write this in terms of the real detuning $\varepsilon - E_0$ thus

$$R = \frac{(q\Gamma/2) + (\varepsilon - E_0)}{(\varepsilon - E_0)^2 + (\Gamma/2)^2} \tag{6.21}$$

which brings out its dependence on Γ, and also its relationship to the normal Lorentz profile formula for $\mid a_\varphi \mid^2$ in equation (6.14).

6.6   The asymmetric lineshape

In fig. 6.2, we show examples of the line profiles which occur as a result of autoionisation, as predicted by Fano's theory.

The quantity q is called the *shape index* and is constant for a Fano profile. In chapter 8, situations in which q varies within an excitation channel will be discussed. However, even in such cases, it can be regarded as a constant over one line.

The shape index determines the symmetry of the line. From equation (6.21), we note that, for a specific value the resonance energy

$$\varepsilon - E_0 = -\frac{q\Gamma}{2} \tag{6.22}$$

of the detuning from the resonance energy, $R = 0$. Thus, for positive values of q, the cross section falls to zero (i.e. the transmission window occurs) below the resonance energy, while for negative values of q, the curves in fig. 6.2 are reversed and the transmission window occurs above the resonance energy.

The important special cases are: (i) if $q = 0$ one has a symmetrical resonance with no absorption maximum and only a minimum (inverted

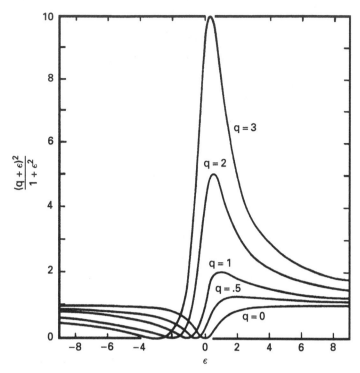

Fig. 6.2. Family of curves generated from the Fano lineshape formula for different values of the shape index q. For negative values of q, reverse the abscissa. (after U. Fano [256]).

Lorentzian shape): such features are called pure *window resonances* or antiresonances; (ii) if $q \to \pm\infty$, then we have a pure, symmetric absorption maximum, again of Lorentzian shape.

The term window resonance is useful for the $q = 0$ case, because it avoids confusion with other kinds of localised minima in the continuum cross section, such as the Seaton–Cooper minimum (see section 4.6.2), whose origin is quite different.

The physical significance of the asymmetric profile is as follows. We may consider that there are two states namely the initial state $\mid i >$ and a final state in which autoionisation has taken place, say $\mid j >$. There are then two paths to reach $\mid j >$: *viz.* one in which the electron is excited directly to the continuum states and another in which the electron is excited into the compound state $\mid E_\varphi >$ and hence into $\mid j >$. In quantum mechanics, whenever two levels are coupled by different paths, interference can occur. This interference is, in general, constructive on one side of E_0

and destructive on the other, hence the asymmetry.[2]

Regarding window resonances, the following picture, though oversimplified, is useful. If the transition to E_φ is unlikely (for example, because double excitations are improbable in the independent electron model), whereas the transition to $\mid \varepsilon >$ is allowed (for example, it is a singly-excited continuum state), then the mixture of the two depresses the intensity in the vicinity of E_0. Conversely, allowed transitions in the prediagonalised scheme tend to yield strong, nearly symmetric maxima in the absorption cross section.

A fundamental difficulty with the theory of autoionisation is that the so-called prediagonalised state does not exist as such. This is a standard situation in perturbation theory, where the unperturbed Hamiltonian is simply a mathematical convenience, because the system cannot physically be 'deperturbed.' It does, however, have the serious consequence that one cannot, at first sight, check the theory in any quantitative sense. Methods of overcoming this limitation will be described in chapter 8

An example of an isolated Fano resonance, recorded in a laser-based experiment is shown in fig. 6.1. We have chosen for this figure a case in which the background continuum does in fact vary outside the range of the autoionising profile, in order to emphasise that, in general, there is no particular reason to assume that continua should remain 'flat' over a wide range of energy. As already stressed, the window in an autoionising line appears as a true zero in the cross section for the simple case of a resonance interacting with a single continuum. This feature, as we shall see in chapter 8, is preserved for a Rydberg series of autoionising resonances interacting with a single continuum (each member of the series has only one associated minimum). It even survives if these resonances are overlapping (i.e. not isolated) and if the Rydberg series is perturbed by intruder states, but the minima are 'filled in' progressively when more underlying continua are present. For this reason, it is desirable to study autoionising resonances and the interactions between them just above the first ionisation threshold.

Techniques which simplify the extraction of the Fano parameters q and Γ from experimental data for isolated resonances are described by Shore [261]. However, the extraction of a single set of Fano parameters q and Γ from the formula for an isolated resonance is not reliable when several resonances overlap in energy [262].

---

[2] This assertion is proved in section 6.16.

6.7   Applicability of the Fano formula

It is important to recognise the limitations of formula (6.20), which are
as follows: (i) the formula applies to a single, isolated resonance; (ii) the
background continuum must be flat, or vary very little over the width
of the resonance; (iii) the radiative width must be much smaller than
the particle width; and (iv) there must be no other channel for particle
broadening than the one considered.

Condition (i) means that, when the autoionisation width approaches
the separation between successive members in a Rydberg series, the formula is not applicable. The same is true if there are other autoionising
lines in the immediate vicinity of the resonance. Condition (ii) means that
the slope of the background absorption or ionisation continuum must be
checked. Condition (iii) is usually satisfied: autoionisation widths are
often in the range $\Gamma \sim 10^{-12}$–10^{-13} cm^{-1}, which is about four orders of
magnitude shorter than typical radiative lifetimes. Condition (iv) is the
least likely to be satisfied. Mostly, autoionising resonances are subject
to several competing mechanisms for the ejection of an electron. Only if
the excited state is built on a stable parent ion state (most usually, the
ground state of the parent ion) and if the resonance is coupled to only one
continuum does the simple Fano formula apply strictly. The energy range
between the first and second ionisation potentials is the one involving the
smallest number of channels for the ejection of one electron. Autoionising
lines then assume their simplest form. For this reason, it is often referred
to as the *first autoionisation range*.

In the presence of several continua, the Fano formula still describes
the general lineshape of autoionising resonances rather well, but the cross
section does not fall to zero near the resonance, i.e. the transmission
window is 'filled in' by the presence of several continua. If the cross
section falls to zero near the resonance, one can in fact deduce that only
one continuum is involved.

Autoionisation is ubiquitous at all energies above the first ionisation
threshold, as evidenced by the broadenings and asymmetries observed.
However, as the excitations extend to higher and higher energies, asymmetries become less and less visible until, in the X-ray range, the structure
tends to become nearly symmetric. There are two reasons for this, one
being Auger broadening (discussed in the next section and also in section 8.32) and the other the increase in radiative widths, due to the ν^3
factor in the Einstein–Milne relations (see section 4.3).

6.8   Auger broadening

After the ejection of an electron from an inner shell, relaxation generally occurs by the emission of a secondary electron. This is known as the *Auger effect*. It can be reasonably well described as a two-step process, leading to double ionisation, because the primary and Auger electron are usually separate. However, if the initial photoelectron is emitted with a very low kinetic energy, then the Auger electron can 'catch up' and interact with it. This process is described as *post-collision interaction* or PCI.

If we consider photoexcitation to very highly excited states, then there are two modes of decay, one of which involves the outer, excited electron, while the other involves reorganisation of the core, or parent ion of the excited state. When the parent ion state lies above the threshold for double ionisation, it is also able to decay by ejecting an electron. Thus, for example, when a deep inner shell is photoexcited, one electron is promoted directly to an outer orbit, whose energy depends on the energy of the incident photon. If the excited electron is transferred into the photoionisation continuum by a radiationless transition, we call this primary process autoionisation, and we expect that the kinetic energy of the ejected photoelectron should be equal to the energy of the incident photon minus the energy of one of the single ionisation thresholds, i.e. *the energy of the autoionising electron varies together with the energy of the incident photon*. If, on the other hand, the ejection of an electron results from a radiationless reorganisation of the core while the Rydberg electron is far away, then it does not matter which particular Rydberg orbit was initially excited: the kinetic energy of the ejected photoelectron, in this case, is not dependent on the energy of the incident photoelectron. We call this an Auger process, and we note that *the energy of an Auger electron does not vary with the energy of the incident photon*. Thus, one can distinguish between autoionisation and the Auger effect in photoelectron spectroscopy by tuning the photon and observing whether the kinetic energy of the ejected electron varies or remains constant.

In photoabsorption spectroscopy, the distinction can also be made in a different way: the probability that an electron on a Rydberg orbit is ejected depends on its penetration of the core, which scales as $1/n^{*3}$. On the other hand, the Auger electron has a constant probability Γ_{auger} of being ejected. Thus, if we observe a Rydberg series of autoionising resonances in photoabsorption, then the total width Γ_n of the nth member is given by

$$\Gamma_n = \Gamma_{auger} + \frac{\Gamma_{auto}}{n^{*3}} \qquad (6.23)$$

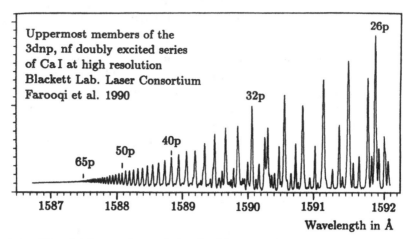

Fig. 6.3.   Very high members of the doubly-excited $3dnp$ autoionising series of Ca. This is in fact the same series as we have used to illustrate many aspects of autoionisation in the present monograph, but it is observed here at much higher resolution, close to the $3d$ series limit, by laser spectroscopy. Since the $3d$ limit is metastable (the ground state of the ion is of the same parity) and since autoionisation falls off as fast as the spacings between successive members, the autoionising series can be followed to very high values of n (after J.-P. Connerade [257]).

The autoionisation width decreases with increasing n as $1/n^{*3}$, whereas the Auger width remains constant. Since the spacing between successsive members of a Rydberg series of high n also decreases as $1/n^{*3}$, we see that *the presence of autoionisation broadening does not shorten a Rydberg series* (see fig. 6.3 for an example). On the other hand, the presence of Auger broadening, which does not decrease when n increases, effectively truncates the Rydberg series.

   In X-ray spectra, when a deep core hole is excited, the excited shell occupies a very small volume and its spatial overlap with other core electrons is large, while with the orbital of the outer excited electron it is small. Under such conditions, the lifetime of the core vacancy becomes short: one speaks of large core-level widths due to Auger broadening, which compete with autoionisation . This is why, in X-ray spectra near absorption edges due to inner-shell excitation, only the first few Rydberg members are observed. We return to this issue in section 11.2.

   Both autoionisation and the Auger effect are often referred to as *radiationless transitions* , because the initial reorganisation of the atom, takes place on very short timescales (typically 10^{-13} s) without the emission of radiation. This does *not*, however, mean that no radiation at all is emitted by the atom during or after either of these processes. It is merely that the branching ratio is usually very unfavourable towards the emission of photons, again because of the longer time involved, which gives ejected electrons the opportunity to carry away all or a large part of the excess energy. For this reason, both effects are most often studied after photoexcitation from the ground state: the timescale for radiationless decay is so short that it is difficult to sustain a high enough electron density to maintain the population of the excited state and observe autoionisation in emission. Nevertheless, it *is* possible to observe autoionisation in the emission spectra of high current arcs [264], and so the emission of photons can sometimes be used for detection.

Despite the unfavourable branching ratio, interference processes can occur between autoionisation and radiative decay. These are discussed in detail in section 8.30. One should also note that, as a result of autoionisation, the remaining parent ion may find itself in an excited state which decays by fluorescence. Thus, the study of fluorescence at wavelengths different from the original excitation may actually be used to detect the presence of autoionisation into a specific channel.

6.9 Auger spectroscopy

Auger spectroscopy is a subject in itself, and is the basis of a general experimental method which serves as a tool in solid state and surface physics as well as in atomic physics, because the initial excitation involves fairly deep-shell vacancies, which survive in solids (see chapter 11).

A more direct way of considering the Auger process than the manner in which it was introduced above is to consider it as a two-step process: we examine how a vacancy can decay by the emission of an electron after the initial photoelectron has been ejected. When the incident photon energy is above the inner-shell ionisation threshold, normal Auger decay takes place. When the Auger emission takes place in the presence of an excited electron, one has a spectator Auger effect, if the excited electron does not participate in the process. If, on the other hand, the spectator electron in the initial and final states of the decay couples with the electrons involved in the decay (spectator–core coupling), then the resulting process is referred to as resonant Auger decay. Resonant Auger spectra of the Kr $3d \rightarrow 5p$ and Xe $4d \rightarrow 6p$ resonances were first reported by Eberhardt *et al.* [265]. More recently, these observations were extended to the $7p$ and $8p$ resonances of Xe [266]. Dramatic changes occur in the intensity distribution between the normal and the resonant Auger processes, which depend sensitively on configuration mixing in both the initial and final states. Auger spectroscopy is thus a useful method of testing atomic theory, and of observing the breakdown of the independent electron approximation [267].

The Auger effect is an important process in solid state spectroscopy. One can use resonant Auger spectra to study the nature of core excitation in ionic solids: by examining the Auger structure, the nature of the core holes can be determined, as well as the splitting of the states by the ligand field.

Auger processes of all kinds (spectator and resonant) are also observed in molecules [268, 269, 270, 271, 272, 273, 274]. In addition, because of the very rapid dissociation of doubly-excited states in molecular species, neutral dissociation can occur before the decay of the core hole, and has been detected in the total ion yield of the H_2S molecule [275].

6.10   Autoionisation by MQDT

MQDT, which was introduced in chapter 3, provides an excellent framework to describe resonances in the first autoionisation range, as one may already infer by comparing the Lu–Fano graphs of section 3.6 with the asymmetric profiles of fig. 6.2. In essence, the Lu–Fano graph pictures interchannel coupling between bound states, but the properties of the bound states go other smoothly into those of the adjoining continua. Thus, for two series limits, the Rydberg members which lie below the lowest of the two limits were used to obtain Lu–Fano graphs in chapter 3, but this neglects all the structure between the two series limits. When channels are coupled, the perturbation analysed by the Lu–Fano graph appears and distorts the energy positions of the bound states: as a result of the same coupling mechanism, the states which lie above the lower ionisation threshold are coupled to the associated continuum, i.e. they autoionise.

It is therefore clear that, if one understands how to extrapolate across ionisation thresholds, the properties of the bound states should enable one to compute autoionising profiles. Indeed, this could be regarded as one of the fundamental aims of MQDT. The principle of this method rests on the continuity which exists between the physical properties of transitions to the uppermost bound states of the atom (high Rydberg members) and the base of the continuum. As already noted, the continuity between bound and continuum states is fundamental to QDT. From equation (3.14), we can write the phase shift as

$$\delta_1 = \pi\mu_1 - \tan^{-1}\left(\frac{R_{12}^2}{\tan\left[\pi(\nu_2 + \mu_2)\right]}\right) \tag{6.24}$$

which describes a recurring series of poles of the argument, one at each zero of the denominator, near which the phase shift changes through π. It will be shown in chapter 8 how a series of poles of this kind defines a series

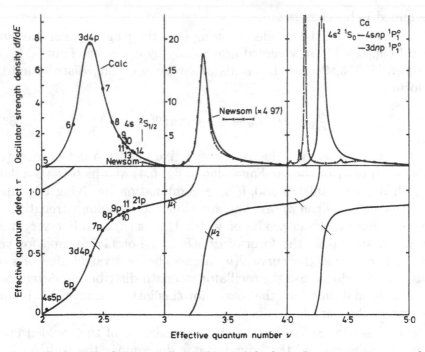

Fig. 6.4. Connection between the analysis of the bound state spectrum and the profiles of autoionising lines, as studied by Geiger, for the $3dnp$ doubly-excited series of Ca. The simple two-channel theory fails at the $3d6p$ resonance (see chapter 8 for a further discussion of this spectrum – after J. Geiger [276]).

of resonances. For the moment, we merely note that the energies of these poles are defined by the zeros in $\tan[\pi(\nu_2 + \mu_2)]$, but that the energies of the resonances are defined by the gradients of $\mu_2(\nu_1)$. For weak coupling $R_{12}^2 \ll 1$, and the resonances, which are very narrow, coincide with energy positions $\nu_2 = n_2 - \mu_2$, which are directly extrapolated from the Rydberg energies of the bound states below the I_1 threshold. For strong coupling $R_{12}^2 \gg 1$, they appear at the half-way positions $\nu_2 = n_2 + \frac{1}{2} - \mu_2$, while for the equal mixing case $R_{12}^2 = 1$, the resonances have ill-defined positions, and correspond to a Lu–Fano graph which is simply a straight line with no obvious resonant position.

A good example of the value of continuity theorems is provided by the work of Geiger [276] who has obtained the autoionised singlet spectrum near the $4s^2$ limit of Ca by using MQDT with two strongly interacting channels. He gives a particularly simple and intuitive picture of the connection between the bound state and autoionising spectra. His results are summarised in fig. 6.4, which shows both the Lu–Fano graph for the bound states of calcium with the oscillator strengths in the principal series and the first autoionising profiles of the doubly-excited channel as determined in his calculation.

The 'strength' of the avoided crossing (i.e. the magnitude of the inter-channel coupling) is represented graphically by d (see the figure). It can be shown [276, 324] that the oscillator strength density distribution has the form

$$\frac{df}{dE} = 2I_0(E_n - E_0)\sin^2\pi(\theta(\nu) - \theta_0) \qquad (6.25)$$

where θ_0 is the specific value of the effective quantum defect $\theta(\nu)$ (the quantity graphed in the Lu–Fano plot of fig. 6.4) at which the oscillator strength density vanishes and I_0 is a combination involving the dipole matrix elements. Then df/dE has, for moderate coupling strengths, the form of a Beutler–Fano profile of width $\Gamma = \pi (d/2)^2$. For very weak interaction strengths, the form of df/dE is a Lorentzian, and for very strong interactions, the curve $\theta(\nu)$ approaches a straight line of slope $d\theta/d\nu = 1$, in which case the oscillator strength distribution degenerates into a sinusoidal function (the reason for this last statement will become apparent in chapter 8).

Thus we see immediately that the magnitude d of the avoided crossing in the spectrum of the bound states determines the widths of the autoionising resonances directly.

QDT allows one to extrapolate directly from the discrete into the first autoionising range: the oscillator strength density distribution occurs unchanged after one unit period of the effective quantum defect, provided only that the parameters do not vary too fast with energy.

Other physical observables which are continuous across thresholds have been discussed by Lane [278], and we return to the general issue of continuity across thresholds in chapter 8.

6.11   Complex quantum defects and superexcited Rydberg states

Yet another way of picturing autoionisation comes about by modifying the Rydberg formula to include a nonradiative lifetime for Rydberg series of autoionising resonances. For bound states, the quantum defect arises from occasional elastic collisions between the Rydberg electron and the core. However, if the core itself is excited, a *superelastic* collision may occur, i.e. one in which the Rydberg electron acquires energy, which may be enough for it to escape from the ion core. This, of course, is none other than the autoionisation process. Such an excited state has a finite lifetime even in the absence of a radiation field. Its energy is therefore complex,[3] with an imaginary part $\Gamma/2$ which is the half-width of the radiationless decay. In the Rydberg formula, this is achieved by introducing a complex quantum defect $q_n = \mu_n + i\gamma_n$, in which case the Rydberg formula becomes:

$$E_\infty - E_n \sim \frac{1}{(n - \mu_n)^2} + \frac{2i\gamma_n}{(n - \mu_n)^3} \qquad (6.26)$$

where the second term is an approximation valid for $\gamma^2 << (n - \mu_n)^2$, and expresses the fact that autoionisation widths fall off at the same rate as the interlevel spacing for high enough n.

Complex QDT has been used to compute autoionising Rydberg series, most notably for the two-electron molecule H_2 which, like He, exhibits autoionising series [279].

6.12   Nonexponential decay and autoionising states

It was implicit in the last section and indeed in all the discussion so far that the decay of an autoionising resonance should be exponential in time. This may seem to be obvious, and is indeed verified with good accuracy in experiments, but there are also fundamental reasons for believing that it is not strictly correct [280]: according to quantum mechanics, the exponential decay law is violated for very long times, where the probability of nonexponential decay eventually prevails, a fact which has been recognised in nuclear and particle theory [282], but has not so far been verified experimentally.

The conditions under which nonexponential decay of autoionising states might appear have been considered phenomenologically [281], and it was suggested that the crucial issue is how close a resonance lies above the ionisation limit. In fact, it is the ratio Γ/E_0, where E_0 the resonance energy is measured from the threshold, which must be significant for effects to appear.[4]

Ab initio calculations by numerical integration of the Schrödinger equation suggest that nonexponential decay, resulting in a slight apparent lengthening of the lifetime, may be observable in some cases [280].

---

[3] The easiest way to see this is simply to substitute a complex energy into the time-dependent Schrödinger equation: the real part of the energy can then be associated with an oscillatory eigenfunction, while the imaginary part is associated with exponential decay.

[4] In this connection, the discussion in chapter 5 about the relationship between linewidths and energy above threshold for giant resonances assumes particular significance: the fact that the ratio, in this case, remains fairly constant shows that the nonexponential terms do not grow as the threshold is approached.

6.13   The f values of autoionising lines

Here, we return to the radiative contribution, and consider how the notion of f value can be extended to a Beutler–Fano profile, by taking account only of the discrete part, and we show that this yields the relevant quantity for studies of the refractive index of an autoionising resonance by MOR.

An autoionising line contains both a discrete and a continuous part. The strength of the discrete part can also be described by an f value, by a slight generalisation in which the integration over all wavelengths is referred to the continuum background, so that a divergent result is avoided. The resulting expression is of use in describing the magneto-optical behaviour of autoionising lines, and is therefore worth writing out. We make use of the identity:

$$\frac{(q+\epsilon)^2}{1+\epsilon^2} \equiv 1 + \frac{2q\epsilon}{1+\epsilon^2} + \frac{q^2-1}{1+\epsilon^2} \tag{6.27}$$

which decomposes the Fano profile into the sum of a constant term, a dispersive profile and a Lorentzian shape.

For a purely Lorentzian lineshape, the definition of the f value is unambiguous, because the integral of the absorption coefficient over the whole profile is independent of the linewidth, as was proved in section 4.4.3. However, this is *not* true for a Fano profile, essentially because it contains a contribution from the continuum, which has been assumed to be 'flat' and therefore continues to all energies. Clearly, the integral in this case diverges. However, what we can do is to define a generalised oscillator strength for Fano profiles, using the formula

$$\overline{f} = \frac{q^2-1}{q^2+1}f \tag{6.28}$$

where f is the oscillator strength which would be expected for a pure Lorentzian profile. This definition is tantamount to measuring oscillator strength *relative to the continuum background*, as can readily be seen by considering the limiting cases. For $q \to \pm\infty$ we find $\overline{f} \to f$ as expected. For $q = \pm 1$, we have $\overline{f} = 0$, again as expected, because the absorption peak on one side of the resonance energy E_0 is exactly counterbalanced by the transmission window on the other. Finally, for $q = 0$ we find $\overline{f} = -f$, which expresses the fact that the resonance then appears entirely as a transmission window.

This generalised definition of the f value for autoionising resonances turns out to be useful in describing the Zeeman and Faraday rotation effects for a Beutler–Fano profile. It also yields a more symmetric form for the profile formula, viz.

$$a(\epsilon) = \frac{Ne^2 f}{mc\Gamma} \frac{(q+\epsilon)^2}{(1+q^2)(1+\epsilon^2)} \tag{6.29}$$

where a is now the absorption coefficient, and the constant has been chosen consistently with the definition of dispersion theory in section 4.11. The factor $1 + q^2$ which ensures that $\bar{f}$ remains finite also appears in the context of scattering theory (see section 8.18 footnote 3) where it fulfills a similar purpose.

6.14 The refractive index of a Beutler–Fano resonance

We can recast equation (6.29) in frequency units ν, to conform with the notation of Mitchell and Zemanski [283] (a standard reference for MOR) as

$$a(x) = \frac{Ne^2 f}{mc} \frac{1}{1+q^2} \left\{ \frac{(q^2-1)\Gamma}{\Gamma^2 + q^2} + \frac{2qx}{\Gamma^2 + q^2} + \frac{1}{q} \right\} \tag{6.30}$$

where, however, we use a for the absorption coefficient (as in chapter 4) to avoid possible confusion with k the wavevector, and we introduce the detuning $x \equiv \nu - \nu_0$ for brevity.

The Fano formula is consistent with scattering theory (see chapter 8), and therefore subject to the requirements of causality which are embodied in the Kramers–Kronig [284, 285] dispersion relations connecting the real and imaginary parts of the complex refractive index $\tilde{n}$. We do not derive the corresponding formula for the refractive index here, but merely outline how it is obtained, giving references for the interested reader. One can show [286] that $\tilde{n}$ satisfies one condition of Titmarsh's theorem [287] and therefore also the Plemelj formulae. By Cauchy integration (see Nussenzweig [286] p 25 equations 1.6 15-16) from the x dependence of equation (6.30), one has, for the real refractive index:

$$n(x) - 1 = \frac{Ne^2 f}{4\pi m\nu_0} \frac{1}{1+q^2} \left\{ \frac{2q\Gamma}{\Gamma^2 + x^2} + \frac{(q^2-1)x}{\Gamma^2 + x^2} \right\} \tag{6.31}$$

For a discussion of the dispersion relations, see, e.g., Feynman [288], and for expressions related to equation (6.31), see Shore [289] who studies possible forms for the refractive index of an autoionising line, but does not give the q dependence of the coefficients in equation (6.31).

In principle, there is a complication concerning constant or slowly-varying background terms which have been omitted from (6.31): in the simple form used here, the Kramers–Kronig relations are valid only for bound rather than free electrons. The refractive index is related to forward scattering through the optical theorem [290] and one should also, when dealing with free electrons, subtract the zero-frequency contribution to forward scattering (see Jackson [291]). Finally, one should account for extraneous transitions at different energies, which contribute to the continuum background as embodied in Sellmeier's formula [292] for the refractive index.

6.15   Magneto-optics of the Beutler–Fano resonance

The expression (6.31) is therefore only suitable for an isolated resonance embedded in an infinite, flat continuum, and we must assume that, in the absence of the resonance, there would be equal refractive indices for left and right hand circularly-polarised radiation in the presence of a magnetic field. In other words, whatever slowly varying term representing the cumulated effects of distant resonances might need to be added to (6.31) to determine the true refractive index would actually cancel out in the evaluation of the difference $n_+ - n_-$, which determines the rotation angle in MOR. In the presence of a magnetic field of strength B, by applying the usual theory (along the same lines as section 4.11), we obtain, after some algebra, the following form for the angle of rotation $\xi(x)$ induced in a singlet-singlet transition to an autoionising state:

$$\xi(x) = \frac{Ne^2 fl\alpha}{mc(1+q^2)} \left\{ \frac{(q^2-1)(x^2-\Delta^2)+4q\Gamma x}{(x^2-\Delta^2)^2+4\Gamma^2 x^2} \right\} \qquad (6.32)$$

where $\alpha \equiv eB/4\pi mc$ and $\Delta^2 \equiv \Gamma^2 + \alpha^2$

Expression (6.32) gives the rotation angle through a Beutler–Fano resonance for different shape indices q and field strengths B. The simplest situation with significant rotation occurs when $\alpha = \Gamma$. The rotation angles for this special case are plotted in fig. 6.5. Profile calculations based on these expressions can be checked by letting $q \to \pm\infty$ and $\Gamma_{fano} \to \Gamma_{natural}$, whereupon the usual MOR patterns for resonance lines are recovered. The plot shows that rotation angles are generally smaller for asymmetric lines than for $q = 0$ or $q \to \pm\infty$. Note also that the sense of rotation reverses for $q = 0$ as opposed to $q \to \pm\infty$. Again, the first autoionising range is the most favourable for the observation of MOR: experiments are easier near the transmission window and rotation generally decreases through coupling to further continua.

For MOR between bound states, it is in principle possible to determine absolute f values, because the number density N can be eliminated if both $a(\nu)$ and $n(\nu)$ are known, as first pointed out by Weingeroff [293]. For a Beutler–Fano profile, this is not possible, essentially because Γ_{fano}

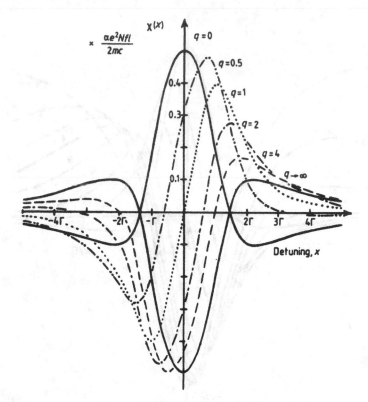

Fig. 6.5.   A plot of the rotation angles in a Beutler–Fano profile as a function of detuning, for several values of the shape index in the special case defined in the text. For negative values of q, reverse the abscissa (after J.-P. Connerade [294]).

is not determined by f. If the profile is Lorentzian, q can be eliminated and one has:

$$\frac{n(\nu) - 1}{a(\nu)} = \frac{c}{4\pi\Gamma} \frac{\nu - \nu_0}{\nu_0} \tag{6.33}$$

so that large rotation angles per unit absorption depth occur only if Γ is small. For transitions between bound states, natural linewidths of the order of 10^{-5} or 10^{-6} occur, so that large rotation angles are encountered, and the MOR patterns extend far out into the wings of the lines, where there is very little absorption. Above threshold, the Fano width dominates and is typically much larger, so that less rotation is experienced, the patterns do not extend as far out into the wings and there is continuous absorption to contend with. Thus, although MOR is the most sensitive method, the refractive index of an autoionising line has proved difficult to determine.

The *visibility* of rotation varies over the profile of a Beutler–Fano res-

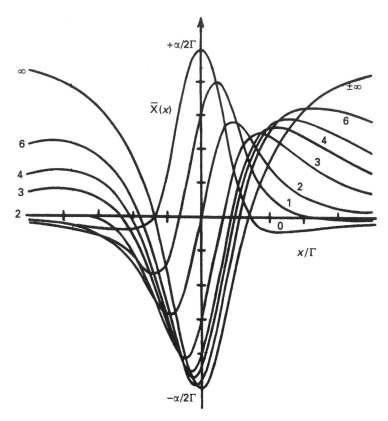

Fig. 6.6. Rotation per unit depth in a Beutler–Fano profile as a function of the detuning, for several values of the shape index q, in the special case described in the text. For negative q, reverse the abscissa (after J.-P. Connerade [294]).

onance. The relevant quantity is the rotation per unit absorption depth $\bar{\xi}(x)$. One can show [294] that

$$\bar{\xi}(x) =$$

$$\frac{\alpha}{2}\left\{\frac{(q^2-1)(x^2-\Delta^2)\Gamma + 4q\Gamma^2 x^2}{(q^2+1)(x^2+\Delta^2)\Gamma^2 + 2qx(x^2+\Gamma^2-\alpha^2)\Gamma + (x^2-\Delta^2) + 4\Gamma^2 x^2}\right\}$$

$$(6.34)$$

If $q \to \pm\infty$, then $\bar{\xi}(x) \to (\alpha/2\Gamma)\left[(x^2-\Delta^2)/(x^2+\Delta^2)\right]$. For zero or for large detuning, the rotation per unit depth rises to an absolute maximum of $\alpha/2\Gamma$. For $q = 0$ we note that the rotation per unit absorption depth in the wings does not tend to a constant value (as it does for $q \to \pm\infty$), but falls off as $\alpha\Gamma/2x^2$ because of continuous absorption, so that any far wing approximation is useless.

Again, the special case $\alpha = \Gamma$ is plotted in fig. 6.6 for different values

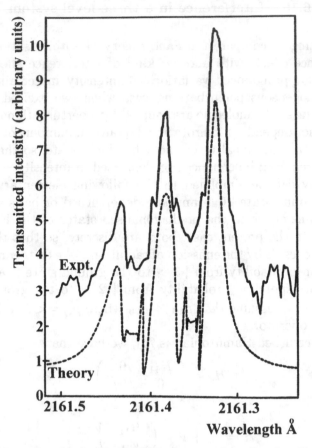

Fig. 6.7. Observed and calculated MOR patterns for a Beutler–Fano resonance in the Ba spectrum (after J.-P. Connerade *et al.* [296]).

of q. Note that, for all q and x, $\bar{\xi}(x)$ is contained between the lower and upper bounds $-\alpha/2\Gamma$ and $+\alpha/2\Gamma$. It follows that, for large rotations, one needs $\alpha \gg \Gamma$, which means both narrow autoionising lines and high field strengths (e.g. $E_0/\Gamma \sim 20\,000$ requires a field of about 9 T).

Since autoionising lines narrow down as $1/n^{*3}$ in the first autoionisation range, the method can be suitable for high Rydberg states. Sharp autoionising lines do exist and have been known for some time [255, 258, 295]. The first observation of an MOR pattern for an autoionising line is shown in fig. 6.7. The observations agree well with the theory, and a pronounced asymmetry is detected.

6.16   Interference in a three-level system

It is interesting to compare the Fano theory of autoionisation, which is an interference effect, with another kind of interference effect which is responsible for pronounced variations of intensity in certain molecular bands. Homogeneous perturbations arise when two excited electronic–vibrational states of a molecule are coupled by a perturbation in a manner which does not depend on the rotational quantum numbers of a molecule, but only on the rotational term and vibrational state, so that complete rotational bands can be enhanced or depressed in intensity.

The theory can be simplified to the following elementary situation: consider a ground state $| \, 0 >$ and two deperturbed or basis states $| \, 1 >$ and $| \, 2 >$ (analogous to the prediagonalised states of the Fano theory, except that, in the present case, both are discrete, so that the only de-excitation process is by the emission of radiation). We further assume that the transition probability from $| \, 1 >$ to the ground state is appreciable, while the corresponding probability from $| \, 2 >$ to the ground state is negligible, i.e. we assume that $\mu_{10} \gg \mu_{20}$ where $\mu_{ij} = < i \, | \, \hat{d} \, | \, j >$ and $\hat{d}$ is the dipole operator.

If the unperturbed Hamiltonian is $\mathcal{H}_0$, we have that

$$\mathcal{H}_0 = \begin{pmatrix} H_{11} & 0 \\ 0 & H_{22} \end{pmatrix}$$

and

$$\mathcal{H} = \mathcal{H}_0 + \mathcal{V} = \begin{pmatrix} H_{11} & V_{12} \\ V_{21} & H_{22} \end{pmatrix}$$

(compare with the structure of the interaction matrix in the Fano theory).

As a result of the perturbation $\mathcal{V}$, the basis states $| \, 1 >$ and $| \, 2 >$ of energies E_1 and E_2 are mixed, and new eigenstates $| + >$ and $| - >$, say, are formed, with eigenvalues Λ_+ and Λ_-.

If c_1 and c_2 are the mixing coefficients, then in general

$$\begin{pmatrix} H_{11} & V_{12} \\ V_{21} & H_{22} \end{pmatrix} \begin{pmatrix} c_1 \\ c_2 \end{pmatrix} = \Lambda \begin{pmatrix} c_1 \\ c_2 \end{pmatrix}$$

or

$$\begin{pmatrix} E_1 - \Lambda & V_{12} \\ V_{21} & E_2 - \Lambda \end{pmatrix} \begin{pmatrix} c_1 \\ c_2 \end{pmatrix} = 0$$

whence

$$\Lambda_{\pm} = E_{av} \pm \left\{ \Delta E^2 + | \, V_{12} \, |^2 \right\}^{\frac{1}{2}}$$

where we define $E_{av} \equiv (E_1 + E_2)/2$ and $\Delta E \equiv (E_1 - E_2)/2$. This equation shows that, for any nonzero perturbation V_{12}, $\Delta\Lambda = \Lambda^+ - \Lambda^-$ is always positive (avoided crossing). We write $| + > = | 1 > c_1^+ + | 2 > c_2^+$ and $| - > = | 1 > c_1^- + | 2 > c_2^-$ so

$$\begin{pmatrix} E_1 & V_{12} \\ V_{21} & E_2 \end{pmatrix} \begin{pmatrix} c_1^+ \\ c_2^+ \end{pmatrix} = \Lambda^+ \begin{pmatrix} c_1^+ \\ c_2^+ \end{pmatrix}$$

whence

$$E_1 - E_2 = -\frac{V_{12}c_2^+}{c_1^+} - \frac{V_{21}c_1^+}{c_2^+} = V_{12} \left(\frac{c_1^+}{c_2^+} - \frac{c_2^+}{c_1^+} \right) \tag{6.35}$$

By writing the equation for Λ^-, we find that

$$\Lambda^+ - \Lambda^- = V_{12} \left(\frac{c_2^+}{c_1^+} - \frac{c_2^-}{c_1^-} \right)$$

By orthogonality, $< - | + > = c_1^- c_1^+ + c_2^- c_2^+$, or

$$\frac{c_2^-}{c_1^-} = -\frac{c_1^+}{c_2^+}$$

and so

$$\Lambda^+ - \Lambda^- = V_{12} \left(\frac{c_2^+}{c_1^+} + \frac{c_1^+}{c_2^+} \right) \tag{6.36}$$

which should be compared with equation (6.35).

Now consider the dipole matrix element for the transition from the ground state $| 0 >$ to the perturbed state $| + >$

$$\mu_{0+} = < 0 | \hat{d} | + > = < 0 | \hat{d} | 1 > c_1^+ + < 0 | \hat{d} | 2 > c_2^+$$

This transition strength will vanish if

$$\frac{c_1^+}{c_2^+} = -\frac{\mu_{02}}{\mu_{01}}$$

which, by equation (6.36) occurs when

$$(\Lambda^+ - \Lambda^-)_{zero\,strength} = -V_{12} \left(\frac{\mu_{02}}{\mu_{01}} + \frac{\mu_{01}}{\mu_{02}} \right)$$

But, from equation (6.16), the avoided crossing occurs when $\Delta E = 0$ or $E_1 = E_2$, in which case, by equation (6.35), $c_1^+ = c_2^+$. Since it was assumed at the outset that $\mu_{02} = \mu_{01}$, the case of zero intensity will generally lie to one side of the avoided crossing. This can be compared with the minimum intensity which generally lies to one side of the Fano profile. The analogy with autoionisation is that, in a Beutler–Fano profile, all the possible values of ΔE are present simultaneously.

In fig. 6.8, we show how the intensities evolve near the avoided crossing for different values of the energy difference ΔE in a typical situation: for

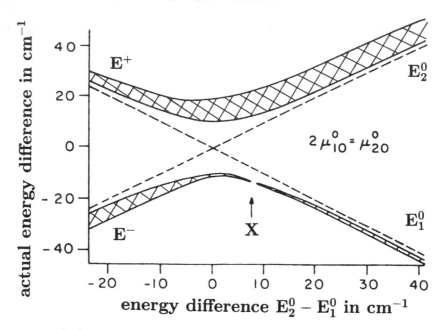

Fig. 6.8. Evolution of the intensities for transitions from the ground state to two bound excited states coupled by a perturbation, in the vicinity of an avoided crossing. The inner side of the hashed area gives the energies of the perturbed eigenvalues, and the hashed area gives approximate intensities. Note the zero in one of the branches at the point marked 'X' in the figure. It is detuned from the avoided crossing and reveals stabilisation against radiation by the perturbation (adapted from K. Dressler [298]).

a given value of ΔE and of V_{12}, the intensity of the transition coupling the ground state to one of the perturbed states falls to zero, i.e. the system behaves as though *one of the states is stabilised against radiation by the perturbation*. Other examples of stabilisation by interference between channels will be presented in chapter 8.

In a homogeneous molecular perturbation, V_{12} is the product of vibrational and electronic overlap elements, but is the same for all the rotational levels of a band. Thus, whenever a cancellation of intensity occurs, a complete rotational band may disappear from view. This situation arises for several $^1\Sigma_u$ bands in the N_2 spectrum [297] which possess the same symmetries but originate from different electronic terms. An analysis based on the same principles as described above has been presented by Dressler [298].

6.17   Alternative derivation of the Fano formula

As promised in the Introduction, we now turn to a third method of deriving the Fano formula, namely diagrammatic MBPT (see section 5.26).

Consider, as before, an initial state $| \, i >$, and let the letter i in Feynman graphs stand for the transition $(i \rightarrow f)$. We can write the single electron (independent electron) dipole transition as the diagram

$$d_i \equiv \times\!-\!-\!\!\prec i \tag{6.37}$$

where the final state f is a correlation-free continuum in the Hartree-Fock potential (which is the usual zero-order basis for MBPT – see e.g. [299]). Now let

$$\mathcal{D}_i \equiv \times\!-\!-\!-\!\!\oslash\!\!\prec i \tag{6.38}$$

be the corresponding transition strength including all the correlations *except* the ones which couple the discrete state $| \, \varphi >$ to the continuum. We can then write a perturbation series for the transition to the prediagonalised state $| \, \phi >$ as

$$\mathcal{D}_\varphi = \times\!-\!-\!-\!\!\oslash\!\!\prec \varphi = \times\!-\!-\!-\!\!\prec \varphi + \times\!-\!-\!\!\bigcirc\!\!\!\!\!\!\!\!\!\!\!\!\!\!\!\!\!\!\!\begin{smallmatrix} i \end{smallmatrix} + \cdots \tag{6.39}$$

and the fully correlated transition to the continuum is then

$$\Delta_i \equiv \times\!-\!\oslash\!\!\prec i + \times\!-\!\oslash\!\!\bigcirc\!\!\varphi\!\!\oslash\!\!\prec i + \times\!-\!\oslash\!\!\bigcirc\!\!\varphi\!\!\oslash\!\!\bigcirc\!\!\varphi\!\!\oslash\!\!\prec i \tag{6.40}$$

$$+ \cdots$$

which we can write as

$$\Delta_i = \mathcal{D}_i + \mathcal{D}_\varphi \frac{N_\varphi}{E - E_\varphi} \Gamma_{\varphi i} + \mathcal{D}_\varphi \frac{N_\varphi}{E - E_\varphi} \Gamma_{\varphi\varphi} \frac{N_\varphi}{E - E_\varphi} \Gamma_{\varphi i} + \cdots$$

$$= \mathcal{D}_i + \mathcal{D}_\varphi \frac{N_\varphi}{E - E_\varphi} \left\{ 1 + \frac{\Gamma_{\varphi\varphi} N_\varphi}{E - E_\varphi} + \frac{(\Gamma_{\varphi\varphi})^2 N_\varphi^2}{(E - E_\varphi)^2} + \cdots \right\} \Gamma_{\varphi i} \tag{6.41}$$

giving a geometric progression, where the widths Γ are just numbers, so that the sum of the series is readily evaluated, giving

$$\Delta_i = \mathcal{D}_i + \mathcal{D}_\varphi \frac{N_\varphi}{E - E_\varphi} \frac{\Gamma_{\varphi i}}{1 - \dfrac{\Gamma_{\varphi\varphi} N_\varphi}{E - E_\varphi}} = \mathcal{D}_i + \frac{\mathcal{D}_\varphi \Gamma_{\varphi i} N_\varphi}{E - E_\varphi - \Gamma_{\varphi\varphi} N_\varphi} \tag{6.42}$$

where N_i, N_φ are just geometrical factors associated with the vertex structure of the perturbation expansion, i.e. $N_i = N_\varphi = 2/3$. We let $\overline{\Gamma}_{\varphi\varphi} = N_\varphi \Gamma_{\varphi\varphi}$ and $\overline{\Gamma}_{\varphi i} = N_\varphi \Gamma_{\varphi i}$ and, noting that the Γ are in general complex, we can write

$$\Delta_i = \mathcal{D}_i + \frac{\mathcal{D}_\varphi \overline{\Gamma}_{\varphi i}}{(E - E_\varphi - \mathcal{R}e\overline{\Gamma}_{\varphi\varphi}) - i\mathcal{J}m\overline{\Gamma}_{\varphi\varphi}}$$

$$= \mathcal{D}_i + \frac{\mathcal{D}_\varphi \overline{\Gamma}_{\varphi i}}{(E - E_0) - i\mathcal{J}m\overline{\Gamma}_{\varphi\varphi}} = \mathcal{D}_i + \mathcal{D}_\varphi \frac{\frac{\overline{\Gamma}_{\varphi i}}{\mathcal{J}m\overline{\Gamma}_{\varphi\varphi}}}{\epsilon - i} \qquad (6.43)$$

where $E_0 \equiv E_\varphi + \mathcal{R}e\overline{\Gamma}_{\varphi\varphi}$ and $\epsilon \equiv (E - E_0)/\mathcal{J}m\overline{\Gamma}_{\varphi\varphi} = (E - E_0)/\Gamma/2$ and Γ is the Fano resonance width.

We note that

$$\Gamma_{\varphi\varphi} \equiv \varphi \rangle\!\!-\!\!\oslash\!\!-\!\!\langle \varphi$$

is the self-interaction of levels $|\varphi\rangle$ mediated by the continuum, so that, by completeness,

$$\mathcal{J}m\Gamma_{\varphi\varphi} = \varphi \rangle\!\!-\!\!\oslash\!\!\left(i \quad i \right)\!\!\oslash\!\!-\!\!\langle \varphi = |\Gamma_{\varphi i}|^2 N_\varphi \quad \text{where} \quad N_i = \left(i \right)$$

gives the width induced by correlations between φ and i. Thus, the total amplitude for transitions to the continuum together with the interference term

$$\Delta_i = \mathcal{D}_i \left\{ 1 + \frac{\mathcal{D}_\varphi \overline{\Gamma}_{\varphi i}}{N_i \mathcal{D}_\rangle |\overline{\Gamma}_{\varphi i}|^2} \frac{1}{\epsilon - i} \right\} = \mathcal{D}_i \left\{ 1 + \frac{\mathcal{D}_\varphi}{\mathcal{D}_i} \frac{\overline{\Gamma}_{\varphi i} N_i}{\epsilon - i} \right\} \qquad (6.44)$$

Noting also that

$$\mathcal{J}m\mathcal{D}_\varphi = \times - - - \oslash\!\!\left(i \right)\!\!\oslash\!\!\langle \varphi = \mathcal{D}_i \Gamma_{\varphi i} N_i$$

we have

$$\Delta_i = \mathcal{D}_i \left\{ 1 + \frac{\mathcal{R}e\mathcal{D}_\varphi + i\mathcal{J}m\mathcal{D}_\varphi}{\mathcal{J}m\mathcal{D}_\varphi(\epsilon - i)} \right\} = \mathcal{D}_i \left\{ 1 + \frac{i - q}{\epsilon - i} \right\} \qquad (6.45)$$

where we define $q \equiv -\mathcal{R}e\mathcal{D}_\varphi/\mathcal{J}m\mathcal{D}_\varphi$ and note that $\Gamma = N_\varphi N_i |\Gamma_{\varphi i}|^2$. Thus, the cross section is

$$\sigma = \sigma_0 |\Delta_i|^2 = \sigma_0 |\mathcal{D}_i|^2 \left| \frac{\epsilon - i - q + i}{\epsilon - i} \right|^2 = \sigma_0 |\mathcal{D}_i|^2 \frac{(\epsilon - q)^2}{1 + \epsilon^2} \qquad (6.46)$$

as before. Note that only one continuum was included, which is why the cross section falls to zero near the resonance. Otherwise, we would have $\mathcal{J}m\mathcal{D}_\varphi = \sum_i (\mathcal{D}_i \Gamma_{\varphi i})$, leading to

$$\sigma = \sigma_c + \sigma_0 \frac{(\epsilon - q)^2}{1 + \epsilon^2}$$

where σ_c is the contribution due to other continua which fill in the transmission window.

6.18 Conclusion

In the present chapter, we have described many aspects of the simplest problem which can arise when an isolated resonance is formed in a single continuum: we have shown that autoionisation is an interference phenomenon and compared it with the behaviour of a discrete three-level system. Two different derivations of the Fano formula have been given, and its connection with MQDT has been described. A third approach will be provided in chapter 8. Beutler–Fano autoionising resonances occur in all many-electron atoms, and a number of examples will be provided in the next two chapters. In chapter 8, the interactions *between* autoionising resonances will be considered, and two further questions will be discussed, namely the influence of coherent light fields on autoionising lines, and the use of lasers to embed autoionising structure in an otherwise featureless continuum.

Finally, we note that autoionisation cannot normally occur in H, because there is only one threshold, which clearly separates bound from continuum states. There is, however, one way of inducing autoionsisation in H: if an external field induces new thresholds, with excited states above them, then autoionisation becomes possible even in H. This situation is discussed in chapter 11.

7

Inner-shell and double-excitation spectra

7.1 Introduction

Many-electron atoms differ from H in an essential respect: when they are excited up to and above the first ionisation potential, they exhibit structure which is not simply due to the excitation of one valence electron. The clearest manifestation of this behaviour occurs in the ionisation continuum. For H, the continuum is clean, i.e. exempt from quasidiscrete features. In any many-electron atom, there will be autoionising resonances of the type discussed in chapter 6. Autoionisation is therefore a clear manifestation of the many-electron character of nonhydrogenic atoms.

In the present chapter, the questions: why does this extra structure occur and how does one set about interpreting it? are addressed. Thus, we will not be so concerned about the lineshapes or even (in first approximation) about interseries perturbations (although they do turn out in some cases to play a crucial role), but rather with the configurations of the inner-shell and doubly-excited states, and their relation in energy to the valence spectrum.

7.2 Inner-shell excitation

Even within the independent electrom approximation, it is obvious that there must exist inner-shell excitation spectra, and that their energy must extend well above the first ionisation potential. This arises from the simple fact that one can choose which electron is excited: it does not necessarily have to be the valence electron, and the inner electrons, being more strongly bound, require photons of higher energy to excite them. Since the valence electron extends furthest out from the atomic core, one is tempted to think that it is always the easiest electron to excite, both because it can more readily interact with an external field (higher transition probability) and because it requires less energy. This, however, is not always true: transitions of valence electrons to high Rydberg states can become extremely weak, and the sum rules may not allow much intensity in the valence spectrum. Inner-shell transitions may sometimes terminate on compact excited states and often come from subshells containing many electrons. In such cases (see chapter 5) the inner-shell spectrum can contain a lot of oscillator strength within a comparatively small energy range. There is a general rule, due to Fano and Cooper [300], which we may regard as a propensity rule (cf section 4.2) to determine which inner-shell spectrum is likely to be most intense: the rule is that the highest intensity is normally associated with the outermost filled subshell of a given shell. The interpretation of this rule does, however, become difficult in dealing with some transition elements, and so it should be regarded as a guide rather than a hard and fast statement.

The most familiar examples of inner-shell spectra are the X-ray spectra, due to the excitation of the deepest shells in the atom. These are produced in an X-ray tube by bombarding the solid anode with fast electrons (of kV energies). Because the inner shells are so deep inside the atom, they are quasiatomic even in the solid, from which they are screened by all the intervening shells. Characteristic X-ray emission is therefore of essentially atomic character. To describe such transitions, Heisenberg introduced the concept of a *hole*, i.e. a closed subshell from which one electron has been removed. A hole behaves like a particle: indeed, it is like an electron, except that it has positive charge. In the context of many-body theory, it is called a *quasiparticle*: a quasiparticle is the name given to a group of particles which conspires to behave like a single particle, by which the group can be replaced. This is never exactly true, but is a convenient and sometimes very good approximation.

If the quasiparticle approximation works well, then this is a sign that the independent particle model is trustworthy. If one is looking for manifestations of the many-electron character of atoms, then the greatest interest attaches to situations where in fact the quasiparticle picture begins to break down. From this perspective, the excitation of very deep inner shells is not the most interesting situation: it is more likely that the concept of a hole will lose validity for the outermost inner shells, although the situations in which this occurs are actually quite specific.

There exist theoretical models which confirm the general argument just presented. Since the details are somewhat complex [301], we summarise their results in graphical form in fig. 7.1.

In case (a) of the figure, we show a typical photoelectron spectrum containing just one line, which corresponds to the excitation of one electron and results in just one ionic state. In case (c), the typical X-ray spectrum is dominated by one line, which represents the quasiparticle, with a few

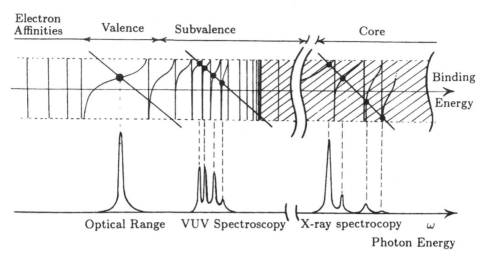

Fig. 7.1. Typical photoelectron spectra as computed from many-body theory: (a) for valence excitation (optical spectra); (b) for subvalence shell excitation (vacuum ultraviolet spectra); and (c) for deep inner-shell excitation (X-ray spectra) (after J.-P. Connerade [355]).

satellites arising from other causes, whose small intensities are a measure of the very slight breakdown of the quasiparticles approximation. In case (b), however, several peaks of comparable strength occur, and it is impossible to say which one might be associated with a quasiparticle: in fact, it is in this energy range that the concept of a quasiparticle is most likely to break down. In section 7.14, we give examples of inner-shell spectra, for which varying degrees of breakdown are observed and can be followed to reveal their causes. First, however, there are some related, general points to be discussed.

7.3   The spectator electrons approximation

The first question which arises when discussing the excitation of an inner electron is: what happens to the *other* electrons in the system, especially those from intervening subshells, which lie between the inner shell being excited and the outer surface of the atom. This question is not quite so pressing for valence excitation, simply because the valence electron is the most external, and therefore usually the *only* electron which can be excited by optical photons. If a fairly deep electron is extracted, however, we may expect more possibilities and that other rearrangements should occur, involving electrons less strongly bound than the one being excited.

In principle, an inner-shell spectrum can be very simple if it happens

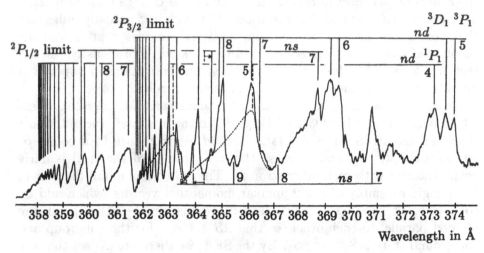

Fig. 7.2. A typical inner-shell excitation spectrum: the 3p spectrum of Ca. Note the wide doublet splitting between the two series limits due to the large spin–orbit interaction of the nearly-closed core, the prominent Rydberg series and the broad, asymmetric autoionising resonances (after J.-P. Connerade *et al.* [302]).

that no double excitations or rearrangements of the core intrude. On the other hand, it may tend to extreme complexity for apparently similar configurations. An example of a rather simple inner-shell spectrum is given in fig. 7.2: the 3p excitation spectrum of Ca [302], which is in principle analogous to the 4p spectrum of Sr and the 5p spectrum of Ba. However, these spectra become rapidly extremely complex with increasing atomic number, for reasons which are associated with breakdown of the quasiparticle approximation and which will be explained in greater detail in chapter 10.

Within the independent particle model, rearrangements which involve a change in the configuration are excluded, because only one electron is allowed to undergo a change in the quantum numbers n and ℓ. This, however, still leaves some possibilities open, because transitions may occur in which the angular momentum couplings between the initial and final states are changed for electrons which do not participate in the transition. Although this is not technically a violation of the independent electron model, one cannot really argue that only one electron was involved in the excitation process when it happens.

The spectator electrons approximation (SEA) puts the most restrictive interpretation on the nature of a single electron excitation. What we suppose is that only one electron is involved in a transition, and that all the others act as spectators, i.e. that they do not change their properties in any way as a result of the transition. Although it is actually quite rare that this approximation works completely, it provides a good starting point, and can be considered as yet another kind of propensity rule (cf section 4.2), which reduces the number of strong transitions in a spectrum below the number allowed by the usual selection rules.

An example is useful to make the situation clear. Consider the $3p$ excitation spectra of Mn [303] and Cr [306]. In the first case, the outermost electrons are arranged as $3d^5 4s^2$. The $3d$ electrons are in a collapsed state (cf chapter 5), and, because the mean radius of the shell is fairly small, interact with each other strongly. They also form a half-filled shell, i.e. a rigid assembly of $\ell = 2$ angular momentum vectors, which add up to $L = 0$, i.e. an S subgroup. The five spins, on the other hand, are aligned parallel to each other, so that $2S + 1 = 6$ for the subgroup and the ground state is $3d^5 4s^2 \, ^6S_5$. By the SEA, we therefore expect that the only strong series due to excitation of a $3p$ electron will converge on limits for which this orientation is preserved, i.e. for which the configuration and term are $(3p^5 \, ^2P_{1/2,3/2})(3d^5 \, ^6S_{5/2})^{5,7}P$. This indeed is confirmed by experiment [303], to the extent that these seem to be the only reported series in photoabsorption spectra of Mn. The SEA, here, is much more restrictive than the independent particle model applied to configurations: if one allows the $3p^5 \, 3d^5$ electrons to form all possible combinations, many more series limits would result.

Now consider the case of Cr: the configuration is then $3d^5 4s \, ^5S_2$ in the ground state, and the transitions allowed both by the dipole selection and the SEA are then built on the ion state $(3p^5 \, ^2P_{1/2,3/2})(3d^5 \, ^6S_{5/2})(4s \, ^2S_{1/2})$. In the ground state, the $4s$ electron has its spin parallel to the five electrons of the $3d$ subgroup. If this orientation were preserved in the excited state, the allowed limits would be $^{6,8}P$, but if the $4s$ electron is allowed to flip its spin so that it no longer points in the same direction relative to the $3d^5$ subgroup as in the ground state, then the allowed limits are $^{4,6}P$. What is observed in this case is that a very weak series appears, converging on the ^4P limit, which demonstrates that, for an isolated $4s$ electron, it is comparatively much easier to 'flip' the spin by exciting an electron from another subshell. This breakdown of the SEA illustrates the fact that the approximation works best for closed subgroups, whose angular momenta are 'locked' together.

It is worthy of note that the SEA is not useful in all cases. For example, if we consider inner-shell excitation from the d shell in an element with outer shells $d^{10} s^2 p$, then the first excited state is $d^9 s^2 p^2$, but the orientation of p was indeterminate in the ground state, so the SEA sets no restriction as compared with the usual selection rules.

7.4 The $\mathcal{Z}+1$ approximation

A very simple way of understanding and analysing the structure observed in inner-shell excitation is to use the $\mathcal{Z}+1$ model, in which the observed inner-shell spectrum is compared with the appropriate spectrum of the *next* element following it in the periodic table. This works well for two reasons. First, as regards the angular part or configuration symmetry, when an inner shell is excited, there is one more electron in the outer reaches of the atom, and it therefore begins to resemble the next element in the table, except for the presence of the open core. Secondly, when a vacancy is created inside the atom, the effect is rather similar to an increase of 1 in the nuclear charge, as far as the outer electrons are concerned.

This idea is as old as inner-shell spectroscopy itself. Already, Beutler [304], in his pioneering experiments to bridge the observational gap between optical and X-ray spectroscopy, used this comparison in a qualitative way to understand the spectral structure he observed. More recently, it has been put on a semiquantitative footing [305].

There are, however, limitations inherent to this approximation: often, the inner-shell spectrum does not appear alone, but is accompanied by *double excitations* in which two external electrons rather than an inner-shell electron are promoted to outer orbits, and the two modes of excitation can and do interact strongly. It is then inappropriate to use the $\mathcal{Z}+1$ approximation [189].

7.5 Double excitations

The simultaneous excitation of two electrons by a single photon is a process rigorously forbidden within the independent particle model. The next stage in the breakdown of the independent electron model goes beyond the slight breakdown of the SEA discussed in the previous section. It arises when the term in the excited state is different from the one expected and one has a configuration in the excited state which is not allowed by the normal dipole selection rules. We then have a *double* or *multiple excitation*, in which more than one electron effects a transition from one configuration to another.

The reason for this breakdown can be understood as follows: as shown in textbooks on classical mechanics,[1] the conservation of angular momentum applies to a particle moving in a central field of force. The extension which is tacitly assumed in atomic physics (following Bohr) is that conservation of angular momentum can be applied *individually* to each electron in the atom, so that the $n\ell$ quantum numbers for each electron can be specified in the configuration assignments of a many-electron atom. This idea was already a subject of argument at the time when it was proposed by Bohr: Langmuir made the very reasonable objection that one should seek 'collective' quantum numbers for groups of electrons, rather than attempt to quantise electrons on an individual basis in a many-electron system.[2] However, the central field approximation and the concept of electronic configurations which results from it were so successful that these objections were largely swept aside.[3]

As a consequence of the breakdown of the independent particle approximation, it then emerged that the quantisation of individual electrons was not completely reliable. This was referred to in the classic texts on the theory of atomic spectra [309] as a breakdown in the ℓ *characterisation*, and it manifests itself in the appearance of extra lines, which could not be classified within the independent electron scheme. The proper solution would, of course, be to revisit the initial theory and correct its inadequacies by a proper understanding of the dynamics of the many-electron problem, including where necessary new quantum numbers to describe the behaviour of correlated groups of electrons. Unfortunately, this plan of action cannot be followed through: it would require a deeper understanding of the many-body problem than exists at present (see, e.g., chapter 10 for some of the difficulties).

Instead, what is usually done is to 'patch up' the independent electron approximation by applying corrections to it using perturbation theory. This purely mechanical approach is referred to as *configuration mixing*, and leads to the following picture or classification of double excitations.

We begin by noting that the $\Delta\ell = \pm1$ selection rule is at issue: if the ℓ value of an electron is uncertain, then there will be corresponding uncertainty in the ℓ values for the excited configuration which, formally, is how doubly-excited configurations can be introduced into the theory.

It is, however, important to note that atoms are symmetrical and have no permanent dipole moment, i.e. that the parity, defined as $(-1)^{\sum_i \ell_i}$, is a good quantum number even in the presence of electron–electron in-

---

[1] See, e.g., Kibble [307].

[2] See [308] for a discussion of this history.

[3] The quest for further quantum numbers to describe electron–electron correlations, as suggested by Langmuir, continues as a topic of current research (see section 7.11 for further comments).

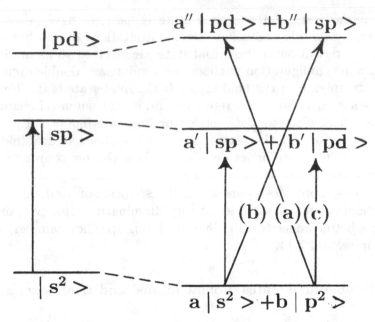

Fig. 7.3. Various mechanisms driving double excitation: (a) ground or initial state mixing; (b) final or excited state mixing; and (c) a combination of both.

teractions (we neglect here any parity nonconserving forces due to the Weinberg–Salam theory as completely negligible in magnitude). Thus, any mixing of ℓ values must preserve this property.

For example, consider an atom whose ground state has an s^2 configuration. We can no longer speak of this configuration being pure if there is a breakdown in the ℓ characterisation. Therefore, we may find that it contains a mixture of s^2 and, say, p^2 character, so that its true eigenvector is $a\,|\,s^2 > + b\,|\,p^2 >$, where a and b are mixing coefficients such that $|\,a\,|^2 + |\,b\,|^2 = 1$ and represent the proportion of each state present in the mixture. In principle, at least for the ground state, one can determine a and b for a given Hamiltonian by performing variational calculations and looking for the combination with the lowest energy. The procedure is, however, arbitrary in many ways, not least because there are infinitely many excited states in the atom, and it is not at all clear whether the right ones have been chosen. Let us, however, suppose that mixing coefficients can be determined in a reliable way.

We can then have the situation illustrated in fig 7.3, from which it is clear that, instead of just one transition to the sp excited configuration, the $\Delta\ell$ selection rule will, in effect, allow transitions to a state one would label pd, since one cannot distinguish quantum mechanically between the mixed vectors of the ground state. Thus, if the ground state label is retained as 's^2', and if the excited state is 'pd', we have generated a mechanism for double excitation. Double excitations due to breakdown in the characterisation of the initial state are referred to as originating by initial state configuration mixing. In a similar way, double excitations can arise by mixing in the final state. If the final state is itself already a mixed state, involving a discrete part and a continuum (cf chapter 6), then either final discrete state mixing or final continuum state mixing can be responsible for the occurrence of a doubly-excited autoionising resonance. Which case arises depends on how the breakdown in the ℓ characterisation of the final state occurs.

This approach provides a purely formal description of the double-excitation mechanism and, as such, is not very illuminating. However, one can build up a better understanding by considering specific examples, as will be done in section 7.14.

7.6   Double-hole states 'hollow atoms' and triple excitation

A *hollow atom* is any atom for which at least one of the inner shells is completely empty. To produce a hollow atom by exciting only two electrons, the inner shell must therefore be $1s^2$. The simplest hollow atom, and the one which has been known for the longest time is He in a doubly-excited state [310]. One can in principle consider partially hollow atoms, in which only a subshell rather than a complete shell is empty.

Once two-electron transitions are introduced, there is no reason why only a single hole should be created inside the atom. Thus, for example, inner-shell excitation from the $4d$ subshell of Ag leads to the excitation $4d^{10} \rightarrow 4d^9 5s5p$, but can also, by final state mixing, lead to $4d^8 5s^2 5p$, which is built on a double $4d$ hole state of slightly higher energy but same parity and perturbs the inner-shell spectrum.

The effect is more dramatic for light atoms: in He, the $1s^2 \rightarrow 2s2p$ excitations are well known (see above); in Li, a similar excitation in the presence of a $2s$ spectator electron leads to the transition $1s^2 2s \rightarrow 2s^2 2p$. Formally, this can also be regarded as a triple excitation, if we consider that the two $1s$ electrons are excited to $2s2p$ as in He, and that the $2s$ electron of Li is excited to $2p$, i.e. that there is no spectator. This transition has been observed [311], and is described as a triple excitation in Li, probably because earlier theoretical papers [312] stress the strong mixing between the lowest triply-excited configurations such as $2s^2 3s$, $2s^2 2p$, $2s2p^2$, which make any other characterisation difficult.

The excited state is hollow, because the $1s^2$ shell is completely empty. It turns out that the observed photoabsorption spectrum of the $1s^2 2s\,^2S \rightarrow 2s^2 2p\,^2P$ transition exhibits a broad and asymmetric Beutler–Fano profile very similar in general appearance to the He doubly-excited transition, or to its isolectronic analogue in Li^+.

Hollow atoms were originally characterised in a different context from double excitations in very simple atoms: when a slow beam of highly stripped ions approaches or penetrates a surface (which is, in effect, an infinite reservoir of electrons), some electrons may be drawn out from the surface and recombine with the ions, producing atoms in hollow excited states [313]. Above the surface, the process feeds electrons from large distances into Rydberg states of the ion, and the resulting hollow atoms are referred to as Rydberg hollow atoms. On the surface or below it, electrons are captured into more compact excited states, and the resulting atoms are called surface hollow atoms. There are further distinctions according to whether the surface is that of a metal, a semiconductor or an insulator. These hollow atom states are different from those previously described, but all hollow atoms share one common feature: the inner shells are unoccupied, while many electrons are in outer orbitals.

The manner in which electrons are extracted from the surface is such that the greatest probability of capture is when the system as a whole remains at the same energy. Thus, electrons are most likely to transfer to the atom in such a way that the binding energy of the electron within the ions is the same as the binding energy of the electron within the surface. It has been shown that hollow atoms produced by collisions with a surface can survive long enough for electrons to be still in high-lying levels as the projectile hits the surface [314].

Apart from their exotic configurations, hollow atoms are interesting, because they can be in both highly correlated and highly symmetric states, whose coupling to the radiation field is unusual. For example, $2s^2$ of He has no dipole coupling to the radiation field: such states lie at the opposite extreme of small principal quantum numbers to the Wannier states, discussed in section 7.9. In fact, they can be regarded as leading members of a Wannier series.

7.7 Inner-shell excitation and Coulomb explosions in molecules

Another interesting situation arises in molecules, when significant double excitation and ionisation occur: if a molecule is doubly charged, then it experiences a very large disruptive force. If two charges q_1 and q_2 are separated by R_{12} Å, then the Coulomb potential energy is given by $U(eV) \sim 14.4 q_1 q_2 / R_{12}$ which is well in excess of the dissociation energy. Thus, the molecules fragment on a very short timescale after the initial excitation, and the kinetic energy of the fragments reveals the nature of the bond and the distribution of the charge in the excited state of the molecule. The fragmentation of small triatomic molecules which can be linear (OCS) or bent (SO_2) provides information on the mechanisms involved. Such processes have been considered by Eland and Sheahan [315], who propose a model in which charges remain delocalised until a separation of about 7 Å, beyond which a Coulomb explosion occurs. This is, as yet, only one of several possible models (sequential fragmentation schemes have also been suggested).

Coulomb explosions are readily induced in collision experiments in which a charged ion of high energy from an ion accelerator collides with a molecular target. Many processes can occur, including Coulomb explosions of the target molecule. If the linear momenta of all the fragments are detected and if the molecule has a simple structure, then each explosion can be reconstructed so that, in principle, the analysis provides an imaging technique for the study of molecular structure. Experiments along these lines have been performed [316], although a high enough spatial resolution remains to be demonstrated.

Because of the inverse square law, the probability of Coulomb explosions decreases rapidly for molecules of larger size. Consequently, in collision experiments, the abundance of doubly-charged ions is far greater for large molecules. This tendency is well known from studies of the heavy hydrocarbons [317], and extends to fullerenes and other clusters (see chapter 12).

7.8 Double-circular states

If two electrons are excited, each one to the maximum allowed value of ℓ for a given n, then one has a situation in which the semiclassical orbit for each electron is circular, whence the name 'double-circular' states. An example is the $4f5g$ configuration of Ba [318], which has been accessed by laser spectroscopy. Note, however, that the n values of these states must be pretty low, or their ℓ values will become extremely large in order to satisfy the definition. For fairly low n values, the concept of a semiclassical orbit does not have much physical validity. On the other hand, at rather large n it would be necessary to increase both n and ℓ together to make a series of such states, and it is hard to conceive a sensible semiclassical limit since there is no obvious recapitulation. Indeed, the $3d$, $4d$, $5g$, etc wavefunctions are mainly characterised by the fact that they possess no nodes. An interesting feature of the $4f5g$ states in Ba (which may be a general property) is their extremely low autoionisation rate: their scaled widths are actually smaller than for higher members of the $4fng$ series (which are not double-circular states). An explanation based on a semiclassical picture has been proposed, but it seems more likely that correlation effects due to interchannel mixing are responsible for the small widths [319], along similar lines to those discussed in the next chapter. Another unusual aspect is the very small spatial overlap between the $4f$ and $5g$ orbitals, which arises because of centrifugal barrier effects: the $4f$ electron is on the verge of orbital collapse, while $5g$ is still held out by a repulsive barrier.

7.9 Two-electron jumps and double Rydberg states

Since doubly-excited states involve the excitation of two electrons, it is natural to consider both excited electrons on an equal footing. To take a model problem, consider the excitation scheme

$$s^2 \longrightarrow n_1 \ell_1 n_2 \ell_2$$

Then, we obtain a Rydberg series by fixing either $n_1 \ell_1$ or $n_2 \ell_2$ to denote an excited state of the ion, and letting the other electron run through all the possible quasibound states until it becomes free, leaving the *parent ion* in the excited state. The Rydberg electron is then referred to as the *running electron*, and it is clear that there is a whole family of excited states of the parent ion which can serve as series limits for doubly-excited Rydberg series. The spectrum of double–excitations is therefore very rich.

The parent ion states themselves form a Rydberg series and, because the ion is singly-charged, the spacings between the parent ion states are a good deal larger (roughly four times as large) as the corresponding spacings in the doubly-excited series. Thus, one observes a manifold of Rydberg series converging to fairly widely spaced limits, which eventually overlap and merge as the double-ionisation threshold is reached. These are the double–excitations as normally observed in photoabsorption spectra.

There exists, in addition, another possibility. It will not have escaped the reader that the two electrons, in the description just given, are not treated completely symmetrically. Were one to do so, it would be necessary to increase *both* n_1 and n_2, so as to form a sequence of levels with $n_1 = n_2$ spaced differently from the normal Rydberg series, converging on the double-ionisation potential. At the limit of this series, both electrons would escape in a completely symmetric way, and so this possibility is often referred to as the 'double-escape problem'.

Double Rydberg states (double because both electrons are running electrons in this case) and the symmetrical double-ionisation problem were considered first by Wannier [320], and so the completely symmetric excited states with $\ell_1 = \ell_2$ and $n_1 = n_2$ as the double-ionisation threshold is approached are called *Wannier states*. These states have very interesting properties theoretically. One of their most intriguing features is that they remain highly correlated as the series limit is approached, since the spatial overlap between the wavefunctions of both electrons does not decrease as in a normal doubly-excited Rydberg series. Because of their symmetry, they are also stable states, and cannot couple directly to the radiation field: in effect, they are monopole excitations. There has been much discussion about how they might be excited in real experiments. Collisions may provide the best route, since the dipole selection rules for excitation can thus be broken. Indeed, transitions of this kind with $n_1 = n_2 \sim 7$ have been observed by electron impact spectroscopy in He. Another possibility is two-photon excitation (see chapter 9) for which the selection rules are different. However, despite many attempts at producing them for very high values of n in a variety of elements, the quest for highly excited Wannier states has not borne fruit so far.

In the present chapter, we therefore confine our attention to doubly-excited series with one running electron, while the other excited electron defines the parent ion limit.

7.10   The special case of He

Before giving examples of the behaviour of many-electron atoms in general, it is worth describing the case of He, which is exceptional in that it contains no inner-shell spectrum, but only a doubly-excited spectrum in addition to the usual valence-shell excitations.

The double-ionisation potential of He lies at an extremely high energy. Not only does He have the highest single-ionisation threshold, but it also has the highest double-ionisation threshold in the periodic table, i.e. the valence spectrum and the spectrum of the doubly-excited states lie further from each other than for any other many-electron atom. This has the consequence that the interaction between the two spectra is at a minimum as compared with other atoms, and that the breakdown of ℓ characterisation due to accidental degeneracies has little role to play. Thus, the doubly-excited spectrum of He comes closer than any other to revealing the dynamical nature of a pure, unmixed doubly-excited spectrum. If correlation quantum numbers exist to describe the excited states of two electrons, then it is probably in He that they stand the best chance of being found. He is in this sense the archetypal system in which to study double excitations. It is the atomic version of the three-body problem.

The history of observations of the doubly-excited spectrum of He is comparatively recent, because of the rather high excitation energies required. The first observation of double excitations in He is due to Silverman and Lassettre [321]. The earliest high resolution spectrum was

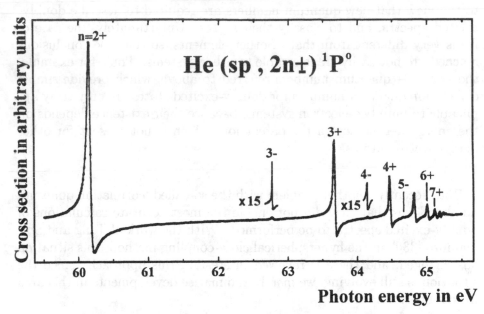

Fig. 7.4.   The doubly-excited spectrum of He: Note in particular the presence of
one strong + and one weak − transition at each of the Rydberg series members
above the leading transition (after M. Domke *et al.*[331]).

obtained by Madden and Codling [322] using synchrotron radiation, and
more recent observations of the same spectrum are shown in fig. 7.4 The
double–excitations in He appear as a Rydberg series of strong autoionising resonances, converging on the $n = 2$ parent state of the hydrogenic
ion: since He$^+$ is hydrogenic, there is no noticeable difference in energy
between the $2s$ and $2p$ limits. Two series occur in this spectrum, one of
which is very intense and gives rise to broad resonances, while the other
is very weak and gives only very narrow resonances. The explanation
is as follows [323]: in view of the degeneracy just mentioned, the true
doubly-excited states are made up from symmetric and antisymmetric
combinations of the two-electron wavefunctions, written as:

$$\left.\begin{array}{l} \Psi_+ = \psi(2snp) + \psi(2pns) \\[2mm] \Psi_- = \psi(2snp) - \psi(2pns) \end{array}\right\} \tag{7.1}$$

so that, in effect, the real or correlated quantum numbers are $n+$ and $n-$
rather $2snp$ or $2pns$. Experimentally, this is revealed by the fact that,
although the selection rules allow both series to occur, one is so markedly
different from the other, in terms of both its intensity and its width. This
points clearly towards the different symmetries involved.

Such observations led Macek [324], Fano [327], Herrick [328] and others to the view that new quantum numbers are required to describe doubly-excited spectra, and to classify their states. Unfortunately, the case of He is very different from that of other elements, so that the conclusions reached are not easily transferrable to other systems. Thus, for example, the + and − quantum numbers referred to above, which provide simple correlation quantum numbers for doubly-excited states are only truly applicable to pure two-electron systems, because their existence depends on the energy degeneracy in the parent ion, which is not present for other many-electron atoms.

Until recently, another problem with the so-called 'correlation quantum numbers' was that they did not enable the most accurate calculations of doubly-excited spectra to be performed. With the work of Tang and Shimamura [330] on the hyperspherical close-coupling method, this situation has changed, and there is renewed interest in this approach: since the situation is still evolving, we merely summarise developments in this area in section 7.11.

Observations of the He spectrum at high resolution have been extended right up to the double-ionisation limit [331, 332], and a vast amount of new data has been obtained, as shown in fig. 7.5. Also, partial cross sections and photoelectron angular distributions in the region of interfering autoionising resonances (cf chapter 8) have been measured [332] . The spectra shown in fig. 7.6 illustrate rather beautifully the fact that autoionising resonances possess widths which are independent of the mode of observation, but shapes which depend on the channel via which they are excited or decay.   The calculations of the full doubly-excited spectrum, also shown in fig. 7.5 alongside the experimental spectrum, were performed, not by using methods involving new quantum numbers, but by the eigenchannel **R**-matrix approach, a method which combines **R**-matrix techniques (see chapter 8) with the methods of MQDT (see chapter 3), and which also proves successful for the calculation of many spectra other than that of He.

Even more recently, the experimental resolution has been still further enhanced, and both partial cross sections and photoelectron angular distributions have been determined [332]. The data shown provide fine examples of interfering autoionising resonances (see chapter 8) and have been analysed in remarkable detail using the hyperspherical close-coupling method: they represent a critical test of the dynamics of double excitation in He.

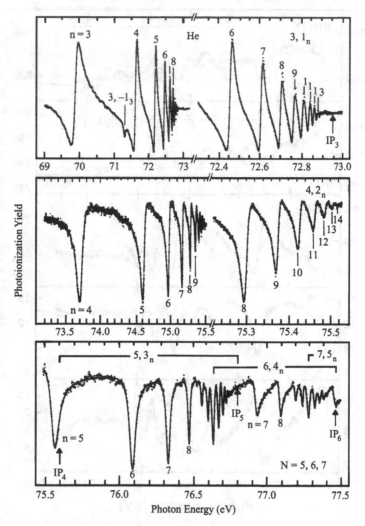

Fig. 7.5. The double–excitation spectrum of He as the double-ionisation threshold is approached, showing a whole family of Rydberg series each one associated with a different parent state of the ion (after M. Domke *et al.* [331]).

7.11 The hyperspherical coordinate method

In the present section, we give brief references to the hyperspherical coordinate method as described by Macek [324] and Fano [327] for the description of doubly excited states and the mechanism which drive double-excitations. A good introduction to the method is given by Fano [327].

The basic idea is a very simple one. Instead of describing doubly-excited states by 'patching up' the independent electron description, we start

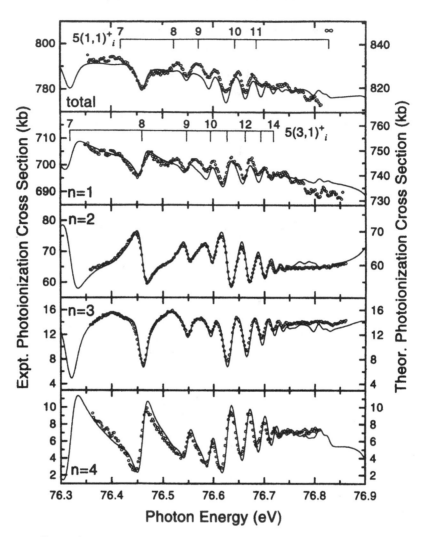

Fig. 7.6. Partial cross sections in the doubly-excited spectrum of He. The total cross section is shown in the top spectrum, and the partial cross sections in the lower spectra. Note how, although the energies and widths of the resonances are the same, their shapes depend markedly on the deexcitation channel (after A. Menzel *et al.* [332]).

from scratch and try to devise a more appropriate system of coordinates to describe the dynamics of the doubly-excited system. Clearly, both electrons should be treated on an equal footing, and both radial and angular coordinates must be selected which involve both of them. For the radial coordinates, one selects $R = [r_1^2 + r_2^2]^{1/2}$, which measures the 'size' of the electron pair. A hyperangle $\alpha = \tan^{-1}(r_2/r_1)$ describes the degree of radial electron–electron correlation, and one thus has two independent variables constructed from r_1 and r_2. There are also further angular quantum numbers which we omit for simplicity.

Hyperspherical methods have the merit of providing a dynamical picture of double excitation and double escape for which the central field approximation is inappropriate. Initially, very accurate calculations were not achieved in this way, and so the hyperspherical method was mainly used as a framework to understand the results obtained by other methods. This situation was transformed by the work of Tang and Shimamura [330] who have performed the most detailed calculations to date on systems with two electrons.

The radial equation in hyperspherical coordinates (we might call it the hyperradial equation) becomes

$$\left\{ \frac{d^2}{dR^2} + 2E - U \right\} (R^{5/2} \sin \alpha \cos \alpha \Psi) = 0 \qquad (7.2)$$

where U, which depends on the angular coordinates, is an effective radial potential and determines radial correlations for each value of R. Just like the normal effective radial potential for the one-electron equation, it consists of a sum of electrostatic and centrifugal terms.

In principle, the hyperspherical method corresponds rather well to the strategy advocated by Langmuir (see [308] and section 7.5): namely that one should seek to quantise the motion of more than just one electron in a many-electron system. The coordinates R and α describe the combined motion of an electron pair, and so the quantum numbers which arise in the solution are radial correlation quantum numbers.

Macek [324] and Lin [326] have solved this equation approximately by different routes, both of which involve an assumption of approximate separability or adiabatic motion (see below). One obtains potential curves rather similar to molecular potential curves for diatomic molecules. Indeed, a similar approach has been followed by Feagin [325] and coworkers, who define their coordinates by analogy with the molecular problem.

The hyperspherical method rests on the approximate separability of the motion in the hyperradius R from motion in the hyperangle α (the so-called adiabatic approximation), which provides an intuitive basis for understanding the dynamics. What Tang and Shimamura have done is to extend the method beyond the adiabatic approximation, by coupling the adiabatic states together using an inner $R < R_{HS}$ and an outer $R > R_{HS}$ region of hyperspherical space, much as the space around the atom is partitioned in the scattering method (chapter 8) or in QDT (chapter 3). The matching condition is then obtained by noting that, in the asymptotic region, as R becomes large, the conventional independent electron picture is more appropriate than the hyperspherical picture. Thus, the external wavefunction is expanded in terms of hydrogenic states. Matching at the boundary $R = R_{HS}$ allows a reaction or **K**-matrix to be defined (cf chapter 8) as in conventional scattering theory. This approach is called the hyperspherical close-coupling method.

Within the hyperspherical method, new quantum numbers K, T and A are introduced to describe two-electron correlations. Both K and T are angular correlation numbers (omitted here for simplicity, see [333]), while $A = 0, \pm 1$ is a radial quantum number, often written as $0,+,-$ because it is related to the $+$ and $-$ classification of Cooper, Fano and Pratts [323] described in section 7.10. Another quantum number which is often used is $v = n - 1 - K - T$, where n is the principal quantum number. The number v turns out to be the vibrational quantum number of the three-body system, or the number of nodes contained between the position vectors $\mathbf{r}_1$ and $\mathbf{r}_2$ of the two electrons [334].

One of the examples of greater understanding of the dynamics achieved within the hyperspherical scheme is embodied in a *propensity rule* (cf section 4.2, proposed by Sadeghpour and Greene [329] which states that $\Delta v = 0$ for the strongest transitions. This is, in fact, only a propensity rule and not a selection rule, as discussed by Tang and Shimamura [330].

The hyperspherical close-coupling method is well adapted to atoms with two external electrons, but should also be capable of treating heavier systems such as the alkaline earths, in which the singly- and doubly-excited spectra interact strongly. So far, not much progress has been accomplished in extending it to systems in which interactions with the valence- or inner-shell spectra become important.

In the simplified summary given here, only the radial part of the problem has been mentioned. Clearly, there are also angular equations and angular correlation quantum numbers to consider. Another, related, approach is to use group theoretical methods to classify doubly-excited states, and this has been pursued mainly by Herrick [328].

7.12   Double excitations in negative ions

Before leaving systems with two electrons and moving on to more complex cases, there is one further interesting situation to consider, namely the case of negative ions with two electrons, beginning with H^-. Negative ions were introduced in chapter 2. The reason why this problem is interesting is that the negative ion has no asymptotic Coulomb potential as one electron is removed. This means that the associated Rydberg series disappear and are no longer mixed with the doubly-excited states. Consequently, one has in a sense 'purer' double excitations. Again, one finds $+$ and $-$ states: the

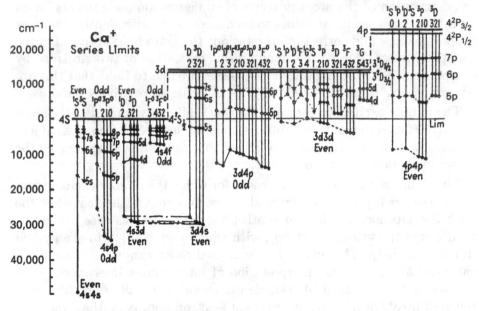

Fig. 7.7. Partial energy level diagram for Ca, showing the relative positions of the singly-excited high Rydberg states and the double excitations near the first ionisation potential (after H.E. White [26]).

+ states are strong and the − states are weak or invisible. The double excitations appear as window resonances, and can be characterised by reference to hyperspherical plots of the wavefunctions and potentials.

Experimental data on H⁻ were reported by Harris *et al.* [335]. There are calculations by a number of authors, including Sadeghpour and Greene [329] and Tang *et al.* [336]. It has also been shown that negative alkali ions possess similar spectra, interpretable along corresponding lines [337].

7.13   Original studies of double excitations

Although He yields the nicest example of a regular, unperturbed doubly-excited manifold, it is, as stated earlier, somewhat untypical, because He is the only element in which single and double excitations are very widely separated in energy from each other. For this reason also, double excitations were actually discovered a long time before the case of He was known. The first example was found in the alkaline-earth spectra, most notably the spectrum of Ca, for which an energy level diagram is given in fig. 7.7. Double excitations appear as intruders within the same energy range as single excitations. The nature of these intruders was at first mysterious [338]. Interestingly, it was Bohr, although he had originated the description of the atom in terms of configuration assignments based on attributing specific $n\ell$ values to each electron independently, who now proposed a new and correct interpretation: the extra lines observed in the Ca spectrum are due to the simultaneous excitation of two electrons by one photon. So many important discoveries are due to Bohr that this important contribution to atomic physics [339] passes relatively unnoticed.

The doubly-excited spectrum of Ca is actually very typical of the cases which are found in many-electron spectra, and has been a frequent subject of investigation. Many further details of this interesting spectrum will be found in chapters 8 and 10.

The significant point, and the reason for which the doubly-excited spectrum was readily observed even at an early date, is the fact that the double–excitations lie close to another spectrum (in this case, single excitations of the valence electron), with which they overlap in energy and interact strongly. This interaction was used as an example of the application of MQDT to the interpretation of interchannel interactions, and is discussed in section 6.10. Double excitations generally owe their existence to breakdown of the independent electron approximation, and thus become particularly prominent when an energy degeneracy with the allowed singly-excited spectrum occurs. It is often stated that they 'borrow oscillator strength' from the singly-excited spectrum, which is an apt description as a number of other examples in section 7.14 will demonstrate.

7.14   Some examples of quasiparticle breakdown

For all atoms with inner shells (i.e. all atoms heavier than He) the extraction of an inner-shell electron requires sufficient energy for double excitations to lie fairly close to the inner-shell spectrum. Double excitations need not involve two valence electrons, as in the spectra of rare earths mentioned above: they can also involve one deep inner-shell excitation, and the excitation of a valence electron, in which case the doubly-excited spectrum lies a little above the associated inner-shell spectrum [340].

Depending on the energy spacings, single and double excitations may also overlap in energy within the same spectrum. When this happens, a great deal of information is gleaned from the interactions between different types of excited configurations. For this reason, it would be artificial to discuss examples of inner-shell and doubly-excited configurations separately. We give here some examples of spectra which demonstrate the interplay between the two modes of excitation. This is a very frequent occurrence, and the present description is far from being exhaustive: it is merely intended to provide examples.

7.14.1 Zn, Cd and Hg

The subvalence shell spectra of Zn, Cd and Hg form a group with very similar characteristics. They were first investigated by Beutler [341]. Later researchers pointed out the need for a special coupling scheme (called J_cK or pair coupling [342]) and investigated them at much higher resolution. The experiments have been repeated using synchrotron radiation [343].

These spectra have already been used in section 2.12 as examples of the extended alkali model. They correspond to the excitation scheme $d^{10}s^2\,^1S_0 \to d^9s^2np, nf(J = 1)$, where s^2 are the valence electrons. Double excitations have also been investigated, especially in Zn [344] and are very significantly enhanced as they approach an inner-shell excited transition. This shows that final state mixing is the dominant mechanism for double excitation.

Because the inner-shell excitation to d^9s^2 always lies below the double-ionisation threshold in these elements (and is of the same parity as the ground state of the ion), it is not broadened. Very long and regular Rydberg series with autoionisation rates decreasing as $1/n^{*3}$ are observed. One further notes that the strong centrifugal barrier effect in the nf excitation channel holds the excited state wavefunctions well outside the core, resulting in very narrow widths and low excitation probabilities as compared with np lines of similar energy. It has been shown [345] that the nf states can be selectively excited as compared with np by a stepwise multiphoton excitation process in which atoms are first prepared in a Rydberg state, with the excited electron outside the centrifugal barrier.

In short, we can say that these atoms exhibit fairly regular subvalence shell spectra, with properties which are well accounted for within the independent particle basis as a first approximation, and double excitations which do not intrude too heavily in the main inner-shell spectrum.

7.14.2 Tℓ

The next element to consider is Tℓ, which is unique and differs from Ga and In from the same column of the periodic table, because its $5d$ subvalence spectrum lies below but within reach of the photo double-ionisation threshold. The $5d$ inner-shell spectrum begins in a fairly simple way at low excitation energies [346], and contains some long, fairly regular Rydberg series. This seems to suggest that the whole inner-shell spectrum will do the same, and contain Rydberg series to allowed limits from the parent ion. However, this turns out not to be true. At higher energy [347], as the photo double-ionisation threshold is approached, the Rydberg series veer off their expected course and terminate as series to quite different limits, which are only allowed if there is double excitation, and which grow very fast in number as the double-ionisation limit is approached. Thus, one finds a very complex doubly excited spectrum (a forest of lines) overlapping in energy with the subvalence spectrum, while the uppermost inner-shell series tend to be lost. We may roughly describe this situation in the following terms: the double excitations acquire oscillator strength at the expense of nearby single excitations from the inner shell.

In addition to this important effect, the inner-shell excitations from the $6s$ and $5d$ subshells also overlap in energy with each other, giving rise to prominent interchannel coupling and examples of the q reversal effect which will be discussed in chapter 8. The $5d$ spectrum of Tℓ is thus unexpectedly rich and interesting.

7.14.3   Ga, In and Pb

The outermost d subshell spectra of Ga [348], In [349] and Pb[350] all have the characteristic that they straddle the double ionisation thresholds. As a result, many of the inner-shell transitions are quenched, while the probability of photo double-ionisation is enhanced. This situation has been discussed [351] in terms of diagrams such as those of fig. 7.8 in which the ionisation potentials for double-ionisation and single-ionisation from an inner-shell threshold are plotted as a function of atomic number. One looks for crossing points, where mixing between the two becomes particularly strong.

7.14.4   Ge and Sn

The final example in this set is the pair of elements Ge and Sn [352], for which the outermost d subshell absorption spectra lie *above* the double-ionisation limit. As a result of Auger broadening of the parent ion core, very few Rydberg members are observed. As already noted in section 6.8, series become rather short when the parent ion state (the core hole) which serves as the series limit is broadened by Auger processes. The resonances arising by inner-shell excitation become very diffuse, and little can be done by way of detailed spectroscopy except to observe the leading series members.

7.14.5   Higher multielectron excitations

It will be noticed that the triple-ionisation threshold is also plotted in fig. 7.8 for completeness. To date, however, triple ionisation has not been found to quench the single-electron spectrum. Likewise, observed triple excitations [311] are generally very weak and higher multiple excitations can be regarded as very improbable. It seems that double excitation and

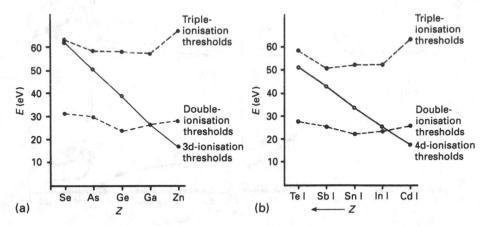

Fig. 7.8. Plots of single-ionisation thresholds from the outermost d subshells and double- and triple-ionisation thresholds from the valence and subvalence shells for different elements in: (a) the first; and (b) the second transition periods (after J.-P. Connerade [351]).

ionisation are much more likely to occur as primary processes. This is further evidence in favour of the Wannier model [320], in which double escape is thought of as a dynamical mechanism in which two electrons tend to be ejected from opposite directions: setting up a stable process in which three electrons escape in a triangular geometry is presumably much more difficult to achieve.

What has been seen in the presence of heavy mixing of the parent ion is that discrete levels can be excited which are *formally* three electron jumps, and also that a significant rate of triple ionisation occurs in the vicinity of a strong resonance. Examples of the latter are shown in fig. 7.9: such occurrences are comparatively rare. While they can be construed as arising from a cascade of deexcitation processes after the primary excitation [354], it is significant [355] that they always seem to occur under conditions where many-body effects are large: one witnesses an evolution in the shape of giant resonances with increasing atomic number for the lanthanide sequence. At the onset of the sequence (towards Xe and Ba), the profiles are as described in section 5.21, and the RPAE method indicates that collective effects are large. At the end of the sequence (towards Yb), the profiles turn into autoionising shapes and collective effects

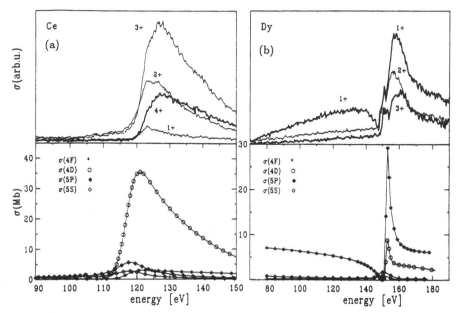

Fig. 7.9. Multiple ionisation in the energy range of the giant resonances of lanthanides, showing the different behaviour in different parts of the lanthanide sequence: (a) before orbital collapse, where multiple ionisation dominates; and (b) after collapse, where single ionisation becomes the dominant process (after P. Zimmermann [353]).

decrease. Similarly, multiple ionisation is most pronounced early in the sequence, and falls off in relative magnitude towards its end.

7.14.6 Orbital collapse and doubly-excited states

There is an interesting connection between orbital collapse and doubly-excited states: as already noted, double excitations acquire oscillator strength from the single-excitation spectrum when they come close together in energy. One of the processes by which this can happen is as follows. Consider excitation from the p^6 shell of Cs, which lies close to the onset of the filling of the $5d$ sequence. When an electron is excited from the inner–shell to an external orbit, screening due to the inner–shell is reduced by one charge, and so the spectrum resembles that of the next element in the periodic table.

Since orbital collapse for unexcited atoms begins before the onset of the long periods, the implication is that the $5d$ orbital, which in the presence of an unexcited core would lie well outside the $6s$ orbital (which is filled), moves inwards to a radius closer to that in Ba once the core is excited. Similarly, if the $5p^6$ subshell of Ba is excited, the $5d$ orbital moves inwards to occupy a similar position to that in La, where $5d$ filling (which is internal) begins. This occurs because of $5d$ orbital collapse, which is a centrifugal barrier effect (chapter 5) and therefore does not occur for $6s$ electrons. Consequently, the $5d$ and $6s$ electrons begin to overlap spatially. Thus, in Cs, the excitation of $5p^5 6sn\ell$ states cannot be separated from that of $5p^5 5dn\ell$ states of the same J, since they are of the same parity, close in energy and overlap spatially as well. Thus, in addition to the normal, inner-shell Rydberg series, one finds additional series, which would not be allowed by one-electron dipole selection rules, and which converge on additional parent ion excited states [356].

These extra, doubly-excited series, are said to arise by parent ion mixing. They are but one example of how centrifugal barrier effects can favour and modify double-excitation spectra [357]

7.15 Two-step autoionisation

The most dramatic of the parent ion mixing effects in $5p$ excited spectra undoubtedly occurs in Ba: instead of series converging to just two limits (the $5p^5 6s^2 \, ^2P_{1/2,3/2}$ states of the parent ion), series to over 14 limits have been detected experimentally [358], and a fully relativistic treatment of configuration mixing was necessary to describe the breakdown of the core hole [359]. In this particular spectrum, it can be said that the parent ion states are very thoroughly mixed and have completely lost their identity.

In the presence of this breakdown, a dramatic enhancement of double ionisation occurs [360]. Up to 2.5 times as many Ba^{2+} as Ba^+ ions can be produced by vacuum ultraviolet irradiation. It was advanced by Hansen [361] that this might come about from a resonant two-step autoionisation process, with the second step driven by a spin flip of the core. Subsequent investigations established that it is indeed a two-step process, but that the driving mechanism actually involves the configuration mixing process described above [363]. This is an example of how resonant multiple ionisation is enhanced by configuration mixing, or driven by the breakdown of the independent electron approximation. A schematic description of the two-step autoionisation process is presented in fig. 7.10. Two-step autoionisation is now a well-established atomic process, and its effects have been confirmed for a wide range of elements [365, 366, 367, 368, 369].

Uncovering two-step autoionisation involved a combination of techniques: the photoabsorption spectrum of Ba was obtained using synchrotron radiation and classical spectroscopy of high dispersion [358]. The analysis was performed using the multiconfigurational Dirac–Fock method, which allowed the limits to be identified [359]. The double-ionisation spectrum was obtained by photoion spectroscopy using syn-

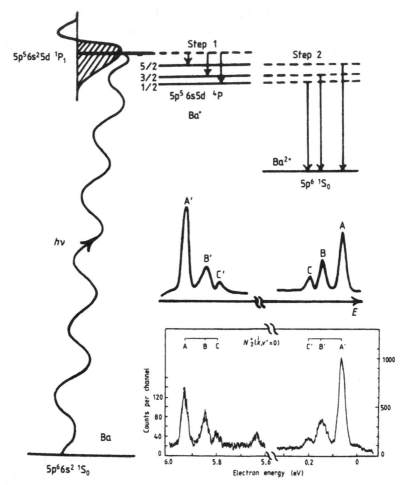

Fig. 7.10. The two-step autoionisation process in the $5p$ spectrum of Ba, and the identification of the intermediate states involved (after J.-P. Connerade [362]).

chrotron radiation [364]. A theoretical mechanism was then proposed [361], and the intermediate state was identified [363], which allowed the two-step process to be finally unravelled.

Two-step autoionisation may be regarded as the lowest energy Auger effect: it sets in as one crosses the double-ionisation threshold towards higher energies.

7.16 Conclusion

In this chapter, we have described many features of subvalence excitation spectra. One aspect, however, has been left aside for discussion in a later chapter. This corresponds to an issue Bohr and Langmuir were both unaware of when the question of quantising groups of electrons first arose [308]: it is a feature of the three-body problem that it is nonlinear. The classical three-body problem, at least in principle, possesses a chaotic regime. It is not, therefore, completely clear whether the usual methods of elementary quantum mechanics describe the quantisation of correlated states completely: by quantising each electron in a many-electron system separately, one reduces the N-electron problem to N one-electron problems, each one of which is 'well behaved', i.e. regular in the semiclassical limit. Thus, in the mean-field approximation, questions concerning 'quantum chaos' never arise. Whether any experimental evidence exists that this approach might fail is a question to which we return in chapter 10.

8
K-matrix theory of
autoionising resonances

8.1   Introduction

The present chapter provides a summary of the basic principles of Wigner scattering or **K**-matrix theory, followed by examples of its application to atomic spectra, and more specifically to the study of interacting autoionising resonances, for which it happens to provide a very suitable analytic framework, within which most of the important effects can be illustrated rather simply. We concentrate on an elementary account of basic principles rather than on the most complete algebraic formulation, because the theory in its full generality becomes rather forbidding. Thus, when only a small number of channels needs to be included in order to illustrate an effect, suitable references are indicated, where the reader can find a fuller treatment. We also make the fullest possible use of analytic methods, which allow one to pick out a number of significant effects without detailed numerical computations: this turns out, rather remarkably, to be possible only for atoms, and this is a consequence of the asymptotic Coulomb potential.

Atoms therefore provide an excellent testing ground for the details of Wigner's theory. Wigner's [370] **S**-matrix theory postulates the existence of a Schrödinger-type equation, but actually requires no explicit knowledge of its solutions. In this sense, it is regarded as the most general formulation of scattering theory (and is more general than MQDT). One can even handle photon decay channels, although no explicit wavefunction can be written for photons. They appear in scattering theory as weakly-coupled radiative channels, and examples will be given in the present chapter.

Wigner's scattering theory revolves around three matrices which couple the incoming and outgoing channels, *viz.*

(i) the **S**-matrix,

(ii) the **R**-matrix,

(iii) the **K**-matrix.

The first part of the chapter contains a brief summary of Wigner's scattering theory, presented so as to emphasise the underlying similarity with the closely related approach of MQDT (chapter 3). This is followed by a discussion of the properties of **S**-, **R**- and **K**-matrices, in which we give the motivation for choosing one or the other, depending on the application in hand. Finally, we turn to some explicit applications of **K**-matrix theory to cases of interacting resonances in atomic physics.

8.2 Scattering theory and MQDT

Scattering theory has very general validity. It is based on the existence of a Schrödinger-type equation and is consistent with causality requirements through the Kramers–Kronig dispersion relations. It is widely used outside atomic physics for the description of all resonance phenomena. There is therefore some advantage in using it in atomic physics: it ensures unification and is more transparent to reseachers from other fields.

In scattering theory, a projectile is imagined to move towards a target from the remote past $(t \to -\infty)$ and from a large distance, where projectile and target are so far separated that the interactions between them tend to zero and the total wavefunction tends to a product of the projectile wavepacket times the basis state of the target, in some definite state, say a. As time increases, an interaction takes place between the projectile and the target and, as $t \to +\infty$, the interaction again tends to zero, leaving the system in another definite state b of the target. Clearly, a and b must both be states for which the system contains a projectile at an infinite separation from the target. These are referred to as *open channel* or continuum states. During a transient interaction, represented by an operator $U(t_1, t_2)$, other states, referred to as *closed channels* or discrete bound states, can become temporarily involved. The closed channels are responsible for the occurrence of resonances. The elements of the scattering or **S**-matrix yield the proportion of b which emerges from a as a result of the interaction

$$S_{ba} = < \psi_b \mid U(t_1, t_2) \mid \psi_a > \qquad (8.1)$$

It is useful to express photoabsorption processes in terms of phase shifts, and to partition the total phase shift Δ into one part which is due to the resonances (Δ_0, say) and one part (say δ) which is the phase shift of the background continuum:

$$\Delta = \Delta_0 + \delta \qquad (8.2)$$

The structure of MQDT involves a compact target state within a radius r_0 and an asymptotic solution, valid at large r, which is joined to the wavefunction of the inner region in order to determine amplitudes and phases. Thus, MQDT is simply a special form of scattering theory, specially adapted to handle bound Rydberg states of atoms and adjacent structure in the base of the ionisation continuum.

One therefore expects a close relationship if not a complete correspondence betweem MQDT and scattering theory. All the interactions involved in atomic physics are expressible in terms of matrix elements, which is the language of MQDT. The equivalence between the two theories was investigated in detail by Lane [373], to whom much of the development of scattering theory and its original application in atomic physics are due [371]. Lane formally proved that MQDT can be derived from Wigner's scattering theory as a special case. Explicit connecting formulae will be found in references [373, 381, 393].[1]

In principle, the correspondence between the two theories is not complete, because scattering theory is the more general formulation. For our purposes, however, the fact that the applications to atomic physics obtained by both methods are quite consistent with each other is an important and useful conclusion. The same result and connections have been obtained independently by Komninos and Nicolaides [378]. Both [373] and [378] noted that the derivation of MQDT from Wigner's scattering theory establishes its basic structure and theorems without special assumptions about the asymptotic forms of wavefunctions. The approach of Komninos and Nicolaides [378] is designed for applications involving Hartree–Fock and multiconfigurational Hartree–Fock bases. In the present exposition, we follow the approach and notation of Lane [379] and others [380, 381], who exploit the analytic **K**-matrix formalism and include photon widths explicitly when interferences occur.

A specific advantage of using the **K**-matrix approach in the present context is that the transition to a numerical treatment of the equations can be postponed until quite a late stage, thus allowing the full algebraic structure of the interacting resonances to be displayed.

8.3   Two well-separated particles

In scattering theory, one seeks to relate two well-defined situations, involving a scattered particle and a particle (or assembly of particles) which

---

[1] For readers interested in the cross connections between various theories, we note that the relation between **K**-matrix and Fano notations is also given explicitly in appendices to [377, 380].

acts as the target. In one situation, the scattered particle and the target are well separated from each other. In another, the scattered particle and the target interact strongly.

The wavefunction for two well-separated particles has the general form:

$$r\psi_E = \exp(-ikr) - S_E \exp(ikr) \tag{8.3}$$

where r is the distance between the particles; $k = Mu^2/\hbar$, with u the relative velocity; M is the relative mass; and both the incoming and outgoing waves are normalised to unit flux, so the conservation of particles requires that $|S_E| = 1$. This is the so-called unitarity condition, which is quite fundamental to scattering theory. Indeed, the whole of the theory can be regarded as an application of the unitarity constraint to excitation and deexcitation processes.

It is convenient to write

$$\{\exp(-ikr) - S_E \exp(ikr)\} = -2i\,(\exp i\delta)\sin(kr + \delta) \tag{8.4}$$

then

$$S_E = \exp 2i\delta \tag{8.5}$$

where S_E depends on the energy with which the two particles collide. In what follows, we also write S instead of S_E for brevity.

8.4   Two particles in a compound state

The other situation we need to consider is the one in which the scattered particle and the target interact strongly, i.e. come close together. They then form a single or compound state. For such a state, by considering a first approximation for that part of configuration space in which both particles are close together, we write:

$$\psi_E = \alpha_E \psi \tag{8.6}$$

where ψ is a normalised wavefunction involving all the coordinates of the compound state, and is independent of E (in other words, the *nature* of the compound state does not depend on the *manner* in which it is excited) and α_E is an amplitude independent of the variables of the compound state, but which does depend on the energy E. It is useful to introduce a complete set of internal states of the target χ_λ, in terms of which we can then expand

$$r\psi_E = \sum_\lambda \alpha_\lambda \chi_\lambda \tag{8.7}$$

in cases where the target has several internal states.

8.5   Wigner's interaction sphere

We now introduce a boundary surface over which the joining conditions can be applied. This is done in a manner quite analogous to the definition of the sphere of radius r_0 in MQDT (cf chapter 3).

Take the origin of coordinates at the point where all the particles coalesce. We then define a sphere centred at the origin, whose radius is as small as possible, consistent with the requirement that ψ_E at the boundary has attained its well-separated form. Let a be the radius of this sphere.

In nuclear physics, because of the nature of the short range interactions involved, this boundary has a definite physical identity. In atomic physics, there is no clear physical value for a, but one introduces the sphere for computational convenience (see, e.g., [371] for further discussion of this point), and it is generally advantageous to define as small a radius a as possible for the problem in hand. We call a the *channel radius*.

For two energies E_1 and E_2, neglecting any small differences in the particle interactions, the joining condition is achieved by applying Green's theorem on the sphere for kinetic energy terms. One obtains

$$-\frac{\hbar^2}{2M}\int_{\substack{surface \\ of sphere}} (\psi_{E_2}^\star \nabla \psi_{E_1} - \psi_{E_1}^\star \nabla \psi_{E_2}) \cdot d\mathbf{S} = (E_1 - E_2)\int_{\substack{volume \\ of sphere}} \psi_{E_1} \psi_{E_2}^\star dV$$

$$(8.8)$$

As discussed by Wigner [370], the left hand side of this equation does not vanish, because of the higher-order dependence of ψ on E.

8.6   Wigner's method of relating S and α

We substitute (8.3) in the left hand side of (8.8) and (8.6) on the right hand side, neglecting the small region where the substitution does not hold, then, for small ka, we find

$$\frac{u_1}{u_2}(1 + S_1)(1 - S_2^\star) + \frac{u_2}{u_1}(1 - S_1)(1 + S_2^\star) = \frac{2}{i\hbar}(E_1 - E_2)\alpha_1\alpha_2^\star \quad (8.9)$$

which, for $E_2 = E_1$, recovers $SS^\star = 1$. Equations (8.9) are ∞^2 relations between twice infinite quantities S and α. Wigner [370] solves the system by introducing

$$S = \frac{1 + ikaR}{1 - ikaR} \tag{8.10}$$

whence

$$kaR = i\frac{1 - S}{1 + S} \tag{8.11}$$

Thus $\mathbf{R}$, like $\mathbf{S}$ to which it is directly related, is a function of energy E and relative velocity u. Since the absolute value of $\mathbf{S}$ is 1, $\mathbf{R}$ is real.

Through (8.3) and (8.4), kaR is the tangent of the phase shift δ, whence

$$\sin^2 \delta = \frac{(kaR)^2}{1 + (kaR)^2} \tag{8.12}$$

and the scattering cross section becomes

$$\sigma(E) = \frac{4\pi(aR)^2}{1 + (kaR)^2}$$

So aR is a length, which turns out to be related to the scattering radius of the colliding particles.

The solution of (8.9) is then

$$- R_2 + R_1 = \frac{2}{\hbar}(E_1 - E_2)\frac{\alpha_1}{u_1(1 + S_1)}\frac{\alpha_2^*}{(1 + S_2^*)} \tag{8.13}$$

8.7 Wigner's resonance scattering formula

To solve (8.13), Wigner [370] notes that the left hand side is real; he sets

$$i\xi_1 = \alpha_1/u_1(1 + S_1)$$

and is able to show that

$$\xi_1 = \frac{\frac{1}{2}\hbar\xi}{E_1 - E_0} \tag{8.14}$$

where ξ and E_0 are both real and independent of energy. Then

$$R_1 = \frac{\frac{1}{2}\hbar\xi^2}{E_1 - E_0} + R_\infty \tag{8.15}$$

where R_∞ is independent of E_1. The cross section is then

$$\sigma(E) = \frac{4\pi}{k^2}\sin^2 \delta = \frac{4\pi}{k^2}\frac{(kaR)^2}{1 + (kaR)^2}$$

Let $\Gamma = \hbar ka\xi^2 / \{1 + (kaR_\infty)^2\}$ then

$$\sigma(E) = \frac{\pi}{k^2}\left\{\frac{4(kaR_\infty)^2}{1 + (kaR_\infty)^2} + \frac{\Gamma^2 - 4\Gamma ka\dfrac{E - E_0}{1 + (kaR_\infty)^2}}{\dfrac{\Gamma^2}{4} + (E - E_0 - \frac{\Gamma}{2}kaR_\infty)^2}\right\} \tag{8.16}$$

where the first term in braces represents the potential scattering or background continuum, and the second, the sum of resonance scattering and interference between resonance and potential scattering.

If $R_\infty = 0$, there is no potential scattering, $\Gamma = \hbar k a \xi^2$ and

$$\sigma(E) = \frac{\pi}{k^2} \frac{\Gamma^2}{\frac{\Gamma^2}{4} + (E - E_0)^2} \tag{8.17}$$

is simply a symmetric profile as expected.

The quantity $\xi^{-2} = kat$ is interpreted as the time taken for the particles separating from the origin to appear at the surface $r = a$.

8.8   The R-matrix for many resonances

Consider the simplest case, with zero angular momentum. We have

$$r\psi = \exp(-ikr) - S \exp(ikr)$$

and we match the logarithmic derivatives at the boundary a, looking first outwards and then inwards, to relate **S** to the internal states.

8.8.1   Outwards construction

The dimensionless logarithmic derivative at the boundary a is given by

$$f \equiv a \left.\frac{r\psi'}{r\psi}\right|_a = -ika \frac{\exp(-2ika) + S}{\exp(-2ika) - S} \tag{8.18}$$

whence

$$S = \exp(-2ika) \left\{ \frac{f + ika}{f - ika} \right\} \tag{8.19}$$

the phase factor $\exp(-2ika)$ which does not appear in (8.10) is unimportant, since $\mathbf{SS}^\dagger = 1$ and ka is small (see (8.9)); it arises from the difference between barred and unbarred quantities discussed in detail in [370] and is of no practical consequence.

8.8.2   Inwards construction

Use the complete set of internal states in (8.7), so that

$$r\psi = \sum_\lambda \alpha_\lambda \chi_\lambda$$

and impose a general boundary condition at the radius a

$$a \left.\frac{\chi_\lambda'}{\chi_\lambda}\right|_a = b \tag{8.20}$$

where b, for the moment, is a free parameter.

8.8.3 Applying the matching condition

As before make use of Green's theorem over the sphere a:

$$\int_a (r\psi)H\chi_\lambda - \chi H(r\psi)dV = (E_\lambda - E)\alpha_\lambda \qquad \text{by eigenvalue equation}$$

$$= \frac{\hbar^2}{2m}\{\chi_\lambda(r\psi)' - (r\psi)\chi_\lambda'\}_a \quad \text{by surface integral}$$

$$= \frac{\hbar^2}{2m}\chi_\lambda(a)(r\psi)_a \left\{ \frac{(r\psi)'}{(r\psi)} - \frac{\chi_\lambda'}{\chi_\lambda} \right\}_a$$

$$= \frac{\hbar^2}{2ma}\chi_\lambda(a)(r\psi)_a (f - b)$$

so

$$\sum_\lambda (E_\lambda - E)\alpha_\lambda = \frac{\hbar^2}{2ma}(f - b)(r\psi)_a \sum_\lambda \chi_\lambda \alpha_\lambda$$

whence

$$1 = \frac{\hbar^2}{2ma}(f - b)\sum_\lambda \frac{\chi_\lambda^2(a)}{E_\lambda - E}$$

We define $\gamma_\lambda^2 \equiv (\hbar^2/2ma)\chi_\lambda^2(a)$ and it turns out that the form of $\mathbf{R}$ is given by

$$\frac{1}{R} = f - b \qquad\qquad (8.21)$$

which yields the canonical or *meromorphic form* of the $\mathbf{R}$-matrix:

$$R = \sum_\lambda \frac{\gamma_\lambda^2}{E_\lambda - E} \qquad\qquad (8.22)$$

8.9 Definition of meromorphy

A function $F(z)$ is said to be meromorphic as a function of the complex variable z in some domain if $F(z)$ is analytic in that domain apart from isolated poles of finite order (see, e.g., [372]).

Meromorphy turns out to be an important mathematical property, especially when resonances overlap in energy: it allows one to connect each physical resonance with a mathematical pole in a one-to-one correspondence even when they interact.

8.10 Dependences on a and b

By construction (equations (8.18) and (8.20)), f is independent of b. Hence, through equation (8.21), it follows that $\mathbf{R}$ does depend on b. Also, through (8.19): it follows that $\mathbf{S}$ does not depend on b. From the initial assumptions, it does not depend on a either. However, $\mathbf{R}$ does depend on both a and b.

8.11 Choice of b

8.11.1 Simple case of one resonance and zero angular momentum

For $\ell = 0$, and dropping the sum over λ, we can write

$$R \approx \frac{\gamma_\lambda^2}{E_\lambda - E}$$

Note that this relation must anyway become valid at small detunings, i.e. sufficiently close to a resonance. Then $\mathbf{S}$ becomes the sum of: (i) a hard sphere or potential scattering term; and (ii) a second term whose denominator is $E_\lambda - E + (b - ika)\gamma_\lambda^2$. If we identify a width $\Gamma_\lambda \equiv ka\gamma_\lambda^2$ to yield the resonance form, and if we choose the shift Δ_λ according to the physical criterion that $E_{res} = E_\lambda$, then we are led to $\Delta_\lambda = 0$ and $b = 0$.

8.11.2 Case of nonzero angular momentum

For $\ell = 0$ one finds

$$S = \exp(-2ika)\frac{f + (U + iV)}{f - (U + iV)}$$

Again, with $R = \gamma_\lambda^2/(E_\lambda - E)$, we find that the denominator in the second term is $E_\lambda - E + [b - (U + iV)]\gamma_\lambda^2$ where

$$U + iV \equiv \left(\frac{G' + iF'}{G + iF}\right)_a ka$$

whence

$$V = \frac{ka}{(F^2 + G^2)_a}$$

One sets $\frac{1}{2}\Gamma_\lambda = V\gamma_\lambda^2$ and $\Delta_\lambda = (b - U)\gamma_\lambda^2$, so that, for $E_{res} = E_\lambda$, we must have $b = U$. In the case $\ell = 0$, $F = \sin ka$ and $G = \cos ka$, but, for $\ell = 0$, one has $F(0) = 0$ and

$$\lim_{r \to \infty} F \sim \sin kr$$

and

$$\lim_{r \to \infty} G \sim \cos kr$$

8.12 Why the K-matrix is introduced

R has the advantage of being real and meromorphic, but the disadvantage that it depends on specific choices of a and b. In reality, this is not always as great a handicap as it may seem. In fact, it may provide useful numerical methods by which **R** can be computed. However, it is a disadvantage if one seeks to establish general properties or theorems, which by their nature should be exact results.

On the other hand, **S** *is* independent of a and b. The problem is that, in general, it does not have a simple pole structure. It is always possible to write **S** as a sum of poles:

$$S = \sum_\lambda \frac{g_\lambda}{E_\lambda - E - \frac{i}{2}\Gamma_\lambda}$$

but, when the levels overlap ($\Gamma_\lambda \geq D_\lambda$, where D_λ is the level spacing), g_λ is complex and bears no simple relation to Γ_λ. Thus, although the **S**-matrix is a rather fundamental quantity, it is not the most suitable with which to analyse interactions between overlapping autoionising resonances.

However, if we define

$$S = \frac{1 + iK}{1 - iK} \tag{8.23}$$

and compare this with the relation between **S** and **R** (for $b = 0$ and $\ell = 0$):

$$S = \exp(-2ika) \left(\frac{1 + ikaR}{1 - ikaR} \right)$$

then it follows that **K** is real and is independent of a and b because it is directly related to **S**. Also

$$K = \tan \delta = \frac{kaR - \tan ka}{1 + kaR \tan ka} \tag{8.24}$$

It is a theorem that, if **R** is a meromorphic function, then the expression $(AR + B)/(CR + D)$, where A, B, C and D are constant is also meromorphic. Hence **K** is real, independent of a and b and meromorphic (i.e. it possesses a simple pole structure).[2]

---

[2] Strictly, **K**, although real, is not completely meromorphic, because it has a branch point at $k = 0$. This, however, is immaterial for the present discussion. Note also that the condition for **K** to be meromorphic in the range of interest requires that $\tan ka$ be at least approximately constant.

In this sense, $\mathbf{K}$ is more fundamental than $\mathbf{R}$. For example, if b is chosen as above to make $\Delta_\lambda = 0$, then all E_λ coincide with the resonance energies in $\mathbf{S}$. It follows that the χ_λ equal the resonance wavefunctions, at least for $r \leq a$. Now consider the dependence of the theory on a, which can normally be omitted since ka is usually small (see, e.g., [370]) and imagine a to be unphysically large. The dependence of $\mathbf{R}$ on a is then simple and explicit: in this case, instead of (8.12),

$$R(a) = \frac{1}{ka} \tan(\delta + ka)$$

for $a \geq a_{min}$. Equivalently, the quantity $K \equiv \tan \delta$, through equation (8.24), displays what the effect of a ridiculously large choice of a would be: while *formally* acceptable, it would make $\tan ka$ oscillate many times within a small range of k, and these oscillations would have to be cancelled by corresponding oscillations of $\mathbf{R}$.

The $\mathbf{K}$-matrix yields the essential part of the cross section as

$$\sigma_{aa} = \frac{4K_{aa}^2}{1 + K_{aa}^2} \tag{8.25}$$

if there is only one open channel, and

$$\sigma_{ab} = \frac{4K_{ab}^2}{\mid (1 - iK_{aa})(1 - iK_{bb}) + K_{ab}^2 \mid^2} \tag{8.26}$$

for two open channels, where $\mathbf{K}$ has especially convenient forms for all of the problems we are interested in here.

8.13   Importance of atomic physics for scattering theory

In addition to the question just raised (see also section 8.5) about the most suitable or natural choice of a, it turns out that there is another important difference between atomic and nuclear physics from the perspective of $\mathbf{K}$-matrix theory: it is a property of the Coulomb field (specifically, the property depends on the fact that Γ_λ/D_λ tends to a constant) that the $\mathbf{K}$-matrix can be expressed in analytic form, so that many of the peculiarities of overlapping resonances can be studied in a very general way for atoms [373, 379, 380, 381]. This serves both as an elegant confirmation of the validity of $\mathbf{K}$-matrix theory and as a new perspective on the subject of interacting resonances in atomic physics: the use of analytic formulae allows one to explore symmetry changes and fluctuations in width for perturbed Rydberg manifolds, and to search for cancellation effects in a rather general way, without any need for explicit calculation of target wavefunctions. Numerical implementations can be postponed until quite a late stage, thus allowing the full algebraic structure of interacting resonances to be investigated.

All the theoretical structure above can be generalised to many channels, which considerably increases the algebraic complexity. For example, equation (14) of [373] is the multichannel generalisation of the present equation (8.24) with $b = 0$ and $\ell = 0$.

8.14 Recapitulation of the properties of S, R and K

(i) The **S**-matrix is the one most directly related to the scattering amplitude, but is not easily connected to specific target states χ_λ since it is independent of a and b. Also, it is unsuitable for studying complicated, e.g. overlapping, resonances, since its pole structure is not simple.

(ii) The **R**-matrix is convenient for numerical applications precisely because it depends on b and is therefore directly related to χ_λ through (8.22). However, because it depends on both b and a, it does not lend itself to a general discussion of the properties of observed resonances.

(iii) Only the **K**-matrix is *both* independent of such nonessential quantities as a and b *and* simple in its pole structure. It therefore provides the most suitable framework within which to discuss the general properties of overlapping resonances, without the need for any specialised knowledge of the target.

8.15 Applications in atomic spectra

Wigner's theory is most useful for studying cases of Rydberg series of interacting features in which an intruder appears. There are several distinct effects in the spectra of interacting resonances which result from perturbations, and are readily described by **K**-matrix theory. We first list them, and then discuss each one in turn.

(i) Changes in shape and in the symmetry of the profiles. These can result in the so-called *q-reversal effect* [382], in which the symmetries of the perturbed states are reversed around some critical value of energy. In an unperturbed Rydberg series of autoionising lines, one finds that the shape index q, as defined from the Fano formula for an autoionising line, is constant, which expresses the fact that the shapes of successive series members are preserved. On the other hand, if the Rydberg series interacts with an intruder state, then **K**-matrix theory predicts a number of interesting situations, all of which are observed. One such case is shown in fig 8.1, where the symmetries of resonances are shown to reverse as the

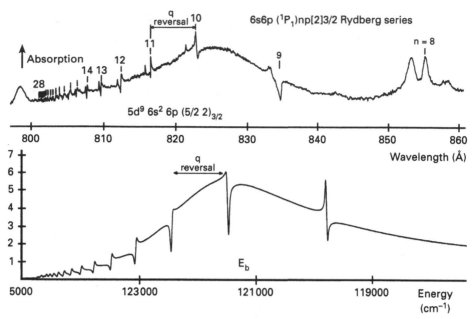

Fig. 8.1. Examples of the *q*-reversal effect: (a) experimental, as observed in the photoabsorption spectrum of Tℓ (note that the *q* reversal, in this case, does not coincide in energy with the maximum in the cross section of the broad perturber); (b) computed, in a rough simulation of the 'skewed *q* reversal effect' based on the simplified equation given in the text (note that this is not a parameter fit, but simply an example – after J.-P. Connerade [382] and J.-P. Connerade and A.M. Lane [381]).

resonance energy of the interloper is traversed. The **K**-matrix equations are able to explain not only the existence of this effect, but also some important details, for example the fact that the critical energy is detuned from the maximum in the cross section of the perturber.

The simplest expression for the cross section $\sigma(\varepsilon)$ in terms of the detuning ε for this case is:

$$\sigma(\varepsilon) = \frac{(\varepsilon + \frac{1}{2}q\Gamma_a)^2}{(\varepsilon + \Sigma)^2 + \left[(\mu_a\Gamma_a^{1/2} + \frac{1}{2}\mu_a^2\varepsilon)\Sigma - \frac{1}{2}\Gamma_a\right]^2} \tag{8.27}$$

where Γ_a is the particle width, q the shape index of the intruder resonance, the sum $\Sigma = \sum_n H_n^2/(E_n - E)$, H_n the coupling strength, E_n, E the resonance and running energies, and $\mu_a \equiv (\Gamma_{na}^{1/2}/H_n)$ is a ratio independent of n. This expression is closely related to the Fano formula, and in particular is skewed from the resonance energy, whence the apparent asymmetry of the *q* reversal. However, it includes a Rydberg manifold of perturbed resonances. See section 8.28 for further details about *q* reversals.

(ii) Fluctuations in the widths of autoionising profiles. Within a channel, the perturbation can induce dramatic changes in the autoionisation widths and, in favourable cases, to a stabilisation of states in the continuum. If a perturbed state appears at the correct energy, a *vanishing particle width* [383] can occur in which one resonance becomes very narrow and resembles a transition between bound states.

Typically, the vanishing particle width occurs at $\varepsilon = -\mu^{-1}\Gamma^{1/2}$ which is related to the width of the broad intruder and the strength of the coupling. This relationship, and the associated effects are discussed in section 8.29.

More generally, one finds many cases where the particle widths of resonances in a Rydberg series, instead of decreasing in proportion to the spacings between successive members, fluctuate dramatically. An example will be given in section 8.33.

(iii) Fluctuations in the radiative lifetimes of bound states. Essentially the same effect as just described may also occur *below* the ionisation threshold, in which case stabilisation against radiative decay can also occur [384]. This is a good example of the need to introduce photon decay channels in the theory: the existence of interferences of this type was originally pointed out by Wigner [370]. Examples from atomic physics are given in section 8.30.

(iv) When the oscillator strength sum rule is exhausted within a given energy range (e.g. if the perturber is a 'giant resonance', then the fine structure appears in the form of *window* resonances (i.e. transmission maxima), which exhibit similar fluctuations of width [375].

(v) Vanishing fluctuations or disappearances of spectral structure may occur: these are due to a vanishing width in one channel in the presence of many open channels. Although the intensity does not go to zero, the fine structure may disappear at a specific energy which is *not* a series limit. This effect is explained in section 8.33.

(vi) The *number* of q reversals within a series and their stability (i.e. whether they travel fast in energy as a function of coupling strength between channels, or whether they remain anchored to the resonance energy of the perturber) can be deduced by the **K**-matrix approach [377]. One can obtain theorems and predict general features of interacting series coupled to a strong interloper. These depend critically on the strength of coupling.

(vii) Because the theory is quite general, the conclusions apply not only to atomic spectra, but also to those of molecules, or indeed of *any* system possessing resonances which can overlap, e.g. atoms in intense electric or magnetic fields.

(viii) In resonantly enhanced multiphoton spectroscopy (REMPI), a phenomenon closely related to q reversals occurs [385]. Indeed, in REMPI it is even easier to produce situations of near-degeneracy in energy which are required for dramatic effects to appear. These are discussed in chapter 9.

(ix) Finally, MQDT was applied, in chapter 3 to Rydberg series of bound states below the first ionisation threshold and to their perturbers, which belong to autoionising series associated with the next higher thresholds. We can, by using **K**-matrix theory, extend the same principles to mutually perturbing series of autoionising resonances. We can show that the principles of two-dimensional Lu–Fano plots (with some care in interpretation) can be extended to autoionising series converging onto distinct thresholds.

8.16   Experimental background

High resolution experiments on interacting resonances in neutral atoms are often conducted in photoabsorption, although thermionic diode detectors (the so-called 'hot wire' systems) and atomic beams are increasingly used. Most of the experiments extend over the vacuum ultraviolet and soft X-ray ranges, above the ionisation potentials of the atoms, but in a range in which instrumental resolution is sufficiently good to allow detailed studies of line profiles. When experiments are conducted with synchrotron radiation below about 1215 Å in wavelength, which is often necessary in view of the high excitation energies involved, there exist no suitable transparent materials for use as windows and a direct vacuum connection between the absorption cell and the accelerator is required. For photoabsorption studies, the pressure differential to be maintained can be as high as seven orders of magnitude between the interior of the cell and the vacuum chamber of the accelerator. Special techniques have been developed for the dynamical containment of atomic vapours in windowless systems, and for the production of vapour columns of even quite refractory elements in high temperature vacuum furnaces (see, e.g., a review on vapour containment techniques [386] for experimental details).

Instead of measuring the attenuation of a beam, one may also count the ions produced with very high efficiency by the use of channelplates or a hot-wire detector [387], an approach which has mainly been applied in laser spectroscopy, where high sensitivity can be achieved by space charge amplification. The principle of the thermionic diode is that the atomic vapour under study is formed within the detector, and a current limited by the space charge is obtained by appropriately biasing a diode, consisting of an external anode (often the outer wall of the vacuum system, formed by a metal tube) and a heated cathode made of a suitable material to emit many electrons (thoriated W is suitable in many cases). A sketch of

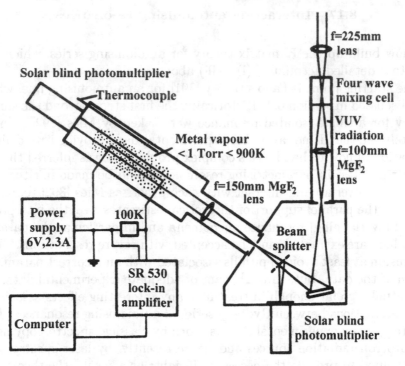

Fig. 8.2. Diagram of a typical thermionic diode arrangement to observe interacting autoionising resonances (after W.G. Kaenders *et al.* [389]).

a typical system is given in fig. 8.2.

The theory of operation of a diode in the space charge limited mode was set up in the early days of vacuum tube technology by Langmuir and is described in standard texts (see, e.g., [388]). When a laser beam is passed through the gas, ions are produced which move slowly as compared with electrons and have the effect of neutralising the space charge. As a result, a large current flows: the ions produced by the laser beam act rather like a grid in a triode valve, and there is a large gain. The technique is described in detail by Niemax [387].

Although most suitable for use with lasers, Thermionic diodes have also been successfully applied to synchrotron radiation studies by using wiggler magnets to enhance the intensity of the beam [390]. Last but not least, one should mention the important category of atomic beam experiments, complemented by the techniques of photoelectron and photoion spectroscopy. All these techniques are suitable for the experimental study of interacting resonances. We turn now to their theoretical description, which will be illustrated by experimental examples.

8.17 Interacting autoionising resonances

We now build up the **K**-matrix theory for autoionising series, which will
lead to a detailed account of (i)–(viii) above.

The starting point is Fano's theory [391] for an autoionising line, which
was discussed in section 6.4. Historically, the first steps beyond the simple
theory for a single isolated resonance were taken by Mies [392]. Unfortunately, at that time, available experimental data provided no definite
indication of how to handle this complex problem. It was inferred that no
further progress in parametrising resonances could be made in situations
where they overlap and interact with each other: as Mies [392] remarked
'without the parallel support of theoretical estimates . . . the interpretation of [overlapping] resonance phenomena and in particular the parameters which are extracted must be accepted with severe reservations.' This
conclusion at least avoided pitfalls associated with an incorrect parametrisation of the data, but, through want of adequate experimental data, did
not actually go far enough in treating many interesting effects associated
with interactions between Rydberg series of autoionising resonances. Fortunately, many experimental studies, both by classical spectroscopy using
synchrotron radiation sources and, more recently, by laser spectroscopy
have helped to provide the necessary insight: as a result, the theory has
been extended.

Having set up the **K**-matrix for an isolated resonance in one 'flat' continuum we build up to further levels of complexity. The first question
one may ask is whether cases occur in which the continuum is not 'flat'
but presents some pronounced modulation. This will lead us to consider
shape or giant resonances. One can also imagine a single resonance 'tuned'
through a broad modulation, a situation clearly equivalent to two resonances of different widths coupled by an interaction. There is, of course,
no reason to stop at two resonances. Indeed, in atomic physics, one usually deals with a full Rydberg manifold comprising an infinite sequence of
resonances which may or may not overlap in energy. It turns out that the
shapes of resonances – their line profiles and symmetries – are markedly
affected by such overlaps and interactions. A study of the lineshapes becomes a powerful method of understanding interchannel interactions, i.e.
the physical processes which couple resonances to each other.

8.18 The isolated autoionising resonance

We use the **K**-matrix formalism, described above, in a form derived for
nuclear scattering by Lane and Thomas [371] and Lane [379, 393]. In the
present context, it provides a convenient and simple description of atomic

resonances in photoabsorption. Expressions for autoionisation in terms of scattering phase shifts were given already in appendix C of Fano's paper ([391]). Later, Shore [394] developed a scattering theory of absorption profiles and refractivity using Wigner's S-matrix formulation. He stresses the interpretation of as a 'half scattering' problem, in which there is no incoming wave, but merely an instantaneously-formed compound state and an outgoing wave.

As noted above (equation (8.2)), one can separate out resonant and non-resonant parts of the phase shift. For a single open channel, the K-matrix reduces to just one element:

$$K = \tan(\delta + \Delta_0) \qquad (8.28)$$

We also begin in the elastic scattering approximation, which has the advantage of being very simple: in this approximation, radiative channels are neglected, and all the observed spectral fluctuations are due to particle widths, i.e. to the decay of the excited state via autoionising channels. Radiative widths are included at a later stage, once the basic effects have been illustrated.

The variation of the phase shift near an isolated resonance has the standard form (see, e.g., [259]) given by:

$$\tan \Delta_0 = \frac{\Gamma/2}{E_0 - E} \qquad (8.29)$$

where Γ and E_0 are the resonance width and resonance energy respectively. Combining equations (8.28) and (8.29), we have that:

$$K = \frac{\tan \delta + \tan \Delta_0}{1 - \tan \delta \tan \Delta_0}$$

$$= \frac{(E_0 - E) \tan \delta + \frac{1}{2}\Gamma}{(E_0 - E) - \frac{1}{2}\Gamma \tan \delta} \qquad (8.30)$$

whence, using equation (8.25)

$$\sigma(E) = 4 \sin^2 \delta \, \frac{\left[(E_0 - E) + \frac{1}{2}\Gamma \cot \delta\right]^2}{(E_0 - E)^2 + (\frac{1}{2}\Gamma)^2} \qquad (8.31)$$

We define

$$q \equiv \cot \delta \quad |\tilde{D}|^2 \equiv 4 \sin^2 \delta \qquad (8.32)$$

One thus obtains the standard form for the Fano formula for an autoionising line, which was obtained by a different route in section 6.4:

$$\sigma(\epsilon) = |\tilde{D}|^2 \, \frac{(q + \epsilon)^2}{1 + \epsilon^2} \qquad (8.33)$$

where ϵ is the rescaled energy variable defined by $\epsilon \equiv (E - E_0)/(\Gamma/2)$. This formula is appropriate for an isolated line in a flat continuum, which means that the phase shift δ of the continuum states is regarded as constant within the range of the resonance and only Δ_0 is allowed to vary. The general shape of isolated autoionising lines for different values of the profile index is displayed in fig. 6.2. In the present chapter, our main concern is how characteristic asymmetric shapes of Beutler–Fano profiles are modified by the proximity of further resonances.

We can also write K as

$$K = \frac{q + \varepsilon}{q\varepsilon - 1} \tag{8.34}$$

which yields a more symmetric version of the Fano formula:

$$\sigma(\varepsilon) = \frac{2K^2}{1 + K^2} = \frac{2(\varepsilon + q)^2}{(q\varepsilon)^2 + (\varepsilon + q)^2} = \frac{2}{1 + q^2}\frac{(\varepsilon + q)^2}{1 + \varepsilon^2}$$

The factor $1/(1 + q^2)$ is important:it appears in the definition of the f-value of an autoionising resonance (see section 6.13), where it ensures the finiteness of $\sigma(\varepsilon)$ even for $q \to \pm\infty$, when $K \to 1/\varepsilon$. For an *antiresonance*, $q = 0$ and $K = -\varepsilon$.

The width Γ is due to ejection of an electron from the compound state, and is therefore referred to as the *particle width*. Note that, whereas the particle width is intrinsic to an autoionising resonance, the shape or symmetry parameter q depends on the mode of excitation through the phase shift δ, which involves the probability of excitation into the continuum, and therefore depends also on the initial state. As remarked by Shore [394] 'although it is possible to predict the position and width of resonance lines [resonances] by considering the decay of prepared states, a full description of asymmetric profiles requires examination of the preparation process'. This point was elegantly made, both experimentally and theoretically, in a paper by Ganz *et al* [395] entitled 'Changing the Beutler–Fano profile of the Ne(ns') autoionising resonances' in which it was demonstrated, both experimentally and by a semiempirical analysis, that the profile symmetry q of autoionising resonances can be varied by altering the initial state, while the resonance width remains constant.

In principle, the particle width is not the only one which needs to be considered, especially if the autoionisation width is small. Narrow autoionising lines are well suited for study by laser spectroscopy. Since they imply the existence of long lived excited states, they can be investigated directly in atomic beams (see, e.g., [396]). However, the earliest examples appear to have been found by Paschen [397], White [398] and Shenstone [399] – the pioneer to whom we owe the very name *autoionisation*. In specific cases, where vanishing widths may occur (see section 8.29) or, alternatively, in deep core hole excitation spectra (see section 8.32), the spontaneous emission coefficient may contribute significantly to observed widths, and must then be included in the theory, as discussed in section 8.30.

A fundamental issue in the description of even the simplest, isolated autoionising resonance in the parametric approach followed by Fano [391] – and further pursued in **K**-matrix theory – is that the atom cannot be deperturbed, that is one cannot access the so-called 'prediagonalised' states which are imagined to exist prior to autoionisation being included as a perturbative interaction, since the effect is anyway internal to the atom and cannot truly be turned off. This has the disadvantage that the parameters, once they have been obtained, must still be calculated from an *ab initio* model of the atom for a full comparison with theory. It might seem that the parametric theory cannot really be checked independently of *ab initio* calculations whose accuracy is hard to ascertain.

A means of overcoming this limitation was suggested by the experimental study [400] of the interaction between a Rydberg series of autoionising lines and a broad interloper. Under the assumption that, with the interaction omitted, the properties of the Rydberg series would follow an otherwise regular behaviour (see section 8.23) for an unperturbed series of autoionising resonances, one could interpolate in energy to deduce the 'deperturbed' positions of the resonances. Such arguments suggest that the study of interacting resonances is a useful step to understand the autoionisation process itself, because one has access to properties of the system 'before' and 'after' the perturbation.

A dramatic improvement occurred a year later, when Safinya and Gallagher [401] published an important study of interferences between autoionising states. Again, interacting resonances were involved but, by using two photons of different colours simultaneously, they excited Ba atoms from $6s^2\,^1S_0$ to $6s15d\,^1D_2$. A laser was tuned through the $6s15d \rightarrow 6p15d$ transition, and ejected electrons were detected. When this third laser was tuned, the weak $6s15d \rightarrow 6pnd(n = 15)$ resonances were also found within the power-broadened linewidth of the intense $6s15d \rightarrow 6p15d$ transition, into which they autoionise as into a broad background continuum. It was possible, by exciting to $6snd(n = 15)$ initially at a greatly reduced laser power, to measure the widths, energies and quantum defects of the $6pnd$ states in the absence of their interaction with the broad background resonance, i.e. in what can be viewed as a 'prediagonalised' condition as far as autoionisation into the power-broadened background is concerned (see fig.8.3).

The novelty of this experiment was that, for the first time, it allowed an experimental determination of prediagonalised states, so that the arbitrariness in the parametrisation was removed and the theory of autoioni-

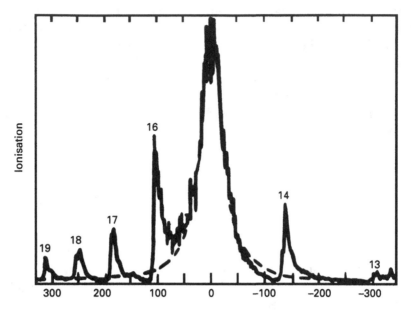

Fig. 8.3. Autoionising resonances coupled to a power-broadened continuum. The pure profile of the power-broadened line was also determined experimentally, and is shown by the broken curve (after A. Safinya and T.F. Gallagher [401]).

sation (chapter 6) became testable, at least in principle.

Thus, laser spectroscopy, and in particular ionisation spectroscopy involving several photons, allow one to study the mechanism of autoionisation itself, as opposed to testing the quality of atomic wavefunctions used to obtain pre-diagonalised states. Thereby, interest in the subject of interacting resonances and their parametric representation is enhanced (see in particular section 8.20).

8.19   Laser excitation and Beutler–Fano resonances

The experiment just described is one of a much more general class of studies in which the spectral properties of atoms are modified by a strong field laser and then probed by another laser – often chosen to have a weak field, so that it does not affect the atom further, although that is by no means essential.

Even if only one laser is involved in probing existing spectral structure, the possibility must always be borne in mind that the strength of the laser field can modify the couplings within the atom, so that the autoionising line may exhibit a different profile from the one of the free atom in the weak field limit. This problem is actually quite involved, because bound and continuum states evolve differently in the presence of a laser field. The whole question of how autoionising states behave under coherent excitation and how their profiles are modified will therefore be discussed in section 9.12 in the general context of multiphoton physics.

However, we pick out one specific aspect here, because its appreciation does not require a detailed preliminary discussion of the underlying high field interactions: the use of a laser to create or embed autoionising structure in an existing continuum is of great significance to the study of how the symmetries of autoionising resonances can be reversed (the so-called *q-reversal effect*, first discovered in the spectrum of an unperturbed neutral atom [382]).

8.20 Laser-induced continuum structure

The embedding of discrete structure in a preexisting continuum by the use of a laser is called *laser-induced continuum structure* or LICS for short. In essence, it is a very simple effect: since excited states of the atom can be coupled to each other resonantly by using laser radiation, it is possible to do so via the continuum. Fano's theory of autoionisation actually places no restriction on which operator V is used to 'turn on' autoionisation, or indeed whether the interaction concerned is internal or external to the atom. The advantage of an external interaction is, of course, that its magnitude can be controlled. The specific additional advantage of the radiation field is that it not only couples states, but does so with an energy difference between them which is adjustable. Thus, one can, in effect transport bound states from another part of the spectrum into a suitable place within the continuum by tuning the wavelength of the embedding laser, and control the magnitude of the coupling by vaying its intensity.

LICS has been the subject of a number of theoretical papers (see, e.g., [402, 403, 404]) and much experimental work ([405, 406, 407, 430]). For a general review, see Knight *et al.* [408].

If the probing conditions are appropriate (for further discussion, see section 9.12), the induced structure is expected to possess a Beutler–Fano profile, with the specific attraction that both the resonance energy and the width are controllable, which gives some experimental meaning to the otherwise somewhat elusive concept of prediagonalisation in Fano's theory (see chapter 6).

Unfortunately the detailed situation in LICS experiments is rather complex, and more processes are involved, even in the simplest situation, than in the simple autoionisation problem. To make this clear, consider the LICS and third harmonic generation experiment illustrated in fig. 8.4. We

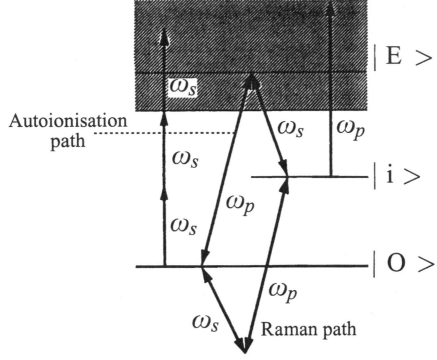

Fig. 8.4.   Energy level and excitation scheme for the simplest LICS situation described in the text.

take two lasers ω_s and ω_p, where the subscript s denotes a strong field laser and the subscript p a weak, probe laser. We apply these lasers in the simplest way for a meaningful LICS experiment, *viz.* to a two-level atom with ground state $\mid 0 >$ and excited state $\mid i >$ and a structureless continuum of states $\mid E >$. Since ω_s is strong, it can induce three photon transitions directly from $\mid 0 >$ into the continuum $\mid E >$. The probe laser, on the other hand, is too weak to do the same, and so its frequency ω_p must be chosen high enough to probe the region surrounding the embedded state in the continuum directly from $\mid 0 >$ by a single photon transition. This embedded state is produced by the action of ω_s on $\mid 1 >$ which effectively raises it into the continuum. However, ω_s can also lead downwards from $\mid 0 >$ to produce a virtual state from which ω_p can also reach $\mid 1 >$ as shown in the diagram. If we consider these processes from the broadenings they produce in the levels $\mid 0 >$ and $\mid 1 >$ by coupling to the continuum, then we see that each one of the bound states suffers two sources of broadening: $\mid 0 >$ is broadened by the one photon transitions driven by ω_p, and by three-photon transitions driven by ω_s, while $\mid 1 >$ is broadened by one photon transitions driven by ω_s and by one photon transitions driven by ω_p, all of which end up in the continuum $| E >$. The combined action of all these broadening mechanisms produces an effective width Γ_{01}, which is the width of the embedded resonance.

Another way of considering the problem which is perhaps physically more meaningful is that in fact the two bound states $| 0 >$ and $| 1 >$ are coupled to each other by the two lasers, in one case via a bound virtual state (the Raman path) and in the other via the continuum (the autoionising path) as marked in fig. 8.4. It turns out that, if the Raman channel dominates, the resulting lineshapes tend to become symmetric, while if the autoionisation channel dominates, the characteristic interference asymmetries of Beutler–Fano resonances emerge .

Since the ground state $| 0 >$ and the excited state $| 1 >$ are both to be coupled via the continuum by dipole-allowed transitions, it follows that they must have the same parity, and that they are coupled *to each other* by a two-photon matrix element which is called the *two-photon Rabi frequency* (for a discussion of the Rabi frequency, see section 9.10).

If we define a resonance energy E_0 somewhere in the continuum, and if we let ω_0 take us from the ground state to E_0 and ω_1 take us from $| 1 >$ to E_0 then, when ω_p is tuned into resonance $\omega_0 - \omega_1 = \omega_p - \omega_s$ connects $| 0 >$ and $| 1 >$. It follows that the detuning of the lasers from the embedded state is $x = \omega_s - \omega_p + \omega_0 - \omega_1$.

We can thus define an energy variable $\epsilon \equiv x/\Gamma_{01}$, while, as usual, q is defined as the ratio of the transition strength to the bound state to the transition strength to the continuum states *viz.* $q = M_{01}/\mu_{0c}\mu_{1c}$, where M_{01} is the two photon transition moment and the μs are single photon transition moments.

There are then three ways of observing the embedded states, namely: (i) by searching the photoionisation signal for the appearance of structure near E_0; (ii) by looking for the enhancement of third harmonic generation; and (iii) by looking for rotation of the plane of polarisation of the probe laser ω_p in the field of the strong laser ω_s, a method closely analogous to Faraday rotation for an autoionising resonance, as discussed in section 6.15.

In any real experiment with strong field lasers, the lasers will be pulsed, and so the time dependence of LICS will need to be considered (this is postponed to chapter 9).

8.21 Optical rotation in LICS

A very interesting technique for observing LICS was used in one of the earliest experiments. It will be called *laser-induced natural rotation* here, because the physical distinction between natural and Faraday rotation is

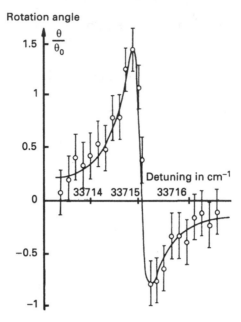

Fig. 8.5. Observation of LICS by optical rotation in Cs. Note the nearly dispersive shape (after Y.I. Heller *et al.* [409]).

in fact that an external magnetic field breaks time-reversal symmetry (see section 10.16) rather than its isolated occurrence in nature. The method consists in selecting the polarisation of the embedding laser to be just circular, say σ^+. An alkali atom is then dressed so that an excited state $ns > {}^2S_{1/2}$ with $M_J = -1/2$ is coupled to the $\varepsilon p > {}^2P_{1/2,3/2}$ $M_J = +1/2$ states, and the $ns > {}^2S_{1/2}$ state with $M_J = +1/2$ is coupled to $\varepsilon p > {}^2P_{3/2}$ $M_J = +3/2$, but there are no embedded states at all in the $\varepsilon p >$ continua with $M_J = -1/2$ and $M_J = -3/2$ originating from $ns > {}^2S_{1/2}$

In this situation, there is then no need for an external magnetic field to induce rotation of the plane of polarisation of a linearly polarised probe beam: in Faraday rotation spectroscopy, the role of the external magnetic field is simply to split the spectral structure in energy according to the sense of circular polarisation, so as to produce a nonzero $n_+ - n_-$ (n being the refractive index), resulting in rotation. In the present example, n_- is constant in the region of interest (no embedded resonance) while n_+ has a large fluctuation (see fig. 8.21). For a nearly symmetric Beutler–Fano resonance, it was shown in section 6.14 that the rotation angle will follow a nearly dispersive shape, which is precisely what Heller *et al.* [409] observed (see fig. 8.5). Again for a purely dispersive refractive index profile, the *maximum* rotation in the profile is one half of the rotation angle available by MOR.

Rotation techniques have the virtue that they can be performed with crossed polarisers, which provide a dark background. The further experimental advantage of using induced natural rotation is that no external magnetic field is required. However, it has the disadvantage that multi-passing the probe beam cannot increase the rotation angle. Thus, Heller *et al.* achieved a maximum rotation angle of 7×10^{-2} radians, whereas an angle of $\pi/2$ could be achieved under the same experimental conditions by multipassing 20 times in the presence of a magnetic field, at which point crossed polarisers become fully transmitting. Also note that the need to polarise the embedding laser circularly to probe laser-induced natural rotation imposes some restrictions on the performance of the embedding system.

The rotation angle is calculated from the product of the density matrix ρ of the system and electric dipole operator d of the atom from the equation for the polarisability

$$P = \operatorname{tr}\{\rho d\} \tag{8.35}$$

while the evolution of the system in time obeys Heisenberg's equation

$$i\hbar \frac{\partial \rho}{\partial t} = [H, \rho] \tag{8.36}$$

In what follows, we neglect this time dependence, which is not an essential aspect of the problem (time dependence is addressed in section 9.12). For small rotations, the angle turns out [410] to be

$$\chi(\epsilon) \propto \frac{|\mathcal{E}_f|^2}{\epsilon_0} \frac{|M_1 2|^2}{\Gamma_{12} q^2} \frac{(q^2-1)\epsilon - 2q}{1+\epsilon^2} \tag{8.37}$$

Polarisation is arguably the most sensitive way to detect LICS and was the method used in the original experiments of [405, 406].

If the ionisation or third harmonic signals are observed, then they turn out to be Fano profiles, with the value of Γ as defined above, and the value of q dependent on the mode of excitation.

In fig. 8.6, we show numerical calculations of the enhancement of third harmonic generation by LICS, as reported by Zhang [410]. The LICS peaks are tuned to several different energies, and move through a smooth curve which is the background spectrum in absence of the coupling laser, near a two-photon resonance which is responsible for the broad modulation. As the Fano profiles arising by LICS are tuned through the structure, their symmetry is changed. This is apparent both in the third harmonic enhancement and in the photoionisation spectra which, furthermore, behave somewhat differently from each other.

It is interesting to compare figs. 8.6 and 8.3: notice how both exhibit the q-reversal effect. However, in LICS, the resonance can be tuned to

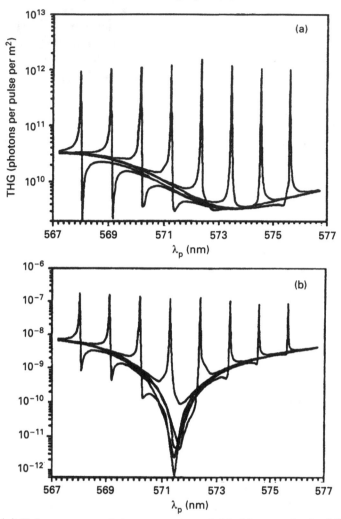

Fig. 8.6.   (a) Enhancement of the generation of third harmonics by LICS. These are theoretical spectra, generated by tuning the strong field laser to various energies; (b) the same as (a) but in the photoionisation spectrum (after J. Zhang [410]).

the desired energy, while its symmetry provides direct information on the phase shift in the continuum.

The change in symmetry or q-reversal effect is accounted for in section 8.28: we note here that it occurs in a Rydberg series of autoionising lines when a broad intruder is present, but that it also occurs when a relatively sharp autoionising line is 'tuned' through a broad resonance. Early theories of strong field laser effects [411] provided this picture, used by Connerade to discuss q reversals [382] when they were first observed, although LICS was then unknown.

8.22   Two overlapping resonances in a flat continuum

Turning now to the case of a flat continuum (constant δ), we can readily construct the case of two overlapping resonances, in which case the phase shift has a resonant part

$$\tan \Delta_0 = \frac{\Gamma_1}{2(E_1 - E)} + \frac{\Gamma_2}{2(E_2 - E)} \tag{8.38}$$

and, by the same procedure as in section 8.18,

$$\sigma(E) = \frac{4K^2}{1 + K^2}$$

$$= 4 \sin^2 \delta \frac{\left(1 + \dfrac{\Gamma_1}{2(E_1 - E)} \cot \delta + \dfrac{\Gamma_2}{2(E_2 - E)} \cot \delta\right)^2}{1 + \left(\dfrac{\Gamma_1}{2(E_1 - E)} + \dfrac{\Gamma_2}{2(E_2 - E)}\right)^2}$$

$$= 4 \sin^2 \delta \frac{\left(1 + \dfrac{q\Gamma_1}{2(E_1 - E)} + \dfrac{q\Gamma_2}{2(E_2 - E)}\right)^2}{1 + \left(\dfrac{\Gamma_1}{2(E_1 - E)} + \dfrac{\Gamma_2}{2(E_2 - E)}\right)^2} \tag{8.39}$$

Such formulae are used to analyse cases in which two Beutler–Fano resonances overlap in energy. For resonances originating from two different channels, different values of q (q_1 and q_2) are introduced in (8.39) (see, e.g., Heinzmann *et al.* [427]). Such situations are also treated by MQDT [428, 429]

8.23   Generalisation to N overlapping resonances with one flat continuum and the connection to QDT

The result in the previous section can be instantly generalised to N resonances, giving

$$\sigma(E) = \frac{4K^2}{1 + K^2} = 4 \sin^2 \delta \frac{\left(1 + q \displaystyle\sum_n \frac{\Gamma_n}{2(E_n - E)}\right)^2}{1 + \left(\displaystyle\sum_n \frac{\Gamma_n}{2(E_n - E)}\right)^2} \tag{8.40}$$

if we assume that all N resonances belong to the same channel, in which case the value of q is the same for all of them. One makes contact with MQDT through a standard replacement:[3]

$$\sum_n \frac{\Gamma_n}{2(E_n - E)} \implies x \cot \pi(\nu + \mu) \equiv x \cot \theta \qquad (8.41)$$

where ν is the energy variable defined by $E_\infty - E \equiv R/\nu^2$ and μ is the quantum defect. Thus

$$\sigma(\theta) = \frac{K^2}{1 + K^2} = 4 \sin^2 \delta \frac{(1 + qx \cot \theta)^2}{1 + (x \cot \theta)^2} \qquad (8.42)$$

After some simple algebra, this equation reduces to:

$$\sigma(\nu) = |\tilde{D}|^2 \frac{\tan^2 \pi\nu + 2B \tan \pi\nu + B^2}{\tan^2 \pi\nu + 2C \tan \pi\nu + D^2} \qquad (8.43)$$

where we have now introduced the three shape parameters

$$B = \frac{qx + \tan \pi\mu}{1 - qx \tan \pi\mu}; \quad C = \frac{(1 - x^2) \tan \pi\mu}{1 + x^2 \tan^2 \pi\mu}; \quad D = \frac{x^2 + \tan^2 \pi\mu}{1 + x^2 \tan^2 \pi\mu} \qquad (8.44)$$

Of these, only B depends on q. From the formula for B

$$q = \frac{1}{x} \tan \pi(\beta - \mu)$$

where $B = \tan \pi\beta$.

For an isolated or Beutler–Fano resonance, $q = 0$ implies a symmetric window. In the present situation, the presence of other resonances arranged as a Rydberg series, with energy intervals which are not the same on both sides of any given resonance means that the lineshapes are not symmetrical even for $q = 0$. However, when $q = 0$ and $B = \tan \pi\mu$, from the first equation (8.44), the transmission maxima coincide with the resonance energies as they do for isolated resonances.

The formula for $\sigma(\nu)$ above [413, 414] is identical to formulae given by Dubau and Seaton [415] and, independently, by Giusti-Suzor and Fano [416]. It provides a single simple expression for a complete and otherwise unperturbed Rydberg series of autoionising resonances in terms of only three shape parameters B, C and D, which are constant for the whole series.

---

[3] Note that this is a *replacement*, not an equation: the sum in **K**-matrix theory is over all the bound states which actually occur, whereas the trigonometric functions used to represent resonances in MQDT continue to repeat even outside the range of physical validity. They therefore include poles in an energy range below the physical resonances, whereas **K**-matrix theory does not. So the replacement should only be used when one is above the lowest resonances in the channel, which is anyway the range of validity of MQDT

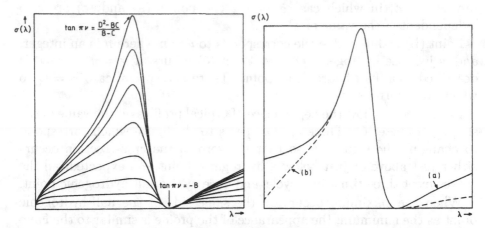

Fig. 8.7. The form of a single resonance in the Rydberg series defined by the Dubau–Seaton formula: (a) plotted with different combinations of parameters so that the maximum and minimum in the absorption cross section remain at fixed energies; and (b) comparing a Dubau–Seaton profile (curve A) with a Beutler–Fano profile of the same shape near the resonance energy (curve B) (after J.-P. Connerade [413, 414]).

8.24 Autoionising Rydberg series in a flat continuum

It is interesting to study how the three shape parameters B, C and D can alter the shapes of the line profiles. Since the numerator in the formula for $\sigma(\nu)$ is a perfect square, there must be a zero for $\tan \pi \nu = -B$. This defines a Rydberg series of transmission windows (one cycle of the series is shown in fig. 8.7).

Likewise, one can readily show that the peak cross section occurs for $\tan \pi \nu = (D^2 - BC)/(B - C)$. There are also bounds on the value of C which one can find by rearranging the formula as

$$\sigma(\nu) = \mid \tilde{D} \mid^2 \frac{(\tan \pi \nu + B)^2}{(\tan \pi \nu + C)^2 + (D^2 - C^2)} \qquad (8.45)$$

Since a cross section is positive definite, by choosing D positive (no loss of generality), we have $-D < C < D$.

Since B is fully determined by a single point on the profile (the zero in the cross section), it is convenient to explore changes in $\sigma(\nu)$ by varying the parameters C and D for fixed B.

We begin with C. Clearly, there exist just three points on the profile within any cycle (a Rydberg cycle is defined by an increase of 2π in the argument $\pi\nu$) which are independent of the values of C, namely: (a) $\tan\pi\nu = -B$; (b) $\tan\pi\nu = 0$, in which case $\sigma = B^2/D^2$; and (c) $\tan\pi\nu \to \infty$, in which case $\sigma = 1$. The points (a) and (c) are also independent of the value of D.

Point (b) within each cycle corresponds to $\nu = n$ where n is an integer, and is just the hydrogenic energy. Point (c) occurs at $\nu = n + \frac{1}{2}$ and is described as a 'half hydrogenic' point. There is a special case $B = C$ to which we return below.

In fig. 8.8, we show the dependence of typical profiles on the value of the shape parameter C. There are two parts to the figure which correspond to changing the sign of B so that the zero in the cross section occurs either just above or just below a hydrogenic point. As expected, all the curves meet three times in a cycle, and once at the half-hydrogenic point.

When the maximum occurs on the same side of the half-hydrogenic point as the minimum, the appearance of the profile is similar to the Fano lineshape. However, if the maximum and minimum become sufficiently separated in energy to straddle the half-hydrogenic point, then the shape of the profile changes to a more sinusoidal form.

When atomic transitions are excited, one is normally dealing with a Rydberg series of autoionising lines, rather than just an isolated line. The profile then lies between two extreme situations: either the linewidth is much smaller than the separation between the hydrogenic and the half-hydrogenic points (type 1 profile), or it is much larger (type 2 profile). Type 1 profiles occur in the *isolated line* limit, and tend to the typical *Fano shape*. Type 2 profiles correspond to the *overlapping line* limit, and tend to *sinusoidal* shapes.

In fig. 8.9, we illustrate how one may distinguish between the two types of profile using experimental data: by drawing tangents at the inflexion points, one can determine whether the lineshape curve rises above one of the tangents near the line centre (Fano shape) or whether it remains below both tangents near the line centre (sinusoidal shape).

It is also of interest to compare a type 1 with the true Fano shape for an isolated resonance, which is done in fig. 8.7(b). The two profiles are nearly coincident close to the resonance energy. They differ in the wings, where the more general formula for a full Rydberg series includes the influence of the adjacent members on either side.

As resonances begin to overlap in a Rydberg series, the original Fano parametrisation becomes inappropriate and individual q parameters become meaningless. In the isolated profile limit of type 1, individual q values are given by:

$$q = \frac{B - C}{(D^2 - C^2)^{1/2}} \tag{8.46}$$

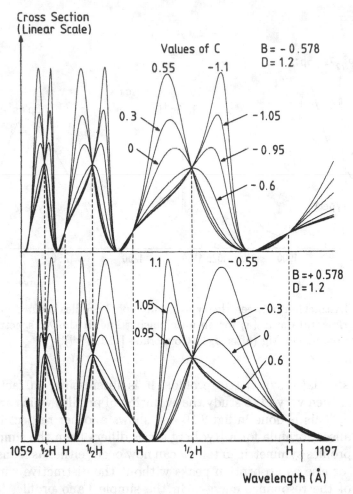

Fig. 8.8. Evolution of a Rydberg series of Dubau–Seaton profiles as a function of the parameter C, for $D = 1.2$ and: (a) $B = -0.578$; (b) $B = +0.578$. Values of C are given in the figure. The narrower resonances tend towards a Fano lineshape, whereas the broader resonances are more sinusoidal in shape. This is illustrated by some experimental examples in fig. 8.9 (after J.-P. Connerade [413, 414]).

and q is therefore (as expected) constant in a series for a given choice of the three shape parameters B, C and D. Note that q is only constant for the series if the background continuum is flat, as assumed here. Other situations are discussed below. Equation 8.46 is useful (see below) in discussing the interaction between a Rydberg series of autoionising lines and a shape or giant resonance.

From equation (8.46), the special case $B = C$ yields $q = 0$ in the isolated profile limit and, more generally, gives a Lorentzian in $\tan \pi \nu$,

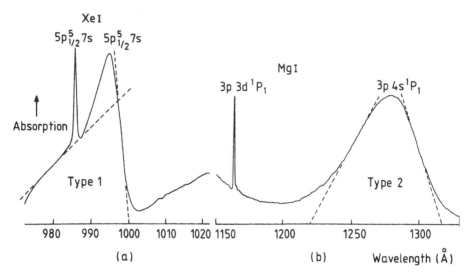

Fig. 8.9. Distinction between the type 1 and type 2 profiles defined in the text using experimental data: (a) for an autoionising line in the Xe spectrum; and (b) for a resonance in Mg (after J.-P. Connerade [413, 414]).

i.e. an essentially symmetric profile. It is interesting, in fact, to set $B = C$ and then vary D to study the evolution of profiles which are nearly symmetric. This is done in fig. 8.10 and shows another respect in which the generalised formula for a Rydberg series differs from the simple Fano formula: profiles symmetric in $\tan \pi \nu$ can now occur either as transmission windows, or else as absorption peaks without the distinctive window on one side of the resonance energy. In the simple Fano profile, for finite values of the shape index q, pronounced absorption maxima are always associated with a 'window' asymmetry within the core of the broadened line.

8.25 Rydberg series in a modulated continuum

In the cases treated so far, the continuum was assumed to be 'flat' as in the original Fano problem. However, experiments tell us that this is usually not the case, and the question arises: how will the series be modified if the continuum itself varies appreciably within its range? To answer this question, we clearly need to know the form of the variation. As an example, consider a series on the continuum background of a shape or giant resonance, which is generally a much broader feature than an autoionising line, but for which the continuum phase shift can be expected

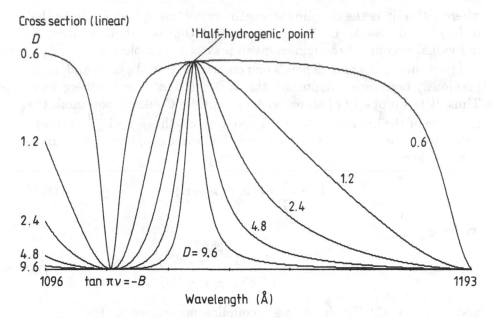

Fig. 8.10. Effect of varying the third lineshape parameter D: note how nearly symmetric peaks turn into nearly symmetric troughs (after J.-P. Connerade [413, 414]).

to change by $O(\pi)$ within the range of the Rydberg series if the two happen to overlap in energy.

We use Fano's notation within our **K**-matrix and, for consistency with the literature [380] emphasise the fact by an additional subscript F: let Γ_{BF} and E_{BF} be, respectively, the particle width and resonance energy of the giant resonance.

We follow the treatment of [375, 377]. As in most cases in this chapter so far, the analysis considers only particle widths: strictly, it applies only to elastic scattering and may or may not extend to photoionisation, depending on the case considered. It does, however, even in this simple form, exhibit all the main features of the problem.

The Rydberg series of narrower autoionising lines is denoted as Γ_{nF} and E_{nF}. The resonant part of the phase shift Δ_0 as defined in previous cases is then given by (see [375] and also section 8.27):

$$\tan \Delta_0 = \sum_n \frac{\Gamma_{nF}}{2(E_nF - E)} + \frac{1}{2} \frac{\left(\Gamma_{nF}^{1/2} - \sum_n \frac{\Gamma_{nF}^{1/2} H_{nF}}{E_{nF} - E} \right)^2}{E_{BF} - E - \sum_n \frac{H_{nF}}{E_{nF} - E}} \tag{8.47}$$

where $\mid H_{nF} \mid^2$ is the coupling strength connecting n and B. Note that, if $H_{nF} \rightarrow 0$, $\tan \Delta_0$ is simply the sum of the contributions from the individual uncoupled resonances, as in previous examples.

The coupling strength depends on penetration into the core, and, as has previously been seen (chapter 3), this varies as $1/n^{*3}$ for Rydberg series. Thus, it is a property of atoms with asymptotic Coulomb potentials that the ratio of the linewidths to the coupling strength $\mu_F \equiv (\Gamma_{nF}^{1/2})/(H_{nF})$ is independent of n for a Rydberg series. For weak coupling of n to B, we can also assume that μ_F^{-1} is small. Writing

$$\Sigma \equiv \sum_n \Gamma_{nF} 2(E_{nF} - E) \tag{8.48}$$

we have

$$\tan \Delta_0 = \Sigma + \frac{1}{2} \frac{(\Gamma_{BF}^{1/2} - 2\mu_F^{-1})^2}{E_{BF} - E - 2\mu_F^{-2}\Sigma} \tag{8.49}$$

and, to order μ_F^{-1} (i.e. in the weak coupling approximation)

$$\tan \Delta_0 \simeq \frac{\Gamma_{BF}}{2(E_{BF} - E)} + \sum_n \frac{\Gamma_{nF}}{2(E_{nF} - E)} - \frac{\mu_F^{-1}\Gamma_{BF}^{1/2}}{2(E_{BF} - E)} \sum_n \frac{\Gamma_{nF}}{2(E_{nF} - E)}$$

$$= \frac{\Gamma_{BF}}{2(E_{BF} - E)} + \sum_n \frac{\Gamma_{nF}}{2(E_{nF} - E)} - \frac{\Gamma_{BF}^{1/2}}{2(E_{BF} - E)} \sum_n \frac{H_{nF}\Gamma_{nF}^{1/2}}{2(E_{nF} - E)} \tag{8.50}$$

i.e. the total resonant phase shift is the sum of the individual phase shifts of all the resonances, plus a mixed term which represents the coupling of n to B. This can be further rearranged (by neglecting terms of second order in the coupling strength) as

$$\tan \Delta_0 \simeq \frac{\Gamma_{BF}}{2(E_{BF} - E)} + \frac{\sum_n \left\{ \Gamma_{nF}^{1/2} - \frac{H_{nF}\Gamma_{BF}^{1/2}}{2(E_{BF} - E)} \right\}^2}{2(E_{nF} - E)} \tag{8.51}$$

$$= \frac{\Gamma_{BF}}{2(E_{BF} - E)} + \left\{ 1 + \frac{\mu_F^{-1}\Gamma_{BF}^{1/2}}{E_{BF} - E} \right\}^2 \sum_n \frac{\Gamma_{nF}}{2(E_{nF} - E)} \tag{8.52}$$

We now follow the connection to MQDT as in section 8.23 and find that Δ_0 can be written

$$\tan \Delta_0 \simeq \overline{\Delta} + x_1 x_2 \cot \theta \tag{8.53}$$

provided we define

$$x_1 \equiv 1 - \frac{\mu_F^{-1}\Gamma_{BF}^{1/2}}{E_{BF} - E}; \quad x_2 \cot\theta \equiv \sum_n \frac{\Gamma_{nF}}{2(E_{nF} - E)}; \quad \overline{\Delta} \equiv \frac{\Gamma_{BF}}{2(E_{BF} - E)}$$

$$(8.54)$$

where the angle θ is as defined in section 8.23.

Again, we write the cross section as the standard form

$$\sigma(\theta) = \frac{2K^2}{1 + K^2}$$

and, after some lengthy but simple algebra, we obtain

$$\sigma(\theta) = \frac{(\overline{\Delta} + \overline{\delta})^2}{(1 + \overline{\Delta}^2)(1 + \overline{\delta}^2)} \frac{\tan^2 \pi\nu + 2B \tan \pi\nu + B^2}{\tan^2 \pi\nu + 2C \tan \pi\nu + D^2} \qquad (8.55)$$

where we now define

$$B \equiv \frac{x_1 x_2}{\overline{\Delta} + \overline{\delta}}; \quad C \equiv \frac{x_1 x_2 \overline{\Delta}}{1 + \overline{\Delta}^2}; \quad C \equiv \frac{x_1^2 x_2^2}{1 + \overline{\Delta}^2} \qquad (8.56)$$

as the new shape parameters in the presence of the modulated continuum.

We instantly recognise the second factor in equation (8.55) as the Dubau–Seaton formula which was obtained above for N resonances in a flat continuum. There is, however, an essential difference: the three shape parameters B, C and D are no longer independent of energy. They become, mainly through the quantity $\overline{\Delta}$, functions of the detuning $(E_{BF} - E)$ from the centre of the giant resonance. Thus, the second factor in (8.55) represents a Rydberg series of autoionising lines, whose shapes (and in particular the profile index q) vary with energy according to the influence of the giant resonance on individual Rydberg members.

On the other hand, the first factor in equation (8.55) is an amplitude term, which depends only on $\overline{\delta}$ and $\overline{\Delta}$, i.e. it contains no information on the Rydberg series as such, but only on the giant resonance and background phases. In fact, it is the profile of the giant resonance itself in absence of any series. In model calculations for Δ, we use the simple formula (5.20) for the phase shift of a giant resonance (see fig. 8.11).

For well-separated Rydberg members, we find from equation (8.46) that

$$q = \frac{B - C}{(D^2 - C^2)^{1/2}} = \frac{1 - \overline{\delta}\overline{\Delta}}{\overline{\delta} + \overline{\Delta}} \qquad (8.57)$$

which means that, in the weak coupling limit (μ_F was assumed to be small at the outset), q is determined entirely by the shape resonance and background phase shifts. The overall situation is illustrated in fig. 8.11. We note that $q = 0$ when $\overline{\Delta} = \overline{\delta}$, which is also the condition for a

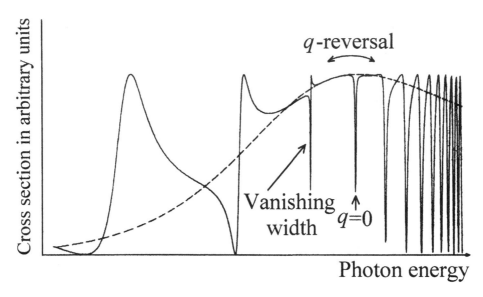

Fig. 8.11.   Calculation of the cross section of a Rydberg series of fairly sharp lines interacting with a giant resonance in the limit of rather weak coupling (full curve). Note that one of the Rydberg members has a very small width. In the limit of vanishingly small coupling strengths, this line coincides with the resonance energy of the background feature, but not with the maximum in its undisturbed cross section (indicated by the dashed curve), because a finite background phase shift has been included, so the giant resonance is asymmetric. However, the $q = 0$ window resonance, as well as the associated q reversal do coincide with the maximum in the cross section, as required by the formula for this case (after J.-P. Connerade and A.M. Lane [375]).

maximum in the first factor of equation (8.55) and occurs when

$$(E_{BF} - E) = \frac{\Gamma_{BF}}{2} \tan \delta \qquad (8.58)$$

i.e. when E is only slightly detuned from the resonance energy E_{BF}, for small values of the background phase shift. Thus $q = 0$ (a window resonance occurs) at the peak intensity of the giant resonance; about this value, $\overline{\Delta}$, and therefore q, change sign.

We can also deduce the widths of Rydberg members from the relation $\Gamma_{nF}\alpha(D^2 - C^2)^{1/2}$ or, more directly, from equation (8.51). Either way, the width tends to zero (a vanishing width occurs) when the detuning is

$$\epsilon \equiv (E - E_{BF}) = -\frac{H_{nF} - \Gamma^{\frac{1}{2}}}{\Gamma_{nF}^{1/2}} \qquad (8.59)$$

irrespective of the value of n. Both the change in symmetry of the profiles (q reversal) about a transmission window and the vanishing width effect are clearly visible in fig. 8.11.

There has been some work on Rydberg series coupled to giant resonances (see, e.g., [417]). Since most examples are at high energies, where the experimental resolution is limited and the influence of many overlapping channels tends to obscure changes of symmetry, q reversals are not very conspicuous. However, it is verified as a general rule that when superposed on a giant resonance, sharper transitions occur as nearly symmetric window resonances near the maximum in the giant resonance cross section. This conclusion is physically reasonable: a giant resonance exhausts the oscillator strength sum rule (see section 4.5) and rises close to the unitary limit, which cannot be exceeded.

8.26 Perturbation by an antiresonance

The opposite case to a giant resonance, which exhausts the oscillator strength within its width, is an antiresonance, or a Fano resonance with $q = 0$. In principle, nothing prevents such a resonance from acting as the intruder. Double excitations appear above the first ionisation potentials of many-electron atoms, and are frequently observed as 'window resonances.' An example where an antiresonance acts as the intruder [418] occurs in the spectrum Ar shown in of fig. 8.12.

In the vicinity of a giant resonance, the continuum background δ is regarded as very small, and, for weak coupling, a q-reversal occurs near the energy E_B of the perturber, at which point $q \to 0$ for the finer structure. Clearly, the assumption of small or negligible transition amplitude to the continuum is not valid if the perturber is an antiresonance. Nevertheless, there is a connection between the present problem and the case of the giant resonance, as will emerge below.

To treat the perturbed series, we use the standard form for the resonant phase shift [419]:

$$\tan \Delta_{Bn} = \sum_n \frac{\Gamma_n/2}{E_n - E} + \frac{1}{2} \frac{(\Gamma_B^{1/2} - \Sigma H \Gamma)^2}{E_B - E - \Sigma H H} \tag{8.60}$$

where we define

$$\Sigma H \Gamma \equiv \sum_n (H_n \Gamma_n^{1/2})/(E_n - E) \quad \text{and} \quad \Sigma H H \equiv \sum_n (H_n^2)/(E_n - E)$$

In (8.60), Γ_B is the width of the broad resonance in absence of the Rydberg series, Γ_n are the widths of the Rydberg members E_n in the absence of the broad interloper E_B, while H_n represents the strength of the interaction

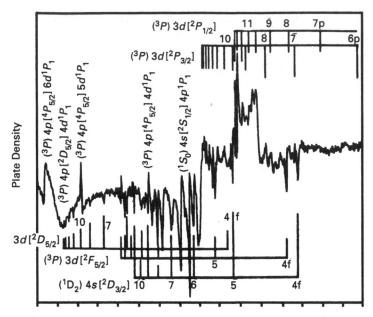

Fig. 8.12.   Doubly-excited series in the absorption spectrum of Ar, showing an example of an antiresonance appearing as an intruder in a Rydberg series (after M.A. Baig *et al.* [418]).

coupling the Rydberg series to the continuum when mediated by the broad level E_B: if $H_n \to 0$, then the E_n are decoupled from E_B and interact only separately with the continuum through the widths Γ_n.

The **K**-matrix has the form:

$$K = \frac{1 + q_B \left\{ \sum_n \dfrac{\Gamma_n/2}{E_n - E} + \dfrac{1}{2} \dfrac{(\Gamma_B^{1/2} - \Sigma H \Gamma)^2}{E_B - E - \Sigma H H} \right\}}{q_B - \left\{ \sum_n \dfrac{\Gamma_n/2}{E_n - E} + \dfrac{1}{2} \dfrac{(\Gamma_B^{1/2} - \Sigma H \Gamma)^2}{E_B - E - \Sigma H H} \right\}} \qquad (8.61)$$

where q_B is the shape parameter of the broad interloper in absence of the Rydberg series. In this expression, radiative widths are neglected, i.e. we use the elastic scattering approximation. Equation (8.61) can be regarded as a simple generalisation of the Dubau–Seaton formula. It describes a perturbed autoionising series in the presence of just one continuum. Calculations from equation (8.61) are shown in fig. 8.13.

The results are given for an antiresonance, but are more general, as are some of the conclusions we now describe. Typical variations in line-shape, previously obtained as a result of arbitrary changes in the three shape parameters B, C and D (see section 8.23) now emerge as a result of changing the coupling strength H_0 (since H_n scales as $n^{-3/2}$, we write

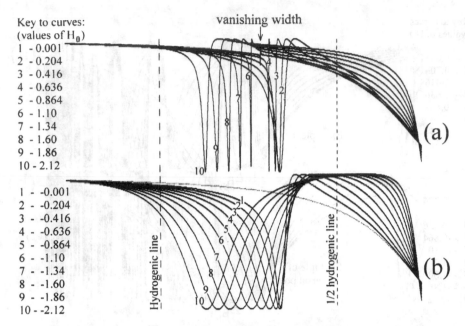

Fig. 8.13. Calculations for perturbations by an antiresonance for: (a) positive; and (b) negative values of the coupling strength parameter near a leading member of the series, showing the vanishing width for positive values of coupling. The dashed curve shows the spectrum in absence of coupling, which is the profile of the antiresonance (after J.-P. Connerade [376]).

$H_n = H_0 n^{-3/2}$). For example, for lines which broaden with increasing coupling strength $|H_0|$, the transition from type 1 profiles (nearly isolated lines) to type 2 profiles (overlapping lines) occurs rather quickly as the coupling strength is increased. This is readily understood: the isolated line condition soon breaks down as the linewidth of the Rydberg series members increases, and the profiles tend towards sinusoidal shapes.

The following theorem applies, as an essential property of the **K**-matrix above: *the cross section $\sigma(E)$ possesses an infinite number of common points independent of H_n, all of which lie on the profile of the undisturbed interloper.*

The existence of one such point associated with each Rydberg member can be inferred by inspection of figs 8.13 and 8.14: all the curves pass through the common points, whatever the value of the coupling strength. They are thus *universal points* and lie on the profile of the interloper E_B, Γ_B, shown as a dashed curve in the figures. A series of similar points was previously found for the unperturbed Dubau–Seaton profiles:

$$\sigma(\nu) = \frac{\tan^2 \pi\nu + 2B \tan \pi\nu + B^2}{\tan^2 \pi\nu + 2C \tan \pi\nu + D^2} \qquad (8.62)$$

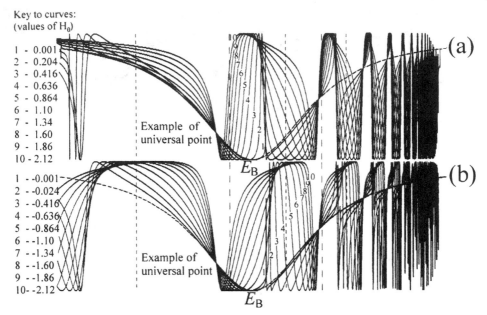

Key to curves:
(values of H_0)

1 - 0.001
2 - 0.204
3 - 0.416
4 - 0.636
5 - 0.864
6 - 1.10
7 - 1.34
8 - 1.60
9 - 1.86
10 - 2.12

1 - -0.001
2 - -0.024
3 - -0.416
4 - -0.636
5 - -0.864
6 - -1.10
7 - -1.34
8 - -1.60
9 - -1.86
10- -2.12

Example of universal point

Example of universal point

E_B

E_B

(a)

(b)

Fig. 8.14.   Portion of the spectrum of a series perturbed by an antiresonance containing several upper members: (a) for positive and (b) for negative values of coupling. Note the universal crossing points (also visible on the right edge of the previous figure). One such point is associated with each series member (after J.-P. Connerade [376]).

in which case the points lie at the intersections with the half-hydrogenic lines, corresponding to the condition $\tan \pi \nu \to \infty$. All the Dubau–Seaton curves pass through these points, whatever the values of the shape parameters B, C or D. Another series of somewhat less general points, through which all the curves pass, whatever the value of the shape parameter C, lies on the hydrogenic lines, where $\tan \pi \nu = 0$. One might at first be tempted to conclude that the more general series of universal points of the Dubau–Seaton formula would survive in the presence of the perturbation by moving away from the half-hydrogenic lines (shown as short-dashed vertical lines in figs. 8.13 and 8.14, while the less general crossing point would dissolve. In fact, as the following proof will show, the reverse is the case: the points associated with the hydrogenic lines in the unperturbed limit are those which survive, and the effect of the change in coupling strength thus turns out to be very similar to a variation in the parameter D for a given line (see fig. 8.10).

   Proof : We note that the only term which depends on H_n in **K** is:

$$\tilde{K} = \frac{1}{2} \frac{(\Gamma_B^{1/2} - \Sigma H \Gamma)^2}{E_B - E - \Sigma H H}$$

(8.63)

This term can only become independent of H_n if there exists an E for which *both*

$$\Sigma HH \equiv \sum_n H_n^2/(E_n - E) = 0$$

and

$$\Sigma H\Gamma \equiv \sum_n H_n\Gamma_n^{1/2}/(E_n - E) = 0$$

together, for any H_n. This is true because of the essential condition for the analytic inversion of the **K**-matrix, namely that all of the ratios[4] $\mu \equiv \Gamma^{1/2}/H_n$ are independent of n, a scaling condition automatically satisfied for asymptotic Coulombic potentials.

We can thus define

$$\Sigma HH \equiv y\tan(\pi\nu + \theta) \quad \text{whence} \quad \Sigma H\Gamma = \mu y\tan(\pi\nu + \theta)$$

and:

$$\tilde{K} = \frac{1}{2}\left\{\frac{[\Gamma_B^{1/2} - y\mu\tan(\pi\nu + \theta)]^2}{(E_B - E - y\tan(\pi + \theta))}\right\} \tag{8.64}$$

If $\tan(\pi\nu+\theta) = 0$, then $\tilde{K} = \Gamma_B/2(E_B-E)$, which is simply the **K**-matrix of the undisturbed perturber. On the other hand, for $\tan(\pi\nu + \theta) \longrightarrow \infty$, $\tilde{K} \longrightarrow \frac{1}{2}\{(y\mu)^2/(E_B - E - y)\}$, which depends on both y and μ, and thus on H_n.

At high n, $\theta \ll n\pi$, and so the universal points lie near the intersections of the hydrogenic lines with the profile of the undisturbed antiresonance, as illustrated in fig. 8.14.

The next question of interest is the presence of vanishing widths, which are strongly dependent on both the magnitude of the coupling strength H_0, and its sign. In general [383], the vanishing width occurs at

$$E_{vw} = E_B - \mu^{-1}\Gamma_B^{1/2} = E_B - H_n\Gamma_B^{1/2}/\Gamma_n^{1/2} \tag{8.65}$$

(cf equation (8.59)). Even if the ratio of the widths $\Gamma_B^{1/2}/\Gamma_n^{1/2}$ is large, E_{vw} can come close to E_B provided the interaction strength H_n is very small. However, as the coupling strength is increased, E_{vw} travels rapidly away from E_B in a direction determined by the sign of H_0.

Since the perturber is symmetric with $q_B = 0$ in the examples shown, a q reversal occurs about the energy E_B for small enough values of the coupling strength. As the coupling strength is increased, the q reversal disappears from sight, and the lines tend, in the strong coupling limit, to

---

[4] it is conventional to use μ to denote the coupling parameter of **K**-matrix theory. It should not be confused with the quantum defect, which is usually clear from the context.

approach the constant value $q = 0$ characteristic of the broad perturber. There are, however, two ways of approaching the weak and the strong coupling limits, depending on the sign of H_0, which is why, in one case, the line may become broader with increasing $| H_0 |$ (as in fig. 8.14(b)), while in the other, it becomes narrower (as in fig. 8.14(a)) until the vanishing width is traversed. Irrespective of the sign of H_0, the minima are shifted in the same direction in energy, *viz.* away from the perturber with increasing magnitude of the coupling strength $| H_0 |$. This is a property known as *spectral repulsion*.

Perturbations by a pure antiresonance are quite rare. A more usual situation is the one shown in fig. 8.15, where the perturber has a broad asymmetric window.

Finally, note a general feature of Rydberg series perturbed by an antiresonance: for all values of the coupling strength, any member occurring close to E_B will appear as an absorption *maximum*, about which, for weak coupling, a q reversal may also occur. This is the converse of the property uncovered for giant resonances: an experimental example occurs for the $n = 12$ member of the autoionising series of Ba [420].

8.27   Perturbation by a broad intruder level

In this section we describe the q-reversal effect and the general **S**-matrix formula for the cross section. The change in symmetry of the profiles in a Rydberg manifold (q-reversal effect) is a fairly common feature of perturbed series. However, it is *not* universal, and the conditions under which a q reversal may occur, as well as the number of q reversals which can exist, are questions of interest. As we shall demonstrate, q reversals may appear or disappear, depending on the coupling strength between channels as well as relative energies of resonances, and so their study can tell us more about the nature of the interactions between resonances. We now treat the problem more generally, and also include the radiative channel.

The q-reversal effect was discovered experimentally [382] through an accidental degeneracy between two subvalence excitation channels in the spectrum of $T\ell$ (see fig. 8.1). Since then, a number of other examples have been found (see [380]), especially where doubly-excited spectra involving two valence electrons and/or the outermost inner valence shell spectra overlap in energy. Another striking example [421] has been found by three-photon laser spectroscopy, and is shown in fig. 8.16. It exhibits a fairly clear asymmetry induced by a perturber, accompanied by fluctuations in the widths of the autoionising lines (which were found to be largest in the immediate vicinity of the perturber, in contrast to some of the examples

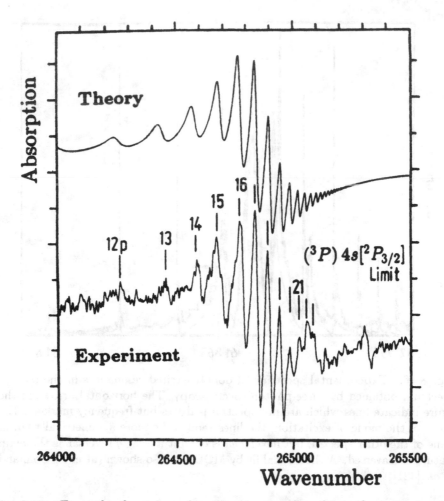

Fig. 8.15.   Example of a series enhancement near a broad intruder resonance in the doubly-excited spectrum of Ar (after M.A. Baig *et al.* [418]).

discussed later in this chapter). Instances have been discovered [435] in which the q-reversal effect can be controlled by an externally applied electric field. We also note that the q-reversal effect is not confined to atoms, and there is every reason to anticipate frequent occurrences in molecules as well (see [422] for further discussion) because the likelihood of appropriate degeneracies is greater in molecules than in atoms. There have been several reports ([423, 424]) of q reversals in spectra of small molecules.

The formal **K** matrix theory for a Rydberg manifold perturbed by a broad intruder state was setup by Lane [393]. The results have been related in detail to the q-reversal problem by Connerade *et al.* [380], whose

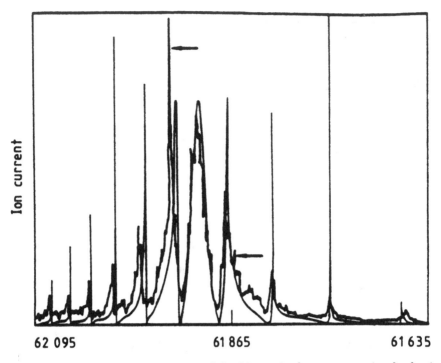

Fig. 8.16. Experimental spectrum of doubly-excited resonances in the barium spectrum obtained by three-photon spectroscopy. The horizontal arrows in the figure indicate lines which are not spectral features but frequency markers. Because of the mode of excitation, the lines tend to be more symmetrical than in some of the other spectra, but nevertheless exhibit a clear q reversal as the main feature is traversed. A theoretical fit by MQDT is also shown (after F. Gounand *et al.* [421]).

treatment we follow here.

When an extra level B is added to a series of levels labelled by n, then the **K**-matrix has the form

$$K_{ab} = \tilde{K}_{ab} + \frac{1}{2}\sum_n \frac{\Gamma_{na}^{1/2}\gamma_{nb}^{1/2}}{E_n - E} + \frac{1}{2}\frac{\hat{\Gamma}_{Ba}^{1/2}\hat{\Gamma}_{Bb}^{1/2}}{E_B - E - \sum_n H_n^2/(E_n - E)} \qquad (8.66)$$

where we define

$$\hat{\Gamma}_{Ba}^{1/2} - \sum \frac{H_n\Gamma_{na}^{1/2}}{E_n - E} \qquad (8.67)$$

and Γ_{na}, Γ_{Ba} are **K**-matrix widths of levels n and B for channel a, other quantities are as before and $\tilde{K}_{ab}$ is a background constant representing underlying continua.

For a general set of levels n which are arbitrarily spaced, as occur in nuclear physics, the inversion $(1 - K)^{-1}$ cannot be performed explicitly to calculate cross sections, and exact analytic expressions cannot be found. Atoms are interesting from this point of view: the ratio $\mu_a \equiv (\Gamma_{na}^{1/2}/H_n)$ is independent of n , because both H_n and $\Gamma_{na}^{1/2}$ are proportional to $n^{*-3/2}$ in an asymptotically Coulombic field. It turns out that an exact result for the inversion can be derived [380] in just this special case.

We now discuss a problem involving two open channels: a strongly coupled particle channel a and a weakly coupled radiative channel γ, so that $K_{a\gamma}$ and $K_{\gamma\gamma}$ are of first and second order of smallness respectively. It turns out that the exact result can be specialised to yield the **S**-matrix:

$$S_{a\gamma} = \frac{(2\overline{K}_{a\gamma}\epsilon - \Gamma_{Ba}^{1/2}\Gamma_{B\gamma}^{1/2}) + (2\overline{K}_{a\gamma} + \mu_a\mu_\gamma\epsilon + \Gamma_{Ba}^{1/2}\mu_\gamma + \Gamma_{B\gamma}^{1/2}\mu_\gamma)\Sigma}{(i + \overline{K}_{aa})\epsilon - \tfrac{1}{2}\Gamma_{Ba} + (i + \overline{K}_{aa} + \mu_a\Gamma_{Ba}^{1/2} + \tfrac{1}{2}\mu_a^2\epsilon)\Sigma}$$

(8.68)

where $\epsilon \equiv E - E_B$ is the detuning from the broad level E_B. It is possible to redefine the quantities so that $\overline{K}_{aa}$ is absorbed, as in Fano's parametrisation, in which the expression for $S_{a\gamma}$ does not include $\overline{K}_{aa}$. One then obtains six real parameters, namely E_B, $\Gamma_{Ba}^{1/2}$, $\Gamma_{B\gamma}^{1/2}$, μ_a, μ_γ and $\overline{K}_{a\gamma}$ to characterise the resonances, the intruder and the interaction between them.

If the radiative widths $\Gamma_{B\gamma}^{1/2}$ and μ_γ are very small and the background constant $\overline{K}_{a\gamma}$ is neglected, then the formula simplifies to the following useful expression:

$$\sigma\epsilon = \frac{(\epsilon + \tfrac{1}{2}q\Gamma_B)^2}{(\epsilon + \Sigma)^2 + \left[(\mu_a\Gamma_B^{1/2} + \tfrac{1}{2}\mu_a^2\epsilon)\Sigma - \tfrac{1}{2}\Gamma_B\right]^2}$$

(8.69)

which is closely related to the standard Fano formula for an isolated level. In fact, the numerator is the same as in the Fano formula for the resonance described by E_B and Γ_B, while an additional Rydberg series of resonances is introduced via Σ in the denominator.

From the simplified formula (8.69), one can represent a number of situations in which radiative widths do not appear to be important. Thus, fig. 8.1(b) shows a 'skewed' q-reversal effect similar to the observed one in fig. 8.1(a). Similarly, fig. 8.17 shows enhancements of upper series members in observed and calculated spectra, and fig. 8.18 shows a q reversal straddling one resonance of $q \to 0$ in both experimental and calculated spectra.

Finally, in fig. 8.19 we show an example in which the perturber has more pronounced asymmetry, but nevertheless induces a q reversal.

An important conclusion from such numerical studies [380] is that q reversals are the hallmark of *weak* coupling, i.e. they tend to disappear

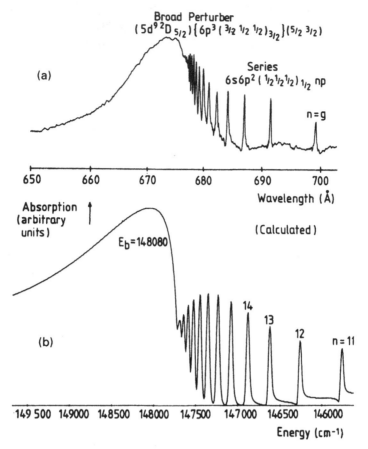

Fig. 8.17. Enhancements of upper series members of a Rydberg series as a broad intruder resonance is approached from below, as obtained: (a) experimentally in the spectrum of Pb; and (b) in an approximate simulation, using the simplified formula in the text (after J.-P. Connerade and A.M. Lane [381]).

as the interaction strength H_n is increased.

8.28 The properties and variation of q

We now consider how many q reversals are expected in a Rydberg series and the conditions under which they occur. Experimental data (such as those in fig. 8.18) demonstrate that there can be more than just one q reversal in a Rydberg series perturbed by an intruder level. One might ask: is this a sign of perturbations due to more than one interloper, or can one interpret the two reversals as having a common origin?

Some insight is gained by asking how many q reversals can be expected

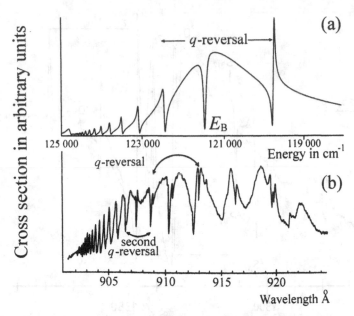

Fig. 8.18. Case of a q reversal straddling one intervening, almost symmetrical window resonance as obtained in: (a) a calculation using identical parameters to those of the first q-reversal calculation above, except that the coupling strength was increased; and (b) an experimental situation which also occurs in the Tℓ spectrum (after J.-P. Connerade and A.M. Lane [381]).

when a Rydberg manifold is perturbed by a single intruder. The answer turns out to be more complicated than one might expect, and depends both on the strength of coupling and on the relative magnitudes of the particle and radiative widths.

General **K**-matrix expressions have been derived for the variation of q as a function of detuning, both in elastic scattering and in photoionisation [375, 377]. It turns out that as many as six q reversals can occur in the general case in photoionisation but that, in elastic scattering, there are at most two.

K-matrix theory is a rather natural framework to analyse the behaviour of the q parameter because, in the case of a single channel, $-q^{-1}$ is already the background **K**-matrix.

For a single channel, the essential result is that

$$- q_a^{-1} = \overline{K}_{aa} + \mu_a(\Gamma_a^{1/2} + \frac{1}{2}\mu_a\epsilon) - \frac{\frac{1}{2}(\epsilon + \Sigma_\infty)(\Gamma_a^{1/2} + \mu_a\epsilon)}{(\epsilon + \Sigma_\infty)^2 + (\pi t)^2} \qquad (8.70)$$

for elastic scattering. In this expression Σ_∞ is defined as follows. We

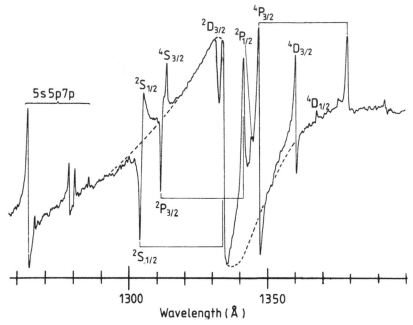

Fig. 8.19. Example of q reversals induced by a highly asymmetric broad perturber, as observed in the spectrum of In. Although the finer structure here does not form a Rydberg series, all the sharper lines originate from the same electronic configuration (after J.-P. Connerade and A.M. Lane [381]).

write

$$\Sigma(E) \equiv \sum_n \frac{H_n^2}{E_n - E} \qquad (8.71)$$

and

$$\Sigma(E + i\delta) = \Sigma_\infty + i\pi t \qquad (8.72)$$

where $t = H_n^2/D_n$ is independent of n, D_n being the level spacing.

The variation of q as a function of energy is plotted in fig. 8.20 for coupling strengths ranging from weak to strong coupling. These curves reveal the following characteristics:

(i) The true form of the variation of q is not a Fano profile as one might expect, but exhibits a pole at each q reversal.

(ii) All curves pass through the background point $q = q_B$ at $E = E_B$ (with $\Sigma_\infty = 0$). However, this is not necessarily observable unless there happens to be a fine level from the series at or near $E = E_B$.

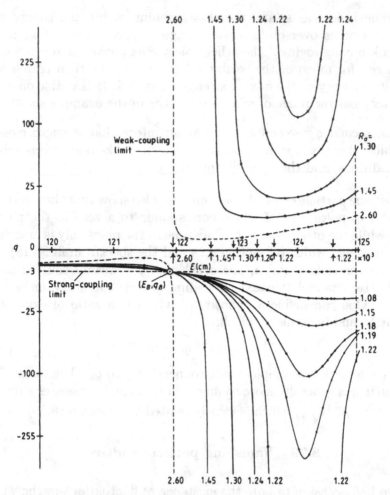

Fig. 8.20. The variation of the shape parameter as a function of energy in the same situation as previously plotted for $T\ell$, for different values of the coupling strength (after J.-P. Connerade ad A.M. Lane [381]).

(iii) All q reversals disappear at large mixing strength, even when there is still appreciable variation in the value of q_a. The sign of q_a then becomes the same as that of q_B. In the strong coupling limit, $q_a = q_B = q$ and is constant.

(iv) With the parameters selected, curves at weak mixing exhibit only one q reversal within the energy range of the Rydberg series. However, within a narrow range of coupling strengths, two q reversals appear. One, which can be regarded as the main one, remains stable in energy and occurs close to E_B. It is only slightly offset from the energy of the intruding level, moving closer to it as the mixing strength decreases, and therefore remains within the energy span of the series over quite a wide range of parameters. This as the weak-mixing regime. The other pole occurs above the series limit in energy for much of the weak-mixing regime, but then travels very fast in energy as the mixing strength is varied. It is visible only over a very narrow range of mixing strengths in the example shown.

(v) The second q reversal appears at an interaction strength close to which there is a sudden switch between weak- and strong-mixing conditions, and the poles disappear.

With just one particle channel open, one can also show that the q reversals occur at two poles, one of which corresponds to a zero in the particle widths, while the other does not. Thus, one of the poles only is associated with a vanishing width (cf section 8.29 and the spectra in fig. 8.18).

For *two* channels γ and a, one of which, γ, is weakly coupled and is the radiative channel (treated to leading order), the situation is more complex. One can define a quantity Q which is a ratio of expressions linear in E, and in terms of which:

$$\cot^{-1} q_\gamma = \cot^{-1} q_a - \cot^{-1} Q \qquad (8.73)$$

which is a remarkably simple result connecting q to q_a. The form of the variation in q is generally more complex, but, again, it possesses a definite pole at $\epsilon = -\mu_a^{-1}\Gamma_{Ba}^{1/2}$, which is clearly related to equation (8.59).

8.29 Vanishing particle widths

We have had occasion to note the existence of fluctuations in the widths of interacting autoionising resonances, with the occasional instance of a near-zero width (cf section 8.25).

As already noted, a very narrow line occurs for a specific detuning from the energy of the intruder level, given by $\epsilon = -\mu_a^{-1}\Gamma_B^{1/2}$ for a Rydberg series of autoionising resonances interacting with a giant resonance. The same result is obtained if the giant resonance is treated explicitly as modulation in the continuum or as an intruder. In the present context, a giant resonance is simply a special case of an intruder which happens to be very broad.

For a general intruder, two quantities are of interest in defining the level sequence, namely:

(i) the shape parameter

$$-q^{-1} = \mathcal{R}eK(E + ie) \qquad (8.74)$$

(ii) the strength function

$$\frac{\pi\Gamma}{2D} = \frac{\mathcal{J}mK(E+ie)}{1 + [\mathcal{R}eK(E+ie)]} \tag{8.75}$$

where D is the average level spacing. The defining property of a giant resonance in this context is its large width. One can show [375] that

$$- q^{-1} = \tan\tilde{\Delta} \tag{8.76}$$

(in fact, the very same equation as (8.32), except that q is no longer constant) where $\tilde{\Delta}$ is now the total background phase shift including the giant resonance. Thus, (8.76) is equivalent to (8.57). Also, the strength function

$$\frac{\pi\Gamma}{2D} = \frac{\frac{1}{2}\pi t(\Gamma_{BF}^{1/2} + \epsilon\mu)^2}{\Sigma_\infty^2 + (\overline{K}\epsilon - \frac{1}{2}\Gamma_{BF})^2} \tag{8.77}$$

vanishes at $E = -\mu_F^{-1}\Gamma_{BF}^{1/2}$, which implies that the particle width vanishes.

The cancellation of particle widths of resonances as a result of an intruder state disturbing a sequence of levels is a well-known effect in nuclear physics [425]. In atomic physics, under suitable conditions, the particle width may go to zero [426] A good example is shown in fig. 8.21, where line narrowing in a doubly-excited series of autoionising resonances is observed, as a result of a perturbation by an intruder state.

Interestingly, the perturber is difficult to detect otherwise than through the dramatic series perturbation for which it is responsible. Indeed, it had been suggested at an early stage in some **R**-matrix calculations that the vanishing width effect in this case might be due, not to this perturbation, but to the occurrence of a Cooper minimum in the underlying continuum, as a result of which the autoionisation matrix element could exhibit phase cancellation. While the latter is undoubtedly an interesting alternative mechanism for producing a vanishing particle width, it does not appear to be relevant for Ca. An analysis by MQDT [429] confirms that the vanishing width is due to this perturbation. The effect has also been calculated by the eigenchannel **R**-matrix method [431].[5]

8.30 Vanishing radiative widths

The vanishing particle width is perhaps a surprise, since it involves stabilising a level in the continuum by introducing a further perturbation. More

---

[5] An interesting feature of the doubly-excited spectrum of Ca is that, at very high resolution, the spectrum becomes extremely complex [263], and displays some of the peculiarities of nearest neighbour distribution which are discussed in chapter 10.

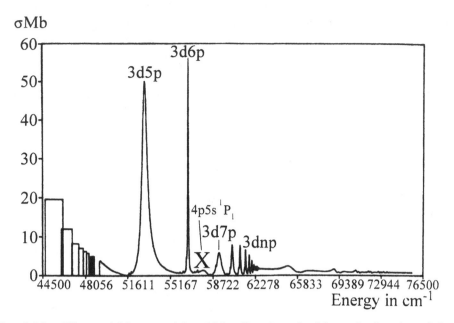

Fig. 8.21. The vanishing particle width effect in a doubly-excited series of the Ca spectrum. Note the narrowing of the $n = 6$ member, although the remainder of the series members are broad. The perturber (hardly visible on account of its breadth and weakness) is indicated by an X in the figure (after U. Griesmann *et al.* [390]).

remarkable still is the *vanishing radiative width*, which involves stabilising a level against the emission of radiation by a perturbation in which the ionisation continuum plays a role. In essence, this is an effect in which the particle and the radiative decay channels interfere destructively, a possibility envisaged by Wigner when he originally formulated his scattering theory. An example is shown in fig. 8.22 This effect provides a beautiful illustration of Seaton's theorem on the continuity of oscillator strength per unit energy across the threshold.

8.31 More than one series of autoionising resonances

In our discussion of interacting resonances, we have so far implied that only one Rydberg manifold was present and that the intruder state was isolated from other resonances of its own channel. In general, this is not be true, and indeed it frequently occurs that two series of autoionising resonances overlap in energy, with several of their members interacting. For a simple case, involving just two such series and one underlying continuum, one can obtain some useful formulae which are summarised here.

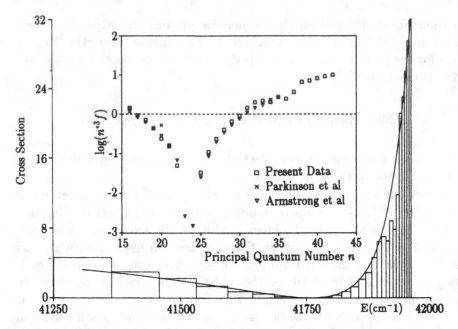

Fig. 8.22. The vanishing radiative width in the Ba spectrum. Experimental data are shown which correspond to oscillator strength measurements below the ionisation threshold by the magneto-optical method described in chapter 4, which also gives an overview of the Ba spectrum including the broad $5d8p$ perturber responsible for the vanishing radiative width. The inset shows a comparison between measured values (squares) and an MQDT extrapolation (triangles) on a logarithmic scale (see also fig. 4.3 – after J.-P. Connerade *et al.* [136]).

We return to this problem from another angle in section 8.35, where we make use of these results.

For elastic scattering (which may or may not apply to photoionisation), we have, as in section 8.25, that

$$\tan \Delta_0 = \Sigma_B + \Sigma_a + \left\{ \frac{-2\mathcal{X} + \mathcal{X}^2(\Sigma_B + \Sigma_n)}{1 - \mathcal{X}^2(\Sigma_B + \Sigma_n)} \right\} \Sigma_B \Sigma_n \qquad (8.78)$$

where the sums are defined as previously, with B denoting a Rydberg manifold of broad interlopers, while the coupling strength

$$\mathcal{X} \equiv \frac{H_n B}{\frac{1}{2}\Gamma_B^{1/2}\Gamma_n^{1/2}} \qquad (8.79)$$

In the special situation of weak coupling (i.e. $\mathcal{X} \ll 1$), we obtain an especially simple formula:

$$\tan \Delta_0 = \Sigma_B + \Sigma_n - 2\mathcal{X}\Sigma_B\Sigma_n \qquad (8.80)$$

which expresses the resonant phase shifts in terms of the sums of the individual phase shifts of all the resonances present and a product term, representing the interaction, weighted by the mixing strength. Expressions for the cross section are readily derived from the procedures we have given previously.

8.32   Extension to the high energy spectrum

In this section, we consider the effect of many channels, Auger broadening and the growth in radiative width.

In previous sections, we have, for simplicity, confined our attention to situations involving one open channel. For isolated Fano profiles, this leads to an exact zero in the cross section at the transmission window, and to a series of exact zeros in the cross section associated with each resonance for a complete Rydberg series. Obviously, if one wishes to study profile shapes in detail and the connection between the bound states and autoionising resonances implied by Seaton's theorem (see previous section) it is desirable to work just above the first ionisation potential, where the number of open continuum channels is at a minimum. In general, this is not possible, and the number of open channels increases rapidly with increasing energy.

Several complications arise from an increase in the number of open channels. First, the zeros just mentioned are 'filled in' by interactions with many continua. Thus, one sees that the asymmetries of the profiles tend to be less pronounced towards higher energies. Other factors conspire to reinforce this general trend.

The series limit on which the autoionising resonances converge may or may not be stable. If the corresponding parent ion state can decay radiatively, then because radiative widths are (for low excitation energies) usually much smaller than autoionisation widths, the series observed is not much affected until high values of n. On the other hand, if the parent ion state itself lies above the double-ionisation threshold, it becomes susceptible to particle decay and acquires its own intrinsic width through autoionisation.

The lowest energy at which this can occur is just above the double-ionisation threshold, in which case particle emission proceeds by autoionisation of the atom followed by a second autoionising step in the ion. This sequence of events is referred to as two-step autoionisation, and was discussed in section 7.15. As the excitation energy is further increased, the processes become more complex, and may involve different shells of the ion. The generic name for this type of particle broadening is Auger broadening (see section 6.8). One distinguishes experimentally between autoionisation and Auger broadening by the fact that the Auger contribution to the width is a property of the excited ion state or core vacancy. It is therefore independent of the principal quantum number of the excited electron. On the other hand, the contribution due to autoionisation should (apart from effects due to perturbations) decrease as the cube of the effective quantum number.

All effects due to broadening of the ion states lead to truncation of the Rydberg series at a finite value of n. Auger broadening also tends to be of rather Lorentzian shape, and as its contribution to the linewidth increases, the profiles become more symmetrical. The two causes of broadening are not entirely independent: they give rise to interference effects which have been studied theoretically, among others, by Combet-Farnoux [432] . The K-matrix formulation is again convenient for handling this problem. One has, for two general channels a and b, that

$$\sigma_{ab} = \sigma_{ab}^0 \left(\frac{1}{1+\epsilon^2}\right) \left| \epsilon + i - 2i \frac{\Gamma_a^{+1/2} \Gamma_b^{+1/2}}{\Gamma \hat{S}_{ab}} \right|^2 \qquad (8.81)$$

(from [377] equations (1)–(18)) where $\Gamma_a^{+1/2} = (i - q_a) \mathcal{J}m\Gamma_a^{1/2}$, in which q_a is the asymmetry parameter for channel a. Now introduce two different particle decay channels b and c, and one weakly coupled radiative channel γ. For small $\sigma_{\gamma c}^0$, one finds

$$\sigma_\gamma \equiv \sigma_{\gamma b} + \sigma_{\gamma c} = \frac{\sigma_{\gamma b}^0}{1+\epsilon^2} \left\{ \frac{|\Gamma_b^+|}{|\Gamma_b^+||\Gamma_c^+|}(q_\gamma^2 - 1 + 2\epsilon q_\gamma) + (1 + \epsilon^2) \right\} \qquad (8.82)$$

Taking b as the autoionisation and c as the Auger channels, we set $\Gamma_b^+ = 2\pi \, | \, H_{auto} \, |^2$ and $\Gamma_c^+ = 2\pi \, | \, H_{auger} \, |^2$ and $\sigma_{\gamma b}^0 = | \, Q_{auto}^0 \, |^2$ whereupon the matrix elements are

$$Q_{auger} = 2\pi \left(\frac{H_{auger} H_{auto}}{\Gamma_{total}}\right) \left(\frac{q - i}{\epsilon + i}\right) \qquad (8.83)$$

and

$$Q_{auto} = Q_{auto}^0 \left(1 + \eta(E)\frac{q - i}{\epsilon + i}\right) \qquad (8.84)$$

where $\eta(E) = (\Gamma_{auto}/\Gamma_{total})$ is called the *branching ratio* and different shapes of profile occur in different channels, depending on whether the dimensionless quantity η is large or small. The main conclusion from such studies is that the asymmetry of resonant lines in Auger channels is due essentially to interchannel effects [432].

If we proceed to still higher energies, then, because of the factor $2h\nu^3/c^2$ in the Einstein–Milne relations, the radiative widths grow very fast and eventually dominate lifetime broadening. The question then arises: how will this interaction affect the discussion of photoabsorption?

We ignore electron emission and assume that the exciting radiative transition ($\ell_i \to n\ell$) is weak compared to other radiative decays ($\ell_j \to \ell_i$) of the excited state. Without assumption, if the exciting transition is w, then the total photoabsorption is $\sigma = 2\mathcal{R}e(1 - S_{ww})$, where

$$\frac{1}{2}(1 - S_{ww}) = \left[1 - (1 - iK)^{-1}\right]_{ww} \tag{8.85}$$

If the **K**-matrix has its general form

$$K = \hat{K} + \frac{1}{2}\frac{\Gamma_\lambda^{1/2}\tilde{\Gamma}_\lambda^{1/2}}{E_\lambda - E} \tag{8.86}$$

then one can show that

$$\sigma = \tilde{\sigma}\left(\frac{(q+\epsilon)^2}{1+\epsilon^2}\rho^2 + (1 - \rho^2)\right) \tag{8.87}$$

i.e. one simply derives the Fano formula, with

$$q = \frac{\mathcal{R}e\Gamma_{\lambda w}^{+1/2}}{\mathcal{J}m\Gamma_{\lambda w}^{+1/2}} \tag{8.88}$$

If there are many electron channels with random phases, then $\Gamma_{\lambda w}^{+1/2}$ is real, implying that $q \to \infty$, which explains why the profiles for the very deep inner shells all tend to become symmetrical.

In summary, the effects of both Auger and radiative broadening generally conspire to make the profiles more symmetrical at high energies. Thus, in all deep core hole spectra, we expect isolated profiles to tend towards symmetrical absorption peaks, and the only asymmetries to result from interchannel couplings when resonances overlap in energy and interact. This general conclusion agrees with observations.

8.33   Vanishing fluctuations

Spectral fluctuations are a distinctive property of atomic spectra, but have rarely been studied in their own right. One example involving fluctuations is the procedure of Gailitis averaging near a series limit (see chapter 3), This involves taking the mean of the maximum and minimum excursions as the series limit is approached, and emphasises the fairly obvious result that spectral fluctuations tend to zero as the limit of a Rydberg series is approached.

Another situation where experimental fluctuations in the cross section tend to zero is when a vanishing width occurs and the instrumental resolution is finite. The fluctuations then pass through zero at an energy which does not coincide with the series limit. However, the vanishing width effect occurs with just one open channel, whereas fluctuations may exhibit variations even if many channels are open. The more general question thus arises: can a point of zero fluctuations be similarly detuned from the associated series limit independently of any instrumental limitations, i.e. do points of vanishing fluctuation exist *per se* as an independent property of atomic spectra, resulting from a perturbation by an intruder state?

In nuclear physics, fluctuations are studied in their own right through the variance:

$$\text{var}\,\sigma \equiv \langle \sigma(E)^2 \rangle - \langle \sigma(E) \rangle^2 \qquad (8.89)$$

where the Dirac brackets denote an average performed over an interval of energy ΔE wide enough to encompass local fluctuations, but sufficiently narrow for gross variations to be preserved.

Spectral fluctuations are an observable, and therefore exhibit the same properties of continuity across thresholds as other obervables in atomic physics. Although the variance is strictly of fourth order in elements of the **S**-matrix, a useful theorem, due to Moldauer [433] states that the fluctuation in the *total* cross section for a given entrance channel a is of second order in **S**-matrix elements:

$$\text{var}\,\sigma_a = 2\left\{ \langle |\,S_{aa}\,|^2 \rangle - |\,\langle S_{aa} \rangle\,|^2 \right\} = 2\langle \sigma_{aa}(fl) \rangle \qquad (8.90)$$

Thus, $\text{var}\,\sigma_a$ behaves like a cross section. For example, it is continuous across thresholds [374] and may possess interesting variations and zeros, just as one finds when considering interferences between Rydberg series of resonances and broad intruder states.

Although inspired by observational data to consider the problem of fluctuations in atomic spectra, Lane [374] was concerned, as a result of his analysis, that the number of parameters needed is too large, so that a zero-point in the fluctuations might prove unlikely. In his own words '... the strong minimum in the fluctuations ... is an exception to general expectation. In view of this, it may be just as well that there is some doubt about the effect ... being genuine....'

However, careful studies by laser spectroscopy have shown that the effect is genuine: a spectrum has been found for which the fluctuations do vanish at a definite energy (see fig. 8.23), while the widths of the resonances tend to zero smoothly as the series limit is approached.

The implication is that the minimum in the fluctuations is *not* due to a vanishing width in the total cross section.[6] One can also observe changes

---

[6] The Doppler width is much smaller than the widths of the lines near the point where fluctuations disappear.

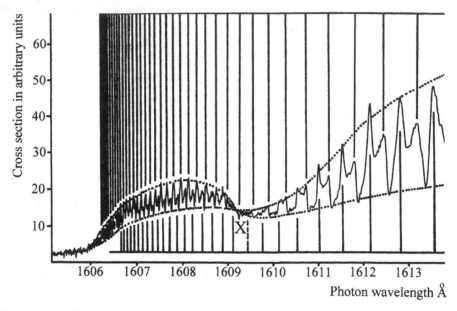

Fig. 8.23.   Vanishing fluctuations in a doubly-excited series of Ba. Note the existence of a point (marked X in the figure) at which spectral structure disappears, and which is not a series limit. Note also the q reversal about a window resonance (after J.-P. Connerade and S.M. Farooqi [442]).

in the shapes of the line profiles as the vanishing point is traversed, which provides further insight.

Using **K**-matrix theory, we now seek the simplest description, bearing in mind that the algebra of the full problem including all possible channels becomes too forbidding to be useful. Assume radiative widths can be neglected, i.e. that the spectrum is dominated by particle widths. We can therefore use the elastic scattering approximation. Referring to the experimental data of fig. 8.23, we note the presence of a transmission resonance, about which a conspicuous q reversal occurs. This is a crucial observation, which enables us to advance the analysis beyond the discussion of Lane [379].

Several instances where a q reversal straddles a $q = 0$ window resonance were described above. The form of the variation of q is then:

$$q = \frac{1 - \bar{\delta}\bar{\Delta}}{\bar{\delta} + \bar{\Delta}} \tag{8.91}$$

where $\bar{\delta}$ is a background, essentially constant phase shift due to the underlying continuum, and $\bar{\Delta} = \Gamma_B/2(E_B - E)$ is the phase shift due to the broad intruder resonance $| B \rangle$.

Hence $q = 0$ when $\bar{\Delta} = 1/\bar{\delta}$, which also turns out to be the condition for a maximum to occur in the cross section the intruder resonance would possess at zero coupling strength. This condition is satisfied at an energy E_0 given by

$$E_0 = E_B - \left(\frac{\Gamma_B}{2}\right) \tan \delta \qquad (8.92)$$

about this value, $\bar{\Delta}$, and therefore also q, changes sign, i.e. a q reversal occurs. This type of q reversal is the one referred to as most stable in section 8.28, and usually remains within the energy range of the intruder level under conditions of weak coupling, because it is tied to the energy E_B. The data exhibit this behaviour quite clearly, close to the point where the fluctuations disappear: one line (marked 'X' in fig. 8.23) has $q \sim 0$ and, about this line, a q reversal of the associated series does indeed occur.

For a single open channel, a vanishing width occurs at an energy

$$E_{vw} = E_B + \mu_a^{-1} \Gamma_B^{1/2} = E_B - H_n \Gamma_B^{1/2} / \Gamma_n^{1/2} \qquad (8.93)$$

Even if the ratio of the widths $\Gamma_B^{1/2}/\Gamma_n^{1/2}$ is large, E_{vw} still comes close to E_B provided the interaction strength H_n is very small. Although there is no vanishing width in the data of fig. 8.23 we have noted above that, for finite instrumental resolution, a vanishing width would also lead to vanishing fluctuations: in the presence of many open channels, if we suppose that only one of them (the dominant channel) contains a vanishing partial width at an energy E_{vw}, the remaining open channels introduce a nonfluctuating width, producing an effect apparently similar to finite instrumental resolution, but entirely internal to the atom.

This connection is explored in [379] section 3. In brief, it can be demonstrated for resonances much narrower than the interlevel spacings, that if only one partial width vanishes as a result of the presence of an intruder, there will be a vanishing of spectral fluctuations at an energy $E_{vf} = E_{vw}$. Lane [379] shows that, in the case of many open channels, an expression for var σ_a is obtained which might plausibly vanish for a single series and a single intruder state in Ba. However, the problem is rendered far more complex if there are two series and two intruders, and a vanishing point then becomes far less likely.

The observations show that the vanishing point lies close to *one* specific intruder energy E_B, signalled by a q reversal in one series (which is the hallmark of weak coupling). It is therefore reasonable to assume that the influence of one intruder state dominates in the vicinity of E_{vf} and that coupling to the other has become negligible close to E_{vf}. As regards the presence of a second series, we note that the ns and nd states built on the same core and with the same total J are anyway strongly mixed: it is therefore entirely reasonable that they should possess parallel properties.

We infer that a vanishing width in the case of a single open channel becomes a vanishing fluctuation in the presence of several open channels The data also demonstrate that the influence of a single perturber is enough for a disappearance of fluctuations to occur, as witnessed by the simultaneous occurrence of a single q reversal.

8.34   Controlling symmetries through external fields

The example of LICS (section 8.20) and other situations in multiphoton spectroscopy generally (see, e.g., [434] or the case treated in section 9.8) show that, spectral symmetries can be controlled in a variety of ways by using applied fields, and this suggests that the same might be achieved for single-photon spectroscopy using external DC fields, a possibility hinted at in [382]. An experimental example in which q reversals are controlled by an externally applied electric field has been reported by Lahaye and Hogervorst [435] in a high resolution laser experiment on bound states of Ba, similar in spirit to the work of Safinya and Gallagher [401]. The q reversal studied in [435] occurs close to the energy of the broad resonance. It is symmetrical about $q = 0$, as can be seen by scanning the applied field and is thus similar to the q reversals in elastic scattering analysed in section 8.25. It is also possible to use magnetic fields for this purpose (e.g. [436]), and there is active research on manipulating atomic spectra through external fields [437, 438]. In chapter 10, the influence of magnetic fields in inducing fine structure in atomic spectra will be described. Here, in order to emphasise the ubiquity of q reversals, we show, in fig. 8.24, yet another example of this effect when fine structure due to the field interacts with a broader resonance.

8.35   Lu–Fano graphs for autoionising series

We now turn to another question involving interacting series of autoionising resonances which can also be tackled by K-matrix theory. The question is whether the two-dimensional graphical representations or Lu–Fano graphs (see chapter 3), used so successfully to represent Rydberg series of bound states and to extrapolate their properties into the first autoionising range, can also be used to represent series consisting *entirely* of autoionising resonances (i.e. with no bound states) converging on a pair of limits well above the first ionisation threshold.

In principle, the answer to this question is contained in the equations of section 8.31, which describe two coupled series of autoionising resonances converging to different limits. We now show how this is achieved, and

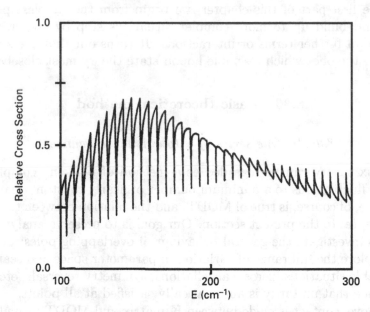

Fig. 8.24. Interaction between fine structure in negative ion photodetachment induced by a strong magnetic field and a broad Feshbach resonance (after C.H. Greene [412]).

what the differences exist between Lu–Fano graphs for bound Rydberg states and for coupled autoionising series.

Experimenters have been prompt in extending the graphical technique to coupled autoionising series because of its convenience, well before its theoretical justification was available (see, e.g., [439]). Two examples of this kind of graph were shown in figs. 3.3 and 3.4.

When a strong interaction between a pair of systems occurs, one can often isolate a pair of limits and argue that all extraneous interactions are weak, or, *a posteriori*, that the the structure of the resulting graph looks the same as for series of bound states, thereby vindicating the original assumption.

One should also examine the conditions under which the standard expressions of Lu and Fano [440] remain applicable when all the lines are autoionising resonances, and the extent to which the usual parametrisation must be reinterpreted or may exhibit differences, bearing in mind potential complications such as symmetry reversals and vanishing fluctuations. Fortunately, even in such cases, the two-dimensional representation remains a valid map.

The treatment we give now also provides a specific example of the connection between MQDT and **K**-matrix theories. Our plan is as follows:

as in the first part of this chapter, we begin from the simplest possible cases, and build up to more complex situations step by step with the inclusion of further terms or interactions. It turns out that the simplest cases are the ones which resemble bound state theory most closely.

8.36   Basic theoretical method

8.36.1   One series of autoionising resonances

K-matrix theory can be regarded as a method of reducing complicated spectral fluctuations to a minimum number of nearly constant quantities. The same, of course, is true of MQDT, and the relation between the two is fundamental to the present section. Our goal is to preserve analyticity in order to investigate the general behaviour of overlapping poles: one may then explore the full range of variation in parameter space and search for remarkable situations (zeros, cancellations, symmetry reversals, etc) with confidence that unitarity is automatically satisfied at all points.

As above, contact is made between **K**-matrix and MQDT formulations through the replacement:

$$\sum_n \frac{\Gamma_n/2}{E_n - E} \rightarrow \mathcal{X} \cot \pi(\nu + \mu) \tag{8.94}$$

This replacement is of key importance: it acts like a change of variables from Γ_n, E_n, E to $\mathcal{X}$, μ, ν.

We begin from the **K**-matrix form for one Rydberg series of resonances:

$$K_{res} = \tan \Delta = \sum_n \frac{\Gamma_n/2}{E_n - E} \equiv \Sigma \tag{8.95}$$

The sum begins with the lowest resonance, and must be taken to infinity at the series limit. In MQDT, the bottom of the channel is not defined, but a sum to infinity does not arise either. The quantity Γ_n (the linewidth) scales as $(n - \mu)^{-3}$ for an asymptotically Coulombic potential. A zero in the cross section occurs between each resonance.

The sign of $\mathcal{X}$ is of interest; it is definite for all ν, as appears from the following argument. Differentiating (8.94) with respect to ν, one has:

$$\mathcal{X} = -\frac{1}{1 + \cot^2 \pi(\nu + \mu)} \frac{2R}{\pi \nu^3} \sum_n \frac{\Gamma_n/2}{(E_n - E)^2} \tag{8.96}$$

whence $\mathcal{X}$ is negative for all $E_1 < E < E_\infty$, since Γ is necessarily positive. Near a pole, the sum in (8.94) collapses to one term, and

$$\mathcal{X} = -\frac{\pi}{4R}(n - \mu)^3 \Gamma_n$$

which is constant provided only that $\Gamma_n \alpha (n-\mu)^{-3}$ (i.e. Auger broadening is excluded from the present theory). The quantity $\mathcal{X}$ is therefore proportional to the ratio of the linewidths to linespacings for an unperturbed autoionising series.

We note from (8.95) that, if $\delta = 0$, $K \to \pm\infty$ when $\nu = n - \mu$ as expected from QDT. If the background continuum cannot be neglected, i.e. δ is nonzero, the Dubau–Seaton formula (section 8.24) results and we find (modulo 1):

$$\nu = \frac{1}{\pi} \tan^{-1}(\mathcal{X} \tan \delta) - \mu = \frac{1}{\pi} \tan^{-1}(\mathcal{X} q^{-1}) - \mu \qquad (8.97)$$

where $q = \cot \delta$ is the shape index of the autoionising line. Equation (8.97) expresses the fact that autoionising resonances are generally asymmetric, so that all the maxima are shifted by an amount which depends both on their width (through $\mathcal{X}$) and on the asymmetry or shape parameter q (for $q \to \pm\infty$, the resonances are symmetric and (8.97) reduces to the standard form for bound states; for $q = 0$, there is of course no maximum but a series of minima). We may thus anticipate that, for two coupled series of resonances (labelled 1 and 2) whose shape is not constant, as in section 8.25, quantum defect plots can be significantly distorted.

In particular, the points in the plot will not be located exactly on the curve

$$I = \frac{R}{\nu_1^2} - \frac{R}{\nu_2^2}$$

where I is the difference in energy between the two series limits and R is the Rydberg, as is the case for bound states, unless the resonances happen to be symmetrical or the lines are very sharp.

Another obvious but significant point follows from the general form of the **K**-matrix with one continuum δ and a series of resonances Δ:

$$K = \tan(\Delta + \delta) = \frac{\tan \Delta + \tan \delta}{1 - \tan \delta \tan \Delta} \qquad (8.98)$$

All the poles of the **K**-matrix are given by the zeros in the denominator of the expanded form for the tangent of the phase shift: infinities in the numerator are cancelled by corresponding infinities in the denominator when either $\delta = \pi/2$ or $\Delta = \pi/2$. This is important in what follows because *all* the branches of two-dimensional graphs plotted from intensity maxima are recovered from the denominator of the expanded **K**-matrix.

8.36.2 *Two series of resonances and one continuum*

For the limiting case of two autoionising series in the presence of one continuum, the meromorphic form again results from adding phase shifts thus:

$$K = \tan(\Delta_1 + \Delta_2 + \delta)$$

$$= \frac{\Sigma_1 + \Sigma_2 + q^{-1} - q^{-1}\Sigma_1\Sigma_2}{1 - \Sigma_1\Sigma_2 - q^{-1}(\Sigma_1 + \Sigma_2)} \tag{8.99}$$

where we define $\Sigma_1 \equiv \tan\Delta_1$, $\Sigma_2 \equiv \tan\Delta_2$, using subscript 1 for the lower and 2 for the upper series. Again, a zero occurs between each resonance. From (8.99) $K \to \pm\infty$ when

$$\Sigma_1\Sigma_2 = 1 - q^{-1}(\Sigma_1 + \Sigma_2) \tag{8.100}$$

whence ν_2 plotted modulo 1 from maxima in the cross section is given by:

$$\nu_2(\nu_1) = \frac{1}{\pi}\tan^{-1}\left\{\frac{\mathcal{X}_2\left[\mathcal{X}_1 + \tan\delta\tan\pi(\nu_1 + \mu_1)\right]}{\tan\pi(\nu_1 + \mu_1) - \mathcal{X}_1\tan\delta}\right\} - \mu_2 \tag{8.101}$$

If we define $\overline{\Sigma}_1 \equiv \tan(\Delta_1 + \delta_1)$ and $\overline{\Sigma}_2 \equiv \tan(\Delta_2 + \delta_2)$, where $\delta_1 + \delta_2 = \delta$, then K has the form:

$$K = \frac{\overline{\Sigma}_1 + \overline{\Sigma}_2}{1 - \overline{\Sigma}_1\overline{\Sigma}_2}$$

$$= \frac{(q_1\Sigma_1 + 1)(q_2 - \Sigma_2) + (q_2\Sigma_2 + 1)(q_1 - \Sigma_1)}{(q_1 - \Sigma_1)(q_2 - \Sigma_2) - (q_1\Sigma_1 + 1)(q_2\Sigma_2 + 1)}$$

$$= \frac{(1 - q_1q_2)(\Sigma_1 + \Sigma_2) - (q_1 + q_2)(1 - \Sigma_1\Sigma_2)}{(1 - q_1q_2)(1 - \Sigma_1\Sigma_2) + (q_1 + q_2)(\Sigma_1 + \Sigma_2)}$$

$$= \frac{(\Sigma_1 + \Sigma_2) + \xi(1 - \Sigma_1\Sigma_2)}{(1 - \Sigma_1\Sigma_2) - \xi(\Sigma_1 + \Sigma_2)} \tag{8.102}$$

where $q_1 = \cot\delta_1$ and $q_2 = \cot\delta_2$ are the Fano shape parameters for series 1 and 2 respectively and $\xi = (q_1 + q_2)/(q_1q_2 - 1)$ is a combined asymmetry parameter for the pair of series. This equation can be regarded as a generalisation of the Dubau–Seaton formula for the case of two autoionising series. The analogue of equation (8.100) is then:

$$\Sigma_1\Sigma_2 = 1 - \xi(\Sigma_1 + \Sigma_2) \tag{8.103}$$

whence, if either $(q_1, q_2) \to 0$ or $(q_1, q_2) \to \infty$ (symmetric resonances), the maxima lie on:

$$\tan\pi(\nu_1 + \mu_1)\tan\pi(\nu_2 + \mu_2) = \mathcal{X}_1\mathcal{X}_2 \tag{8.104}$$

(bearing in mind that $\mathcal{X}_1$ and $\mathcal{X}_2$ are *both* negative – see (8.96) and the associated discussion, the product $\mathcal{X}_1\mathcal{X}_2$ is positive). The significance of these equations is readily understood: for asymmetric resonances, the maxima in the cross section do not coincide with the poles. Equation (8.104) is instantly recognisable in form because it has the same structure as a standard equation of *bound state* MQDT:

$$\tan\left[\pi(\nu_1 + \mu_1)\right]\tan\left[\pi(\nu_2 + \mu_2)\right] = R_{12}^2 \qquad (8.105)$$

(cf equation (3.14)) Despite their similarity, equations (8.104) and (8.105) should not be confused, because they do not convey quite the same meaning: in MQDT, the quantity R_{12}^2 (which is always positive) stands for the strength of coupling *between* two series of bound states. It is not related to autoionisation widths, except indirectly by an extrapolation across the lowest series limit into the adjoining continuum, where it yields widths for series 2. In equation (8.104), the quantities $\mathcal{X}_1$ and $\mathcal{X}_2$ do *not* represent an interseries coupling strength, but are related instead (cf equation (8.95)) to the widths Γ_{n_1} and Γ_{n_2} of the *unperturbed* autoionising series. What this expresses physically is the following interesting fact: the two-dimensional plot of two autoionising series in the presence of one continuum *when the resonances are symmetric* possesses the same branch structure and general appearance as the MQDT graph of two *interacting Rydberg series of bound states*, but exhibits avoided crossings whose widths depend on the breadths of the *unperturbed* resonances.

When the resonances are *not* symmetric, the maxima are shifted, so that they no longer occur at positions defined only by μ_1 and μ_2, and equation (8.103) should be used; it contains an additional term involving $\Sigma_1 + \Sigma_2$, and thus, in general, describes two-dimensional plots with a structure slightly different from the usual MQDT pattern described by equation (8.105).

The significance of the combined asymmetry parameter ξ is as follows. If $\xi = 0$ (symmetric case), then the maxima in the cross section lie on the curve $\Sigma_1\Sigma_2 = 1$, while the minima lie on the curve $\Sigma_1 + \Sigma_2 = 0$. If, however, $\xi \to \pm\infty$, then the roles are reversed. Note in particular that if $\mathcal{X}_1 = \mathcal{X}_2$, then large absolute values of ξ yield a straight line graph equivalent to the strong coupling graph of MQDT.

In general, the full range of variation of ξ has the effect shown in fig. 8.25 for a case in which $\mathcal{X}_1 = \mathcal{X}_2$: changes in the combined asymmetry parameter alter the appearance of the avoided crossing in a manner which can fill the whole of the available area in the graph, while preserving positive slope.

We may further consider the case in which the fine structure due to series 1 overlaps with a broad resonance $\Gamma_B/2(E_B - E) \equiv \varepsilon_B^{-1}$ which dominates Σ_2 in the energy range of interest. From equation (8.103) we

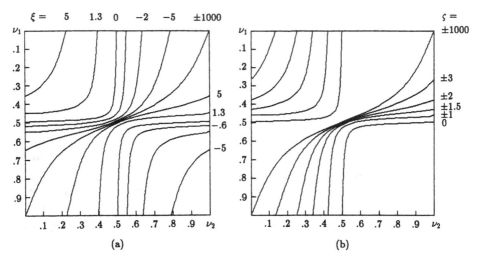

Fig. 8.25. Influence of the combined asymmetry parameters on a typical Lu–Fano graph: (a) for zero direct coupling and real values of the combined asymmetry parameter; (b) for zero combined asymmetry and various values of the coupling strength. Other parameters: $E_{\infty 1} = 71800$, $E_{\infty 2} = 71890$, $\mathcal{X}_1 = -0.1$, $\mathcal{X}_2 = -0.2$, $\mu_1 = 0.4$, $\mu_2 = 0.6$ (after J.-P. Connerade [444]).

have (modulo 1):

$$\nu_1 = \frac{1}{\pi} \tan^{-1} \left(\mathcal{X}_1 \frac{1 - \epsilon_B \xi}{\epsilon_B + \xi} \right) - \mu_1$$

So that the two-dimensional graph of ν_1 against E, plotted modulo 1, contains a single discontinuity shifted from the pole at $E = E_B$ to an energy near $E = E_B - \xi \Gamma_B / 2$, which is clearly related to the profile of the broad resonance. In fact, the absolute value of the term multiplying $\mathcal{X}_1$ is simply the K-matrix the resonance E_B, Γ_B would possess if its shape parameter were ξ.

Equations (8.103) and (8.104) explain the empirical fact that two-dimensional plots 'apply' to autoionising series: they yield closely similar graphs to those for bound states. Even so, one should be careful about interpreting avoided crossings: they cannot be strictly zero for autoionising lines even in the absence of interseries coupling and they also depend on the combined asymmetry parameter ξ.

In connection with asymmetry, an interesting experimental point arises: its influence on the two-dimensional graphs can in principle be probed directly in two-colour laser experiments, in which the same excited Rydberg states are accessed via different intermediate states, so that all parameters are the same *except* q_1 and q_2 [441].

8.36.3 Directly coupled series of autoionising resonances

When two series of resonances are coupled directly, the **K**-matrix contains off-diagonal terms, and a mixed product containing poles of both series results. For the coupled system:

$$K = \Sigma_1 + \Sigma_2 + F(\zeta, \Sigma_1, \Sigma_2) \cdot \Sigma_1 \Sigma_2 \qquad (8.106)$$

with

$$\Sigma_1 \equiv \sum_{n_1} \frac{\Gamma_{n_1}/2}{E_{n_1} - E}$$

$$\Sigma_2 \equiv \sum_{n_2} \frac{\Gamma_{n_2}/2}{E_{n_2} - E}$$

and

$$\zeta \equiv \frac{2H_{n_1 n_2}}{\Gamma_{n_1}^{1/2} \Gamma_{n_2}^{1/2}}$$

(see [381] for the definition of $H_{n_1 n_2}$); ζ is a dimensionless coupling strength which is independent of both n_1 and n_2 by construction and the function F depends both on ζ and on the series Σ_1 and Σ_2. An explicit form for F neglecting radiative widths (which are usually, though not always, much smaller than particle widths) is the elastic scattering formula [381]:

$$F(\zeta, \Sigma_1, \Sigma_2) = \left\{ \frac{-2\zeta + \zeta^2(\Sigma_1 + \Sigma_2)}{1 - \zeta^2(\Sigma_1 + \Sigma_2)} \right\} \qquad (8.107)$$

where ζ, ζ^2 are of first and second order of smallness. In general, this is a complicated expression, but reduces to a simple form in the weak coupling limit $\zeta \to 0$, when $F(\zeta, \Sigma_1, \Sigma_2) \to -2\zeta$. The effects of interseries coupling are described by the product term $\Sigma_1 \Sigma_2$ in the numerator of K. When the coefficient of this term is zero, the series are uncoupled otherwise than indirectly, via the continuum.

The analogue of equation (8.100) then reads

$$(\Sigma_1 + \Sigma_2) = 1/\zeta^2 \qquad (8.108)$$

which is similar on the left hand side to the extra term present in 8.103 but not in (8.104). Again, we may explore the meaning of this expression if we replace Σ_2 by a single broad resonance $\Gamma_B/2(E_B-E) = \varepsilon_B^{-1}$, yielding

$$\nu_1 = \frac{1}{\pi}\tan^{-1}\left(\frac{\mathcal{X}_1\zeta^2\varepsilon_B}{\varepsilon_B-\zeta^2}\right) - \mu_1$$

The resulting curve of ν_1 against energy, plotted modulo 1, exhibits a break near the energy $E = E_B - \zeta^2\Gamma_B/2$, where its slope is infinite, and re enters the graph on the opposite face.

The expressions just given do not contain asymmetries due to the background continuum, which further increase the complexity of the problem. Their inclusion yields:

$$F(\zeta,\overline{\Sigma}_1,\overline{\Sigma}_2) = \frac{-2\zeta(1-\overline{\Sigma}_1\overline{\Sigma}_2) + \zeta^2\overline{\Sigma}_1 + \overline{\Sigma}_2}{(1-\overline{\Sigma}_1\overline{\Sigma}_2) - \zeta^2(\overline{\Sigma}_1+\overline{\Sigma}_2)}$$

and

$$K = \frac{\overline{\Sigma}_1 + \overline{\Sigma}_2 + F(\zeta,\overline{\Sigma}_1,\overline{\Sigma}_2)\overline{\Sigma}_1\overline{\Sigma}_2}{\left(1-\overline{\Sigma}_1\overline{\Sigma}_2\right) - \zeta^2\left(\overline{\Sigma}_1+\overline{\Sigma}_2\right)} \tag{8.109}$$

which reduces to (8.102) for $\zeta \to 0$, and yields maxima[7] when

$$\overline{\Sigma}_1\overline{\Sigma}_2 = 1 - \zeta^2\left(\overline{\Sigma}_1+\overline{\Sigma}_2\right) \tag{8.110}$$

which is clearly a further generalisation of equation 8.103. After a little algebra, equation (8.110) becomes

$$(1 - \zeta^2\xi)\Sigma_1\Sigma_2 + (\zeta^2 + \xi)(\Sigma_1 + \Sigma_2) = 1 - \zeta^2\xi \tag{8.111}$$

In particular, if $\xi = -\zeta^2$, this equation reduces to the same form as (8.104). Since ξ can be either positive or negative depending on the values of q_1 and q_2, *provided* $\zeta = 0$, an identical graph to that for bound states may be recovered despite asymmetries. This arises because the shifts in the positions of maxima due to inter series coupling may occur in a direction opposite to those resulting from profile asymmetries, leading to varying degrees of cancellation.

From equation (8.109), we deduce (modulo 1) that *provided* $\Sigma_1, \Sigma_2 = 0$:

$$\nu_2 = \frac{1}{\pi}\tan^{-1}\left\{\mathcal{X}_2\frac{\mathcal{X}_1(1-\zeta^2\xi) + (\zeta^2+\xi)\tan\pi(\nu_1+\mu_1)}{(1-\zeta^2\xi)\tan\pi(\nu_1+\mu_1) - (\zeta^2+\xi)\mathcal{X}_1}\right\} - \mu_2 \tag{8.112}$$

which is the generalisation of (8.101) We can now consider the effect of ζ^2 (direct inter series coupling) on the two-dimensional plots. First,

---

[7] Note that the poles in the numerator of (8.109) which occur when either $\overline{\Sigma}_1 + \overline{\Sigma}_2 \to \pm\infty$ or $\overline{\Sigma}_1\overline{\Sigma}_2 \to \infty$ are cancelled by corresponding poles in the denominator as noted above for equation (8.98) so that all the branches of the two-dimensional plot are described by equation (8.110).

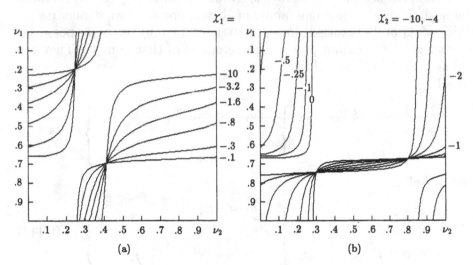

Fig. 8.26. Lu–Fano plots, showing showing the effect of variations in the parameter $\mathcal{X}$ for each of the two series in 8.25: (a) $\mathcal{X}_2 = -0.245$, $\mu_1 = 0.3$, $\mu_2 = 0.7$; (b) $\mathcal{X}_1 = -0.1$, $\mu_1 = 0.3$, $\mu_2 = 0.7$ (after J.-P. Connerade [444]).

fig. 8.25(b) shows the effect of varying ζ^2 when $\xi = 0$. This is similar to the effect of varying ξ shown in fig. 8.25(a), except that, for real interaction strengths, ζ^2 is positive, so that only a part of the two-dimensional square is explored. An increase in ζ^2 then has a very similar effect to an increase of the interaction strength R_{12}^2 in bound state MQDT.

Since equation (8.109) is fairly complex in form, but includes the other cases for suitable choices of the quantities $\mathcal{X}_1$, $\mathcal{X}_2$, μ_1, μ_2, q_1, q_2, plus the interlimit spacing and the interseries coupling strength, all the situations enumerated in the present chapter can be represented.

8.36.4   *Special points in the spectra and in the $\nu_1\nu_2$ plane*

In the present section, we analyse the structure of the two-dimensional graphs. In particular, we consider special or invariant points in the $\nu_1\nu_2$ plane with the property that they lie on many different curves for different values of either the interseries coupling strength or the autoionisation widths, and we examine the position of the avoided crossing. We also consider special points in the spectra.

Equation (8.105) has the well-known consequence that, for all values of the inter series coupling strength R_{12}, the family of $\nu_2(\nu_1)$ curves pass through two invariant points. The same is true from (8.104) for all values of $\mathcal{X}_1\mathcal{X}_2$, and for the same pair of invariant points, defined (modulo 1) by $\nu_1 = -\mu_1$, $\nu_2 = \frac{1}{2} - \mu_2$ and $\nu_1 = \frac{1}{2} - \mu_1$, $\nu_2 = -\mu_2$.

In the presence of a nonzero background δ, however, the situation is changed. Instead of just one family of curves, one has two, depending on which of $\mathcal{X}_1$ or $\mathcal{X}_2$ is varied. From (8.103), each family of curves possesses its own pair of invariant points, characteristic of that family.[8] They are defined (modulo 1) as follows:

for all $\mathcal{X}_2$:

$$\left. \begin{array}{l} \nu_1 = \frac{1}{\pi}\tan^{-1}(\xi\mathcal{X}_1) - \mu_1 \\[2ex] \nu_2 = \frac{1}{2} - \mu_2 \end{array} \right\} \qquad \left. \begin{array}{l} \nu_1 = \frac{1}{\pi}\tan^{-1}\left(\frac{\mathcal{X}_1}{\xi}\right) - \mu_1 \\[2ex] \nu_2 = -\mu_2 \end{array} \right\} \qquad (8.113)$$

and, for all $\mathcal{X}_1$:

$$\left. \begin{array}{l} \nu_1 = \frac{1}{2} - \mu_1 \\[2ex] \nu_2 = \frac{1}{\pi}\tan^{-1}(\xi\mathcal{X}_2) - \mu_2 \end{array} \right\} \qquad \left. \begin{array}{l} \nu_1 = -\mu_1 \\[2ex] \nu_2 = \frac{1}{\pi}\tan^{-1}\left(\frac{\mathcal{X}_2}{\xi}\right) - \mu_2 \end{array} \right\} \qquad (8.114)$$

from which we see that the two pairs of invariant points lose their remaining dependence on $\mathcal{X}_1$ or $\mathcal{X}_2$ when $\xi = 0$ (symmetric resonances) and become the usual pair of invariant points associated with equation (8.105).

Turning now to directly coupled series of resonances, it is interesting to consider whether pairs of invariant points still arise in the presence of inter series coupling as described by equation (8.109). From (8.111), we see that, if $\Sigma_2 \to \infty$, then

$$\Sigma_1 = \frac{\xi + \zeta^2}{1 - \zeta^2\xi} = \tan(\delta + \delta_{\mathrm{coup}}) \qquad (8.115)$$

where $\tan\delta_{\mathrm{coup}} = \zeta^2$. If $\nu_2 = \frac{1}{2} - \mu_2$, then

$$\nu_1 = \frac{1}{\pi}\tan^{-1}\left[\frac{\xi + \zeta^2}{1 - \zeta^2\xi}\mathcal{X}_1\right] - \mu_1$$

for all $\mathcal{X}_2$, which is the analogue of equation (8.113). Thus, the pairs of invariant points persist, with separate dependences on $\mathcal{X}_1$ and $\mathcal{X}_2$, as previously noted. This is consistent with the fact that the effects of asymmetry and of inter series coupling are additive, and do not change the general form of the two-dimensional graphs.

In figs. 8.26(a) and (b), we show, for a typical set of parameters, the families of graphs which result from separate variations of $\mathcal{X}_1$ and $\mathcal{X}_2$

---

[8] These points are related to the hydrogenic and 'half-hydrogenic' points discussed in section 8.24.

respectively. Note that the pairs of special points are distinct from each other in each graph.

The avoided crossing in the ν_1, ν_2 plane may be located from equation (8.112), whence:

$$\frac{d\nu_2}{d\nu_1} = -\frac{1 + \tan^2 \pi(\nu_1 + \mu_1)}{1 + \tan^2 \pi(\nu_2 + \mu_2)} \frac{(\zeta^2 + \xi)\mathcal{X}_2 - \tan \pi(\nu_2 + \mu_2)}{(\zeta^2 + \xi)\mathcal{X}_1 - \tan \pi(\nu_1 + \mu_1)} \quad (8.116)$$

yielding one avoided crossing about the point:

$$\left.\begin{array}{l} \nu_1 = \frac{1}{\pi} \tan^{-1}\{\mathcal{X}_1(\zeta^2 + \xi)\} - \mu_1 \\[2mm] \nu_2 = \frac{1}{\pi} \tan^{-1}\{\mathcal{X}_2(\zeta^2 + \xi)\} - \mu_2 \end{array}\right\} \quad (8.117)$$

(the first factor in (8.116) yields no avoided crossing: if either tangent is infinite, equation (8.111) becomes inapplicable, as noted above equation (8.112)).

When one series is broader than the other but the background δ is very weak, a possible occurrence for directly coupled series is a cancellation such that the widths of a sequence of poles progressively decrease, pass through a zero point and increase again, leading to a vanishing point in the spectral fluctuations. We may see this from equation (8.106): if an energy is approached at which $\Sigma_1 + F(\zeta, \Sigma_1, \Sigma_2)\Sigma_1\Sigma_2 = 0$, the dependence of K on Σ_1 will disappear. What remains to be shown is that such an energy E_{vf} exists within the range of the series.

Near a broad resonance Γ_B, E_B belonging to series 2, the sum over poles Σ_2 is dominated by one term. Then

$$K_{res} = \frac{\Gamma_B/2}{E_B - E} + \sum_{n_1} \frac{\Gamma_{n_1}/2}{E_{n_1} - E} + F(\zeta, \Sigma_1, \Sigma_2)\frac{\Gamma_B/2}{E_B - E} \sum_{n_2} \frac{\Gamma_{n_2}/2}{E_{n_2} - E}$$

$$= \left(\sum_{n_1} \frac{\Gamma_{n_1}/2}{E_{n_1} - E}\right)\left(1 + F(\zeta, \Sigma_1, \Sigma_2)\frac{\Gamma_B/2}{E_B - E}\right) + \frac{\Gamma_B/2}{E_B - E}$$

$$(8.118)$$

whence, under conditions of weak coupling and at the specific energy

$$E_{vf} = E_B + F(\zeta, \Sigma_1, \Sigma_2)\Gamma_B/2 \approx E_B - \zeta\Gamma_B$$

K becomes independent of the sum over poles for series 1. Near this energy, the spectral fluctuations disappear, but no change is induced in the appearance of the two-dimensional graph. The existence of vanishing fluctuations in atomic spectra at energies distinct from series limits was suggested by Lane [379] and occurrences have been confirmed experimentally in the spectra of Ba [442] and Sr [443] (see section 8.33).

For very weak coupling, as noted above, $F(\zeta, \Sigma_1, \Sigma_2) \to -2\zeta$ and $E_{vf} \to E_B - \zeta\Gamma_B$, which tends towards E_B as $\zeta \to 0$ (a relevant example is given in section 8.37). Thus, E_{vf} moves in energy as a function of coupling. At

$E = E_B$, there is a pole. For very weak coupling, the point of vanishing fluctuations coincides with this pole, which may, however, be detuned from the maximum in the broad resonance.

In the presence of continua into which series 1 and 2 do not autoionise, E_{vf} is unchanged, but vanishing fluctuations tend to disappear when a background continuum δ into which series can autoionise becomes conspicuous, as one sees from equation (8.109).

Other remarkable points in the spectra are the q reversal energies. Since these have been discussed at length in the present chapter, we merely note that they are due to sign changes which do not affect the energies of the poles in the **K**-matrix, and therefore do not induce new structure in two-dimensional graphs.

8.37   Numerical studies of two-dimensional plots

In this section, we illustrate the theory by some numerical examples. Each one involves a simulated spectrum and the corresponding two-dimensional plot. We confine ourselves to the energy range below both series limits, where two-dimensional plots are appropriate.

The positions of local maxima in the cross section and therefore in the square of the **K**-matrix, are used to determine resonances. Experimenters often argue that, for very asymmetric peaks, one should measure the 'edge' within the resonance because, in many cases it is more narrowly defined by the data. Also, the edge should lie closer to the true resonance energy. This procedure may sometimes yield energies closer to the energies E_{n_1}, E_{n_2}, but is unsatisfactory for several reasons: (i) it only applies to asymmetric resonances of Fano form, or to type 1 profiles which are isolated, and so can only be used if the widths of the resonances are all much smaller than the spacings between them; (ii) the point on the line profile which this measurement aims for is in principle a point of inflexion, but is not well defined theoretically for profiles of complex shape; (iii) for perturbed, or coupled series, the shapes of resonances do not recapitulate with increasing n but may fluctuate widely so that a common method of measurement, applicable to many different lineshapes is desirable.

A pure or perfect $q = 0$ window resonance would of course be missed in a search for maxima in the cross section. Very few resonances can satisfy this condition exactly for interacting series. In practice, one misses at most one resonance, and this does not change the shapes of the graphs.

In the two-dimensional plots, the full curves through the points are obtained from equation (8.112). The dashed lines in figs. 8.27–8.34 show the values of μ_1 and μ_2.

Figs. 8.27, 8.28 and 8.29 show the influence of asymmetry on the avoided

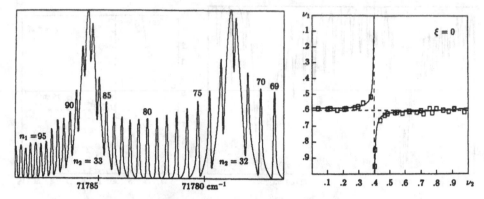

Fig. 8.27. The sequence comprising of this and the next three figures shows the influence of the q or shape parameters of the individual unperturbed series on the Lu–Fano graphs. Note the dramatic changes in magnitude and position of the avoided crossing. $\zeta, \xi = 0$ $E_{\infty 1} = 71800$, $E_{\infty 2} = 71890$, $\mathcal{X}_1 = -0.1$, $\mathcal{X}_2 = -0.2$, $\mu_1 = 0.4$, $\mu_2 = 0.6$, $q_1 = q_2 = 1000$ (after J.-P. Connerade [444]).

crossing. Note that the series and widths are the same in all these figures: only the symmetries of the lines are altered, which has a dramatic effect on the avoided crossings.

We begin, in fig. 8.27, with a case for which q_1 and q_2 are both large, i.e. the unperturbed resonances are symmetric. We display a range with two broad perturbers ($n_2 = 32$ and $n_2 = 33$ from series 2 and 31 members of series 1 (from $n_1 = 69$ to $n_1 = 99$). In fig. 8.27, a fairly narrow avoided crossing (corresponding to $\xi \sim 0$) occurs, which is nevertheless nonzero, since the widths of the resonances are quite large. Notice the presence of q reversals [382]: (i) about the two symmetric perturbers; and (ii) about a position roughly half-way between them (at high principal quantum number). As noted in section 8.28, some q reversals tend to migrate quite rapidly in energy as a function of coupling strength.

Figs. 8.28 and 8.29 show examples with parameters identical to those of fig. 8.27, except that ξ has been increased in absolute value (positive and negative respectively), which leads to large avoided crossings.

From these three calculations, we see that, for series of weakly coupled, sharp autoionising lines, the same principles apply as in the analysis of bound states, except that anticrossings should be interpreted with caution as they approach the linewidths of one of the autoionising series. An example occurs in the spectrum of Cr, which possesses a more complex

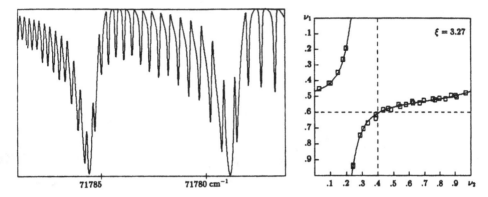

Fig. 8.28.   As 8.27, with $q = 0.1$ for the first series and $q = -0.5$ for the second one (after J.-P. Connerade [444]).

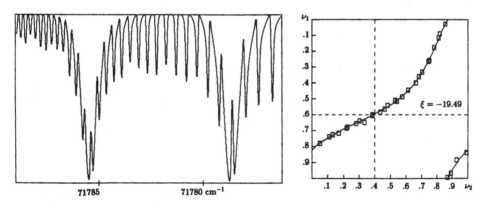

Fig. 8.29.   As above, with $q = 0.95$ for both series (after J.-P. Connerade [444]).

two-dimensional plot, with more interacting channels than in our model calculation, but which comprises very narrow autoionising lines yielding a two-dimensional plot characteristic of weak coupling [439].

   We turn now to the case of a somewhat broader perturber, whose influence is felt over a wider range and affects all the fine structure in the calculated spectrum. In fig. 8.30, we show a case in which one series is

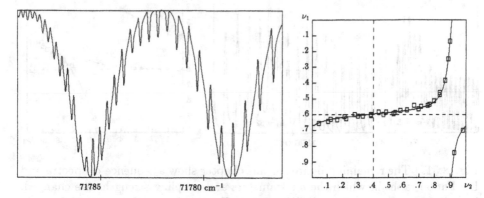

Fig. 8.30. Case of very broad perturbers interacting with a sharp series; $\chi_1 = -0.05$, $\chi_2 = -0.06$, and $\xi = 10^6$ (after J.-P. Connerade [444]).

much broader than the other so that the avoided crossing is tilted substantially to one side of the position defined by μ_1 and μ_2.

This calculation shows that, for a broader interloper and zero or very weak interseries coupling, the structure of the two-dimensional plots emulates that for two series of bound states, but with a quite broad anticrossing, which would be interpreted as the consequence of a stronger interseries interaction in the latter instance whereas, in the former, it results from the breadth of the interloper. Such seems to be the situation in the spectrum of Yb, for the series interaction analysed by Baig and Connerade [445], and may also be true in more complex plots reported by Baig *et al.* [446] for other series of Yb: although the rules of MQDT are well respected by the empirical two-dimensional plots, the interpretation of the parameters is not the same. This applies not only to R_{12}, but also to the apparent quantum defects μ_1 and μ_2, which no longer yield the locations of the undisturbed poles in the presence of pronounced asymmetries.

In this connection (as already hinted above), an interesting experimental possibility arises. The symmetry of an autoionising resonance (but not its width) depends on the mode of excitation. Thus, for example, for an atom with an s^2 ground state, one may excite the same autoionising resonances either directly by single-photon spectroscopy, or by stepwise excitation via an intermediate, excited $msns\,^1S_0$ state, which can be reached by a two-photon transition [441]. In the latter instance, the symmetries of the lines can be dramatically altered, and in the present theory, this

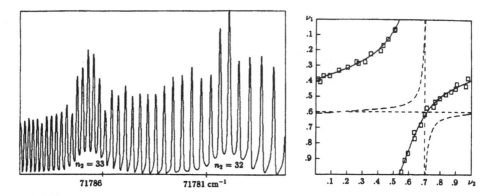

Fig. 8.31.   The remaining figures in this chapter show a sequence of spectra and Lu–Fano graphs in which the q parameters and coupling strengths are changed, but all other parameters are held constant. For this figure, the $q = 1000$ for both series, and the coupling strength $\zeta = 2$. A zero coupling strength, zero combined asymmetry plot is shown as a dashed curve for reference. $\mathcal{X}_1 = 0.3$, $\mathcal{X}_2 = -0.2$, $\mu_1 = 0.4$, $\mu_2 = 0.3$, $q_1 = q_2 = 1000$ and $\zeta = 2$ (after J.-P. Connerade [444]).

leads to changes in q_1, q_2, and ξ, while $\mathcal{X}_1$, $\mathcal{X}_2$ and ζ remain unaffected. Thus, the two-dimensional graphs are changed, but in a manner which depends *only* on the asymmetries of the lines.

In figs. 8.31–8.34, we now demonstrate the influence of inter series coupling, and of its interplay with asymmetries.

Fig. 8.31 shows the effect of quite a large interseries coupling on a series containing symmetric resonances. In the presence of interseries coupling, the two broad interlopers are 'diluted' into the overlapping fine series and the avoided crossing is broadened (the dashed curve in the two-dimensional plot shows the corresponding case with $\zeta = 0$).

In the presence of opposite asymmetries (q_1 and q_2 of different sign), the intensities of the overlapping series may reflect those of the perturbers, as shown in fig. 8.32. This does not, however, lead to any dramatic changes in the corresponding two-dimensional graph. A good experimental example of this behaviour appears in the $3p$-spectrum of Ca [447], where the presence of perturbers from the upper series can be detected through intensity modulations in the lower one.

One of the most important effects of direct inter series coupling is the existence of vanishing fluctuations for an appropriate combination of parameters, as discussed in the previous section. Fig. 8.33 shows such a case, and demonstrates that the presence of vanishing fluctuations does *not* lead to any change in structure of the two-dimensional graph. In-

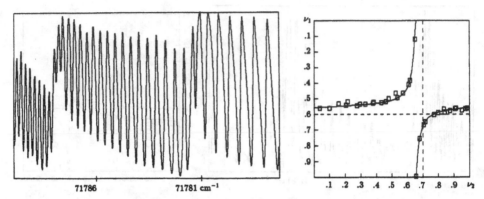

Fig. 8.32. For this figure, $q_1 = 1$ for series 1, $q_2 = -1$ for series 2, and the coupling strength $\zeta = -0.9$ (after J.-P. Connerade [444]).

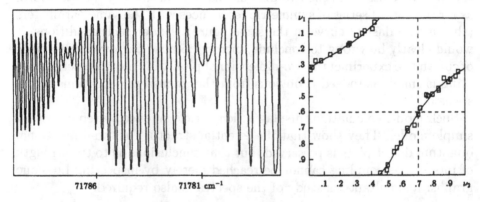

Fig. 8.33. For this figure, $q_1 = .5$, $q_2 = -3$, and the coupling strength $\zeta = 0.9$ (after J.-P. Connerade [444]).

deed, this is consistent with the work of Brown and Ginter [448], which has not indicated any new branches in two-dimensional plots as a result of vanishing fluctuations.

Finally, in fig. 8.34, we show how the combined effect of interseries coupling and asymmetries may lead to cancellation: although a comparatively large value has been selected for the interseries coupling, its effects are mostly cancelled by the influence of ξ. Therefore, in spite of a greater dilution of the perturbers into the finer structure and the appearance of a vanishing point in the fluctuations to the right of each broad interloper,

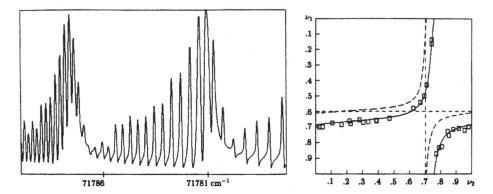

Fig. 8.34.   For this figure, $q_1 = 0.1$, $q_2 = -5$ and the coupling strength $\zeta = 1.8$. A zero coupling, zero combined asymmetry plot is shown as a dashed curve for reference (after J.-P. Connerade [444]).

the avoided crossing in the two-dimensional plot has nearly the same form as for uncoupled series of symmetric resonances with the same parameters (shown as a dashed curve in the same figure). Under such conditions, it would clearly be wrong to conclude, merely because narrow anticrossings occur an the experimental two-dimensional plot, that the inter series coupling is small, as indeed is obvious from the intensity fluctuations in the spectra.

Such model calculations display a rich variety of effects even on a very simple model. They show that the essential structure of two-dimensional quantum defect plots is preserved, but that conclusions as to the strength of inter-series coupling cannot be reached merely by inspecting Lu–Fano graphs: a simultaneous study of the spectra is also required.

8.38   Conclusion

In this chapter, Wigner's scattering theory has been presented and applied to a wide variety of effects which occur when Rydberg series of autoionising resonances interact with one another. Overlapping resonances occur in many atomic spectra: they become more and more frequent as the number of subshells increases. They can also be 'manufactured' by multiphoton spectroscopy, a theme we return to in the following chapter. The effects we have described (symmetry reversals, width fluctuations and disappearances of structure) are also expected to occur in photoelectron spectroscopy [449].

9

Atoms in strong laser fields

9.1   Multiphoton spectroscopy

The subject of multiphoton excitation spectroscopy began in 1931 when Göppert-Mayer [450] wrote a theoretical paper in which she calculated the transition rate for an atom in the presence of two photons rather than just one. At the time, the process seemed rather exotic, and it was reassuring that the calculated rate was so low as to guarantee that it could not readily be observed in the laboratory with conventional sources. This conclusion was reassuring because it implies that a simple perturbative theory (one photon per transition is the weak-field limit) is adequate for most purposes.

The subject came to life with the advent of lasers, when it became easy to create intense beams of light. Since the probability of excitation by two photons grows as the square of the photon density, whereas the probability of single-photon excitation grows only linearly with photon density, two-photon transitions gain in relative strength with increasing intensity despite the small value of the rate coefficient.

The development of multiphoton spectroscopy has followed that of lasers: as the available power has increased, so has the number of photons involved in individual transitions. More significantly, it has become apparent that the physics of the interaction between radiation and matter is not the same at high laser powers as under weak illumination, i.e. that there is a qualitative change which sets in at strong laser fields. This is normally expressed by saying that perturbative approximations break down. A more direct (and equally accurate) statement is that new effects are observed, which are not present in conventional spectroscopy, even where the latter is extended to include, say, two- and three-photon transitions.

We begin with a brief summary of some aspects of multiphoton physics involving just a few photons per transition, before turning to the new subject of 'super strong' laser excitation. This last topic will, however, be considered not from the standpoint of what maximum laser field strengths are achievable, but whether and what one can learn about highly-excited atoms and their properties from this new class of experiments.

9.2 The different regimes of excitation

Strong field excitation may be divided into three different regimes, according to the value of a parameter γ_K introduced by Keldysh [494], defined by

$$\gamma_K = \frac{\epsilon_I}{2\Phi_L}$$

where ϵ_I is the ionisation potential and Φ_L is the field of the laser (see section 9.16 for the exact definition of Φ_L).

If $\gamma_K < 1$ with more than one photon involved per transition, then one has the *multiphoton regime*, which can be handled using the methods of perturbation theory.

If $\gamma_K < 1$ and perturbative methods break down, a simple description may still exist, in which the valence electron sees a field which, in one dimension, can be written as

$$V(x) = -\frac{Ze^2}{x} - e\mathcal{E}x$$

where $\mathcal{E}$ is the electric field due to the laser. This is essentially a Coulomb field, with an electric field term which induces a barrier. To a first approximation, the electron still behaves as a bound electron, but may ionise by tunnelling through the barrier, and so this is often called the *tunnelling regime* [451, 452].

Finally, if $\gamma_K > 1$ and there are no bound states, even in the first approximation, one might imagine that there cease to be any atomic aspects to consider. This turns out not to be quite true, because of dynamical effects. This third regime is sometimes referred to as the *'over the barrier ionisation' regime* .

The classification just described has the merit of being simple, and leads to rules which agree well with observations. However, it neglects some important issues of the dynamics and the time dependence of laser pulses which also need to be envisaged.

In the present chapter, we provide and discuss examples from all of these regimes.

Table 9.1. Selection rules for two-photon transitions

General Rules			
$\|\Delta J\| \leq 2$			$J_i + J_f = $ integer
Particular Polarization Rules			
Polarization of ω_1	ω_2	Allowed Transitions	Forbidden Transitions
σ^+	σ^-	$\Delta M = 0$	
σ^+	π	$\Delta M = -1$	$\Delta J : 0 \rightarrow 0$
σ^-	π	$\Delta M = 1$	$\Delta J : 0 \rightarrow 0$
π	σ	$\Delta M = \pm 1$	$\Delta J : 0 \rightarrow 0$
π	π	$\Delta M = 0$	$\Delta J : 0 \leftrightarrow 1$
σ^+	σ^+	$\Delta M = -2$	$\Delta J : 0 \leftrightarrow 1, 0 \rightarrow 0, 1/2 \rightarrow 1/2$
σ^-	σ^-	$\Delta M = 2$	$\Delta J : 0 \leftrightarrow 1, 0 \rightarrow 0, 1/2 \rightarrow 1/2$
Additional Rules for Equal Frequency Photons			
$\Delta J : 0 \leftrightarrow 1$		Forbidden for all polarizations	
If $\|\Delta J\| = 1$	$\Delta M : 0 \rightarrow 0$	Forbidden for all polarizations	

9.3 Selection rules for two-photon excitation

The selection rules for multiphoton transitions are clearly different from the usual dipole selection rules, since each photon carries an angular momentum ± 1 Thus, for two-photon transitions, one rule is $\Delta J = 0 \pm 2$, but further control can be exercised by selecting the polarisation of the light: the $\Delta J = 0$ transitions are only possible if the laser light is linearly polarised (i.e. contains both σ^+ and σ^- circular polarisation), while the choice of either circular polarisation results in an increase or a decrease of J. Detailed discussion of the selection rules for two-photon transitions can be found in several papers [453, 455, 459]. For multiphoton transitions, the same principles apply, and the role of polarisation is still more significant. A general reference is [460], in which selection rules are derived from first principles, and a list of selection rules for two-photon transitions is given in table 9.1.

Note that these rules are somewhat different from the rules for two-step excitation, in which the intermediate state coincides with a real level, which can be viewed as a special case. This process was first discussed by Javan *et al.* [461], and complete solutions are given by other authors [462, 463].

9.4 The two-photon transition rate

Suppose the light field is the sum of two waves, of frequencies ω_1 and ω_2:

$$\vec{\mathcal{E}}(t) = v_1 \mathcal{E}_1 \cos(\omega_1 t - \mathbf{k}_1 \cdot \mathbf{z}) + v_2 \mathcal{E}_2 \cos(\omega_2 t - \mathbf{k}_2 \cdot \mathbf{z}) \qquad (9.1)$$

and suppose that the sum of the photon energies satisfies the resonance condition, i.e. $\hbar\omega_1 + \hbar\omega_2 = E_f - E_i$. Then the two-photon transition rate is proportional to the intensity of each of the two fields, and is written as

$$W_{if} = \frac{|A_{if}|^2 \, \mathcal{E}_1^2 \mathcal{E}_2^2}{2\hbar^4 \Gamma} \qquad (9.2)$$

where Γ is half the width at half-maximum of the natural line profile and A_{if} is a two-photon matrix element given by

$$A_{if} = \sum_n \left\{ \frac{(\mu \cdot v_1)_{in}(\mu \cdot v_2)_{nf}}{\omega_{in} - \omega_1} + \frac{(\mu \cdot v_2)_{in}(\mu \cdot v_1)_{nf}}{\omega_{in} - \omega_2} \right\} \qquad (9.3)$$

where the sum is taken over all possible intermediate states connecting $|\,i>$ with $|\,f>$. The assumption in this expression is that states $|\,i>$ and $|\,f>$ are nondegenerate and that none of the intermediate states is resonant with the light fields ω_1 and ω_2. One can then use perturbation theory to estimate the two-photon transition probability [464]. Two-photon transitions, which are inherently very sharp, have been used as wavelength markers to calibrate spectra of autoionising resonances observed by three-photon spectroscopy [454]. They are of practical importance in atomic physics as a means of cancelling the longitudinal Doppler effect. By propagating two photons in opposite directions (counterpropagation), one ensures that the resonance condition is only satisfied by a specific class of atoms, whose velocity component parallel to the laser beams is zero. This allows the spectral resolution of laser experiments to be enormously increased [455]. Finally, two-photon transitions have been used in fundamental experiments to test the validity of quantum mechanics [456, 457].

The excitation of an atomic transition simultaneously by two photons is not the only possibility for combined excitation by two particles: it is also possible for a single transition to be excited simultaneously by one photon and one electron, neither of which have sufficient energy separately to excite the transition [458].

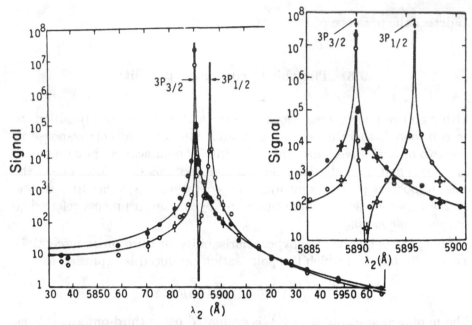

Fig. 9.1. Resonantly enhanced two-photon ionisation in Na. Two lasers are used: the sum of the two frequencies is fixed, and the intermediate energy is tuned through the 3p doublet. The inset shows the region close to resonance (after J.E. Bjorkholm and P.F. Liao [466]).

9.5   Resonant enhancement

As equation (9.3) indicates, the transition rate increases dramatically when either of the two fields becomes resonant with an intermediate state.   This resonant enhancement of the two-photon transition by a near-degeneracy has been observed in the two-photon spectrum of Na by Bjorkholm and Liao [466] and is shown in fig. 9.1. Resonant enhancement is a valuable way of boosting the transition probability for any multiphoton process, and brings out the complexity of this kind of spectroscopy: it is no longer sufficient to consider just two states of the atom, coupled by the radiation field: many other near-resonant possibilities may have to be considered as well, and can dramatically affect the probability of excitation. Another example of the resonant enhancement of two-photon transitions occurs in a quite different area of physics: as was noted in section 2.22, stable excitonic quasimolecules can be formed in non metallic solids. Hanamura [467] has shown that two-photon absorption in the excitonic molecule is strongly enhanced, because of a giant resonance in the exciton, so that the two-photon cross section is actually as large as the single-photon cross section at the same wavelength, and can be used to detect the presence of the biexcitonic state.

9.6   The third-order susceptibility

Although two-photon transitions were the first multiphoton transitions to
be considered, there is no reason to limit studies of atomic response to
just two optical waves, once the possibility of nonlinear coupling between
the atoms and the radiation field has been recognised. Indeed, one of the
most important processes involves the generation of a higher frequency ω
by the superposition of three waves ω_1, ω_2 and ω_3, a process referred to
as *four-wave mixing*.

   This process involves the hyperpolarisability, or third-order susceptibility $\chi^{(3)}$, in terms of which the polarisation produced is expressed as

$$P = \chi^{(3)}(-\omega; \omega_1, \omega_2, \omega_3)\mathcal{E}_1\mathcal{E}_2\mathcal{E}_3$$

The nonlinear susceptibility $\chi^{(3)}$ is evaluated using third-order perturbation theory, and resonant enhancement is readily demonstrated to occur.
Four-wave mixing is a useful experimental technique to extend the energy range available to tunable dye lasers [468]. It is also of interest that
processes involving excitation by three photons allow transitions between
even and odd parity states to be excited, as do single-photon transitions.

9.7   Measurements of atomic f value by nonlinear optics

When more than one wavelength is present in a medium and it is driven
hard by the intense oscillating field of the laser, radiation is generated
at the sum or difference frequencies corresponding to certain harmonics
[469] of the fundamental driving frequencies. This mixing effect is a large
subject in itself and concerns nonlinear optics rather than atomic physics,
but it has one interesting application for the measurement of oscillator
strengths: the efficiency of the conversion process depends on whether the
harmonics travel in phase with the fundamental wave inside the medium.
There is a phase-matching condition, dependent on the refractive index,
which must be satisfied for maximum intensity to be obtained in the
harmonic. Since the refractive index itself depends on the f value (a
connection exploited in the Faraday rotation method of section 4.9), it
is actually possible to use the nonlinear conversion signal to measure the
absolute f value of an atomic transition, a method exploited by Wynne
and Beigang [470].

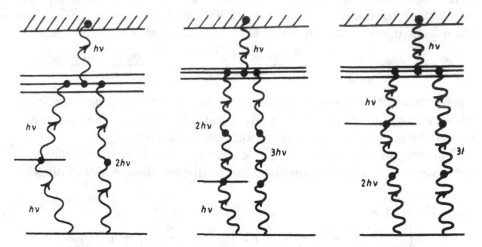

Fig. 9.2. Excitations schemes for resonantly enhanced multiphoton ionisation using three and four photons (schematic) (after J.-P. Connerade and A.M. Lane [385]).

9.8 Symmetry reversals in resonant enhancement

Resonant enhancement can also be interpreted as a perturbation of the multiphoton spectrum. In a sense, the enhancing resonance acts like an interloper 'lifted' into an otherwise unperturbed Rydberg series, which suggests that symmetry reversals should occur, as in chapter 8. This is indeed observed, and allows for new possibilities not available in single-photon spectroscopy. When only a single photon is used, one can only probe existing atomic structure, but, with the advent of intense tunable laser radiation, spectroscopy can be extended beyond this, as already demonstrated in the case of LICS (section 8.20). By using several photons, the influence of a perturber can be shifted in energy by a selected interval when the atom is dressed by the radiation field. The principle of this 'controllable perturbation' is illustrated in fig. 9.2 for three- and four-photon ionisation: under n-photon ionisation, we arrange for an energy degeneracy between the $(n-1)$-photon states and an $(n-2)$-photon allowed transition. The resulting interference is then probed by n-photon ionisation.

The disturbance which results from superposing this new perturber on a Rydberg manifold is extremely similar to the disturbance caused when an intruder perturbs a Rydberg series in single-photon spectroscopy, but there is now the possibility to exercise external control over the nature and strength of the interaction. For example, consider three-photon ionisation of K vapour using circularly polarised light with energies $\hbar\omega$ corresponding to the range between $12\,500$ and $16\,300$ cm^{-1}. To lowest order in the atom–laser interaction, the essential term in the ionisation amplitude is

$$T(\omega) = (e\mathcal{E})^3 \sum_{n_1,n_2} \frac{<\epsilon\,|\,\mathcal{Z}\,|\,n_2><n_2\,|\,\mathcal{Z}\,|\,n_1><n_2\,|\,\mathcal{Z}\,|\,0>}{(E_{n_2} - E_0 - 2\hbar\omega)(E_{n_1} - E_0 - \hbar\omega)} \quad (9.4)$$

where $|\,0>$ is the ground state of the atom, $|\,\epsilon>$ is the ionisation state whose energy $(\epsilon - E_0)$ is $3\hbar\omega$ and n_1, n_2 are, for example, p and d states excited by one- and two-photon absorption respectively.

Following [471], we choose $\hbar\omega$ to lie near a p state $n_1 = N$ and $2\hbar\omega$ near a d state $n_2 = n$ and then rewrite $T(\omega)$ to display these states explicitly:

$$T(\omega) = (e\mathcal{E})^3 \times$$
$$<\epsilon\,|\,\mathcal{Z}\left(\frac{|\,n><n\,|}{2\hbar\omega - (E_n - E_0)} + \mathcal{B}_2\right)\mathcal{Z}\left(\frac{|\,N><N\,|}{\hbar\omega - (E_N - E_0)} + \mathcal{B}_1\right)\mathcal{Z}\,|\,0>$$
$$(9.5)$$

where the background operators $\mathcal{B}_1$ and $\mathcal{B}_2$ are defined by

$$\left.\begin{array}{l} \mathcal{B}_1 = \sum_{n_1=N} \dfrac{|\,n_1><n_1\,|}{\hbar\omega - (E_{n_1} - E_0)} \\[2em] \mathcal{B}_2 = \sum_{n_1=n} \dfrac{|\,n_2><n_2\,|}{2\hbar\omega - (E_{n_2} - E_0)} \end{array}\right\} \quad (9.6)$$

We can introduce the detunings

$$\left.\begin{array}{l} \epsilon_N \equiv \hbar\omega - (E_N - E_0) \\[0.5em] \epsilon_n \equiv 2\hbar\omega - (E_n - E_0) \end{array}\right\} \quad (9.7)$$

and the ionisation yield $W(\omega)$ then has the form

$$W(\omega) = |<\epsilon\,|\,\mathcal{Z}\mathcal{B}_2\mathcal{Z}\mathcal{B}_1\mathcal{Z}\,|\,0>|^2 \left(\frac{\epsilon_n + Q_n}{\epsilon_n}\right)^2 \left(\frac{\epsilon_N + Q_N}{\epsilon_N}\right)^2 \quad (9.8)$$

where Q_n and Q_N are generalised asymmetry parameters, which turn out to have the dimensions of energy.

We can further generalise these expressions to include an explicit summation over many levels of the Rydberg manifold n. To display energy variations more directly, it is convenient also to set $E_0 = 0$ and $E = \hbar\omega$ and to write E_∞ for the series limit. Finally, we have

$$T(E) = \left(1 + \frac{Q_N}{\epsilon_N}\right)\left(1 + \frac{\epsilon_N + \hat{\mathcal{Z}}}{\epsilon_N + Q_N}\frac{Q_{n_0}^0}{(E_\infty - E_{n_0})^{3/2}} \sum_{n=n_0}^{n/0} \frac{(E_\infty - E_n)^{3/2}}{2E - E_n}\right)$$
$$(9.9)$$

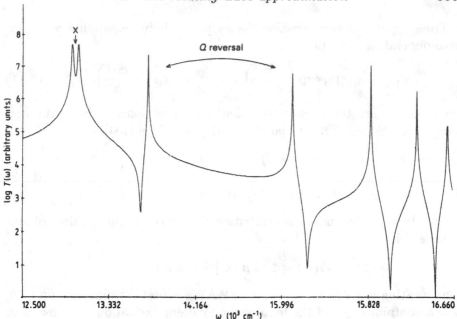

Fig. 9.3. The Q-reversal effect in multiphoton ionisation. Note that Q stands for the generalised asymmetry parameter in multiphoton spectroscopy, as opposed to q in single-photon spectroscopy. The choice of parameters corresponds to REMPI of K (see text – after J.-P. Connerade and A.M. Lane [385]).

where

$$\hat{\mathcal{Z}} = \frac{<n \mid \mathcal{Z} \mid N><N \mid \mathcal{Z} \mid 0>}{<n \mid \mathcal{Z}\mathcal{B}_1\mathcal{Z} \mid 0>}$$

A typical calculation based on equation (9.9) is shown in fig. 9.3, with parameters appropriate for three-photon ionisation of K I [472]. Note the Q_n-reversal effect due to the influence of the perturber, displaced by the laser field into the energy range of the Rydberg manifold.

9.9 The rotating wave approximation

For near-resonant excitation by a coherent light wave of frequency ω, the frequency difference $\Delta\omega = \omega - \omega_0$ (where ω_0 is the frequency of the atomic transition) is very small, and indeed much less than the transition frequency. Thus, the maximum value of the interaction energy which induces the transition ($\mu\mathcal{E}$) is expected to be much smaller than the transition energy:

$$\mu\mathcal{E} \ll \hbar\omega_0 \qquad (9.10)$$

since otherwise the radiation field would be 'pulled' out of resonance.

Time-dependent perturbation theory proceeds by expanding a general time-dependent wavefunction

$$\Psi(t) = a_i(t)u_i \exp\left(-\frac{E_i t}{\hbar}\right) + a_f(t)u_f \exp\left(-\frac{E_f t}{\hbar}\right) \tag{9.11}$$

where u_i, u_f are the time-independent eigenfunctions of the initial and final states E_i and E_f. By the standard procedures [464]

$$\left.\begin{aligned} i\hbar \, \dot{a}_i &= V_{fi}(t)a_j \exp\left(-i\omega_0 t\right) \\[2ex] i\hbar \, \dot{a}_j &= V_{if}(t)a_j \exp\left(+i\omega_0 t\right) \end{aligned}\right\} \tag{9.12}$$

where the time-dependent perturbation due to the radiation field of amplitude $\mathcal{E}$ is

$$V_{if}(t) = <i \mid \mu \cdot v \mid f> \mathcal{E} \cos \omega t \tag{9.13}$$

The rotating wave approximation (RWA) is a useful simplification [465] which contains most of the features of coherent excitation: it consists in assuming that only $\exp(i\omega t)$ terms are present (counter rotating terms are neglected), in which case the amplitudes $a_i(t)$ and $a_f(t)$ experience only slow oscillations at the frequency $\Delta\omega$, according to the differential equations:

$$\left.\begin{aligned} i\hbar \, \dot{a}_i &= -\tfrac{1}{2}\mu\mathcal{E} \exp\left(+i\Delta\omega t\right) a_f \\[2ex] i\hbar \, \dot{a}_f &= -\tfrac{1}{2}\mu\mathcal{E} \exp\left(-i\Delta\omega t\right) a_i \end{aligned}\right\} \tag{9.14}$$

where the factor $1/2$ comes about from $\cos \omega t = \tfrac{1}{2}\left(\exp(i\omega t) + \exp(-i\omega t)\right)$. If the atom is initially in state $\mid i>$ with probability 1 and a light field of constant amplitude $\mathcal{E}$ is turned on then the solution at time t is

$$\left.\begin{aligned} a_i(t) &= \left\{\cos\left(\frac{\gamma t}{2}\right) - i\frac{\Delta\omega}{\gamma}\sin\left(\frac{\gamma t}{2}\right)\right\} \exp\left(i\frac{\Delta\omega t}{2}\right) \\[2ex] a_f(t) &= i\frac{\mu}{\hbar\gamma}\sin\left(\frac{\gamma}{2}t\right)\exp\left(-i\frac{\Delta\omega t}{2}\right) \end{aligned}\right\} \tag{9.15}$$

where

$$\gamma = \Delta\omega^2 + \Omega^2 \tag{9.16}$$

is the *Rabi oscillation frequency* in general, and

$$\Omega = \left(\frac{\mu\mathcal{E}}{\hbar}\right)^2$$

is the Rabi oscillation frequency on resonance.

On resonance, $\Delta\omega = 0$ so

$$\left.\begin{aligned} a_i(t) &= \cos\frac{\Omega t}{2} \\[2mm] a_f(t) &= i\Omega\sin\frac{\Omega t}{2} \end{aligned}\right\} \qquad (9.17)$$

Thus, the populations are cycled or optically pumped by the coherent radiation field at a frequency Ω, an effect which is not observed for incoherent excitation.

9.10 The Autler–Townes satellites or AC Stark splitting

The Rabi oscillation frequency depends on both the transition matrix element from $\mid i >$ to $\mid f >$ and the intensity of the radiation field. As a result of this cycling of population, the observed fluorescence no longer consists of a single line, but exhibits sidebands at $\omega_0 - \Omega$ and $\omega_0 + \Omega$. The height of the central fluorescence peak is twice as great as the height of the sidebands. This cycling is completely *reversible*, since the population remains in the bound states: when the laser is turned off, all the population returns to the ground state.

The two satellites on either side of the main transition are called an Autler–Townes doublet after the names of those who first observed them [473]. They were, however, predicted at an earlier date by Mollow [474]. An elegant method of calculating the fluorescence profile was described by Cohen-Tannoudji and Avan [475]. A very full discussion of how to represent this problem, including the central elastic component, the inelastic contributions and the sidebands, can be found in lecture notes by Cohen-Tannoudji [476]

The AC Stark effect is relevant, not only in atomic spectroscopy, but also in solid state physics. The biexciton state (or excitonic molecule), where two Wannier excitons are bound by the exchange interaction between electrons, occurs in various semiconductors (see section 2.22). Various experiments on the AC Stark effect of excitons have been reported, but the clearest example to date is probably the observation of the Rabi splitting of the biexciton line in CuCℓ reported by Shimano and Kuwata-Gonokami [477]. It is very interesting to consider how Bloch states in solids, which themselves are delocalised and periodic, are dressed or modified by the electromagnetic field, since their properties are rather different from those of purely atomic states, which are by definition completely localised.

9.11   Coherent excitation of continuum states

If the excited state is not a bound, but a continuum state, then the situation is fundamentally different, because of photoionisation. Once an electron is excited into the continuum, it escapes rapidly from the influence of the laser field (except at the highest laser field strengths), so that cycling of the population can no longer occur. This process of escape is *irreversible*, so that Rabi nutation can no longer occur. The time dependence of the laser and ionisation amplitudes must then be considered.

Suppose the ionisation probability per unit time of an atom is dP/dt. The number of atoms left in the ground state $\mid 0 >$ after a time t is

$$n_0(t) = \exp\left(-\int_0^t \frac{dP}{dt}dt\right) \tag{9.18}$$

The rate at which population is transferred into the continuum by a coherent field is determined by the Rabi frequency $\hbar\Omega_{0\varepsilon}(t) = < 0 \mid \hat{V}(t) \mid \varepsilon >$, where the time dependence must now be included explicitly. The exciting laser is a coherent source of light, so we suppose that the field $\hat{V}(t) = \hat{V}_0(t)\cos(\omega t + \phi(t))$ where both $\hat{V}_0(t)$ and $\phi(t)$ vary only slowly with time. We then have

$$n_0(t) = \exp\left(-\int_0^t \frac{\pi\hbar}{2} \mid \Omega_{0\varepsilon}(t') \mid^2 dt'\right) \tag{9.19}$$

which is the result predicted by the golden rule, because $\hat{V} = \hbar\Omega_{0\varepsilon}/2$ in the RWA. The decay rate is $\Gamma(t) = (1/t)\pi\hbar \mid \Omega_{0\varepsilon}(t) \mid^2$, which of course varies with time during the laser excitation. Several approximations have been made on the way to obtain this result, which do not concern us here. Further discussion and references can be found in [478].

9.12   Coherent excitation of an autoionising state

From the two previous sections, it is clear that the dynamics of exciting an autoionising resonance, which has a mixture of bound state and continuum character, will be rather complex, involving both reversible and irreversible contributions, i.e. a combination of Rabi oscillations and damping.

In fact, if we think of oscillations as involving a factor $\exp(i\Omega t)$, then if the Rabi frequency is taken as complex, its real part will be the 'bound state' contribution, and the imaginary part can be used to represent loss of population into the continuum.

An interesting example is the case of LICS, which was discussed in section 8.20. In this case, one can show that the complex Rabi frequency

Ω is

$$\omega = \left\{ \left[D - \frac{i}{2}(\Gamma_1 - \Gamma_0) \right]^2 + \Gamma_0 \gamma_1 (q - i)^2 \right\}^{1/2} \qquad (9.20)$$

where D is the detuning of the probe laser from the LICS centre, and Γ_0, Γ_1 are the ionisation widths. The real part of the Rabi frequency

$$\mathcal{R}e \{\Omega\} = D^2 + (q^2 - 1)\Gamma_0 \Gamma_1 - \hat{\Gamma}_{01}^2 \qquad (9.21)$$

represents the amplitude of nutations, and the imaginary part

$$\mathcal{I}m \{\Omega\} = -iD(\Gamma_1 - \Gamma_0) - 2iq\Gamma_0\Gamma_1 \qquad (9.22)$$

represents the amplitude of ionisation, while the total

$$\mid \Omega \mid^2 = \left[(D^2 - \hat{\Gamma}_{01}^2)^2 + 2\Gamma_0\Gamma_1 \left\{ d^2(q^2 - 1) - q\hat{\Gamma}_{01}(\Gamma_{01} + D) \right\} \right]^{\frac{1}{2}} \qquad (9.23)$$

where we define $\hat{\Gamma}_{01} \equiv \Gamma_0 \Gamma_1 / 2$. Consider, for example, the case $\Gamma_1 \gg \Gamma_0$ and equal to Γ, say. Then if the reduced detuning $x = 2D/\Gamma$, we have $\mid \Omega \mid^2 = 1 + x^2$, and we find that

$$I = \frac{(\mathcal{I}m \{\Omega\})^2}{\mid \Omega \mid^2} \rightarrow \frac{-D\Gamma - 2q\Gamma\Gamma_0}{D^2 + (\Gamma/2)^2} \qquad (9.24)$$

whence

$$I = \frac{(\bar{q} - x)^2}{1 + x^2} \qquad (9.25)$$

by defining $\bar{q} = 4\Gamma_0/\Gamma$, i.e. we recover a Fano shape for ionisation under conditions where one autoionisation width dominates the coupling and the other is small. Likewise, if D is large and Γ dominates, there are no Rabi oscillations at the centre of the LICS profile where autoionisation dominates, but they reappear at large detunings, where one state is decoupled from the other in the autoionisation channel.

In general, however, the lineshapes are completely different from a Fano profile if the laser intensity is large. Thus [478] gives examples of profiles which may even contain two minima.

9.13 Laser-induced mixing of autoionising states

While on the theme of modifying the profiles of autoionising states using a strong electromagnetic coupling, one should also mention resonant laser-induced mixing achieved by using a strong field visible laser to couple two autoionising states together, and then probing the change in the structure by a two-photon transition using a weaker laser and detecting ionisation.

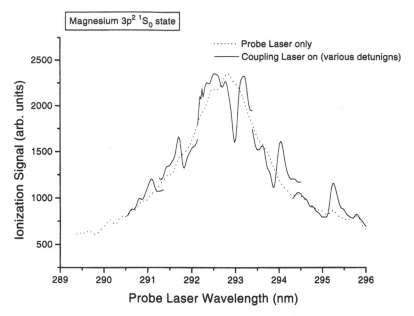

Fig. 9.4. Example of the changes in shape which are observed, as a function of the detuning of the embedding laser, when two autoionising resonances, one broad and one narrower, are coupled together. Note the symmetry reversals. which are strongly reminiscent of those described in chapter 8 (by courtesy of N.E. Karapanagioti 1995).

This type of experiment has been performed on Mg atoms by Karapanagioti *et al.* [479]: the experimental data indicate a drastic change in the ionisation rate, while the coupling enables many different lineshapes to be produced by varying the detuning of the coupling laser, as shown in fig. 9.4 These results are interesting, because they demonstrate precisely the same kind of symmetry reversal effects as were described in chapter 8 (see fig. 8.6).

9.14   Laser-induced orbital collapse

Another property of atoms which is sensitive to the conditions in the outer reaches of the atomic field is of course orbital collapse, which can be controlled as described in section 5.23. This has led Golovinskiy *et al.* [480] to consider whether a strong laser field could be used to precipitate orbital collapse, and to propose an experiment in which dynamic collapse at the Rabi frequency could be detected by X-ray spectroscopy of the irradiated sample.

9.15 Influence of the laser intensity on spectra

The previous few examples show that, under coherent illumination, one can no longer consider observed spectral intensities as purely characteristic of the atom, since the properties of the Autler–Townes doublet depend on the laser intensity, and the profile of a laser-excited autoionising line will in general possess a shape unrelated to that of an atom excited by a weak source.

The implication is that laser spectroscopy brings out new phenomena which would not otherwise have been apparent, but also distorts the information so that it is not always simple to recover atomic parameters from observations of the spectra. This increased complexity becomes more and more apparent as the intensity of laser light is increased.

9.16 The ponderomotive potential

Consider a free electron in a laser field. The Hamiltonian has the form

$$\mathcal{H}(t) = \frac{\left(\mathbf{p} - \frac{e}{c}\mathbf{A(t)}\right)^2}{2m} \tag{9.26}$$

There are two time dependences to be considered: first, there is the oscillation period of the radiation, and then there is the duration of the laser pulse. If the second period is much longer than the first, we can neglect it and average over one period of oscillation. The result is

$$< \mathcal{H} >= \frac{p^2}{2m} + \frac{e^2 A_0^2}{4mc^2} = \frac{p^2}{2m} + \frac{e^2 \mathcal{E}_0^2}{4m\omega^2} \tag{9.27}$$

where p is no longer the instantaneous momentum of the particle, but is the average or drift momentum. The term $\Phi_L = e^2 \mathcal{E}_0^2 / 4m\omega^2$ is the ponderomotive potential, or average energy due to the wiggling motion in the field. At very strong laser fields, the amplitude of this motion is large, and the ponderomotive energy becomes very important. A simple working formula for the ponderomotive potential due to a laser of wavelength λ is

$$\Phi_L = 10^{-13} I \lambda^2$$

, where I is the intensity in $W\,cm^{-2}$ and λ is the wavelength in μm. For a focussed laser beam giving about $10^{14}\,W\,cm^{-2}$ of photon energy $\sim 1\,eV$ (Nd YAG), the ponderomotive energy is about 10 eV. This wiggle is not simply an up-and-down motion, because the influence of magnetic fields must also be considered: in reality, the ponderomotive motion is a 'figure of eight' trajectory which lies in the v–k plane. A difficulty in the interpretation of real experiments is that the ponderomotive potential within the laser focus is constant neither in space nor in time.

9.17   Volkov states

In order to discuss the tunnelling regime, we begin by considering the extreme situation, where the field of the atom is negligible, and the dominant interaction is with the laser field. A high AC field (with $\gamma_K < 1$) is one so high that it becomes comparable to or dominates the atomic field strength (a definition comparable to the one for the strong magnetic field problem – see section 10.14). The atomic unit of magnetic field strength is $e^2/4\pi\epsilon_0 a_0^2$, where a_0 is the Bohr radius. This turns out to be about 5×10^9 V cm^{-1} and corresponds to a laser intensity of 3.5×10^{16} W cm^{-2}.

For an electron in a high AC field, a solution of the Schrödinger equation exists which is analytic, and therefore plays a fundamental role, rather like the hydrogenic solution in relation to atomic physics. If we consider a plane polarised, homogeneous and infinite classical wave[1]

$$\vec{\mathcal{E}}(t) = \vec{\mathcal{E}}_0 \sin \omega t \tag{9.28}$$

or alternatively a vector potential $\mathbf{A}(t) = \mathbf{A}_0 \cos \omega t$ where $\mathbf{A}_0 = \dfrac{c}{\omega}\vec{\mathcal{E}}_0$ and write the time-dependent Schrödinger equation

$$i\frac{\partial}{\partial t}\psi(r,t) = \left[\frac{1}{2}\left(\mathbf{p}+\frac{1}{c}\mathbf{A}\right)^2 + V - \Phi\right]\psi(r,t) \tag{9.29}$$

where Φ is the electrostatic potential, then it is invariant under the gauge transformations

$$\left.\begin{array}{c} \mathbf{A}\prime = \mathbf{A} + \nabla f \\[2ex] \Phi\prime = \Phi - \dfrac{1}{c}\dfrac{\partial f}{\partial t} \\[2ex] \psi(r,t) = \exp\left(-\dfrac{i}{c}f\right)\psi(r,t) \end{array}\right\} \tag{9.30}$$

By a suitable choice of the gauge, the A^2 term in (9.29) can be eliminated. This is called the $\mathbf{A}\cdot\mathbf{p}$ or velocity gauge. One obtains

$$i\frac{\partial}{\partial t}\chi(r,t) = \left[\frac{p^2}{2} + \frac{1}{c}\mathbf{A}(t)\cdot\mathbf{p}\right]\chi(r,t) \tag{9.31}$$

---

[1] We use $\mathcal{E}$ for electric field strength, to avoid confusion with E the energy. Consequently, we use $\vec{\mathcal{E}}$ for the electric field vector.

with

$$\chi(r, t) = \phi(t) \exp\left(i\mathbf{k} \cdot \mathbf{r}\right) \tag{9.32}$$

and

$$\phi(t) = C \exp -i \left(\frac{k^2}{2} + \frac{1}{c}\mathbf{k} \cdot \mathbf{A}_0 \sin \omega t \right) \tag{9.33}$$

which is the Volkov solution [481] for an electron moving in the field of a plane wave, and is exact to all orders. This solution will be used in section 9.23.3.

Notice that, in the limit of a strong laser field, where atomic effects are small, the solution has a universal and simple form, whose main properties may therefore be sought in observed spectra. As will be shown below, several features of experimental data for atoms in strong fields can be explained from this very simple starting point.

9.18 The nonperturbative regime

With the advent of very intense, pulsed laser radiation, a new area of physics was born, namely the study of the interaction between radiation and matter in a regime in which the normal perturbative dipole approximations of atomic physics are invalid. This field of research is actively being pursued in many laboratories today. It involves not only the response of single atoms or molecules placed in a strong field, but also the propagation of intense radiation in gaseous media which may exhibit partial ionisation. To remain within the theme of the present monograph, we restrict our attention to the single-atom response. Even with this restriction, one may well ask whether the new subject falls within the area of 'highly-excited atoms', i.e. whether the atomic character of the targets can usefully be probed at radiation field strengths attainable with modern lasers ($>10^{16}$ W cm^{-2}). It may seem strange to raise this point: we now endeavour to show why this question is a fruitful one to ask.

Several different points of view are arguable:

(i) One can assume that all the physics of the atom in a strong laser field is basically that of a single bound electron in a superintense oscillating wave, and that all one needs to know of the atomic physics is essentially the binding energy of that electron, since the interaction of one electron with the strong laser field dominates all other interactions. This is sometimes called the *single active electron approximation*. The model presupposes that electron–electron correlations are somehow 'turned off' by the strong laser field, and that a one-electron model is adequate. We will refer to this as the *minimum atomic physics option*.

(ii) If we start by noting that an external field is only strong in relation to the internal ionisation or binding energy of the atom, and that any experimentally generated applied field must grow from a low value, we may be pessimistic about reaching the high field regime. We may argue that ionisation is likely to occur sequentially, before the maximum field is reached, and that the 'next' electrons available for study have higher binding energies. Since an external field is only strong with respect to the binding field (cf the definition of γ_K) strong field physics of the atom would become an unattainable quest. We will refer to this as the *zero atomic physics* option, because, according to this picture, atoms in superstrong laser fields are basically unobservable.

(iii) We might, however, be more optimistic, and suppose that, by some subtle means, the atom can organise itself to survive the intense pulse for long enough to experience the full strength of the laser field. Bearing in mind that the experiment will always involve a pulsed field in order to achieve high power, the implication is that the rise-time of the pulse becomes an important consideration: if the field rises too slowly, we might revert to situation (ii) above. A proper description of the dynamics then requires a detailed understanding of properties specific to a given atom. We will call this the *maximum atomic physics option*.

(iv) Finally, still assuming that the atom somehow survives to experience the full strength of the laser pulse, it might be that all the electrons are excited coherently and, so to speak, correlated by the external laser field. In this option, atomic physics would again be 'washed out' at the strongest laser fields: the spherical closed shell structure of the atom would disappear from view, and the only role of conventional atomic physics would be to *allow* the atom to survive into the strong field regime. We will call this the *laser-correlated option*.

There is no clear evidence that any one of these models is sufficient or complete. Examples exist of situations in which one or other of these pictures applies, and which model is appropriate depends on: (a) the atom; (b) the intensity of the laser pulse; and (c) its duration.

9.19   The role of atomic forces

To an atomic physicist, it would be a disappointment if the minimum atomic physics option turned out to be all that is required to achieve a satisfactory description of experimental data. The interaction between strong radiation fields and matter might still turn out to be an important problem of modern physics but would not reveal new properies of specific atoms.

Interactions involving short laser pulses have occasionally been compared to collisions, which also depend impulsively on time. Collision problems require at the outset a rather good description of the correlated atom, without which realistic predictions cannot be made. This analogy suggests that a maximum atomic physics option is required.

While considerable theoretical progress has been made, calculations to date do not involve a full many-electron atom, but only one-electron systems. Thus, there is as yet no answer to the question: what is the effect of the laser field on electron–electron correlations? and we cannot decide between (iii) and (iv). The results of calculations using one-electron models are compared with experiments involving many-electron systems[2]. Indeed, the present situation is incomplete in the following respect: in order to avoid situation (ii) where strong field atomic physics is hindered by ionisation, experiments are done mainly on the elements with the highest ionisation potentials, i.e. the rare gases, which possess many equivalent electrons in the outer subshell. However, the only theory available at present is for one-electron systems or, as is said, in the *single-active-electron* approximation.

9.20   Atoms in very strong laser fields

The remainder of this chapter is organised as follows. First, we briefly describe a typical system which has been used to achieve intense fields, the purpose being to stress the experimental connection between high laser fields and short pulses. We move on to describe various high field effects, together with simple physical explanations which have been used to interpret them. We then return to our main discussion about the role of atomic physics in these effects and, finally, to the description of a theoretical picture which holds promise to describe the response of many-electron atoms in strong fields.

---

[2] We stress that a one-electron model, in this context, does not mean an independent particle SCF system, but merely a one-electron-in-a-potential calculation.

9.21   Strong field physics and quantum optics

The subject of 'quantum optics' is concerned with the quantum properties
of the radiation field, i.e. the properties of photons. Since the word
'multiphoton' has been used, it might seem that strong laser fields are in
some way relevant to quantum optics. However, the word 'multiphoton' is
something of a misnomer in the strong field regime. In fact, if very many
photons are involved, quantisation of the radiation field is more or less
irrelevant: the intense, coherent laser pulse tends to a quasiclassical beam
of light. Indeed, it has been pointed out by several authors [483] that the
use of the word 'photon' in the context of laser physics is of questionable
validity.

The condition for a beam to be classical is that the number of photons
per cubic wavelength should be much larger than 1. For modern strong
field lasers, this number is of the order of 10^{14} or more.

9.22   The generation of strong laser fields

The generation of strong laser fields is a very specialised activity in which
much ingenuity has been deployed. It is a large area of research in its
own right. To illustrate some of the important points, we describe here
one fairly typical experimental system, which has been developed in the
laboratory at Imperial College London [484].

There are two approaches to making strong field lasers: the first is to
increase the energy within the pulse, and the second is to shorten the
duration of the laser pulse. If the aim is to develop a good system for
atomic physics then, as pointed out above, a long pulse laser is a poor
choice because ionisation will soon produce situation (ii). Therefore, it
is advantageous to reduce the duration of the laser pulse, which also
has the advantage that a smaller laser system can be built since the energy required is lower. The exact shape of the pulse and in particular
its time structure are important experimental parameters if any attempt
at theoretical interpretation is to be made, and they must therefore be
determined experimentally.

The reason why high power lasers involve short pulses comes from fundamental limitations in the performance of amplifiers. For a given energy,
the most effective way to raise the power is to compress the pulse in time
after amplification has taken place, provided that a broad bandwidth can
be preserved during the initial amplification, so that subsequent compression to a short timescale is not impaired. Indeed, to achieve a high
enough initial amplification, it is advantageous to stretch out the pulse
in time, so as to avoid damaging solid state amplifiers. Pulse stretching

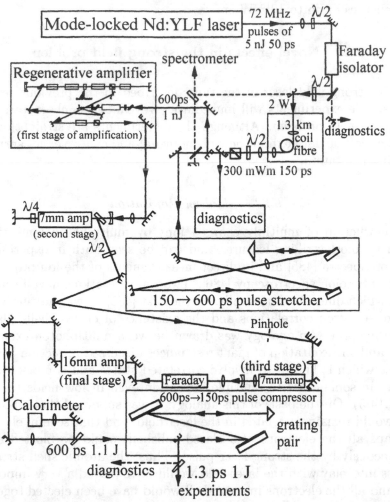

Fig. 9.5. Example of a CPA laser built at Imperial College, which provides 1 J pulses of about 1 ps duration (1 TW of power – after J.-P. Connerade *et al.* [482]).

can be achieved by using either an optical fibre or a pair of gratings. It is important that the stretched pulse should also be 'chirped' linearly – i.e. that each time in the pulse should be associated with a different frequency according to a linear law, so that pulse compression can be effected by using a pair of plane gratings between which each frequency will traverse a different path length. In fig. 9.5, we show an outline diagram of a typical system of chirped pulse amplification (CPA), which allows a pulse to be stretched, amplified and finally recompressed, to achieve a total power of about a terawatt. A lens can then be used to focus down the laser beam to intensities of up to 10^{18} W cm^{-2}.

9.23   Novel effects in the strong field problem

Given a strong field laser of the type described in the previous section, quite simple experiments will immediately reveal effects characteristic of the strong field problem. Although the experiments are straightforward in principle, the data exhibit new phenomena, and controversies surround their interpretation.

9.23.1   Multiple ionisation

The production of multiply charged ions by multiphoton ionisation is one of its most striking features, and can be seen both in experimental emission spectra [485] and in direct measurements of the ion production rate by photoion spectroscopy [486]. Two interpretations have been proposed which differ fundamentally. In the first [485], it was advanced that electron–electron correlations and the existence of closed shells play an important part. An analogy was drawn between multiphoton excitation of Xe, and the excitation of giant resonances discussed in section 5.16 – a process which is known to involve correlated states and complete shells, leading in some cases to multiple ionisation even in low fields (see section 7.14.5). On the basis of this analogy, it was supposed that an atomic shell would remain correlated in the laser field, and that several electrons (perhaps all the electrons of a closed subshell) could be blasted off simultaneously by the strong laser pulse. Thus, the atomic shell structure and its interplay with the laser field would have played a very important role, and all the electrons in a subshell would have been ejected together, leading to direct multiple ionisation. We recognise this model as a fully atomic picture in the sense of section 9.18 (iii), but there is the difficulty of understanding how the closed subshells respond to the laser field.

This picture was subsequently criticised [486] on the basis that several steps of sequential ionisation are more likely, because of section 9.18 (ii). In this case, ionisation would not not result from the strong field or as a cooperative process. Indeed, the strong field conditions are never probed. Obviously, these two interpretations are totally different and mutually exclusive.

In principle, the matter could readily be sorted out by time-resolved detection, but there are no detectors fast enough for this purpose. Thus, one cannot distinguish experimentally between the two interpretations.

An analogy has therefore been drawn between multiple ionisation of atoms and Coulomb explosions of molecules in a laser field (see section 9.23.5): the experiment on molecules favours alternative (ii) of section 9.18 and, when applied to atoms, suggests that ionisation proceeds sequentially so that the strong field regime would never be reached [487].

This last conclusion has in turn been challenged. The point was made [489] that molecules should evolve quite differently from atoms, precisely because fragmentation in the molecules is so fast that collective oscillation modes cannot build up during the laser pulse. The experiment for molecules would then bear little relation to the situation for atoms.[3]

A new conceptual framework for the ionisation of many-electron atoms in strong laser fields has been introduced [490], based on different principles (see section 9.26). According to this picture, multiple ionisation does not proceed sequentially in a superstrong field: instead, the mechanism for excitation and ionisation depends explicitly on the duration of the laser pulse. This model suggests a situation closer to option (iv) of section 9.18: multiple ionisation is not intrinsically dependent on correlations or on the existence of closed shells, both of which are considered to be 'washed out' in the presence of the strong field. However, it must occur simultaneously rather than sequentially, because of the nature of the interaction.

As a final point concerning ionisation in a strong laser field, we note that the ionisation threshold is itself Stark-shifted, and is not the same as for the free atom (see the discussion in section 9.23.2).

9.23.2 *Multiphoton absorption by quasifree electrons*

The next strong field effect is usually termed *above threshold ionisation* or ATI. Of course, all ionisation processes, whether at low or high fields, involve excitation to states above threshold, so this is a misnomer.[4] We may think of it in the following way. It is a well-known elementary result that a completely free electron can neither absorb nor emit a photon, since it would not satisfy the energy–momentum conservation relations. When an atom is ionised by a strong laser field, the escaping electron does not have time to escape from the field of the parent ion before encountering another photon. In this case, momentum–energy balance still can be satisfied, and it may absorb another photon, or indeed several further photons. Thus, the usual, smooth distribution in the photoionisation continuum is lost. Instead, one finds a 'comb' of peaks separated in energy

---

[3] The dissociative ionisation of molecules in intense laser fields is an interesting problem in itself: experiments show that the fragment ions emerge with dissociation energies which are definite fractions of the Coulomb explosion energy, specific to a given molecule but independent of the charge state. A model has been proposed to explain this [488].

[4] Other names have been suggested, such as excess photon ionisation, but the simple designation ATI has now been adopted and is difficult to change.

by $\hbar\omega_L$, where ω_L is the frequency of the laser. We can say that the photoionisation continuum has been structured or dressed by the laser field.

This effect can be thought of as a generalisation of Einstein's equation for the photoelectric effect. For the kinetic energy of the escaping electron, we can now wite:

$$E_{kin} = n\hbar\omega + m\hbar\omega_L - I_p$$

where n photons are required to promote an electron into the continuum, and a further m photons are absorbed by the escaping photoelectron.

Again, in the description just given, one might think of the absorption processes in the continuum occurring sequentially or, alternatively, that all the ATI peaks are produced simultaneously inside the laser pulse (which is the picture developed below).

Freeman [491] has proposed an experimental test of which picture is correct: he argues that, if ATI peaks are observed even for negative ions, where the escaping electron cannot readily interact since the core is not ionic, then the process should be simultaneous rather than sequential. Indeed, several ATI peaks are observed [492] even for negative ions, which suggests that the process as a whole is simultaneous.

We now address the question: how much atomic physics needs to be included in order to account for ATI? In fig. 9.6, we show experimental data for ATI from [493], obtained at several laser intensities. One of the important properties of ATI peaks, referred to as *peak suppression*, is that the relative intensity of the first ATI peaks above threshold does not increase uniformly with laser field strength, but actually begins to *decrease* in intensity relative to higher energy peaks as the laser field strength increases. Such behaviour cannot be explained in a perturbative scheme, in which interactions must decrease monotonically order by order as the number of photons involved increases, but can be accounted for in terms of the AC Stark shift of the ionisation potential in the presence of the laser field. In ATI experiments, the ionisation potential appears to shift by an average amount nearly equal to the ponderomotive potential, so that prominent, discrete ATI peaks are seen despite the many different intensities present during the laser pulse. However, ATI peaks closest to the ionisation limit become suppressed as the amplitude of the laser field oscillations increases and the ionisation threshold sweeps past them (a different effect which also suppresses ionisation near threshold is discussed in section 9.24.1).

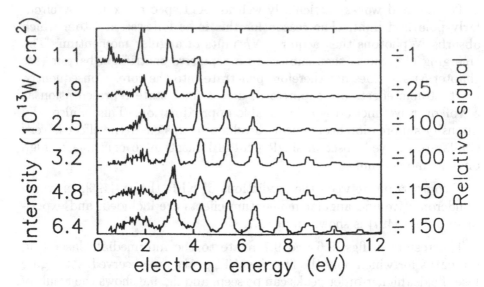

Fig. 9.6. Experimental data, showing the first few ATI peaks above threshold obtained at several different laser field strengths. Note the *reduction* in relative intensity of the lowest energy peaks (peak suppression) which occurs as the laser intensity is *increased*. This is a nonperturbative effect (after T.J. McIlrath *et al.* [493]).

9.23.3 Computation of ATI spectra: Keldysh–Faisal–Rees theory

Keldysh [494] obtains the transition rate in a strong field from the ansatz

$$W_{if}(t) = -\frac{i}{\hbar} \int_0^t < \phi_f(t') \mid V(t') \mid \psi_i(t') > dt' \qquad (9.34)$$

using, for $V(t)$, the dipole approximation. Other authors ([495, 496]) have stressed the importance of the A^2 term for the atom in a strong electromagnetic field (see equation (4.1)) and have used the Coulomb gauge interaction

$$V(\mathbf{r}, t) = -\frac{e}{m}\mathbf{p} \cdot \mathbf{A}(t) + \frac{e^2}{2m}A(t)^2$$

These equations allow the main features of observed ATI spectra to be computed in a simple way. The method is referred to as the Keldysh–Faisal–Reiss (KFR) theory. Note that the final states are Volkov states, and therefore contain no information about the atom. In this sense, we are dealing with a model of type (i) in section 9.18. Very little atomic physics is required: only the ionisation potential of the atom, the angular momentum acquired by the emerging electron and the properties of the laser wave need to be known.

The method works particularly well for ATI spectra excited by circularly polarised light. The reason for this is as follows: an atom which absorbs N photons then acquires $N\hbar$ units of angular momentum. The emerging electron is then subject to a repulsive centrifugal barrier (see chapter 5) and does not therefore penetrate into the core. Consequently, most atomic effects are suppressed, and a final state representation as a Volkov wavefunction is a reasonable approximation. This is also why intensity suppression occurs near threshold in this case: the effect is very similar to delayed onset in single-photon ionisation to continua of high angular momentum.

A comparison between the predictions of KFR theory [496], in which ponderomotive and angular momentum effects are included, and experimental data [497] is shown in fig. 9.7.

The results in figs. 9.6 and 9.7 relate to the intermediate laser field strengths for which only the first few ATI peaks are observed. At higher laser fields, higher-order peaks can be seen, and fig. 9.8 shows the result of a systematic study of the energies of higher order peaks by photoelectron spectroscopy. No qualitative change occurs: their general behaviour is similar to the properties outlined above.

The ATI spectra described above are found for pulse durations of about 1 ps or longer. For much shorter pulses, there is a qualitative change which is further discussed in section 9.25: it turns out that the minimum atomic physics option is then no longer sufficient.

9.23.4 Angular distributions

A further step in sophistication is to study the angular distributions of the escaping photoelectrons, emitted within each of the ATI peaks. On general grounds, one expects a 'propensity rule' (cf section 4.2) that should favour increasingly elongated distributions with increasing numbers of photons, i.e. higher angular momenta. Experimentally [498], however, unexpected 'scattering rings' in the ATI angular distributions are found, which scale as $9U_p$. The proposed explanation is that these are produced by backscattering of the emerging electrons from excited states of the ion core during ionisation. Thus, even under ps excitation, atomic features begin to emerge when the process of ATI is studied in finer detail, but come about from secondary processes rather than from the direct excitation mechanism.

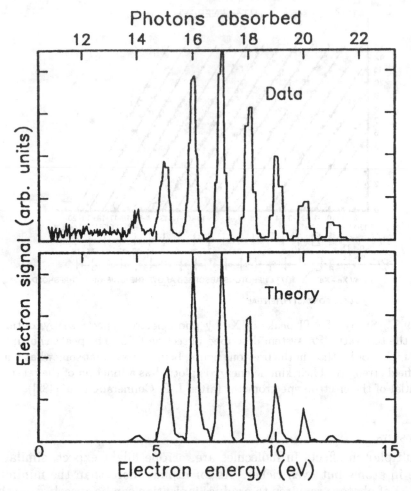

Fig. 9.7. Comparison between experimental data and KFR calculations, which demonstrates that the main features of ATI for circularly polarised light (including angular momentum barrier effects) are well accounted for in a non-perturbative model using Volkov final states (after H. Reiss [496]).

9.23.5 Bond softening and Coulomb explosions

As already remarked in section 2.31, H_2 has an exceptionally high ionisation potential amongst molecules, and a Rydberg series extending far into the vacuum ultraviolet range of the spectrum. Consequently, it is very suitable for strong field studies, especially as its stability means it is a permanent gas.

Thus, the harmonic spectrum of H_2 in a strong laser field has been studied (Tisch, private communication), and there have been similar experiments on other diatomics [500], although few to date on polyatomics

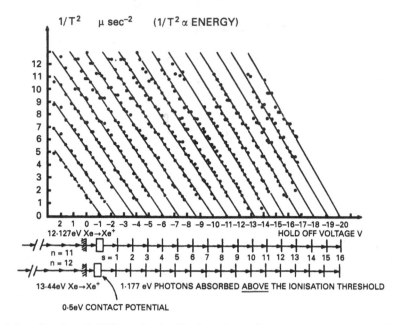

Fig. 9.8. Study of ATI peaks in Xe by photoelectron spectroscopy, obtained using the terawatt CPA system described in section 9.22. The peaks shown here are of higher order than in the previous figure, being recorded at somewhat higher laser field strength. Their kinetic energy is plotted as a function of the retarding potential of the electron spectrometer (after J.-P. Connerade *et al* [482]).

or clusters.

Multiphoton effects in molecules are, as one might expect, similar to those in atoms but even richer. As for atoms, more than the minimum number of photons required to produce ionisation can be absorbed, so that ATI is observed [499, 501]. Many other effects are possible, for example, Coulomb explosions [502] which arise when two charged fragments fly apart.

When the laser field becomes comparable to the internuclear binding field of a molecule, which clearly depends on the orientation of the internuclear axis with respect to the electric field vector of the radiation, the molecular potential is deformed: potential wells become more shallow and, as a consequence, the number of vibrational states which are observable is reduced. This effect is called *molecular bond softening* [503].

Another effect is found near and above the photodissociation thresholds. Just as in the case of ionisation, molecules may continue to absorb excess photons over and above the minimum required, and this results in extra peaks in the spectra: this effect is called ATD by analogy with ATI.

Since the internuclear potential is typically a few volts per angstrom, peak laser fields of 10^{13}–10^{14} W cm^{-2} are enough to produce strong field effects in molecules, this being a somewhat smaller number than for atoms.

Another interesting area is the study of Coulomb explosions in very strong laser fields. Coulomb explosions have already been described in section 7.7: there, it was explained that multiply charged molecules are unstable because the Coulomb potential energy is higher than the dissociation threshold. They fly apart on a very short timescale, which is of the order of 20 fs or less, and is thus much shorter than a ps laser pulse. This means that the fragmentation of a molecule in a strong laser field carries information on the redistribution of energy well within the timescale of the pulse.

Since different fragmentation paths originate at different points on the potential surface diagram, one can predict which fragments will appear if fragmentation proceeds sequentially and one can distiguish between sequential break up and multiple fragmentation in a single step.

It has been shown in this way that, for molecules, fragmentation and ionisation tend to occur sequentially. In part, this occurs precisely because the timescale for the fragmentation of molecules is so short. The study of molecular processes in strong fields is thus a separate subject of study: it provides new information on their dynamical properties, which are somewhat different from those of atoms [487].

9.23.6 High harmonic generation

High harmonic generation (HHG) is another new effect which occurs when an atom experiences a strong laser field. The ponderomotive force drives both bound and free electrons up and down (or in a figure of eight, if relativistic effects become important) with a large amplitude. Anharmonicities result in the emission of harmonics of the fundamental driving frequency, which extend up to very high orders (often in excess of 100 times the driving frequency).

An example of HHG in He, obtained using the laser system of fig. 9.5 is shown in fig. 9.9: the driving laser has a photon energy of ~ 1 eV, and so the harmonics extend up to over 100 eV in energy.

While the valence electron remains attached to the atom, it is an efficient radiator of energy. However, free electrons also experience the strong laser field, and thus both contribute to HHG. In practice, however, when ionisation occurs, propagation effects due to changes in the refractive index of light in the presence of free electrons make it more difficult to generate high harmonics, and so the observed spectra usually come from neutral atoms with high ionisation potentials.

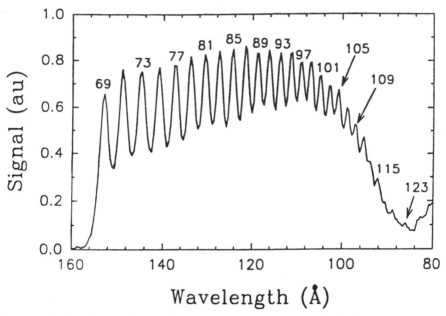

Fig. 9.9. High harmonics in the spectrum of He. Note the high energy cut-off, which is real (the low energy cut-off is due to the optical arrangement), and how the harmonic spectrum remains essentially flat in intensity up to very high orders, which is a nonperturbative effect (data averaged over 18 shots) (after J.-P. Connerade *et al.* [482]).

The laser field which is responsible for HHG acts similarly in some respects to the oscillating fields experienced by an electron in a synchrotron: high harmonics are emitted under semiclassical conditions at many times the fundamental frequency, which in this case is the frequency of the laser. For reasons of symmetry, only the odd harmonics are observed. Also (for reasons mentioned below) the dominant contribution comes from initially bound electrons.

In a similar way to ATI, HHG exhibits nonperturbative behaviour, which, for HHG, concerns the fall off of intensities with increasing order. After an initial drop for the first few harmonics, the intensities settle down to a plateau region, where they remain fairly constant until a cut-off is reached. The cut-off varies somewhat from atom to atom, but follows a general rule [505]. One supposes that HHG with good intensity is produced as long as the electrons remain attached to the atomic sites. In the tunnelling regime, this is the case, and so electrons return to their respective sites with a maximum kinetic energy given by an average over the optical cycle, which turns out to be $3\Phi_L$. This needs to be added to the ionisation potential, and so the cut-off is given by

$$E_{max} = \epsilon_I + 3\Phi_L$$

This rule is only approximate because it misses out many important physical considerations. For example, ϵ_I is taken as the field-free ionisation limit for a one-electron system, and the rule contains nothing explicit about pulse duration. Nevertheless, it has been found to work reasonably well. From this rule, we can see that a well-developed spectrum of high harmonics is only expected in species with a high ionisation threshold.

Even the existence of a plateau turns out to be a very general feature: calculations with a square well potential containing only one bound state of suitable binding energy reproduce a plateau, so that, in this case, a minimal atomic physics description (option (i) of section 9.18) proves adequate to interpret most HHG experiments.

Observations prove that high harmonics may generate radiation of much higher photon energy than the binding energy of the valence electrons. For example, in fig. 9.9 we can follow them out to about the 121st order (i.e. radiation up to about 121 eV photon energy), which lies well above the field-free *double*-ionisation potential, for which both electrons would be stripped off in absence of the laser field.

Experimental studies of the time dependence of harmonic generation [506] also show that harmonic generation becomes very weak once ionisation occurs, so that shorter pulses are preferable to generate harmonics in elements of lower ionisation potential.

The reason why ionisation quenches the harmonics is more subtle than studies of the single-atom response can explain. As noted above, free electrons possess a high refractive index, which can easily break the phase matching between the harmonic and the driving field, and which can also spoil the focussing conditions. In principle, there is no reason why ionisation should quench HHG because: (i) a many-electron atom contains more tightly bound electrons which can also 'wiggle' and (ii) free electrons can also radiate when placed in an very strong laser field.

A fuller understanding of HHG is complex, because of these propagation effects which modify the observed spectra [511, 512]. One way of studying the dynamics in more detail has been suggested namely to apply an external magnetic field. Laboratory fields can be generated which approach the magnitude of the magnetic component of laser fields. It would then become possible to break the symmetry so that even harmonics are also generated. Such spectra have been computed [513], but not yet observed.

9.24   General discussion of strong field effects

At very strong fields, simple approximations, such as RWA and the tunnelling approximation are not viable approaches, and so one returns to solving (numerically or otherwise) the time-dependent Schrödinger equation for an atom subjected to a strong pulse of coherent light. This involves at least three different times: the natural response time of the system or its internal frequency(ies); the period of the driving oscillation, and the rise-time of the laser pulse. Thus, the dynamics of the system are in general extremely complex.

We now describe different ways of visualising this problem. In the presence of an intense, oscillating field, the energy levels of the atom and its ionisation potential are no longer the same as for a field-free atom. They become characteristic of an atom *dressed by the radiation field*. It proves convenient to solve the resulting differential equation with time-dependent coefficients and cyclic boundary conditions by a mathematical method due to Floquet. In the Floquet picture, the dressed energy levels of the atom appear as *quasienergies* (energies of the atom in the presence of the field) which can be calculated. Experimentally, there is no simple way of recovering them, but Floquet theory describes many aspects of strong field physics very well, and leads in particular to a useful description of the dynamics of strongly driven quantum systems. We now review some of the theoretical background.

9.24.1   The Kramers–Henneberger frame

One approach to solving the Schrödinger equation in a strong field is to make a transformation to the so-called Kramers–Henneberger or 'wiggling' frame. Starting from the time-dependent Schrödinger equation in the form:

$$\left\{ \frac{\left(\mathbf{p} - \frac{e}{c}\mathbf{A}\right)^2}{2m} + V + e\varphi \right\} \Psi(\mathbf{r},t) = i\hbar \frac{\partial \Psi(\mathbf{r},t)}{\partial t} \qquad (9.35)$$

The scalar potential φ can be removed by moving to the $\mathbf{A} \cdot \mathbf{p}$ gauge. We perform a unitary transformation:

$$\Psi(\mathbf{r},t) = \exp\left(-\frac{ie^2}{mc^2} \int_{-\infty}^{t} \mathbf{A}(t')^2 dt' - \frac{ie}{mc} \int_{-\infty}^{t} \mathbf{A}(t') \cdot \mathbf{p} dt' \right) \psi_{KH}(\mathbf{r},t) \qquad (9.36)$$

where the first term in the argument of the exponential eliminates the $\mathbf{A}^2$ term and the second, the $\mathbf{A} \cdot \mathbf{p}$ term. With this transformation, the

Schrödinger equation assumes a new and simpler form which can be solved for the functions $\psi(\mathbf{r}, t)$:

$$\left\{ \frac{\mathbf{p}^2}{2m} + V(\mathbf{r} + \alpha(t)) \right\} \psi_{KH}(\mathbf{r}, t) = i\hbar \frac{\partial \psi(\mathbf{rt})_{KH}}{\partial t} \qquad (9.37)$$

where the quantity

$$\alpha(t) \equiv \frac{e}{mc} \int_{-\infty}^{t} \mathbf{A}(t') dt' = \alpha_0 \sin \omega t \qquad (9.38)$$

defines an oscillating space translation since $\mathbf{A}(t)$ is periodic. The oscillation is referred to as the *quiver motion*. This frame was introduced by Pauli and Fierz [514], and was later used by Kramers [515] and Henneberger [516]. The Kramers–Henneberger frame can be combined with Floquet analysis (see the next section).

One effect of the wiggling motion is to produce a so-called 'dichotomy' of the wavefunctions [517]. The idea is quite straightforward: the time-averaged potential in the Kramers–Henneberger frame has the general form shown in fig. 9.10, which tends towards two wells located near the classical turning points. For large fields, the wavefunction is split between the two wells (this is the effect called 'dichotomy'), with little amplitude in between, and little spatial overlap with the ground state. Dichotomy is one of several schemes which lead to the notion of 'stabilisation' induced by the strong field: once the atom is in such an excited state, it has little probability of radiating down to the ground state, and one also finds that it has increased survival at high intensity. Models of this kind suggest that ionisation is not a necessary outcome in the presence of a very intense laser field, provided that the 'turn-on' time is sufficiently rapid for the atom to be excited directly into wavepackets of this or an analogous kind. There are now several different models which predict the survival of excited atomic states as they enter the strong field regime.

9.24.2 Floquet theory

The RWA is appropriate when the Rabi frequency is much smaller than the other frequencies in the problem, *viz.* the transition frequency and the detuning. The first of these conditions limits the intensity, and so ultimately RWA breaks down: it then becomes necessary to include the effects of the counterrotating terms, and one returns to solving the time-dependent Schrödinger equation, with a Hamiltonian which is periodic in time: this is done in a more general way by applying Floquet's theorem for differential equations with periodically varying coefficients.

We now describe Floquet theory, as it has been applied to the strong

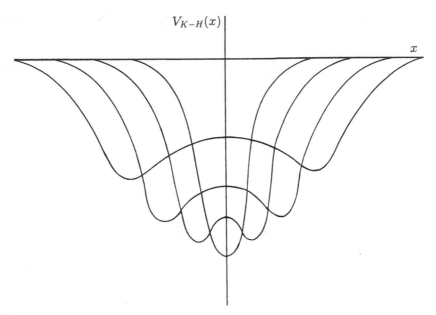

Fig. 9.10. The appearance of the time-averaged effective potential in the Kramers–Henneberger frame for several different amplitudes of the oscillating wave.

field laser problem by Dietz and his school [506], which provides a convenient conceptual framework for the rest of this chapter.

The time-dependent Schrödinger equation for a one-electron atom in a coherent field of strength Λ and frequency ω reads:

$$-i\hbar\frac{d\Psi(\mathbf{r}, t)}{dt} + (\mathcal{H}_0 + \Lambda\tilde{D}\sin\omega t)\Psi(\mathbf{r}, t) = 0 \qquad (9.39)$$

where $\mathcal{H}_0$ is the field-free Hamiltonian, and $\tilde{D}$ is the usual dipole operator. An alternative to RWA is to attempt an exact solution in a periodic framework: Floquet's theorem [507] states that there exists a complete set of solutions of equation (9.39)

$$\Psi_n(\mathbf{r}, t) = u_n(\mathbf{r}, t)\exp\left(-i\epsilon_n t\right) \qquad (9.40)$$

where the u_n are themselves solutions of the eigenvalue equation:

$$-i\hbar\frac{du_n(\mathbf{r}, t)}{dt} + (\mathcal{H}_0 + \Lambda\tilde{D}\sin\omega t)u_n(\mathbf{r}, t) = \epsilon_n u_n(\mathbf{r}, t) \qquad (9.41)$$

which satisfy the periodic boundary condition

$$u_n\left(\mathbf{r}, t + \frac{2\pi}{\omega}\right) = u_n(\mathbf{r}, t) \qquad (9.42)$$

and the ϵ_α, which are independent of time, are called the *quasienergies*.

An important feature of the Floquet solution is that it can be obtained equivalently from other solutions within the same class: the same solution is obtained if $u_n(\mathbf{r}, t)$ is replaced by $u_n(\mathbf{r}, t)\exp(imt)$ where m and n are any integers, and ϵ_n is replaced by $\epsilon_n + m\hbar\omega$. This means that, if we introduce the convention that the suffix α represents a double index n, m, the quasienergies ϵ_α are not defined uniquely: they are, in fact, only determined *modulo* $\hbar\omega$, i.e. the quasienergy spectrum forms Brillouin zones of width $\hbar\omega$. Exactly as in solid state physics, where the band structure in a crystal can be 'folded over' itself so as to require only one finite spectral region for its representation, the spectrum of an atom in a strong laser field can be plotted in a quasienergy space which is finite and branches for different values of m are folded over into a single diagram.

9.25   Avoided crossings and Landau–Zener transitions

The quasienergy spectrum ϵ_α can be plotted as a function of Λ, the field strength of the laser, and this generally will lead to a number of 'avoided crossings' (rather like the avoided crossings in the Lu–Fano graphs of chapter 3). An important difference between these these plots (in quasienergy space) and the Lu–Fano graphs (in another pseudoenergy space, defined modulo of the principal quantum number) is that the Floquet quasienergy diagram may contain branches which, although close to each other in quasienergy, differ from each other in real energy by any multiple of $\hbar\omega$, where ω is the frequency of the strong field laser. If the system suddenly jumps from one branch on the diagram to another (a 'diabatic' transition), this can represent what is loosely termed a 'multiphoton' transition (despite the facts that quantisation of the electromagnetic field is not involved, and that the perturbation is a purely classical wave). We see that the classification of 'multiphoton' transitions in this perspective is merely a matter of accounting for energy balance.

In any problem involving avoided crossings, there are two modes of variation. A physical system will evolve *adiabatically*, i.e. it will remain on one or other branch if the intensity Λ is varied slowly. If, however, the variation is rapid then *diabatic* or Landau–Zener transitions may occur [508]. The probability of a transition from one branch to the other at the avoided crossing is given by

$$P(\eta) = \exp\left(-\frac{\pi\eta}{2}\right) \tag{9.43}$$

with

$$\eta = \frac{\delta\epsilon \cdot \delta\Lambda}{\dfrac{d\Lambda}{dt}}$$

where η is evaluated at the avoided crossing. This formula is actually an expression of the uncertainty principle: for short times, the system does not 'know' on which branch it is, and can hop from one to the other. Formally, it appears to be identical to the Landau–Zener formula [508] as used in solid state physics, *but it is not*: the important generalisation is that the distance $\delta\epsilon$ between the branches is not an energy difference but is a difference *inside the Brillouin zone*. $P(\eta)$ therefore expresses the probability of a multiphoton rather than a single-photon transition.

Thus, Floquet theory provides a natural means of studying the dynamics of the response of an atomic system to a pulsed laser field. Indeed, it yields a general framwork for discussing the adiabatic motion of any periodically driven quantum system. If we define a characteristic time $\delta\tau = \delta\Lambda/(d\Lambda/dt)$ which is the time taken to sweep through the avoided crossing, then one can distinguish between three regimes:

(i) $\delta\epsilon\delta\tau \ll 0$: The system evolves adiabatically past the avoided crossing without noticeable change in its wavefunction.

(ii) $\delta\epsilon\delta\tau \sim \hbar$: The wavefunction of the system splits at the avoided crossing into a superposition of states involving both branches of the avoided crossing.

(iii) $\delta\epsilon\delta\tau \gg \hbar$: The system again evolves adiabatically, remaining on one branch before and after the avoided crossing.

9.26   Effects due to laser pulse duration

A simple picture of multiphoton excitation by short pulses [509] will now be described. This picture is arrived at by a large number of *ab initio* numerical integrations of the time-dependent Schrödinger equation in both quantum and semiclassical conditions and for a wide variety of laser field strengths, pulse durations, etc. From this experience, a model has been developed. The model should have more general validity than the calculations which have suggested it, because it is based on fundamental principles which should apply even in more complex systems. Consider first the dressed atom levels plotted schematically in fig. 9.11(a) as a function of the field strength Λ. As remarked in the previous section, quasienergies are available from Floquet calculations in any model potential, and a map of quasienergies against laser intensities can be drawn. It will usually reveal a complex manifold, with avoided crossings, of which there can be many. At each of these avoided crossings, generalised Landau–Zener transitions can occur.

Next, consider excitation by a pulse $\Lambda(t)$ of laser radiation (fig. 9.11(b)). If the laser pulse is long on a timescale set by the magnitude of the avoided

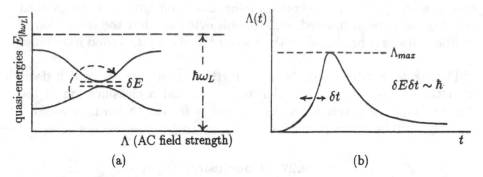

Fig. 9.11. (a) Schematic map of the quasienergies, plotted modulo laser photon energy, as a function of the laser intensity, showing avoided crossings. (b) The shape of the laser pulse as a function of time. Under appropriate conditions (as described in the text) a generalised form of Landau–Zener transition occurs which is what one commonly calls a multiphoton transition (after J.-P. Connerade *et al.* [482]).

crossing, then the system will travel outwards and return along the same branch adiabatically, i.e. no excitation occurs. If, on the other hand, the duration of the pulse matches the timescale defined by the avoided crossing, then a multiphoton transition can occur, in the sense described above. The diagram demonstrates that the field strength is equally important, and that the properties of the atom or molecule, together with the laser field strength, determine the positions and numbers of the avoided crossings. In general, therefore, multiphoton excitation by an intense pulse is a very complex problem. As the ionisation threshold is approached, the density of avoided crossings increases, and so there are many pathways to ionisation. Within this complexity, however, one can, for a given process, select a subset of relevant avoided crossings once the laser pulse duration is known. Indeed, there exists experimental information which confirms that excitation mechanisms change according to laser pulse duration. For example, as mentioned above, ATI peaks have been observed to break up and reveal resonance structure when the pulse length is shortened below 1 ps [504]. A particularly significant manifold of avoided crossings occurs near a Rydberg series limit, where the density of curves is large, and they tend to become similar. As a consequence, under sufficiently short pulse excitation, a whole family of transitions to high Rydberg states can occur: since these lie close in energy to the lowest ATI peaks, the latter have been observed to break up into a fine structure of Rydberg character.

It is also known that the atoms may be left in highly excited states after irradiation by a short pulse. A second laser pulse can be used to photoionise the excited states, which are then detected by photoelectron spectroscopy [510]. This provides information on which of the avoided crossings have been bridged, and there is evidence that the excitation of Rydberg states is produced by the rising edge of a femtosecond laser pulse [518].

The model outlined here is taken further in chapter 10, which deals with the order-to-chaos transition in atoms, and a classification of the different kinds of transition which can occur for excitation near avoided crossings will be given.

9.27 Conclusion

In the present chapter, the rapidly growing subject of atoms in strong laser fields has been briefly described, the main emphasis being on the novel effects which have been observed. Several aspects of the problem have not been discussed: for example, above 10^{20} W cm^{-2}, relativistic effects will become important, although these have not yet been observed. Similarly, we have omitted any discussion of coherent control in two-colour excitation and femtosecond chemistry.

The importance of the rise-time of the pulse has also been stressed: it is at least as significant as effects due to the strength of the field. The extent to which details of atomic physics enter into the description depends on the situation, but a detailed knowledge of quasienergy maps for a given atom would be needed for optimum pulse durations to be selected to produce a given excitation. The interplay between strong laser fields and the internal correlations of the atom is likely to remain one of the most challenging areas of research in atomic physics for some time to come.

10

Statistical methods and 'quantum chaology'

10.1 Introduction

A system is considered as chaotic in classical mechanics if the orbits, instead of remaining confined to a specific region, invade the whole of available phase space. A simple example is a pendulumn with a magnet below the bob: if a sheet of paper is placed between the magnet and the bob and a pen is attached, the pendulumn will write all over the page within the range accessible to it. More exactly, if we examine phase space, it will seem completely disordered, with interwoven tracks throughout its volume. If we magnify the volume, the disorder will persist, and so on *ad infinitum* no matter how great the magnification, because classical mechanics imposes no limit on the resolution which can be achieved.

Chaotic behaviour can arise in any system whose motion is described by a nonlinear differential equation. Whether or not it is prevalent depends on the details of the problem, but it is a general theorem that *any* system described by a nonlinear differential equation possesses some chaotic regime.

In quantum mechanics, by contrast, chaos does not occur. We may see this in several ways. First, note that we cannot magnify *ad infinitum* the volume to be analysed in phase space: eventually, we reach the elementary volume $\hbar^3$ within which trajectories lose their meaning. Another way of reaching the same conclusion is to note that any Schrödinger type equation is linear: its solutions obey the superposition theorem. Under these circumstances genuine chaos is excluded by fundamental principles.

Nonetheless, traces of chaotic behaviour persist in some quantum systems. They are found when calculations are extended into the semiclassical limit and the underlying classical dynamics of the corresponding classical system is chaotic. This is anyway an interesting situation, because it relates to the correspondence principle. To some degree, it challenges

363

the strong view expressed by the Copenhagen school (and articulated most clearly by Landau) that a problem in quantum mechanics cannot be solved unless the solution of the corresponding classical problem can also be specified. If the classical system is chaotic, then there is some the difficulty of understanding how the correspondence principle should be applied.

This general area of investigation is described as *quantum chaology*. Quantum chaology can be defined as the study of a quantum system in its semiclassical limit, when the corresponding classical system is chaotic. Its definition would seem rather contorted without the opening statements made previously. The best model problems of quantum chaology (i.e. those which are most completely understood) involve highly-excited atoms. In addition, there are good reasons for an atomic physicist to study this subject: it was shown by Poincaré [519] that the classical three-body problem is chaotic. In principle, therefore, even as simple an atom as He possesses a regime of chaotic underlying classical dynamics.

Nevertheless, until relatively recently, chaos has not figured prominently in the preoccupations of atomic physicists. The *independent* electron, Hartree–Fock model, replaces the N-electron equation by N one-electron equations, each one of which contains well-ordered Rydberg series, which seems to preclude the possibility of 'quantum chaos', at least until many-electron correlations are introduced. This suggests that a transition to 'quantum chaos' might appear in the presence of strong configuration mixing.

In the present chapter, we examine some of these issues, drawing examples from current research. We begin by considering strongly interacting, many-electron systems, which are not necessarily in the semiclassical limit and therefore not true examples of 'quantum chaos', but which share many of its statistical properties, and we move on to highly-excited atoms in external fields, which can be followed to the semiclassical limit and are therefore very good systems for the study of 'quantum chaos'.

Since chaos is only defined for a classical system, the word will be used in quotes when applied to quantum systems: these quotes should be taken to imply all the caveats of the present introduction, and will allow us to use the words 'chaos' or 'quantum chaos' for brevity without implying more than the loose connection we have described. The subject of chaos has a vast literature. For further information, the reader may consult [521, 522] which provide more insight into its mathematical foundations.

10.2   Total breakdown of the $n\ell$ characterisation

As long as n and ℓ remain good quantum numbers, the independent particle model and the central field approximation both apply, and 'quantum chaos' does not arise. We can thus identify two situations where 'chaos' could emerge: the first is a complete breakdown in the independent electron approximation (due, for example, to strong correlations) and the second is a distortion of the central field approximation (due, for example, to a strong external field).

Thus far, it has been tacitly assumed that, although some breakdown in the $n\ell$ quantum numbers might occur for many-electron systems, the basis of their quantisation is essentially the same as for one-electron systems. In other words, n and ℓ might cease to be good or correct quantum numbers, but the principles of Bohr–Sommerfeld quantisation on which the quantum numbers are based have been accepted as the correct zero-order description.

10.3   The statistics of spectra

One of the consequences one might expect in a highly correlated atom, as a result of 'quantum chaos' is an explosion in the number of interacting states. This would give some kind of random distribution within an energy range where, in absence of correlations, there would be but a few levels and a well-ordered spectrum. Thus, quantisation would be persist in the form of discrete energy levels, but incipient chaos would nevertheless appear through disorder in their spacings.

Two assumptions are implicit in this description:(i) the levels must be *interacting*, so that the analogy with classical mechanics, in which the same trajectory fills the whole of phase space, is as complete as possible; and (ii) it must be established, at least in principle, that regularity would reappear if the interactions coupling the levels were turned off.

Thus, one is interested in a rather special kind of statistics, viz. the statistics of a dense population of interacting levels. This is the fundamental distinction between *chaos* and *complexity* : there may arise situations in which levels do not necessarily all interact (they might have different quantum numbers) but are simply present in large numbers, so that their analysis is not possible in practice but could be performed in principle. These are called *unresolved transition arrays* (UTAs). One can develop [526] a theory of UTAs which yields general theorems about them as a whole. Such theories are a statistical approach to the interpretation of spectra, but are not related to the problem of quantum chaology.

In the case of strongly interacting levels, it must be the case that the

quantum numbers which might be used to distinguish one level from another within the set have all broken down. In the example of the independent particle model, excited states differ from each other by at least
one quantum number of at least one electron. After strong mixing has
occurred, this will cease to be true for levels of the same J and parity.

It is interesting to note that the quantum numbers which lose their
meaning in this case are the n and ℓ of the individual electrons. These
quantum numbers are essential in achieving the great simplification of
the many-body problem which results from the Pauli principle and the
occurrence of closed shells. Consequently, the simplicity of the subshell
structure implicit in the periodic table can be lost when 'quantum chaos'
emerges. This is not altogether suprising, since the many-body problem
in classical physics has no analogous simplification.

10.4   Semiclassical quantisation and the harmonic oscillator

The simplest way to introduce semiclassical quantisation is via the harmonic oscillator. Consider the Hamiltonian:

$$\mathcal{H} = \frac{p^2}{2m} + \frac{k^2}{2}. \tag{10.1}$$

Since the total energy E is a constant of the motion, we can rewrite (10.1)
as:

$$\frac{p^2}{2mE} + \frac{q^2}{2E/k} = \frac{p^2}{a^2} + \frac{q^2}{b^2} = 1 \tag{10.2}$$

which is an ellipse in p, q space (phase space), since both a and b are
constants.

The variables p and q are canonically conjugate, so that the Bohr–
Sommerfeld quantisation condition yields:

$$\oint p\,dq = \pi ab = \pi\left\{(2mE)\cdot(2E/k)\right\}^{1/2} = 2\pi\frac{E}{\omega} = \left(n+\frac{1}{2}\right)h \tag{10.3}$$

whence:

$$E = \left(n+\frac{1}{2}\right)\hbar\omega \tag{10.4}$$

10.5   Closed orbits

Periodic or closed orbits in phase space are especially convenient, because the quantisation conditions are then particularly straightforward,
and the correspondence principle takes one over smoothly from classical
to quantum physics.

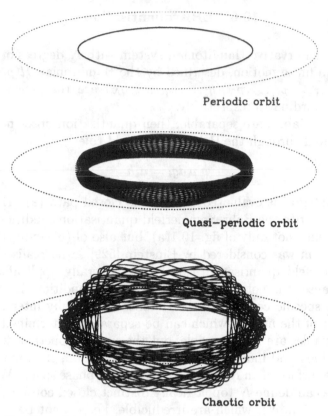

Fig. 10.1. Showing: (a) a closed orbit; (b) multiperiodic orbits on an invariant torus; and (c) chaotic trajectories (after D. Pfenniger [527]).

Indeed, they are so convenient that quantum physicists initially treated *all* systems as though they could be quantised in this fashion, including systems with $N > 1$ degrees of freedom. Many of the difficulties associated with this problem can be appreciated by considering fig. 10.1, which shows three situations one might anticipate, corresponding to different degrees of complexity in a more general situation, involving more dimensions. Fig. 10.1(a) is the one just considered, namely a closed orbit, and is actually the least likely; fig. 10.1(b) also represents a closed orbit, but of much greater complexity, in which the particle describes a very involved motion on the surface of a torus before the orbit repeats itself. Finally, in fig. 10.1(c), the orbit never repeats, and a portion of phase space eventually becomes filled with trajectories (one can no longer even call them orbits). This last example is of a classically chaotic system.

10.6   EBK quantisation

Consider a conservative Hamiltonian system with N degrees of freedom, performing a finite motion, described by the Hamiltonian $\mathcal{H}(p,q)$, where $(p,q) \equiv (p_1, \ldots, p_i, \ldots, p_N)(q_1, \ldots, q_i, \ldots, q_N)$ are the generalised momenta and coordinates respectively.

If all the variables are separable, then quantisation poses no problem and is achieved through the Bohr–Sommerfeld rule:

$$\oint p_i dq_i = n_i h \tag{10.5}$$

where the integral is evaluated around the closed path (a). How, then, should one rewrite the Bohr–Sommerfeld quantisation condition in order to take account, not only of fig. 10.1(a), but also of (b) and (c)?

This problem was considered by Einstein [529] , who realised that the Bohr–Sommerfeld quantisation rule is not generally applicable, to any system because it is not a canonically invariant condition, i.e. it is only valid in the specific case where the Hamiltonian is fully integrable, with N constants of the motion which can be separately determined.

If the variables are *not* separable, but the system nonetheless possesses N single-valued and independent integrals of motion, then motion takes place on the surface of an N-dimensional torus in phase space. Within this surface, one can define N topologically distinct closed contours (labelled C_k, with $k = 1,...,N$), which are irreducible, i.e. cannot be turned into each other by continuous deformations. Examples are shown in fig. 10.2.

The EBK quantisation rule ([529, 530, 531]) (after Einstein, Brillouin and Keller) states that:

$$\oint_{C_k} \sum_{i=1}^{N} p_i dq_i = n_k h \tag{10.6}$$

and allows one to quantise motion on such a surface, because the N quantum numbers n_k are invariant, i.e. independent of the path for reducible contours. Interestingly, EBK quantisation does *not* depend on detailed knowledge of the full trajectory in phase space.

We then come to the difficult question: how is quantisation to be applied for a system whose Hamiltonian contains *fewer* than N independent and single-valued integrals of the motion? Classically, such a system will exhibit chaotic or irregular orbits, and the question is then whether and how this property will persist in the corresponding quantised system.

It was suggested by Percival [532] that a semiclassical system with N degrees of freedom executing bounded motion possesses either: (i) a regular spectrum of bound states labelled by N quantum numbers for the case where there are N independent constants of the motion; or (ii) an

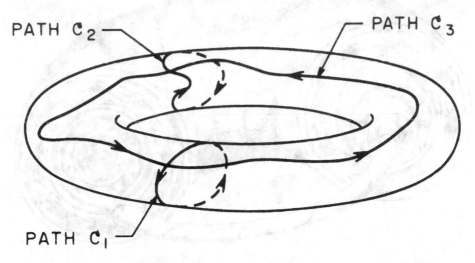

PATH C_2

PATH C_3

PATH C_1

Fig. 10.2. The paths involved in EBK quantisation: paths C_1 and C_2 are not topologically distinct from each other, but paths C_1 and C_3 are (after W.P. Reinhardt and D. Farelly [528]).

irregular spectrum in which it is not possible to find satisfactory quantum labels in the case where there are fewer than N first integrals of the motion.

10.7 The many-electron atom

How the correspondence principle should be applied to an atomic system thus depends critically on whether or not there exists a multiperiodic representation of the classical trajectories – the question first raised by Einstein. If the system possesses multiperiodic orbits, then its motion becomes separable, i.e. it becomes equivalent to as many independent modes as there are degrees of freedom. Dynamical separability is assumed in all independent particle and perturbative models of the many-electron atom. It is, however, not strictly applicable and the successes of simple quantum theory for many-electron systems are, to say the least, surprising. It was pointed out by Einstein, who based his arguments on the work of Poincaré [519], that there exists no true separation of the three-body problem.

In general, a Hamiltonian is neither completely integrable, nore purely chaotic, which means that, in practice, phase space fragments itself into *islands of stability*, inside which the motion is quasiperiodic, and regions,

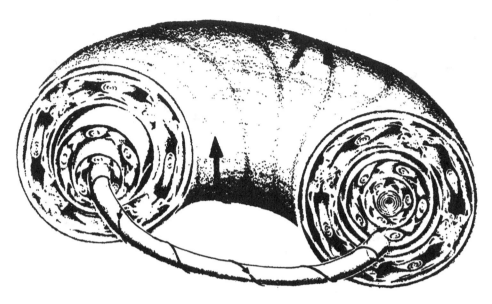

Fig. 10.3. Illustrating the deformation of invariant tori predicted by the KAM theorem. Chaotic or irregular trajectories grow between the tori, at so called separatrix points on a Poincaré surface of section or cut normal to the tori (after A. Hacinlyian [520]).

called stochastic layers, which become densely filled by single trajectories. These different regions are usually visualised by making cuts known as Poincaré sections, across the invariant tori, as shown in fig. 10.3 and further described in section 10.8. The volume occupied by these layers grows rapidly with increasing energy around a critical energy which is characteristic of the system.

Thus, the unperturbed H atom becomes the unique example of perfectly regular quantisation for atoms. For many-electron atoms, complete separability cannot be assumed, and Bohr–Sommerfeld quantisation cannot apply exactly. In the semiclassical limit, one may expect to find at least some situations where the corresponding classical system will exhibit chaos, at least in some domain of parameter space, and where, for the quantum system, some related complications due to this breakdown will persist.

As a practical guide, we may follow the line that any gross breakdown in $n\ell$ characterisation signals difficulties in applying the Bohr–Sommerfeld quantisation condition and should be investigated. It is important (as indicated above) to select groups of levels which interact with each other, since otherwise, their behaviour may be ruled by complexity rather than by chaos.

We therefore seek systems in which the independent particle model breaks down and many different excitation channels are strongly coupled together.

10.8   The Kolmogorov–Arnold–Moser theorem

As already noted, the strength of the interaction is a crucial parameter: the KAM theorem (after Kolmogorov, Arnold and Moser [523, 524, 525] expresses the fact that the invariant tori on which the EBK quantisation (see above) is based are, in the classical system, replaced by very complex fractal structures in which only residues of the tori can survive. Thus, a progressively increasing volume of phase space is lost to chaotic orbits as the strength of the dynamical coupling between the modes is increased.

The KAM theorem states that, if a nondegenerate unperturbed Hamiltonian is subjected to a sufficiently small, conservative, Hamiltonian perturbation, most of the invariant tori survive, being only slightly deformed, in the phase space of the perturbed system. These invariant tori become densely filled with the trajectories of the perturbed system, which wind around them conditionally-periodically, the number of independent frequencies being equal to the number of degrees of freedom.

The KAM theorem, it should be noted, has nothing to say about what happens when the strength of the perturbation increases. However, a considerable amount of experience has accumulated from detailed numerical calculations performed for many systems. One can visualise the results by studying Poincaré sections: if a cut is made across an invariant torus (see fig. 10.3) and a numerical calculation of trajectories is performed over a sufficiently long time, the stable orbits fill the deformed tori densely, and so result in closed curves in the two-dimensional cut, whereas the irregular or chaotic orbits yield a random speckle.

As the magnitude of the coupling increases, it is found that the regular region breaks up into smaller and smaller fragments, and that the irregular or chaotic region grows in size. Examples of such behaviour will be given below (see section 10.24). When dealing with many-electron atoms and the influence of perturbations, there is the difficulty that the strength of the perturbation cannot be varied, since it is intrinsic to the system. The approximate validity of $n\ell$ quantisation, which is derived for an independent electron system is in a sense the limit in which the problem becomes regular.

10.9   The quest for chaos in many-electron atoms

In view of the fact that the periodic table is so well accounted for using an independent electron model, there is not much room for quantum chaos to develop in the spectra of unperturbed many-electron atoms.

The strongest pieces of evidence complex atoms provide *in favour* of independent electron modes and simple Bohr–Sommerfeld quantisation are: (i) the existence of Rydberg series; and (ii) the regularity of the periodic table of the elements. As a corollary, we should look for quantum chaos (if it occurs) in atoms for which there is some breakdown in the quality of the shell structure, combined with prolific and heavily perturbed overlapping series of interacting levels. These conditions are most readily met, as will be shown below, in the spectra of the alkaline-earth elements, as a result of *d*-orbital collapse.

This problem is in a sense analogous to the classical problem of anharmonically coupled oscillators, which is an example of a chaotic system.

10.10   A connection to nuclear physics: random matrices

One of the surprising results of 'quantum chaology' is the finding that the statistics of random matrix theory apply equally well to two apparently unrelated problems: the spectra of complex systems with many coupled degrees of freedom and those of relatively simple systems with few degrees of freedom whose underlying classical dynamics is chaotic. Examples will be given below. We now describe the origin of random matrix theory.

Experiments performed by nuclear physicists have led to the collection of a vast amount of data for excited states of nuclei [533, 534]. While the ground states are fairly well explained in an independent particle scheme with an average potential [535, 536], the same cannot be said for excited states, basically because the static mean field approximation breaks down more easily than it does for atoms (the very small mass of electrons and the long range interaction they are subject to are both favourable to mean field theory in atomic physics). For excited states in nuclear physics, the interactions become so complex that the Hamiltonian is essentially unknown. On the other hand, it is observed experimentally that the level structure is very dense. While it would be quite impossible to explain each and every state in the spectrum under such conditions, there do exist general statistical rules which have been developed by nuclear physicists to account for the average or statistical properties of discrete levels, such as the separations between adjacent interacting levels, the distribution of particle and radiative widths, etc, all of which can be measured with good precision if a large enough ensemble of states is available.

One assumes that, according to the general laws of quantum mechanics, the energy levels are all eigenvalues of a Hermitian operator, the Hamiltonian. Since the system possesses a spectrum consisting in general of a fairly large number of bound states and a continuum, the space spanned by the eigenstates of the Hamiltonian is infinite-dimensional. In practice, however, this is too difficult to handle, and so, for the bound part of the spectrum, the space is truncated by working with a very large but nevertheless finite number of dimensions. Within this approximation, the Hamiltonian is a very large, finite-dimensional matrix.

At low excitation energies, the energy levels are far apart, but as the energy is raised, level densities increase, and all approximate quantum numbers are destroyed by the perturbations and interactions between levels. Consequently, the *only* quantum numbers which are significant are those which correspond to *exact* integrals of the motion such as parity or the total J value. The Hamiltonian matrix thus reduces to diagonal blocks labelled by these quantum numbers. All off-diagonal elements coupling different blocks must be zero, and level sets coming from different blocks are statistically uncorrelated. Thus, the first important task is to sort or *bin* levels according to the exact quantum numbers. Levels within each block are then subject to such complex interactions that any local regularities which might have been expected on the basis of partial diagonalisations are washed out.

The matrix elements then appear as random variables, and can be allowed the maximum statistical variations consistent with global symmetry requirements imposed on the ensemble of operators. Thus, distributions are only limited by general properties of the system: statistical theory does not even attempt to describe the details of a level sequence, but can represent its general features and the degree of fluctuation.

One must distinguish between this situation and the statistical theories described under the general heading of 'statistical mechanics': in classical statistical mechanics, the laws which govern the evolution of the system are extremely well known, but the path is so complex that it cannot be followed in detail. Here, the number of states is large, but finite, and the randomness arises from an ensemble of possible Hamiltonians.

The suggestion that the local statistical properties of a sequence of levels all possessing the same exact quantum numbers (known as an interacting or simple sequence) should be the same as those of the eigenvalues of a random matrix is originally due to Wigner (Wigner's hypothesis). It is as fundamental to the statistical theory of levels as the ergodic hypothesis is to classical statistical mechanics. He also conjectured [537] that the probability distribution for spacings x between adjacent levels of a simple sequence would be given by

$$P_{wigner}(x) = \frac{\pi}{2} x \exp\left(-\frac{\pi x^2}{4}\right) \qquad (10.7)$$

where x is made dimensionless by dividing the actual spacing ΔE by the mean spacing $\overline{\Delta E}$ (a matter we return to below). This conjecture is called Wigner's surmise, and the distribution (10.7) is known as a Wigner distribution.

There is a simple way [537, 538] of arriving at (10.7): one may assume (as is indeed reasonable) that, for a given level of energy E, the probability that the next level occurs at $E + \Delta E$ is simply proportional to ΔE, for small enough ΔE. Now extend this to all ΔE by subdividing the energy interval ΔE into N equal intervals, and by assuming that the probabilities in each interval of length $\Delta E/N$ are mutually independent. This leads to:

$$P_{wigner}(x)dx = \lim_{N\to\infty} \prod_{r=0}^{N-1}\left(1 - \frac{rx}{N}\frac{1}{N}a\right) axdx = ax\exp\left(-a\frac{x^2}{2}\right)dx \qquad (10.8)$$

where the constant a is then determined by requiring that the *average* value of x is unity, since x itself is the ratio of ΔE to $\overline{\Delta E}$.

This simple argument turns out to be an oversimplification, but the result (10.7) nevertheless turns out to be pretty close to the exact distribution. An extensive discussion of random matrix theory in nuclear physics is given by Mehta [539].

Under similar circumstances, particle widths γ follow the distribution:

$$P(\gamma^{1/2})d\gamma^{1/2} = \left(\frac{2}{\pi}\right)^{\frac{1}{2}} \exp\left(-\frac{1}{2}\gamma\right) d\gamma^{1/2} \qquad (10.9)$$

where $\gamma = \Gamma/\overline{\Gamma}$ is again dimensionless. This is known as a Thomas–Porter distribution [542].

An early attempt [543] was made to apply statistical methods to the interpretation of the spectra of complex atoms. For reasons to be discussed in the next section the data are not truly convincing. Another problem to which random matrices are applicable is the electronic structure of metallic clusters when the interlevel spacing becomes larger than the thermal energy, as first shown by Frölich [545]: such systems are discussed in chapter 12.

10.11   Level spacings in complex atoms

We now turn to some experimental examples. The first attempt to study the statistics of levels in complex atoms is due to Rosenzweig and Porter [544, 543] who noticed that elements with incomplete $4f$ subshells have a very rich line spectrum, and therefore tried to select groups of levels of the same J by 'binning', and then analysing them statistically. There are several problems with their approach. First, the levels originate from tables which contain data from different sources, which poses questions about spectral resolution, sensitivity, etc. Second, there is no physical parameter which could (even in principle) be turned off and which would provide an ordered spectrum for comparison. Third, there may even be doubts about the reliability of the binning procedure, which results in a rather small number of levels for each J.

A much better example was mentioned in section 7.14.6: it arises by parent ion mixing in $5p$-excited Ba. As was described, the subshell characterisation breaks down in the presence of a $5p$ hole, because of the overlap in energy between $6s$ and $5d$ states. As a consequence, the $6s^2$ subshell is completely shattered, and breaks into a $6s^2 \otimes 5d^2 \otimes 5d6s \otimes 6p^2$ manifold, all the $J = 1/2$ and $J = 3/2$ levels of which can serve as limits to $J = 1$ series accessible in photoabsorption from the ground state. No fewer than 14 overlapping Rydberg series have been detected, all of which are strongly perturbed. A particularly nice feature of this example is that no binning is required since the ground state is $^1S_0$ and, because of the dipole selection rules *only* the $J = 1$ upper states are accessible.

The Ba spectrum forms part of the homologous sequence Ca, Sr, Ba: the $3p$ spectrum of Ca [546] and the $4p$ spectrum of Sr [547] have also been investigated. The Ca subvalence spectrum possesses a fairly regular Rydberg structure, shown in fig. 7.2, consisting of rather clear Rydberg series of autoionising resonances. This structure has been interpreted with the help of *ab initio* multiconfigurational Dirac–Fock calculations: while some configuration mixing is present, there is no evidence that the $n\ell$ designations have fully broken down. The Sr spectrum is a good deal more complex, and no longer follows anything like the simple pattern expected from the perturbed independent particle model. As for Ba, it is so complex and so heavily perturbed that there is little prospect of making assignments.

It turns out that there are three regions of the spectrum: (i) a comparatively simple region where Rydberg spacings are large; (ii) a very complex region where all the series overlap and spacings are small; and (iii) another simpler region in which high series members to the highest limits stand clear of all the lower series. An overall view of this spectrum

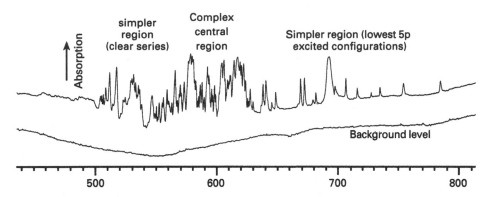

Fig. 10.4. Overall view of the 5p spectrum of Ba. Three regions of differing complexity are indicated (see text). The Wigner distribution is found in the complex central region (after J.-P. Connerade *et al.* [358]).

is shown in fig. 10.4. In the central region, the complexity is so great that detailed configuration assignments turn out to be impossible.

The statistical approach has therefore been followed to interpret the Ba data [547]: we show, in fig. 10.5, a plot of the accumulated density of states alongside a portion of the Ba 5p spectrum. This figure demonstrates that there is a large enough density of levels to draw a meaningful average, i.e. that the *rigidity* of the spectrum is small enough.

The accumulated density of states or *spectral staircase* is a convenient representation of the overall spectral density. It is defined mathematically as

$$\mathcal{N}(E) \equiv \sum_n \Theta(E - E_n) \qquad (10.10)$$

where $E_1, E_2, \dots E_n$ is the sequence of eigenvalues and Θ is the unit step function. The spectral density $d(E)$ is given by

$$d(E) \equiv \frac{d\mathcal{N}(E)}{dE} = \sum_n \delta(E - E_n) \qquad (10.11)$$

The *rigidity* $\Delta(L)$, introduced by Dyson and Mehta [548], is the local or sliding average of the rms deviation of the spectral staircase from the best fitting straight line in a given energy range corresponding to L mean level spacings. From the same data, a plot of the distribution of nearest level spacings can be made. This is shown in fig. 10.6. There is some evidence that their distribution tends towards a law of Wigner type: otherwise, the histogram would peak at the smallest level spacings.

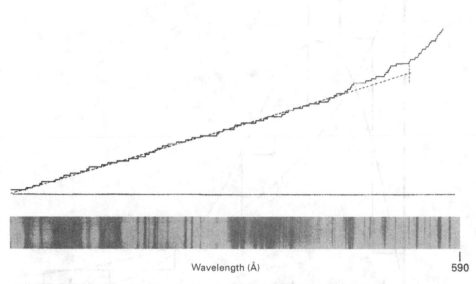

Wavelength (Å)   590

Fig. 10.5. A portion of the $5p$ spectrum of Ba, together with a plot of the accumulated density of states (after J.-P. Connerade *et al.* [547]).

10.12   The influence of linewidths

For chaos to be revealed in the spectrum of a many-electron system, a test such as the one just described is not quite enough. One must also show that a semiclassical limit exists in which the underlying classical dynamics would be chaotic. The semiclassical limit is obtained by following a Rydberg series to very high n. In practice, this means that any broadening mechanism which scales more slowly than $1/n^3$ (cf section 6.8) and therefore truncates the Rydberg series is an obstacle to the characterisation of chaos. Linewidths which do *not* scale as $1/n^3$ include core-level broadening, either radiative or Auger, and the latter, which usually dominates, emerges as the main difficulty.

Physically, this is readily understood in the following way: in a classical system, chaos requires a finite time to develop. If the excitation can decay by some alternative path in a quantum system well before that time is reached, then chaos will be quenched. In the example of the previous section, two-step autoionisation occurs (cf section 7.15), and is an obstacle to the observation of chaos because it is an Auger effect and involves widths which do not scale as $1/n^3$. For quantum chaos to be possible, one must choose both the atom and the energy range such that: (i) the excited core of the parent ion which defines the series limit is stable

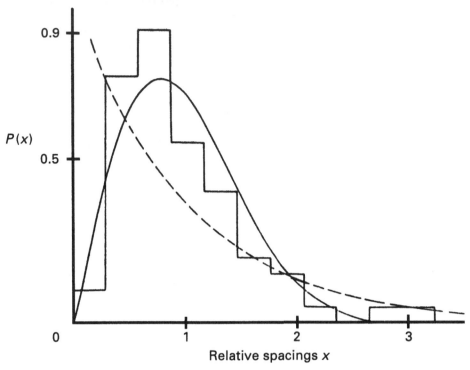

Fig. 10.6.   Nearest neighbour distribution in the $5p$ spectrum of Ba. Overlapping Rydberg series with $J = 1$ converging to more than 14 thresholds occur in this strongly interacting spectrum (after J.-P. Connerade *et al.* [547]).

against radiative and nonradiative decay; and (ii) dynamical motions of different symmetries coexist within an energy range where the density of interacting states is high.

Such a situation exists in the doubly-excited spectrum of Ca (see section 7.13) which has served repeatedly as an example in the present book. The single and double excitation spectra come very close to each other in energy, and prominent doubly-excited series can be followed to very high members (see fig. 6.3) at the top end of the energy range concerned, which shows that the semiclassical limit is attainable. Ca is therefore a good candidate for the emergence of chaos in the spectrum [549].

10.13   Multiconfigurational calculations

The emergence of Wigner statistics can be approached from the point of view of *ab initio* multiconfigurational Dirac–Fock theory. Although a full calculation for the $5p$ spectrum of Ba presents a formidable challenge and has not yet proved possible, fairly extensive calculations have proved fea-

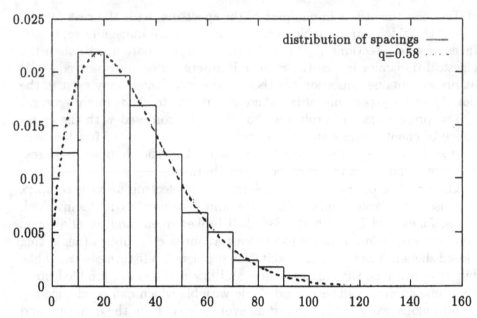

Fig. 10.7. Nearest neighbour spacings in multiconfigurational Dirac–Fock calculations for the $4p$ spectrum of strontium (after J.-P. Connerade *et al.* [550]).

sible for the $4p$ spectrum of Sr [550] and the data from these computations can be treated statistically. This has the merit that, since the number of calculated levels turns out to be rather large (several hundred), more statistical tests can be performed than on the experimental data to check the reliability of the conclusion. We show, in fig. 10.7, a plot of the nearest level distribution for Sr deduced from *ab initio* calculations. The results show a distribution which does not quite conform to a Wigner law but is evolving towards it: a good measure is to use the Brody distribution [551]:

$$P_q(x) = \alpha x^q \exp\left(-\beta x^{q+1}\right) \tag{10.12}$$

with $\alpha = (1 + q)\beta$. The Brody parameter varies from 0 for a Poisson distribution to 1 for a Wigner function, and is therefore a measure of spectral repulsion. For Sr, it was found to be 0.59 with a probability of 0.96.

We stress that level spacings alone do not suffice to make statements about 'quantum chaos': integrable systems are known which give Wigner distributions [561, 562]. Simple systems are also known whose classical dynamics is chaotic, but whose nearest level spacings differ markedly from the Wigner distribution [563]. For atoms, such systems can have the property that one series is dominant, i.e. has a much higher density of states than the others in one part of the spectrum, as is the case for He, where the single- and double-excitation spectra are unusually separated in energy (see sections 7.10 and 7.13). In this situation, although the classical dynamics is chaotic, regular Rydberg series are observed. It is at present an open question whether some energy range may exist in the doubly-excited spectrum of He where effects due to chaos might emerge.

The present class of problems should not be confused with the emergence of chaotic behaviour in systems with few degrees of freedom (see next section below). The connection is merely that they happen to possess the same kind of nearest neighbour distribution.

The problem of the highly-correlated, many-electron atom is complex because of the large number of particles and dynamical variables involved. Thus, $5p$-excited Ba with the $6s^2$ shell broken open, and an additional Rydberg excitation, behaves like at least a four-body problem (neglecting closed shells). The Ca double-excitation problem is a little more tractable, but it is still true that high n states and low n states are jumbled up in the same spectrum. There is no single variable which can be changed in a continuous way so that the system evolves away from the quantum and into the semiclassical limit.[1] Under these conditions, one cannot pick out a single semiclassical limit, because the high n states of one channel are 'scrambled' with the low n states of another. Thus, the notion of 'quantum chaos' and its elegant connection with the correspondence principle are then lost.

Before leaving this subject, we note that the energy range in which the spectrum is dense in our example and a Wigner distribution begins to emerge is bounded above and below. Towards very low energy, the Rydberg spacings become much larger than the spacings between parent ion thresholds, and n emerges as a good quantum number. At very high energies, above all the lowest parent ion thresholds, one eventually encounters one isolated limit which lies above all the others: close to this limit, n is again a good quantum number and the spectrum becomes simple again. This feature of a spectrum which is simple at low and at high energies, but complex in between, will also reappear in the next example.

10.14   The high magnetic field problem

The full complexity of the many-body problem appeared in the previous example. A simpler situation would clearly be to restrict ourselves to a a

---

[1] In this respect, the highly-correlated, many-electron atom is similar to the situation encountered in nuclear physics, where there are many dynamical variables and there exists no simple semiclassical limit. Random matrix theory can be a useful framework, but the word 'chaos' is not easy to define in this context.

single Rydberg series and apply one perturbation capable of breaking its fundamental symmetry, so as to induce chaos in the underlying dynamics.

A hint that such a possibility exists was provided by the example of molecular spectra (see section 2.32) where the breaking of symmetry was shown to have a dramatic effect on the apparent complexity of spectra (see in particular figs. 2.16 and 2.17).

The search for a simpler situation is not the only motivation for seeking another example. In the many-electron atom, little or no external control can be exercised on the couplings. In the absence of correlations, the spectrum would be ordered as regular Rydberg series. One can show experimentally that *analogous*, less correlated spectra exist in which the spectrum exhibits clean Rydberg series, rather than a Wigner distribution of interacting levels. However, one cannot actually control the magnitude of correlations for a given spectrum since they are internal to the atom.

These difficulties are removed in the problem of a Rydberg atom in a strong magnetic field. A 'strong' magnetic field will now be defined analogously to a strong laser field in the context of multiphoton physics (see chapter 9): it must be strong enough to induce effects (shifts, splittings) comparable in magnitude to the binding energy. In a semiclassical context, we can also think in terms of the curvature of orbits: a strong external field is one which can bend the orbit of the electron to a curvature comparable to the curvature produced by the field of the nucleus (or ionic core). Since curvature and energy are proportional to each other, these two definitions are, in fact, equivalent. A 'superstrong' magnetic field regime also exists, where the coulombic forces of the atom become a small perturbation. This regime will be discussed in section 11.16.

We introduce an index of the field strength γ^B, conventionally defined as the ratio of one half the cyclotron energy $\hbar\omega_L$ to the Rydberg energy R. Its magnitude $\gamma^B = B/B_c = B/(2.35 \times 10^5)$ (B_c is the atomic unit of magnetic field) is such that the effective field strength comparable to the binding energy of the ground state of H is $\sim 2.35 \times 10^5$ T, which is unattainable in the laboratory. Another useful measure is the index γ_n^B which compares the cyclotron energy to the spacing between two successive Rydberg members at high n: $\gamma_n^B = n^3 \hbar\omega_L/R$.

We note in passing that this definition of a 'strong' field is not the one found in textbooks, which relates to the relative strengths of the Zeeman splitting and multiplet structure (the Paschen–Back effect). A field capable of inducing angular-momentum uncoupling can still be a weak field under the present definition.[2]

The first experiment in which the strong magnetic field regime was

---

[2] An exception occurs in the spectrum of In, where both are similar in magnitude.

probed at adequate resolution to reveal new effects was performed by
Garton and Tomkins in 1969 [552]. This was a very important experiment,
and probably the most significant development since the work of Zeeman
[553]. Prior to this experiment, it can fairly be said that the existence
of a strong field regime with different quantisation rules had not been
suspected.

The spectrum obtained by Garton and Tomkins is shown in fig. 10.8.
As will be shown in section 10.15, this spectrum divides into three regions: on the right, one notes a persistence of Rydberg states, i.e. one
has dominant spherical symmetry with the symmetry of the **B** field as
a perturbation. On the left, one has equally-spaced structures, characteristic of a two-dimensional harmonic oscillator, which denotes motion
on a cylinder, or the symmetry of the **B** field perturbed by the spherical
symmetry of the atom. Between the two, and at the boundary between
these two irreconcilable symmetries, the spectrum displays a region of
high complexity which grows rapidly with increasing field strength. Here,
the underlying dynamics becomes chaotic.

The experiment was performed in photoabsorption using σ^- circularly
polarised light, propagating parallel to the magnetic field direction. Since
Ba has a $^1S_0$ ground state and exhibits Zeeman doublets in the principal
series (triplet states hardly intrude), this means that only one line is seen
for each Rydberg member at low magnetic field strengths, i.e. that the
normal Zeeman spectrum has been removed from view. Consequently,
any new structure which appears near a given Rydberg member as the
magnetic field strength is increased (moving upwards in fig. 10.8) is 'new,'
i.e. unexpected according to the simple theory of the linear Zeeman effect.
Starting at the bottom right hand corner of the figure, i.e. at the lowest
n and moving upwards to high fields, we see satellite lines appear. They
are at first weak, but gain intensity as the magnetic field is increased.
The gradual emergence of these lines can be explained in the following
way: in a strong magnetic field, the atom is distorted away from central
symmetry to some kind of spheroidal or cigar shape, i.e. the conservation
of angular momentum breaks down, and it becomes necessary, in order
to account for the distortion of the orbitals, to expand them as a series
of spherical harmonics of different ℓ. This effect is called ℓ-mixing, and
can be represented perturbatively: as a result, instead of a single angular
momentum state, one sees all the odd ℓ states for a given n (parity is
conserved) in proportions which reflect the degree of mixing.

This mixing can increase in two different ways: first, one can increase
the field strength B for a given n, but secondly, one can increase the n
value for a given B, which increases the size of the excited state, and
therefore decreases the effect of the nuclear field. The strong field effects
are quadratic, i.e. they depend on terms in B^2 which are neglected in the

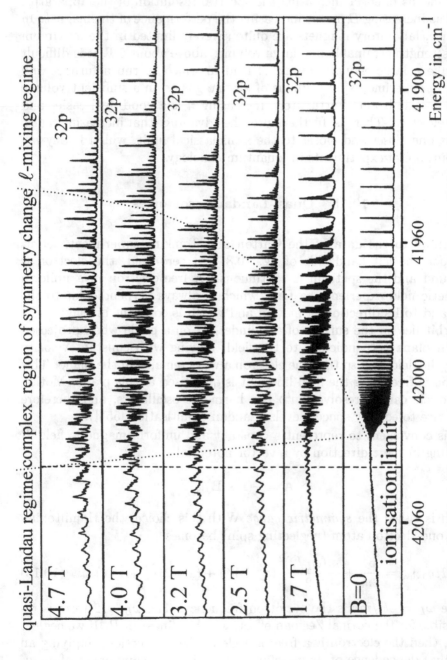

Fig. 10.8. The photoabsorption spectrum of Ba in strong magnetic fields as obtained by Garton and Tomkins (after W.R.S. Garton and F.S. Tomkins [420]).

theory of the linear Zeeman effect. They turn out to scale as $B^2 n^4$ (as long as n remains a good quantum number). One can therefore choose, in performing high field experiments, to increase either B or n. Increasing n means in effect increasing the spectral resolution of the apparatus, whereas increasing B means increasing the performance of the magnet. In practice, laboratory magnets are quite severely limited in the maximum field strength attainable in a large volume: above about 6 T, it is difficult to make a large enough volume of field in which to run a furnace and perform experiments on columns of atomic gas or on a sufficient volume of an atomic beam. Fortunately, increasing n is a good deal easier and more effective. There is, furthermore, the advantage that increasing n also brings one closer and closer to the semiclassical limit, which is anyway the purpose of experiments in 'quantum chaology.'

10.15   Quasi-Landau resonances

Returning to the data of the Garton and Tomkins experiment, let us consider the left hand side of fig. 10.8. At zero field, the electron is unbound and the spctrum is continuous. A free electron in a uniform magnetic field experiences a force which is always perpendicular to the field and to its direction of motion and which is constant in magnitude. Its orbit lies on the surface of a cylinder, or on a circle when projected onto a plane perpendicular to the field, moving with constant velocity along the field line, so that the electron always remains in the plane. This motion is free along the field line, but is quantised in this plane. Motion in a circle can be resolved into two harmonic oscillators, and therefore gives rise to equally spaced levels (Landau quantisation [554]).

It is convenient to represent a constant and uniform magnetic field B pointing in the z direction by a vector potential

$$\mathbf{A} = \frac{1}{2}(\mathbf{r} \times \mathbf{B})$$

which is called the *symmetric gauge* With this choice, the Hamiltonian for a one-electron atom (neglecting spin) becomes

$$\mathcal{H}_{landau} = \frac{\mathbf{p}^2}{2m} + \frac{m(\omega_L/2)^2}{2}(x^2 + y^2) + (\omega_L/2)\hat{L}_z + V(\mathbf{r}) \qquad (10.13)$$

where $\omega_L = eB/mc$ is the cyclotron frequency. The $\omega_L \hat{L}_z/2$ term is responsible for the normal Zeeman effect, which is linear in B. If we neglect $V(\mathbf{r})$, then the electron is a free particle in the z direction, implying an $\exp(ikz)$ dependence of its wavefunction. It has a z component of angular momentum m, implying also an $\exp(im\phi)$ dependence, and the energy eigenvalue equation yields

$$E_{\eta mk} = \left(\eta + \frac{m + \mid m \mid + 1}{2}\right)\hbar\omega_L + \frac{\hbar^2}{2m} \qquad (10.14)$$

The second term is the energy of a free electron, while the first gives the eigenstates of a two-dimensional Harmonic oscillator, with principal quantum number η and azimuthal quantum number m. The spacing between successive levels of the same m in this picture is just the cyclotron frequency. Actually, the spacing is $3/2\hbar\omega_c$ near the ionisation threshold, and we return to this point in section 10.21. These equally spaced levels appear as modulations in the continuum, with a frequency which slowly increases as the field strength B becomes larger. The modulations correspond to states which are not quite free, since they are constrained to orbit in a plane, but which are not bound states either, since the electron can escape from the field of the atom along the z direction. From this point of view, one can understand why the modulations are broad. However, one should not forget that atomic electrons (even in the continuum) are not completely free: they experience the field of the parent ion as well as the magnetic field, so that the states are modified as compared with pure harmonic oscillator levels. For this reason, they are called *quasi-Landau resonances*.

We thus arrive at the following simple description: at the low energy end of fig. 10.8, the electron moves in the Coulomb field with the magnetic field as a perturbation, giving *ℓ-mixed* Rydberg states. At the high energy side of fig. 10.8, the electrons move in a magnetic field, with the atomic field as a perturbation, giving quasi-Landau resonances, Between these two regions, there is clearly a problem, since the classical motions are incomensurate and the quantum-mechanical laws governing the energy spacings are also irreconcilable.

It is in this intermediate region (roughly outlined with dashed curves in fig. 10.8) that 'quantum chaos' develops. The classical analogue of this motion is a pendulum-like motion (the Rydberg orbit) in an applied magnetic field, and is chaotic.

It has since been realised that the experiment of Garton and Tomkins was not performed with a pure magnetic field: the Maxwellian velocity distribution of the atoms gave rise to a motional Stark field (see section 10.20). This field has been eliminated in experiments performed with atomic beams, by applying a compensating electric field, and the spectral resolution has been much enhanced [555]. Data from these more modern experiments are shown in Fig. 10.9, which shows spectra in the energy range where ℓ mixing ceases, and manifolds of different n begin to overlap.

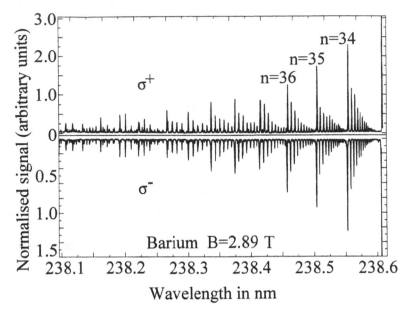

Fig. 10.9. The Ba spectrum in a high magnetic field with the motional Stark field compensated. The data are obtained in atomic beam experiments, and the relative intensities do not suffer from opacity or saturation effects. Both circular polarisations are separated experimentally, and are found to have the same structure (shown by presenting them as though reflected in the axis) when displaced in energy by the linear Zeeman splitting.

10.16   Time-reversal invariance and magnetic fields

As pointed out above, the Wigner surmise and the analysis of statistical distributions require a careful study of which quantum numbers are exact, i.e. which rules a system must obey precisely. A rule which has not been considered so far is *time-reversal invariance*, which applies when a system evolves backwards in time in precisely the same way as it evolved forward, i.e. when the sign of time can be changed without affecting the basic equations.

Not all systems are time-reversal invariant. In particular, time-reversal invariance is broken by an external magnetic field. This is most easily understood by considering the physical difference between natural rotation and Faraday rotation: in naturally occurring rotation (in a crystal or a sugar solution) the plane of polarisation of the radiation comes back onto itself when the light beam is reflected backwards along the same path. This does not occur in Faraday rotation, where the angle of rotation continues to accumulate for the reflected beam. In fact, the Faraday effect is the only effect in optics which destroys time-reversal symmetry.

It is possible to define a time-reversal operator T as follows [556]:

$$T = UC \qquad (10.15)$$

where the operator U is unitary, i.e. preserves the normalisation of eigen-vectors, while the operator C is complex conjugation, i.e. reverses the sign of the $i\omega t$ arguments of oscillating exponentials.[3]

Preservation of normalisation follows from the transformation $| \, n' >= U \, | \, n >$ so that $< m' \, |=< m \, | \, U^\dagger$, where $U^\dagger$ is the hermitian transpose of U. Thus, for $< m' \, | \, n' >=< m \, | \, U^\dagger U \, | \, n >$ to be the same as $< m \, | \, n >$, we require that $U^\dagger U \equiv 1$, i.e. U is unitary. Complex conjugation, of course, does not affect normalisation either.

Operating twice with T must bring the system back onto itself and preserve normalisation, so

$$TT \, | \, n >= UCUC = UU^\star CC \, | \, n >= UU^\star \, | \, n >= \alpha \, | \, n > \qquad (10.16)$$

where $| \, \alpha \, |= 1$. Since U is also unitary, it follows (simply by taking the transpose of $UU^\dagger = 1$) that

$$U^\star U^T = 1 \qquad (10.17)$$

Combining (10.16) and (10.17) we have that

$$U = \alpha U^T = \alpha(\alpha U^T)^T = \alpha^2 U$$

whence $\alpha^2 = 1$ and $\alpha = \pm 1$. This means that U must be either symmetric or antisymmetric, i.e. we have either $UU^\star = 1$ or $UU^\star = -1$. The former applies to integral angular momentum J and the latter to half-integral J (even spin and odd spin, respectively). For even-spin systems with time-reversal, it is always possible to choose a representation such that the Hamiltonian is real and symmetric. In the special case where a system is invariant under space rotations, then for either even spin or odd spin the system is represented by a real and symmetric Hamiltonian.

10.17 Gaussian orthogonal ensemble

For either of the situations just considered the appropriate level distribution is the *Gaussian orthogonal ensemble* or GOE distribution,[4] defined

---

[3] In quantum mechanics, the operation $t \to -t$ (time-reversal) is to be accompanied by complex conjugation ($i \to -i$) so that the Schrödinger equation remains invariant. This operation is called *Wigner time-reversal*.

[4] Ensembles of N-dimensional Gaussian random matrices (GRMs) with invariances under orthogonal or unitary rotation groups correspond to the GOE or GUE; they were introduced by Wigner and developed by Porter, Dyson, Mehta and others [539, 540, 541]. Wigner guessed that the statistics of these GRMs can be used to model the statistical properties of the observed spectra of compound nuclei.

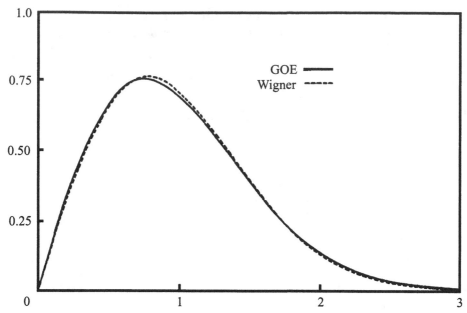

Fig. 10.10. Comparison between the GOE and Wigner distribution functions

in the space of real symmetric matrices by two requirements:

(i) the ensemble is invariant under any transformation

$$\mathcal{H} \rightarrow \mathcal{H}\prime = O^T \mathcal{H} O \tag{10.18}$$

where O is a real orthogonal matrix; and

(ii) the elements $< m \mid \mathcal{H} \mid n >$ are statistically independent.

Under these conditions, one can show that the probability distribution is

$$P_{GOE}(x) = \frac{\pi^2}{6}x - \frac{\pi^4}{60}x^3 + \frac{\pi^4}{270}x^4 + \frac{\pi^6}{1680} + ... \tag{10.19}$$

This differs slightly from the Wigner surmise: in particular, its slope at the origin is $\pi^2/6$, whereas the slope given by (10.7) is $\pi/2$. Thus, the Wigner surmise is not exact, although the approximation is extremely close. fig. 10.10 shows a comparison between $P_{GOE}(x)$ and $P_{wigner}(x)$.

10.18   Level statistics for H in a high magnetic field

Although Garton and Tomkins discovered ℓ-mixing, n-mixing and the quasi-Landau resonances in the spectrum of Ba, from the standpoint of quantum chaology the asymptotically Coulombic nature of the atomic field (i.e. the situation in absence of any external field) remains relevant, because it guarantees the regularity of the semiclassical problem for the field-free atom. The simplest atom in which 'quantum chaos' can be studied is H, which has been investigated in great detail, both experimentally [557] and theoretically [559].

From the theoretical point of view, as stated at the outset, there is no problem of quantisation if the variables of the problem are separable. In fact, the problem of H in a strong magnetic field, taking even the simplest possible Hamiltonian with only Coulomb and diamagnetic terms

$$\mathcal{H} = \frac{p^2}{2} \frac{(\gamma^B)^2 (x^2 + y^2)}{8} - \frac{1}{r} \qquad (10.20)$$

is not separable, which is why the issue of 'quantum chaos' arises. We can classify the spectrum into three broad regimes as follows:

(i) where $\gamma_n \ll 1$, the Coulomb energy dominates, and the spectrum is that of Rydberg states perturbed by ℓ-mixing effects;

(ii) where $\gamma_n \sim 1$, Coulomb and magnetic field energies are of the same order: n-mixing occurs and, when this occurs near the series limit, the spectrum becomes 'chaotic';

(iii) above the field-free ionisation limit, the magnetic field energy dominates, and the atomic field is the perturbation, so that quasi-Landau resonances emerge.

The beauty of the magnetic field problem is the degree of control over its structure which can be achieved. In particular, increasing a single parameter (the energy or n value) allows the system to be raised to the semiclassical limit. Varying the field strength, on the other hand, allows the point where the 'order-to-chaos' transition occurs to be modified.

In an intermediate range, considered as 'chaotic', strong inter-n-mixing occurs. The structure of the spectrum is thus simple at low energies, simple at high energies, while an intermediate region of high complexity exists where both n and ℓ cease to be valid quantum numbers. *This is exactly the same overall structure as in highly correlated spectra*, discussed in section 10.13. It is an important feature of 'quantum chaos' that there are 'chaotic regions' and islands of stability which are interleaved (this was discussed in section 10.8 and illustrated in fig. 10.3). Thus, one needs to recognise where the boundaries between them occur. In the high magnetic field problem, looking for GOE distributions of interacting levels is one method which has been used to search for the emergence of chaos. An example is shown in fig. 10.11.

There are, however, difficulties in this approach. These difficulties are not connected with the high magnetic field problem: rather, they are

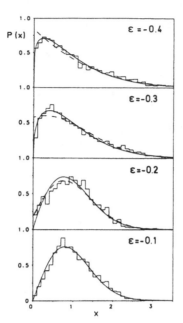

Fig. 10.11.   Emergence of a GOE distribution in the histogram of nearest level spacings for an atom in a high magnetic field. Plots are given for different values of the scaled energy. As the series limit is approached, the probability of zero spacings tends to zero, in accordance with the GOE distribution (after H. Friedrich [112]).

intrinsic to quantum chaology. The problem is really the extent to which one can assert that the emergence of a GOE or Wigner type distribution is the spectral signature of 'quantum chaos.' Many examples of quantum systems whose underlying classical dynamics is chaotic have been found to exhibit similar distributions. However, this is not a rigorous theorem and so the doubt persists that it might not be universal in atomic physics [563]. Also, the inverse proposition (that a system whose spectrum exhibits a GOE or Wigner-type distribution is necessarily chaotic) may well be untrue. Indeed, counterexamples have been advanced [561, 583].

10.19   Scaled energies and Fourier transforms

The form of the Hamiltonian for the atom in a strong magnetic field suggests a scaling transformation $r' = (\gamma^B)^{2/3}r$ and $p' = (\gamma^B)^{-1/3}p$ [564]. With this transformation, the EBK quantisation condition becomes

$$\oint p dq = (\gamma^B)^{1/3} nh \tag{10.21}$$

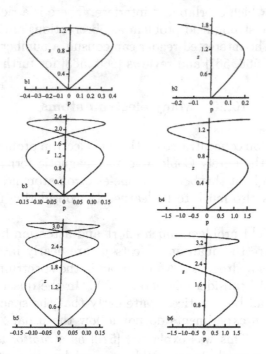

Fig. 10.12. Some examples of quasiperiodic orbits outside the plane perpendicular to the magnetic field (after P.F. O'Mahony *et al.* [558]).

Thus, for quasi-Landau resonances with $m=0$, equidistant resonances are obtained when varying the magnetic field in steps proportional to $(\gamma^B)^{-1/3}$ at a fixed scaled energy $\tilde{E} = E(\gamma^B)^{2/3}$. Hence, the Fourier transform of a spectrum recorded at a constant scaled energy and plotted against $(\gamma^B)^{-1/3}$ yields a single frequency for this whole set of resonances. This is a more rigorous way of using the Fourier transform technique than the method of [565]. By searching for peaks in the Fourier transform spectrum, a number of different classes of quasiperiodic orbits, corresponding to excitations out of the plane perpendicular to the magnetic field have been discovered [557]. These trajectories are quasiperiodic in the sense that the electron returns once to the nucleus, but they are not closed, i.e. they are not classical orbits in the real sense. Some examples are shown in fig. 10.12.

Extra peaks in the Fourier transform spectrum have been found to correspond to many such orbits. Thus, a new understanding of the underlying classical dynamics has evolved.

For large fields, the separation between Coulomb states becomes much smaller than the separation between quasi-Landau resonances, and one observes sequences of overlapping Rydberg channels distinguished by the

Landau index η, which overlap and interfere. There is a rich structure of interactions in the strong field problem as observations are extended into the continuum. The interested reader can consult a number of conference proceedings [438, 566, 567] and reviews [560, 568] for further details.

10.20   Many-electron atoms

The previous section concentrated on the one-electron problem, i.e. on the balance between the purely Coulombic and magnetic terms. As stressed continually throughout this book, the one-electron atom is an exceptional case, and there is also much to be learned by considering many-electron systems.

In the strong field problem, much effort and attention has been given to situations where the quantum defects μ_ℓ are nearly integral, i.e. the solutions are quasihydrogenic. An example is the spectrum of Li, which was investigated in considerable detail [569]. In the quasi-Landau resonance range, it had been noticed quite early that the spacings observed in photoabsorption experiments do not follow the usual $\frac{3}{2}\hbar\omega_L$ rule but are $\frac{1}{2}\hbar\omega_L$ instead. This was explained [570] as a *motional Stark effect*, resulting from a crossed electric field, induced by the motion of atoms across the magnetic field lines. It happens to be particularly large for Li, which is a light atom and requires a rather high evaporation temperature. The explanation was confirmed by beam experiments [565] in which the atoms were constrained to move parallel to the field and the $\frac{3}{2}\hbar\omega_L$ splitting was restored. In this experiment, regular splittings were searched for by taking the Fourier transform of the spectrum, a technique which has also led to theoretical developments [571].

In the ℓ-mixing region, Li has the peculiarity that the p electrons have a rather low quantum defect ($\mu_p \sim 0.053$). Care is needed to exclude nonhydrogenic effects. In particular, they can intrude in the presence of electric fields because the μ_ℓ are not all hydrogenic. In an important experiment [569], the ℓ-mixed manifold was observed with no motional Stark effect, and the result is shown in fig. 10.13.

It turns out that there are two regimes of motion which can be understood from the classical dynamics of the system. One is a rotational motion in which the electron is mainly localised in the plane perpendicular to the field lines and the other is a vibrational motion corresponding to stretching along the field lines. These two motions have distinct symmetries and therefore occupy separate regions of phase space. The vibrator states, for which the mean value of $x^2 + y^2$ is small, correspond to the lower energy states in the manifold, while the rotator states, for which the mean value of $x^2 + y^2$ is increased by centrifugal forces, lie higher

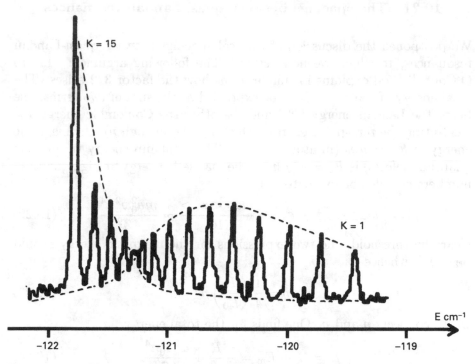

Fig. 10.13. The $n = 30$ manifold of Li (odd diamagnetic spectrum) showing two different regimes of motion – the vibrator and rotator states (after J. Pinard *et al.* [569]).

in the manifold. The surface separating them in phase space is called a *separatrix*. It is not crossed by trajectories, and lies in a region where the density of paths is very small. It has been shown [572] that this surface is defined by the properties of an approximate constant of the motion, related to the Runge–Lenz vector (a vector pointing along the major axis of a Kepler orbit, and whose length is proportional to its eccentricity). The existence of a separatrix is intimately linked with the KAM theorem, and with the manner in which chaos grows in phase space. Separatrix crossing will also be discussed in section 10.24.

In general, for many-electron atoms, the μ_ℓ values are all different, according to the nature of the core, the proximity of double excitations, etc. The appearance of the ℓ-mixed region is therefore strongly dependent on nonhydrogenic effects, and differs greatly from one atom to another. A good example is provided by the work of O'Mahony and Taylor [573], who have studied theoretically the spectra of Sr and Ba.

10.21   The spacings between quasi-Landau resonances

We postponed the discussion of the real spacings between quasi-Landau resonances, to which we now return. The following argument, due to O'Connell [574] explains in simple terms how the factor 3/2 arises. The total energy of the system can be expressed as the sum of two terms, one being the Landau energy (E_η) and the other the Coulombic energy E_c. Neglecting the zero-point energy (which is valid at high η), the magnetic energy is $E_m = \eta \omega_c$ (in atomic units). The Coulomb energy (neglecting quantum defects) is $E_c = -1/2n^2$. The magnetic energy at high quantum numbers can also be evaluated as

$$E_m = \frac{1}{8}m\omega_c^2 <x^2 + y^2> \sim \frac{m\omega_c <r^2>}{10} \sim \frac{ma_0^2\omega_c^2 n^4}{4Z^2} \tag{10.22}$$

Near the threshold, the two expressions for the magnetic energy should be equal, whence

$$\eta = \frac{1}{4}\left(\frac{B}{B_0}\right)n^4 \tag{10.23}$$

which connects η and n. One finds for the total energy

$$E = E_c + E_m = -\frac{R}{n^2} + \frac{\omega^2 a_0^2 n^4}{4Z^2} \tag{10.24}$$

whence

$$\frac{\delta E}{\delta n} = -\frac{2E_c}{n} + \frac{4E_m}{n} = -\frac{2E}{n} + \frac{6E_m}{n} \tag{10.25}$$

and, using (10.23)

$$\frac{\delta E}{\delta \eta} = -\frac{1}{2}\frac{E}{\eta} + \frac{3}{2}\frac{E_m}{\eta} \tag{10.26}$$

Near the continuum edge, $E \sim 0$, and so the separation is $\frac{3}{2}\hbar\omega_c$.

A more rigorous argument is given by Rau [575]. Along $z = 0$ the potential is $V(\rho) = -Ze^2/(4\pi\epsilon_0\rho) + \frac{1}{8}m\omega^2\rho^2$, where $\rho^2 = x^2 + y^2$. The quantisation condition is

$$2(2m)^{1/2}\int_\rho^{\rho_0}(E - V(\rho))^{1/2}d\rho = \left(\eta + \frac{1}{2}\right)h \tag{10.27}$$

where ρ_0 is obtained from the condition $E - V(\rho) = 0$. Differentiating the quantisation condition and setting $E = 0$,

$$\frac{dE}{dn} = \frac{h}{(2m)^{1/2}\int_\rho^{\rho_0}(E - V(\rho))^{1/2}d\rho} \tag{10.28}$$

The integration can be performed explicitly, yielding $dE/d\eta = \frac{3}{2}\hbar\omega$. Thus, provided V is of such a form that it vanishes at some finite ρ_0

and the effective quantum number η is a large enough number, one has equally spaced states.

10.22 Strong electric fields

By contrast, the strong electric field problem has appeared (perhaps prematurely) to be well understood. This view is reinforced by the fact that the Schrödinger equation for an atom in a strong electric field, although nonseparable in spherical polar coordinates (n and ℓ are not good quantum numbers) does turn out to be separable in parabolic cylinder coordinates, given by

$$\xi = r + z \quad \eta = r - z \quad \phi = \tan^{-1}(y/x) \tag{10.29}$$

(note: this is not the same η as in the previous section). The wavefunction is written as $\psi = \psi_1(\xi)\psi_2(\eta)\exp(im\phi)$ and the differential equations for motion in the ξ and η coordinates then separate as

$$\left. \begin{aligned} \frac{d}{d\xi}\left(\xi\frac{d\psi_1}{d\xi}\right) + \left(\frac{E}{2}\eta + Z_1 - \frac{m^2}{4\xi} - \frac{F\xi^2}{4}\right)\psi_1 = 0 \\[2mm] \frac{d}{d\eta}\left(\eta\frac{d\psi_1}{d\eta}\right) + \left(\frac{E}{2}\eta + Z_2 - \frac{m^2}{4\eta} - \frac{F\eta^2}{4}\right)\psi_2 = 0 \end{aligned} \right\} \tag{10.30}$$

in atomic units, where F is the external electric field and the separation constants Z_1 and Z_2 sum to 1.

Since they are separable, these equations describe regular motion, which is nonchaotic in the classical limit. A number of experiments have been performed [576] [577] and calculations have reproduced the data very well [578].

There is, however, one area of uncertainty: the region very close to $E = 0$ may require more detailed investigation, since it has been argued [579] that it could contain a very large number of close resonances.

It has also been noted [580] that the arguments given above on the separability of the Schrödinger equation might not suffice: electric and magnetic field problems are related through the Lorentz transformations, and there is no such thing as a purely electric strong field problem for atomic electrons. In principle, the Dirac equation should be used: the Schrödinger equation is at best an approximation when dealing with the dynamics of charged particles in strong fields. Unfortunately, there has, it seems, been no treatment so far of the strong electric field problem using the Dirac equation, so further comment is not possible.

Rau [575] gives the near threshold spacings for the $m = 0$ case by the same method as in section 10.21: in the ξ coordinate, the motion is one-dimensional, with a binding potential, while in the η coordinate there is the possibility of escape. Consequently, one finds resonances, which are equally spaced in energy near the series threshold. The spacings turn out to scale as $F^{3/4}$.

10.23   Microwave ionisation

The problems discussed above all relate to spectral properties and statistics. In fact, the classical definition of chaos is more concerned with *dynamical* properties: one can define classical chaos in terms of the instability of trajectories under infinitesimal displacements, which leads to the exponential divergence of neighbouring trajectories in phase space. The Liapounov exponent is the argument of the exponential which determines the rate of this divergence, and is often taken as a measure.

One problem in which dynamics comes to the fore is the ionisation of a Rydberg atom exposed to an unspecified number of oscillations of a microwave beam. It turns out that this problem can be treated semiclassically and in one dimension: since the atom is in a high Rydberg state, the microwave field can be a strong perturbation, and the time dependence of the Hamiltonian becomes important. In its simplest form (in atomic units)

$$\mathcal{H} = \frac{p^2}{2} - \frac{1}{r} + \mathcal{E}z \cos \omega t \qquad (10.31)$$

where $\mathcal{E}$ is the strength of the electric field. It is conventional to represent the resulting motion in terms of action-angle variables (I, Θ) [581]. These are defined such that the action $I = \oint p \, dq / 2\pi$, while the canonically conjugate angle θ ranges from 0 to 2π over one period of oscillation. As $\mathcal{E}$ tends to zero, the trajectories in the Poincaré section are simply lines of constant I. As $\mathcal{E}$ increases, experiments show [582] that ionisation becomes possible for lower and lower values of n, the principal quantum number. The angle-action plots then become distorted from their low field structure. For any given n and ω, ionisation can be measured as a function of the field strength. It is found [583] that there is a threshold for transition to chaotic motion given by $n^4 \mathcal{E} \sim 1/49 n \omega^{1/3}$. Above this field strength, strong excitation and ionisation occur in a classical model. The quantum excitation model follows the classical model closely, but interference terms are found, which lead to the suppression of classically chaotic excitation, and this has been interpreted as a suppression of chaos. Further aspects of this problem are discussed in [583]. Examples of angle-action plots for a related problem are given in the next section.

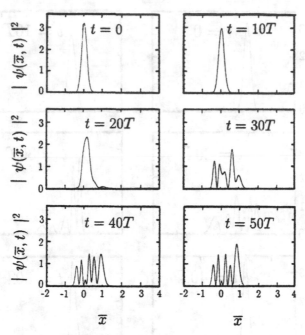

Fig. 10.14. Evolution of the wavefunction of a strongly driven Morse oscillator, whose parameters are those of the HF molecule for different times during a laser pulse of total duration $50T$, where T is one optical cycle. A snapshot is taken every ten cycles, and the laser intensity is 0.05 atomic units. Note how the wavefunction becomes more complex, showing that excitation has been achieved towards the end of the pulse (after J.-P. Connerade *et al.* [584]).

10.24   Atoms in high AC fields

We return now to a problem introduced in section 9.26, namely the dynamics of excitation by a very short pulse laser of high intensity. This has been studied [584] by two different theoretical approaches. First, quantum calculations plotted frame by frame at different times during the laser pulse reveal how the composition of the eigenstates varies and how the system may re-emerge either in an excited state or return to the original composition depending on the laser parameters (see figs. 10.14 and 10.15).

Second, calculations can be performed in a semiclassical regime, and the results plotted on a Poincaré section in action-angle (I, θ) coordinates. Such diagrams may seem complicated (see figs. 10.16 and 10.17), but are at least in principle readily understood: a near-horizontal line across the (I, θ) plot corresponds to a torus in ordinary phase space. When periodically extended in the time coordinate, each line corresponds to a vortex tube embedded in the extended phase space of the periodically

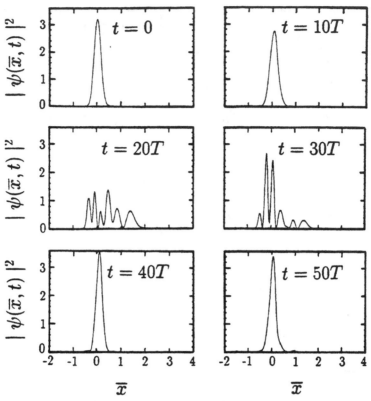

Fig. 10.15. As in the previous figure and with all parameters as before, except that the field strength of the laser has been turned up to 0.1 atomic units. Note how the system evolves through a complex admixture, but then returns almost exactly to the initial state at the end of the pulse, showing that stabilisation has occurred, i.e. that the system preserves its integrity despite exposure to the strong field. Calculations show that, if the intensity is turned up still further, one returns to a condition where the system is highly excited at the end of the pulse (after J.-P. Connerade *et al.* [584]).

driven system, which in turn corresponds to a quasistationary Floquet state. A set of closed curves within the diagram describes a classical libration, i.e. a resonance.

Excitation is achieved as follows: one considers a sequence of such (I, θ) plots taken at various times during the pulse as the amplitude $\Lambda(t)$ varies. For large enough Λ_{max}, an initial ensemble of tori is progressively deformed until they cross the separatrix and are caught by the resonance. As $\Lambda(t)$ decreases again, the resonance shrinks, and the captured ensemble blows up again, touches the shrinking separatrix symmetrically and divides into two deformed tori which appear in the last (I, θ) plot – after

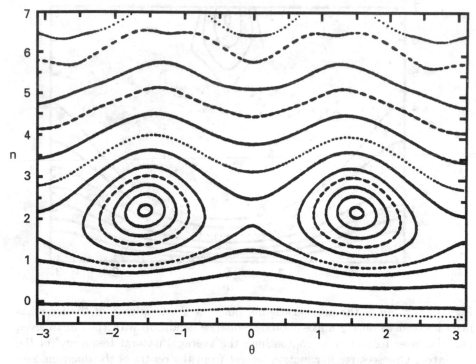

Fig. 10.16. Poincaré surface of section of a strongly driven Morse oscillator at a laser field strength of 0.1 atomic units, with the frequency tuned to a two-photon resonance condition with excitation from the ground state to the fourth excited level. The plot indicates the presence of a resonance structure, away from which the motion lies on regular tori (after J.-P. Connerade *et al.* [584]).

the pulse – as two separate sets of horizontal lines. It is interesting that separatrix crossing, which can be described geometrically in the semi-classical picture, becomes a very complicated reordering of wavefunctions in the quantum limit.

These calculations reveal another interesting property. As the complexity of the plots increases, chaotic trajectories appear, which grow (as expected) from the separatrix outwards. The presence of such orbits heralds new paths for ionisation, which may eventually dominate the whole of phase space.

It is clear from such calculations that the dynamics are very complex, but that they do depend on the nature of the system being studied. The major difficulty is to unravel the map of avoided crossings for a given system. It is at this level that one may expect an interplay between the correlating effects of the ponderomotive field itself and electron–electron

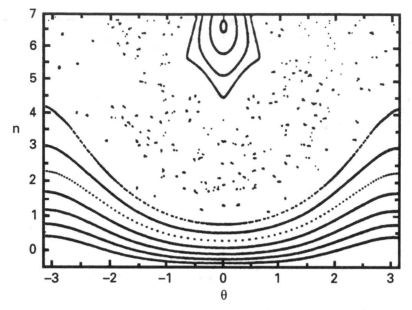

Fig. 10.17.   As in the previous figure, but now the frequency of the laser has been lowered to satisfy a five-photon resonance condition with the fourth level, i.e. the laser frequency is approaching the average 'natural frequency' of the oscillator. One sees regular motion retreat from the centre of the diagram, and chaotic paths invade phase space (after J.-P. Connerade *et al.* [584]).

correlations intrinsic to the system. Unfortunately, there is no easy way to probe the quasienergy map experimentally. The challenge is to make laser systems with equivalent field strength, but different pulse lengths, and then study the effect of pulse length on excitation.

10.25   Classification of excitation/ionisation processes

We can now summarise the situations which occur after irradiation of an atom or molecule by a strong field laser pulse:

(i) The state of the system does not change. If the system was initially in the ground state, then it is found again in the ground state after the pulse. In the quantum calculations, we recognise the same ground state wavefunction reemerging towards the end of the laser pulse, while, in the semiclassical picture, the system remains on a strictly adiabatically moving torus in phase space throughout the pulse.

(ii) Excitation to only one energy level occurs. A superposition of wavefunctions is then excited after the laser pulse. In the semiclassical picture, this is viewed as a strong classical primary resonance splitting into two, symmetrically lying, stable tori via separatrix crossing. Transition probabilities close to 1 are induced by this process.

(iii) Excitation to many bound states occurs. This is best understood in the semiclassical picture: for this to happen, chaos must loom large in the corresponding Poincaré section. This picture immediately brings up the question of how many excitations have paths leading into the continuum, and thus, the last case.

(iv) Ionization is the dominant process. This situation will obtain when the Poincaré section is dominated by chaotic trajectories [585].

10.26   Conclusion

If we regard as one of the main purposes of atomic physics the study of the many-body problem, i.e. the search for the breakdown of the independent particle model, then 'quantum chaos' is an important possibility.

The full complexity of the many-electron atom is, however, so great that simpler examples are needed in order to understand the underlying physics. Remembering the theme of chapter 2 that Rydberg series represent simplicity (the independent electron model) in atomic physics, one searches for the simplest example of 'quantum chaos' in the form of a single Rydberg series whose fundamental (spherical) symmetry is broken by applying a (cylindrical) magnetic field.

We have given examples of 'quantum chaos' in both spectral and dynamical properties of highly excited atoms. As regards spectral properties, the most significant result is perhaps the emergence of a new class of 'highly correlated' spectra, when the level density, the number of dynamical variables and the interchannel interactions are large enough for Wigner nearest level spacings to appear. As regards dynamical properties, several problems exist for which the correspondence with chaotic underlying classical dynamics can be explored using highly excited atoms.

As regards the high magnetic field problem, several aspects were not discussed in the present chapter, which the reader may wish to explore. There is the problem of structure within the continuum, giving rise to an abundance of sharp structure [586], and to many overlapping Rydberg series converging onto quasi-Landau resonance thresholds [587]. Then, one may combine high magnetic fields with parallel and crossed electric fields [588].

Despite its importance of principle, one should not overstate the rôle of chaos in the spectroscopy of highly excited atoms: although favourable circumstances can arise, they are rare. There are two fundamental reasons for this. The first is the Pauli principle: as noted in chapter 1, the shell structure of atoms restores spherical symmetry to the many-electron atom at each new row of the periodic table, and spherical symmetry, which helps the independent particle model, inhibits chaos. Secondly, as the excitation energy is increased, autoionisation and the Auger effect also become obstacles to the emergence of chaos, because the lifetimes are so short that instabilities in the underlying classical dynamics do not have time to develop.

Situations favourable to the emergence of chaos are those in which a dense manifold of bound Rydberg states is subject to perturbations of a strength comparable to the level spacing. This can occur mainly close to the first ionisation threshold and in the first autoionising range.

As regards the high laser field problem, extensions include the so-called 'two-colour' experiments [589] in which coherent mixtures of two laser fields of different frequency are used to tailor the excitation. The present chapter merely provides an introduction to research on atoms in strong fields, both oscillatory and constant in time, which are rapidly developing areas of atomic and molecular physics.

11

Atomic effects in solids

11.1 Introduction

There are many connections between the physics of free atoms and that
of solids which have been noted, in passing, several times already in the
present volume. One should add that many-body theory and, especially,
the concept of excitations as quasiparticles in free atoms, owe much to
the theory of excitations in solids [590].

The theme of the present chapter is rather more specific: the intention
is to present a number of effects which are counterparts of those we have
studied in previous chapters, but for atoms in the solid rather than in
the gaseous phase. Also, the intention is to set the scene for the last
chapter, in which atomic clusters will be used in an attempt to bridge
the gap from the atom to the solid experimentally. A highly excited
atom in a solid will be taken as an atom excited close to or above the
Fermi energy (including, of course, core excitation). There are some solid
state systems for which electrons with energies close to the Fermi level
behave like those in atoms. X-ray absorption and electron energy loss
spectroscopy involving core excitation to empty electronic states can then
be described by initial and final states possessing L, S and J quantum
numbers, and the allowed transitions follow strict dipole selection rules.
Examples include the $d \rightarrow f$ transitions of Ba in high T_c superconductors,
and many instances involving transition metals and lanthanides. The
term *quasiatomic effects* has been coined [591] to describe this kind of
behaviour: it is to be contrasted with situations in which the electrons
follow a band-like behaviour, and electron correlations assume a more
complex form, no longer characteristic of single atoms.

The crudest model for a one-dimensional solid is that of atoms brought
together in a regular line, as would occur in a highly simplified ideal crystal. When this is done, the electrons from any given subshell are treated

as moving in the field of positive ions, and of all the other electrons, averaged in the sense of static mean field theory. The resulting field is, of course, periodic. For a periodic potential $V(x)$ with period a, it is readily shown that the solutions of the Schrödinger equation form a series of *bands*, i.e that all the solutions have the form $\psi(x) = u(x)\exp(ikx)$ where $u(x)$ is periodic with period a and $-\pi/a < k < \pi/a$. For each allowed band of energy values, k varies continuously from $-\pi/a$ to $+\pi/a$ The lowest state of each band is purely periodic, while the higher states consist of plane waves modulated by the periodic field [592].

A simple proof is as follows: let $\psi_1(x)$ and $\psi_2(x)$ be two solutions for the same enegy E. Then $\psi_1(x + a)$ is also a solution and so $\psi(x + a) = A\psi_1(x) + B\psi_2(x)$. Two linear combinations of $\psi_1(x)$ and $\psi_2(x)$ can thus be formed (call them ϕ_1 and ϕ_2), satisfying

$$\phi_1(x + a) = \lambda_1\phi_1(x) \quad \text{and} \quad \phi_2(x + a) = \lambda_2\phi_2(x) \qquad (11.1)$$

Then, differentiating each equation and multiplying by the other

$$(\phi_1'\phi_2(x) - \phi_2'\phi_1(x))_{x+a} = \lambda_1\lambda_2(\phi_1'\phi_2(x) - \phi_2'\phi_1(x))_x \qquad (11.2)$$

However, for two independent solutions of the Schrödinger equation,

$$\phi_1'\phi_2(x) - \phi_2'\phi_1(x)$$

is constant, whence $\lambda_1\lambda_2 = 1$. We can thus write $\lambda_1 = \exp(ik)$ and $\lambda_2 = \exp(-ik)$ and the two solutions ϕ_1 and ϕ_2 are $u_k(x)\exp(\pm ikx)$, where $u_k(x)$ is periodic, with period a, while $\exp(\pm ikx)$ is, of course, a plane wave. This solution is a Bloch wave: k, the wavevector, is continuous within the band, and the lowest energy solution, which occurs when $k = 0$, is purely periodic.

For the inner electrons, the wavefunctions in the solid will not differ much from those of free atoms, but from the orbitals of the outer electrons, broad valence and conduction bands are formed. Elementary solid state theory then goes on to describe how electronic motion can occur throughout the solid if the conduction band is half-full, giving rise to conductors, while insulators are materials in which bands are either full or empty. The top of the filled states at zero temperature is called the Fermi energy. The magnitude of the band gap between the valence and conduction band determines the properties of nonmetallic materials: a large value makes it difficult for an electron to jump from the valence to the conduction band (insulator) while a small value makes it easier (semiconductor). Finally, one needs the Fermi–Dirac distribution law to describe the behaviour at a general temperature of electrons close to the Fermi energy. Thus, from the outset, the elementary theory of solids distinguishes between *localised states* which are attached to a given atomic site and *itinerant states* which are delocalised. The theory also suggests that localised states are of essentially atomic character, becoming more so the more deeply they are bound to an individual nucleus, while the others reveal the periodic structure of the solid.

Solid state spectroscopy usually revolves around the determination of properties characteristic of a solid, such as the density of states in the conduction band, the position of the Fermi energy, etc. These can be probed by X-ray photoabsorption or more directly by photoemission (either angle-resolved [593] or inverse [594]) spectroscopies, in which case the initial quasiatomic state is reasonably well known since it is close to the wavefunction of a free atom, while the final state is the continuum characteristic of the solid, so that the dispersion of the wavevector **k** can be determined.

However, solids do not always behave according to the idealised band model. When photoabsorption spectra are studied in the X-ray range, it can happen that the observations reveal quasiatomic properties only slightly modified by solid state effects. These atomic effects in solids must be distinguished from purely solid state properties, but are also of intrinsic interest: once properly understood, they provide an alternative route to the study of atoms in condensed matter.

There are two reasons for which X-rays are ideal probes for condensed matter. First, they can traverse thin films or penetrate a finite depth into bulk solids. Second, they excite deep core holes, whose properties are quasiatomic: these are readily related to the atomic constituents, while at the same time revealing, through subtle changes in the spectra, properties characteristic of the solid.

11.2 Localisation of excited states

In the spectroscopy of solids, questions of localisation or delocalisation of particles with respect to each other or else with respect to individual sites play a crucial role. Thus, for example, one distinguishes between two types of luminescence, *viz.* excitonic type and electron–hole type, depending on whether the electron and hole are bound together to form an exciton, or whether they are essentially independent. In the former case, the yield increases with incident energy, while in the latter it is essentially constant.

In many situations, X-ray absorption spectroscopy is site specific, i.e. it provides information local to the atom whose X-ray edge is being probed. The questions arising are then: how well localised are the excited states in the vicinity of these particular atoms and to what extent do their properties differ from those of free atoms?

We have already considered some examples of how the solid state envi-

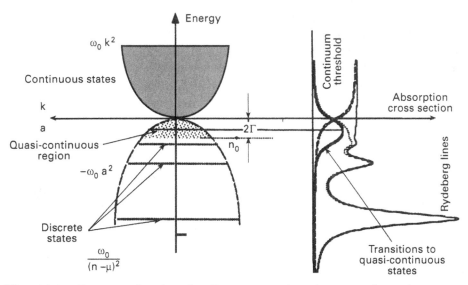

Fig. 11.1.   Diagram showing the discrete, quasicontinuous and continuous regions of an inner-shell spectrum as compared with the resulting photoabsorption spectrum (after C.M. Teodorescu *et al.* [596]).

ronment can modify excited states. One was in section 2.14 where solutions of the Schrödinger equation inside a hollow sphere were described: the lowest member of a Rydberg series may just fit inside a sufficiently open structure, whereas the upper states are destroyed.

It is a characteristic feature of inner-shell excitation in free atoms that the series limits are broadened by core-lifetime or Auger effects (see section 8.32), which truncate the series beyond the first few members, giving rise to intrinsically unresolvable upper members and a lowering of the ionisation threshold below the series limit. This purely atomic effect, which results in a quasicontinuous region below the onset of the true continuum associated with a deep inner shell excitation threshold, allows an internal density of states to be constructed, whose properties (illustrated in fig. 11.1) have been studied both theoretically and experimentally for free atoms [596].

It is clear that, for deep-shell excitation, there is not much difference between the destruction of Rydberg states due to the finite volume available around the atom and the termination of series due to Auger broadening. Furthermore, since the ionisation threshold no longer concides with the series limit even for the free atom, one has to be careful to distinguish between Auger depression and the Fermi edge.

As already noted in section 6.9, Auger spectroscopy is a general method based on the persistence of the Auger effect in core-excited atoms in solids and in surfaces: this allows it to be used as a probe of local electronic structure inside the solid. There is a vast literature in this subject [595] to which the reader is referred.

11.3 Continuum effects: Seaton–Cooper minima in solids

The persistence of atomic multiplet structure from core states is not the only quasiatomic feature to be found in the spectra of solids. Even properties associated with the base of the atomic continuum can survive in the solid, as will be explained here and in the next section. We saw, in subsection 4.6.2, how a Seaton–Cooper minimum can occur in the photoabsorption continuum of an atom if the excited state wavefunction contains a node, and that this minimum, which involves a phase cancellation between two parts of the cross section integral, depends sensitively on the wavefunctions. The implication is that it can be used as a probe, to reveal modifications of these wavefunctions due to solid state effects. Thus, the $4d$-photoionisation cross section of Ag impurities in Aℓ exhibit a quasiatomic Seaton–Cooper minimum [597] which reflects the high degree of localisation of these orbitals. When diluted into the d-band metal Cd, the Seaton–Cooper minimum of $4d$-excited Ag persists but is strongly suppressed (by a factor ~ 20 from the atomic case). Calculations for the impurity wavefunction reveal that this effect is due to hybridisation of the impurity wavefunction with the host $4d$ states. The order of magnitude variation of the Seaton–Cooper minimum with chemical environment makes it a uniquely sensitive experimental probe of changes in the radial wavefunction, providing new insight into the electronic structure and chemical bonding of molecules and solids [598].

11.4 Centrifugal effects

A still more dramatic size variations occurs as a result of the centrifugal barrier effects discussed in chapter 5: since orbital collapse results in deep filling within the atom, in which the outermost electrons are not involved, it is possible for excited states and resonances involving collapsed orbitals in the final states to survive in the solid.

This is particularly the case for giant resonances. In the example of the previous section, it may have seemed surprising that quasiatomic continuum states should be able to persist in solids. The key point is that the dipole matrix element which determines the excitation is due to the overlap between a localised initial state function, corresponding to a core state, and a final state which extends outwards, becoming oscillatory at large r. Since the overlap integral has significant magnitude only within the volume of the atom, the form of the excited state wavefunction at large r can be of little significance. We say that the transition only samples the inner reaches of the atom. In the presence of a centrifugal barrier, the inner part of the excited state is not much modified by the field of the solid, and thus giant resonances remain quasiatomic even within the solid, which accounts for the early observations of Zimkina (section 5.6). There are, however, some fine details in the spectra which reveal the influence of the solid, and thus the comparison between spectra obtained in different phases becomes an important exercise.

11.5   The persistence of multiplet structure

One of the sure signs of quasiatomic behaviour is the persistence of multiplet structure in the spectrum of the solid. A very good example occurs in the $3d \rightarrow f$ spectra of La: the triplet excited states are bound states localised in the inner well of a double-valley potential, while the singlet excited state is resonantly localised in the inner well of the slightly shallower singlet potential. All these states survive in the solid, and one thus observes the full $^3D_1$, $^3P_1$ and $^1P_1$ multiplet structure very clearly in total electron yield spectra obtained from the surface of the solid (see fig. 11.2). There is even evidence of asymmetry in one of the resonances, which shows that continuum effects other than Auger broadening of the core states are involved, as one expects for atoms.

11.6   The quasiperiodic table

We have emphasised the distinction between localised and itinerant states. Under certain circumstances (governed by atomic properties) a given orbital is poised at the critical point where it can become either one or the other for small changes in the environment of the atom. In solid state physics, this gives rise to a first-order Mott transition. In the present context, such a situation is closely related to the problem of controlled orbital collapse (section 5.23): if a solid is built up from free atoms with a double well potential and the corresponding orbitals in the outer well, these may hybridise easily, the external part of the orbital going into itinerant states. If one forms a solid from atoms with collapsed orbitals, then they remain localised.

   In solid state physics, the first-order Mott transition is a phase transition between the metallic (delocalised or itinerant) and insulator (lo-

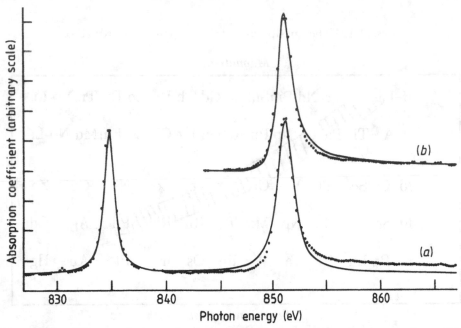

Fig. 11.2. The total photoelectron yield spectrum of lanthanum in the solid phase, as obtained by synchrotron radiation spectroscopy in the neighbourhood of the 3d threshold. Curve (a) shows the experimental data points with the sum of two Lorentzian profiles fitted to them. Curve (b) shows the same data points for the higher of the two resonances, with a Fano profile fitted to it. The improvement in the fit is clear evidence for a slight asymmetry due to autoionisation (after R.C. Karnatak *et al.* [599]).

calised) states which occurs as a function of density in a disordered system [600], when the average interparticle distance becomes of the order of a few times the particle size (the Mott criterion). In this context, what one means by 'particle' is simply the constituent objects. For example, in section 2.21, the particle would be a Rydberg atom. In solid state physics, the usual example involves excitons in a doped semiconductor.

Another example of this kind of transition is shown in table 11.1, taken from the work of Smith and Kmetko [601]. It is a *quasiperiodic table* of all the transition elements and lanthanides in the periodic table, arranged in order of mean localised radius in the vertical direction, and adjusted horizontally so that filled and empty *d* and *f* subshells coincide. What Smith and Kmetko discovered is that a broad diagonal sweep across this table separates metals with localised electron properties (magnets) from those with itinerant electron properties (conductors). This boundary (shown as a shaded curve in the figure) is the locus of the Mott transition. Metals lying along this curve are sensitive to pressure effects (Ce has an isomorphic phase transition from the α to the γ phase at about 1 kbar, U becomes

Table 11.1.   The quasiperiodic table of Smith and Kmetko.

Magnetism

4f–La	Ce	Pr	Nd	Pm	Sm	Eu	Gd	Tb	Dy	Ho	Er	Tm	Yb–Lu
5f–Ac–Th	Pa	U	Np	Pu	Am	Cm	Bk	Cf	Es	Fm	Md	No–Lr	
3d	Ca	Sc	Ti	V	Cr	Mn	Fe	Co	Ni	Cu	Zn		
4d	Sr	Y	Zr	Nb	Mo	Tc	Ru	Rh	Pd	Ag	Cd		
5d	Ba	Lu	Hf	Ta	W	Re	Os	Ir	Pt	Au	Hg		

Superconductivity

This table was derived from empirical data on conductivity and magnetism in solids. The shaded area maps the first order Mott transition between localised and itinerant behaviour in the solid, and the elements which lie on it have sensitive (e.g. pressure-dependent) properties. However, they are also remarkable in atomic physics for their giant resonances, are noted catalysts, or provide good materials for H storage, and then exhibit photon-stimulated desorption peaks which replicate the giant resonance profiles. Many of these properties seem to depend on the critical localisation of f and d electrons.

superconducting under pressure, etc). Many of them are good catalysts or good H storage materials.

It is believed that these properties are connected. To the atomic physicist, the Smith and Kmetko table is also significant: the broad sweep across the table is related to the emergence of giant resonances, i.e. to the properties of the inner well of the double-well potential.

11.7   Photon-stimulated desorption

The interrelation between several of these properties is well illustrated by the study of photon-stimulated desorption of H and radicals containing H from a solid surface of CeO_2, illustrated in fig. 11.3. When a surface of CeO_2 on which H_2 has been adsorbed is exposed to radiation, desorption

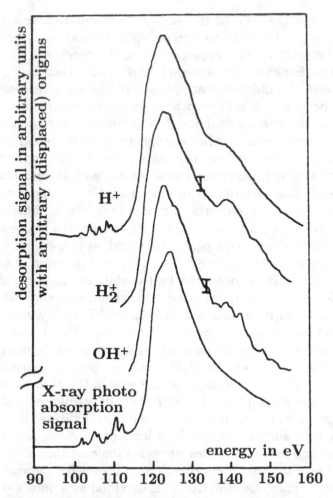

Fig. 11.3. Photon stimulated desorption from a surface of CeO_2, as a function of photon energy, in the range of the $4d \rightarrow f$ giant resonance. The top three curves show the desorption signal for several molecular ions containing H, while the bottom curve shows the photoabsorption cross section for comparison (after B.E. Koel *et al.* [602]).

is stimulated, and the products can be studied by mass spectroscopy. If the photon energy is tuned in the soft X-ray range, through the giant resonance in Ce, the resulting spectrum follows closely the profile of the photoabsorption peak. Photon stimulated desorption is a complicated process, but this example nonetheless demonstrates the importance of underlying quasiatomic processes.

The adsorption of atoms or molecules onto a solid surface can occur in two ways: if the wavefunctions of the atom or molecule are profoundly modified by the presence of the surface and acquire a quasimolecular character, one speaks of *chemisorption*, while if the atom or molecule preserves the integrity of its occupied orbitals, the process is referred to as *physisorption*. For example, benzene adsorbed on a cleaved (111) surface of Si possesses two different states, separated by an activation barrier.[1] At low temperatures, it is physisorbed. At room temperature, it exhibits a strongly bound chemisorbed state. X-ray absorption spectroscopy allows the molecule–metal hybrid wavefunctions to be recognised, and is therefore a useful probe of chemisorption [603]. Returning to the example of CeO_2, it is well understood as an example of physisorption: CeO_2 crystallizes in a fluorite structure in which each Ce atom is coordinated by eight O neighbours. When ceria (a mixture containing different oxides) is reduced under an atmosphere of pure H_2, neither the hexagonal Ce_2O_3 nor the rhombohedral $CeO_{1.8}$ phases are found, as is apparent from X-ray diffraction peaks. Rather, there occurs a lattice expansion of the fluorite structure, which can be measured by the shift in the X-ray diffraction peaks of CeO_2. This expanded phase depends on the temperature of the reduction, and can reach values as high as 2%. Thus, hydrogenation occurs *with no change of the crystalline structure*. This is described as a *topotactic* process, and is similar in some respects to the isomorphic phase transition discussed in section 11.11. Such properties are consistent, as will be shown, with radial changes at the level of the atom. Clearly, if no chemical change takes place, the occupation of angular orbitals remains the same, and overall expansion of the solid can indicate that the bonds of the host have adapted slightly by a local compression and stretching, to accommodate extra ions. Measurements indicate that there are three distinct zones of topotactic insertion [604]. In zone (i), there is very little change in the lattice size: this can be understood as a purely geometrical insertion range, in which the ions move to fill all available sites which require no distortion of the host. In the zone (ii), a slow expansion occurs, while in zone (iii), the expansion occurs more rapidly. It is tempting to relate the sudden transition between zones (ii) and (iii) to properties of an inner and outer range of the potential for the Ce atom in the solid.

Reversible H_2 storage is a technically important problem to resolve if a nonpolluting motor based on burning H_2 is to be developed, since it is then necessaryy to find a safe means of transporting and storing the fuel. It has been found that quite large quantities of H_2 can be safely and

---

[1] The concept of an activation barrier arises in the following way. The rate constant of a chemical reaction is conventionally described by an Arrhenius temperature dependence, namely $k_r = A\exp(-E_a/kT)$, where A is a constant, and E_a is called the activation energy. There are many important reactions which do not have this simple form for the rate constants, and these are said to exhibit barriers.

reversibly stored in a semiporous material called misch-metal, which is an alloy containing a substantial quantity of La (note how close La lies to Ce in table 11.1). This fuel 'tank' can be placed above the exhaust pipe of a vehicle, and the heat generated while running the motor assists in desorbing more H_2 during operation.

11.8 Intercalation and the rechargeable Li battery

H_2 and positive ion storage in solids are related to a process called *intercalation* which is exploited in current research for the development of composite electrodes for Li rechargeable batteries.

To a chemist, the presence of a greater quantity of H in a system, or of a larger amount of any electropositive ion are both examples of 'reduction,' while the opposite is called 'oxidation.' Thus, reversible storage of H_2 or Li^+ are both examples of the 'redox' process. Li^+ is the best ion to use in rechargeable batteries because of its low equivalent weight and high standard potential. Energy densities in laboratory cells of over 150 Wh/kg are reported, considerably higher than any other rechargeable room temperature system. The Li^+ battery is thus a contender, not only for miniature coin type cells, or rechargeable batteries used in portable equipment, but also for very large storage cells to power electric vehicles and for other heavy duty applications.

Cathodes for Li^+ batteries are usually made from composite materials into which the Li^+ ions can diffuse, and lodge themselves within the solid structure, inside which they are then stored or intercalated. The anodes one might expect to be made of Li metal, and this indeed was historically the first choice. It turns out, however, that metallic Li anodes pose severe problems under cyclic recharging. Deposition of the metal by cycling leads to irreversible loss of power, and such batteries may even become hazardous [605]. For these reasons, attention has shifted to cells based entirely on intercalation for both the anode and the cathode to store Li^+ ions within solid electrodes, so that no deposition of metal is required. Indeed, this is one route towards the all solid state battery, from which the liquid electrolyte is also eliminated [606]. Intercalation, sometimes described as 'solid solution' electrodes were first discussed by Steele [607] and Armand [608]. The charging and discharging process is assumed to proceed without the formation of any new phase. The Li^+ ions then rock back and forth between two electrodes (the 'rocking chair' battery principle), and the resulting cell is made both safe and reversible. Most intercalation compounds remain single-phase only over a relatively restricted range of lithiation but in order to maintain reversibility, the compounds should not be cycled over a phase transition. Interest concentrates on materials which are capable of expanding to accommodate lithiation with the minimum structural change. The term 'soft chemistry' is sometimes used to describe the appropriate redox conditions: we consider the implication of this terminology below. Reversible, metal-free Li^+ batteries are also called 'Li ion' or 'swing' batteries. The difference in potential is achieved by using two host electrodes which have different chemical potentials with respect to Li^+. Various electrolytes are used, e.g. a solution of $LiC\ell O_4$ in propylene carbonate is suitable. The anode is usually made of specially prepared forms of graphite, care being needed to achieve good reversibility. Another choice is fullerite, which holds much promise, for physical reasons which will be explained in section 12.5. We now turn to Li^+ intercalation cathodes, for which a variety of materials are being developed, based on the oxides of metals which occupy strategic positions in table 11.1.

Cathode materials for Li^+ batteries are made from mixed oxides of Li and transition metals. These range from layered or folliated oxides such as $LiNiO_2$ or $LiCoO_2$ (where the layers are held together by weak van der Waals forces and the gap between them can expand on lithiation) to spinel structures of more complex formulae, which swell isotropically while retaining their structure (these are two forms of topotactic insertion see section 11.7). Isotropic expansion is considered preferable, because it gives greater mechanical stability.

Kozawa [609] was among the first to discuss MnO_2–Li intercallation clearly as a one-phase process, and to recognise the full significance of this concept in electro-chemistry: he argued explicitly that reversibility is linked to the existence of a single-phase redox cycle.

The mixed oxides have high potential differences with respect to a Li anode (more than 3 V), are stable to air in the fully lithiated state and exhibit good reversibility. Favoured combinations are $Li_xCo_{1-x}O_2$, $Li_xNi_{1-x}O_2$ and $Li_xMn_{2(1-x)}O_4$, where x is the degree of lithiation, although combinations involving vanadium oxide (V_6O_{13}) are also of interest. They can be prepared as thin films (less than $10\mu m$ thickness) by various deposition techniques, with excellent adhesion properties on many substrates. In general, $Li_xCo_{1-x}O_2$ is reported to have better cycling properties than Mn-based oxides.

Other, more complex composite materials suitable for battery electrodes, such as $LiNi_{1-y}Co_yO_2$ are discussed by Ménétrier *et al.* [610] and by Delmas *et al.* [611]. They offer the opportunity to optimise the host structure by varying the composition. There exists a wide variety of such materials of both amorphous and crystalline forms [612] as well as more open spinel structures [613] which possess well-interconnected interstitial sites. These are important to ensure the highest mobility of the Li^+ ions over the whole of the redox cycle. As already mentioned, a common property of all the hosts used as reversible cathodes is the presence of transition metals. This suggests that such materials can expand and contract or 'breathe' in the redox cycle because they involve elements with 'soft' effective radial potentials, related to the centrifugal barrier phenomena discussed in chapter 5.

Not only are Li^+ intercalation and H^+ intercalation in these solids closely similar processes, but topotactic Li^+/H^+ exchange has been observed on complex oxides [614]: this is further evidence that 'soft chemistry' is related to properties of the radial Schrödinger equation, the occupied angular wavefunctions being the same throughout the reversible redox cycle, so that bonds are not broken and bond directions, in particular, are hardly altered. In this respect, soft chemistry resembles physisorption, as opposed to chemisorption (see section 11.3).

11.9   Brief account of intermediate valence

In section 5.18, we described how an atom with a strong centrifugal barrier might be poised with an orbital shared evenly between two wells, and the specific example of Ba^+ was given. If an atom close to this condition is used to build a solid, then the questions arise: how many of the electrons remain attached to the atom, and how many delocalise and join those of the conduction band?

Instabilities of valence (*viz.* atoms 'flipping' from one state of valence to another as a function of changes in the environment) and mixed valence (an atom exhibits simultaneously two valences, or two valence states coexist on the same site) are both related to intermediate valence (the atom in the condensed phase exhibits some mean, nonintegral valence). The effects are usually encountered when dealing with $4f$ and $5f$ electrons, and it is therefore very relevant to determine the f count, or effective number of f electrons on a given site. Various core-level spectroscopies have been used to probe f electron occupancies, and there is a vast literature on this field (see the review by Fuggle [615]).

The most popular model to describe the situation uses a Hubbard Hamiltonian to represent the localisation–delocalisation of an f electron in the presence of the potential of a d core hole, thus:

$$\hat{\mathcal{H}} = \sum_{\substack{\text{all possible} \\ \text{symmetries}}} \left\{ \begin{array}{c} n_f \varepsilon_f + n_d \varepsilon_d + \int n_\varepsilon \varepsilon d\varepsilon + \frac{1}{2} U_{ff} \sum_{i=j} n_{fi} n_{fj} \\[2mm] + U_{fc} n_f (n_d - 1) + \int V_\varepsilon (\alpha_f^\dagger \alpha_\varepsilon + \ldots) d\varepsilon \end{array} \right\} \qquad (11.3)$$

where the first three of the terms in braces represent, respectively, the valence energy, the core-electron energy and the conduction band energy, while the last three (bilinear or 'mixed' terms) stand for the Coulomb repulsion between f electrons, the interaction between an f electron and the core hole and, finally, the hybridisation energy of an f electron into the conduction band, represented by the continuous energy variable ε. The three mixed terms, which between them drive the localisation–delocalisation transition of the f electrons, are weighted by three parameters U_{ff}, U_{fc} and V_ε, which are adjusted to achieve the optimum representation of experimental spectra.

Unfortunately, the model does not predict the magnitudes of these parameters, on which the transition depends. It is even conceivable in the presence of atomic multiplet structure, or of term-dependent localisation effects internal to the atom, etc., that the semiempirical parametrisation might absorb effects not really intended in the model, such as atomic multiplet splittings or term dependence of the orbitals.

Such models, applied by Gunnarsson and Schönhammer [616], or by Jo and Kotani [618] have provided a very successful framework for reconciling results from a wide variety of different spectroscopies (X-ray photoabsorption (XAS), photoemission (PES), bremstrahlung isochromat spectroscopy (BIS), etc.) by the use of only a few adjustable parameters.

One is also able to deduce f counts directly from the analysis of the spectral peaks. In fact, the peaks in some spectra are interpreted directly in terms of the occupancy. For example, in materials containing trivalent Ce, when the photoelectron spectrum splits into three peaks, these are interpreted as $4f^0$, $4f^1$ and $4f^2$ final states. One has to be careful that such designations do not become confused with any persistence of multiplet structure in solids.

11.10   Hybrid quasiatomic models

The implicit assumption in the previous model is that atomic structure has little influence on the degree of f electron localisation in the solid. There are difficulties in reconciling this view with any persistence of atomic multiplet structure, and with the fact that changes of localisation are clearly associated with certain regions of the Periodic Table. For this reason, various attempts have been made to explore what properties of atoms might survive in solids and, perhaps, provide a quasiatomic mechanism to drive changes of valence.

Provided it can be interpreted unambiguously and separated from the localisation effects (which, unfortunately, are not unambiguous), quasiatomic multiplet structure is useful: it yields the electronic configuration, and hence provides the signature of a specific valence state in a case of uncertain valence. Thus, there is a powerful motivation to compare atomic structure calculations of core-hole spectra with experimental data for solids and compounds as a first step in understanding.

In this sense, atomic multiplet theory provides complementary information about the valence state. One can take the view that this information should be used, and then blended in some way with the conceptual framework of the Anderson single-impurity model, so that the matrix elements coupling the f electrons to the conduction band can continue to play the decisive role in determining the extent of f electron localisation.

A good example of this hybrid approach is the work by Thole *et al.* [619] on the $3d$ subshell X-ray absorption spectra of the lanthanides. While a substantial measure of agreement is achieved for a whole sequence of elements, there are areas of uncertainty in the atomic structure calculations themselves, which turn out to be most troublesome precisely for the elements which exhibit valence fluctuations .

In order to achieve good agreement between experimental data and atomic structure calculations, Thole *et al.* [619] were obliged to scale a number of parameters empirically, whereby the main feature of atomic theory, *viz.* its ability to make *ab initio* predictions, is compromised. With an open $4f$ subshell, the number of accessible levels can run into many hundreds in some cases. They tend to merge into just a few observed features in experimental soft X-ray spectra, and both the energies and the relative intensities must be accurately computed for a large number of possible transitions in order to achieve a good interpretation. When parameters are scaled empirically in order to achieve good agreement for the energies, the wavefunction overlaps which determine the oscillator strengths are not necessarily consistent with the scaled energy calculations, and some control is lost over the accuracy of atomic theory.

In a specific instance, that of Sm (further discussed below), the discrepancy between theory and observations for the condensed phase could not be resolved even by using empirically scaled atomic structure calculations.

There have been several attempts to combine information from atomic physics with the Anderson impurity scheme outlined above, so as to calculate multiplet structure consistently with the occurrence of hybridisation. From the atomic physics standpoint, the possibility of correcting for the loss of localised $4f$-electron density by reducing the Slater–Condon integrals by 10 or 20% respectively, depending on whether one or two $4f$ electrons are involved, has been suggested [620]. The precise amount of the correction required is of course borrowed from parametric studies, because atomic physics *per se* can yield no information whatever on the reduction of charge density due to hybridisation. This is the reverse path to the one outlined above, i.e. it feeds back information from the parametric studies into atomic calculations, whose *ab initio* character must of course be sacrificed.

It is hard to reach any definite conclusion from such calculations, because the sensitivity of the results to variations in the empirical reduction is uncertain, and also because a very important consequence of hybridisation, *viz.* the *broadening* of the atomic structure, cannot readily be included.

From the opposite end, the desire to blend the Anderson approach with some kind of atomic structure calculation is a fundamental motivation [617]. Attempts have been made [618] to include some atomic structure effects in the calculation as follows. One writes:

$$\hat{\mathcal{H}}_{TOTAL} = \hat{\mathcal{H}}_{PARAMETRIC} + \hat{\mathcal{H}}_{ATOM} \qquad (11.4)$$

where $\hat{\mathcal{H}}_{PARAMETRIC}$ is as defined above, and $\hat{\mathcal{H}}_{ATOM}$ contains all the Slater–Condon parameters, i.e. the problem is treated additively in a configuration averaged scheme, with Slater–Condon parameters adjusted to represent the multiplet structure of the states concerned, as determined from an analysis of the free atom or ion.

However, it is again difficult to extract a full description without sacrificing the predictive ability of the theory.

11.11   Centrifugal barrier models

Another approach is to abandon the full description of solid state effects, and to study more closely the atomic description, in the hope of identifying mechanisms which might then be adapted and included in a model of the solid.

As emphasised above, lanthanide and actinide elements play a central role in the study of intermediate valence. A purely parametric theory alone does not account for this, and the dependence on atomic number (as summarised, for example, in the quasiperiodic table of section 11.6) suggests an atomic, centrifugal origin, along the lines of chapter 5. The implications are that: (i) the contraction must be sensitively dependent not only on the environment, but also on the binding strength of the inner atomic well; and (ii) once contraction has occurred, the orbital becomes essentially atomic, even in the environment of the solid. Taking (i) first, one can use the sensitive dependence on the properties of the inner well to 'mock up' from atomic theory the transfer of an electron from the outer to the inner well, and deduce from first principles the change in external conditions which would be required to produce it.

A well-known example of critical phenomena encountered in solid state physics is the pressure-sensitive isomorphic phase transition in the amorphous solid SmS [621]. A similar effect is the α–γ transition on solid Ce. In both cases, the structure of the solid does not change, but there is a significant alteration in its volume, and many physical properties are suddenly altered as expected in a phase transition. It has proved possible [622] , by using the change ΔE in Hartree–Fock energy for the atom, and the change ΔV in its volume, together with the thermodynamic relation $\Delta E = P\Delta V$, to calculate the pressure required to 'flip' the solution from one well to the other in a quasiatomic model, without introducing any adjustable parameters. The result is about 1 kbar, which is of the correct order of magnitude, and suggests that a mechanism internal to the atom is involved in the isomorphic transition. Such a prediction is not possible using parametrised theories of the solid.

It has further been argued by Band and Fomichev [623] who have performed Dirac–Fock calculations of the transition from collapsed to 'decollapsed' orbitals that a coexistence of states in the same atom, due to the formation of hybrid states, can occur One thus forms a picture in which it could also happen, in cases of fluctuating valence, that two different valence states would coexist on the same site [624].

It has also been speculated on the basis of Thomas–Fermi calculations for the atom in a solid [625] that a double-well structure might exist, and that a coexistence of localised and diffuse f orbitals for states nearly degenerate in energy might explain the occurrence of intermediate valence. Against this, Bringer [626] has argued that there is in fact no double-well structure for an atom in a solid when treated by a Hartree–Fock model with a Wigner-Seitz boundary condition, and therefore that the centrifugal barrier model for valence changes is incorrect.

However, Bringer's conclusion is at variance with the results of Band *et al.* [624]. It is also important to note that the centrifugal model does *not* require the occurrence of two distinct potential wells for orbital contraction to take place: it is sufficient that there should be an inner potential well of finite range, connected to an outer well of much longer range. The main influence of the barrier (when it occurs) is to ensure a fuller spatial segregation of the collapsed and decollapsed states. This is how the Hartree–Fock version of the Mayer–Fermi theory for free atoms accounts for d-electron contraction, in which case a repulsive barrier actually does not form, and localisation becomes even more term- and configuration-dependent.

11.12 Survey of the lanthanides

A classification of lanthanide or rare-earth elements based on the number of $4f$ electrons in the atomic and condensed phases is presented in table 11.2.

We also present, in table 11.3, the changes Δn_f in the number of $4f$

Table 11.2. Classification of the lanthanides according to the number of $4f$ electrons in the atom and the solid. Values of p and q in the electronic configuration $\bar{X}e4f^p(5d6s)^q$ are given, where p is the number of $4f$ electrons and q is the number of valence electrons; R stands for a rare earth.

	R	La	Ce	Pr	Nd	Pm	Sm	Eu
	$Z =$	57	58	59	60	61	62	63
Atom	$p =$	0	2	3	4	5	6	7
	$q =$	3	2	2	2	2	2	2
Solid	$p =$	0	1	2	3	4	6	7
			0	1			5	6
	$q =$	3	3	3	3	3	2	2
			4	4			3	3

	R	Gd	Tb	Dy	Ho	Er	Tm	Yb
	$Z =$	64	65	66	67	68	69	70
Atom	$p =$	7	9	10	11	12	13	14
	$q =$	3	2	2	2	2	2	2
Solid	$p =$	7	8	9	10	11	13	14
			7				12	13
	$q =$	3	3	3	3	3	2	2
			4				3	3

Table 11.3. Changes of valence Δn_f from the free atom to the solid.

Δn	Rare earths
0	La, Gd
-1	Nd, Pm, Dy, Ho, Er
0 or -1	Sm, Eu, Tm, Yb
-1 or -2	Ce, Pr, Tb

electrons for these elements between the atom and the solid. It is interesting to note that, in table 11.3, seven out of fourteen rare-earth elements show valence changes in various environments of the solid. This in itself is a strong indication that valence instabilities are related to the quantum

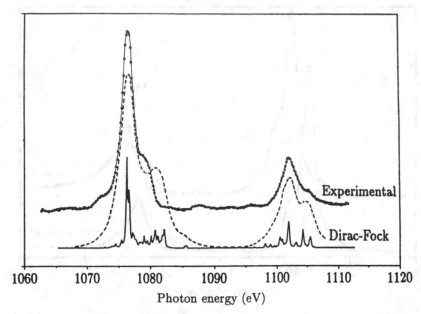

Fig. 11.4. The inner-shell absorption spectrum of Sm vapour. Also included are results of multiconfigurational Dirac–Fock calculations for the free atom (after B.K. Sarpal *et al.* [627]).

chemistry of this part of the periodic table. We can classify the elements presenting valence problems into two broad groups, based on the knowledge of their electronic configuration and known chemical valence.

11.12.1 *Divalent Sm, Eu, Tm and Yb*

The elements Sm, Eu, Tm and Yb, in which $|\Delta n_f|$ may be 1 or remain zero, exhibit characteristic spectroscopic features. Both divalent and trivalent compounds of Sm, Eu, Tm and Yb exist. The $3d \rightarrow 4f$ spectra of these compounds are found to be characteristic of the number of $4f$ subshell electrons present. The change from divalent to trivalent forms is accompanied by the delocalisation of the 'last' electron, which enters into ionic bonding of the $3d$–$4f$ multiplet with one $4f$ electron removed from the $4f$ subshell.

These elements possess well-developed multiplet structure in the soft X-ray spectra of the condensed phase, which can be compared with the spectrum of the free atom: by performing *ab initio* Dirac–Fock calculations, the actual $4f$ occupancy in the solid can be deduced. An example is shown in fig. 11.4 The first point to note is that the multiplet structure of the atom survives in the solid, because of the strong localisation of the $4f$ electrons (see section 5.6), so that soft X-ray spectroscopy provides a

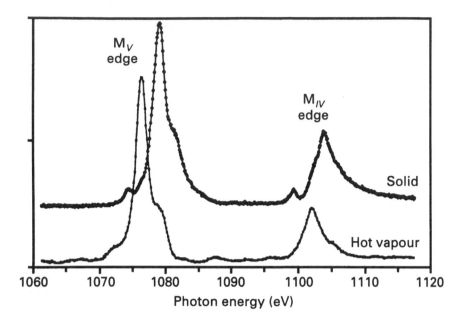

Fig. 11.5.   The 3*d* spectra of Sm in the solid and hot vapour phases. Note the close correspondence of structure and the slight energy shift between them (after J.-P. Connerade and R.C. Karnatak [637]).

useful probe of the $4f$ occupancy. To emphasise this point, we show, in fig. 11.5, a comparison between spectra of Sm vapour and Sm in the solid phase, which reveals the great similarity of structure between them. Also included in fig. 11.5 is a multiconfigurational atomic structure calculation [627] of the atomic multiplet structure which demonstrates that, even in this complex situation, it is possible to deduce from *ab initio* atomic structure calculations what the $4f$ occupancy is.

This technique of analysis is very useful in determining valence changes, as will be illustrated in section 12.11. Note in particular that no adjustable parameters were used in the calculation of multiplet structure, which is therefore the purely atomic result.

11.12.2   *The Ce intermetallics*

The elements Ce, Pr and Tb belong to another group in which $\mid \Delta n_f \mid$ is found to be 1 or 2. Among this group, the case of Ce intermetallics, in which the Ce valence remains very close to 3 ($\mid \Delta n_f \mid \sim 1$) is unique and very interesting. The $3d$ multiplet structure of Ce intermetallics exhibits two main lines with further structure, separated by the spin–orbit splitting, and weaker structures about $5\,\mathrm{eV}$ above the main lines. The multiplet is clearly recognisable as due to a $4f^2$ final configuration.

The intensity of the weaker structure is observed to depend on the partner element in the intermetallic. The manifestation of intermediate valence in the $3d \longrightarrow 4f$ XAS and other core and outer spectra of Ce intermetallics has been interpreted in the framework of the Anderson impurity model [628].

Such calculations are able to reconcile the solid state properties (magnetic susceptibilities, etc) with the high energy spectra by using a single set of adjusted parameters [628, 629].

On the other hand, although spectra of Ce and Yb have been thus interpreted, no impurity model calculations of core- and outer-level spectra of Sm, Eu and Tm appear to have been reported so far. Indeed, the Anderson Hamiltonian is reputedly unsuitable for the interpretation of such systems. In the interesting cases of YbP [630] and YbAℓ_3 [631], impurity model calculations have been performed by considering Yb as the hole-analogue of Ce.

11.12.3 Rare-earth dioxides – covalent insulators

The rare-earth dioxides RO_2 (R=Ce, Pr, Tb) and other tetravalent insulating compounds belong to a system where $\mid \Delta n_f \mid = 2$. These insulators show a high degree of covalence. Among these compounds, CeO_2 has attracted most interest with regard to intermediate valence in the Ce intermetallic compounds. The controversy about its mixed f^0 and f^1 ground state and its classification as an intermediate valence material remained a subject of debate for some time. Now, it is accepted as a covalent insulator on the grounds of its $3d$ X-ray absorption spectrum [632, 633] and optical data [634].

The $3d \rightarrow 4f$ spectra of rare-earth dioxides provide some insight as to the nature of the $4f$ states involved in the transition [635]. The selected examples of M_V spectra of CeO_2, PrO_2, TbO_2 and $CeNi_2$ (an intermediate valence material) are characterised by two different types of feature. The intense peaks exhibiting broad multiplet structure are found to resemble those observed in corresponding La, Ce and Gd metals and trivalent compounds [636], which indicates a reduction in the local f count (in fact, the multiplets correspond to transitions from f^0, f^1 and f^7 ground configurations respectively in CeO_2, PrO_2 and TbO_2, while in $CeNi_2$ such an identification yields a mainly f^1 ground configuration).

In addition to the main multiplet lines, one additional weak feature appears systematically several eV above each M_{IV} and M_V multiplet. Its separation from the main lines (3–5 eV) is too large as compared to the observed separation (1.7 eV) between tetravalent and trivalent $3d^9 4f^{n+1}$ final atomic multiplets for any explanation in terms of multiplet splittings. It has been suggested that the weak feature originates from excitation to *extended orbitals* of f symmetry [637].

Another interesting fact about RO_2 is the observation of unusually wide multiplets. They are found to be 40–50% wider in span than similar multiplets observed in the corresponding $Z - 1$ elements [638] and the increase cannot be accounted for merely by invoking an increase in the $3d$ spin–orbit separation for higher Z elements.

The information one gathers from observations of $3d \rightarrow 4f$ transitions in RO_2 and the Ce materials is:

(i) In both types of materials, the localised and extended characters of the last $4f$ electron *coexist*, as evidenced by $3d$ XAS multiplet analysis.

(ii) Ce intermetallics are further characterised by the presence of nearly one f electron in a localised level. Hybridisation of the $4f$ electron with the valence band leads to a fractional delocalisation, and the vacant states thus created are found to be of an extended nature, owing to which they also give rise to spectral features in other core-level and BIS spectra of the same materials. In the context of Ce materials, such states should not be confused with f^0 *localised* states, and the $4f$ occupancy in each type of state can be precisely distinguished.

(iii) In RO_2 compounds, the delocalisation mechanism of the last $4f$ electron is quite different from that in Ce intermetallics. For the latter, the vacant states created exhibit localised f character corresponding to trivalent compounds of $Z - 1$ elements, and extended states similar to those in Ce intermetallics. The unusual line broadening in the excited multiplets is thought to be due to the interaction between the excited f electron and the finite continuum of f symmetry states [639].

(iv) In fluctuating valence Sm, Eu and Tm materials, atomic $3d \rightarrow 4f$ multiplets due to two distinct valence states are observed.

The above description of the ground state of RO_2 deduced from M_{IV-V} spectra is further supported by L_{III}-edge measurements [640].

Within a quasiatomic interpretation, these results can be understood if two types of $4f$ orbital are allowed to coexist in the solid. These are denoted as $4f$ (the usual, localised type), and a new kind of orbital, denoted $\overline{4f}$, which can coexist with the ground state, as originally suggested by Band and Fomichev [623] for atoms of the $4f$ sequence. An interpretation along such lines has been pursued by performing *ab initio* relativistic Dirac–Fock calculations for atoms with modified boundary conditions at large r to represent the effect of the Wigner–Seitz cell [624]: two solutions exist for f electrons, and the influence of the external boundary conditions can be studied.

However Allen [641] and others have argued against an atomic interpretation. It is probably fair to say that atomic multiplet theory and the condensed matter models have not yet been completely reconciled. This question is important: high temperature superconductors of the form $RBa_2Cu_3O_6$, where R can be any member of the rare-earth series of elements except Ce and Pr, have attracted much attention. A fluctuation in valence involving $4f$ electrons is likely to be involved in the emergence of high temperature superconductivity for these materials.

Experimental data on the free atom provide an essential benchmark, without which interpretation of data for the solid must remain somewhat speculative.

11.13 X-ray dichroism

The core-level spectra of solids have been observed to exhibit magnetic circular dichroism in the X-ray range. There are sometimes large effects in individual peaks, but the integrated magnetic circular dichroism tends to be zero, because there is no preference for left or right circular polarisation in dipole transitions from a completely filled core to an empty continuum state (this is similar to the situation for Faraday rotation in a continuum, discussed in section 6.14). The situation is different for emission from a partly filled or open shell, when a nonstatistical distribution of the population over magnetic sublevels results in a difference in the integrated photoemission for left and right hand circularly polarised radiation. General sum rules have been derived which show that these integrated intensities are proportional to ground state expectation values of the spin and orbital momentum products. Thus, polarised photoemission becomes a useful probe for solids.

The most interesting case is photoemission of $4f$ electrons in the rare earths: as noted in the previous section, because of the collapsed nature of the $4f$ orbitals, the photoemission spectrum can be interpreted completely even in the solid by atomic multiplet theory, and this applies also to magnetic circular dichroism. Thole and van der Laan [642] have derived sum rules for magnetic dichroism in rare-earth $4f$ photoemission. They have shown that the integrated intensity is simply the sum over each sublevel of its occupation number times the total transition probability from that sublevel to the continuum shell. Polarisation effects in the $4f$ photoemission spectra of rare earths are very large, and this tool based on quasiatomic analysis is of considerable significance: it provides a new

means of characterising the magnetic properties of rare-earth surfaces
and thin films, which is an application of technological importance. This
completes our discussion of rare-earth spectra in solids.

11.14   Extended X-ray absorption fine structure

Site selectivity in many applications of X-ray spectroscopy was stressed
in section 11.2: the different absorption edges are often widely separated
in energy by a broad band of continuum states containing rather little
atomic structure. This allows one to study the immediate surroundings
of specific atoms.

For excitation from core states of atoms to a final continuum state
in the solid, an important observation is the occurrence of quasiperiodic
modulations in the intensity of the observed continuous spectrum. These
modulations are readily explained as arising from interference between
the wave of the escaping electron, and a backscattering off the nearest
neighbour atoms. They are called extended X-ray absorption fine structure or EXAFS for short. They are very important, because they can be
rather simply analysed to yield the distance between the atom on which
the core excitation has taken place and its nearest neihbours. There is an
extensive literature on this technique [643] which is a very powerful one,
since it can be applied to all kinds of compounds: each atom has its own
particular, quasiatomic core states, and so the method is atom-specific.
The only complication is when the same atom can occupy two different
sites with different nearest neighbours.

The amplitude of an EXAFS signal is given by

$$\xi(E) =$$
$$-\sum_j \frac{N_j}{kR_j^2} \mid f_j(k) \mid \sin\left\{2kR_j + \delta(k) + \psi(k)\right\} \exp(-2\sigma_j k^2) \exp\left\{\frac{-R_j}{\lambda(k)}\right\}$$

$$(11.5)$$

where:

> $\mid f_j(k) \mid$ is the amplitude reflection from a single atom;
> $\sin 2kR_j$ is the Fourier transform of the interatom separation
> R_j (gives the density);
> $\psi(k)$ is the central atom phase shift, associated with the elec-
> tron leaving the first atom;
> $\delta(k)$ is the phase shift associated with a reflected wave from
> the nearest neighbour atoms;
> σ_j^2 is the mean motion associated with the temperature of the
> atoms; and
> $\lambda(k)$ is the mean free path of the electrons.

The last two result in damping terms, while the interatom separation can be determined very accurately, essentially by a Fourier transform of the oscillatory part of the pattern. Although there may be very little atomic structure in the continuum, it cannot be completely neglected. There is one situation in which care is needed in the interpretation: doubly-excited configurations may extend far into the continuum above an inner-shell threshold and may coincide in energy with EXAFS for some atoms. They are then not easily separated from each other, which may become a source of error in the interpretation.

11.15 Deep inner-shell experiments

For very deep inner shells, an interesting situation arises, where pure atomic physics experiments can actually be better performed on the solid: very deep shells have a very small excitation cross section, because their effective radius is so small. This difficulty can be overcome by an enormous increase in the density of absorbers, i.e. by using thin metallic films. Another advantage of this approach is that absolute cross sections are then easily measured, merely by determining the absorption coefficients and weighing the sample, from which its density can be determined much more accurately than for atomic vapours.

Experiments along these lines have been performed by Nakel and coworkers [644]. By using very high energy electrons for the primary excitation, they have measured triply differential cross sections for deep-shell ionization processes in which one fast and one slower electron are produced in a coplanar geometry. One can also measure cross sections for the production of two electrons of similar energies in a symmetric coplanar geometry. Such experiments at very high energies are quasiatomic even in the solid, and can be compared with *ab initio* Dirac–Fock calculations [645]. A new effect which has been found is the presence of a 'recoil peak' in the angular distribution of the ejected electrons. This peak is not predicted by fully relativistic plane wave calculations, but does appear if interference effects between the initial and the target polarisation channels are included [645]. This is an example of how relativistic effects can be probed for atoms by using solid targets: so far, experiments of this kind have been performed using high energy electrons. Nothing in principle prevents such work being performed with very high energy photons, which remains an open area of research.

11.16   Strong magnetic fields

Finally, we give an example of a problem of atomic physics which cannot be explored in the laboratory by using atoms, but only by using the 'pseudoatoms' described in section 2.34. The problem is that of the atom in an even stronger magnetic field than discussed in section 10.14. One seeks to make the magnetic field strong enough to produce effects comparable to or larger than the binding energy of the atom in its ground state (superstrong magnetic fields). This requires field strengths in excess of the atomic unit of magnetic field, i.e. greater than 2.35×10^5 T, which is unattainable, except in the atmospheres of white dwarf or neutron stars [646, 647]. However, if one replaces a normal atom by a pseudoatom, consisting of a shallow donor impurity in a semiconductor such as InP, GaAs or InP, then giant quasihydrogenic orbits are produced, and even the ground state is very wide. Taking GaAs as our example, the effective mass m^* of the electron is $0.065m$, so the orbits are some 200 times wider than for the atom. As a result, with an external magnetic field of about 6.5 T (readily achieved in the laboratory), the superstrong magnetic field condition can be reached [648]. This system has been investigated experimentally in fields as high as 14 T [649], and can be accounted for within a simple theoretical model. In this unique situation (known as the Landau limit) the term quadratic in B dominates the Hamiltonian, while the Coulomb field becomes a small perturbation. Under these conditions, the spectrum becomes a series of equally spaced Landau levels, to each one of which a Rydberg series converges. Above the lowest Landau limit, a continuum occurs, and therefore, even in hydrogen, excited states above this limit experience autoionisation in the superstrong field condition [650]. While the autoionisation rate for hydrogen has been calculated theoretically, there is no way of performing controlled experiments in the laboratory to test the theory for a real atom. The test has been carried out by making use of pseudohydrogenic states in GaAs, and scaling the results [651]. This provides a beautiful example of the unity of physics in the study of very unusual highly-excited atomic states.

12

Atomic clusters

12.1   Introduction

The present chapter is devoted to the comparatively new and rapidly developing subject of clusters, a field intermediate between atomic physics, chemistry and solid state physics, in which concepts borrowed from nuclear physics have also proved very important. Although the field is new, it has expanded very rapidly, and there are many different aspects beyond the scope of the present book. We therefore confine our attention to: (i) a general introduction and (ii) some aspects of cluster physics which are specifically connected with material already presented in the previous chapters.

A cluster is an assembly of identical objects whose total number can be chosen at will. An atomic cluster is therefore an assembly of atoms in which the total number is adjustable. Just as, in solid state physics, one distinguishes between cases in which the valence electrons become mobile and those in which they remain localised on individual atomic sites, so one finds different kinds of clusters, depending on the degree of localisation of the valence electrons. Broadly speaking, these differences are dictated by the periodic table: at one extreme, one has the rare-gas clusters, in which electrons remain localised, while at the other, one finds the alkali clusters, which are metallic in the sense that the valence electrons can move throughout the cluster.

The subject of atomic clusters arose only recently because it was not appreciated in earlier times that identical atoms could hang together in this way. The existence of dimers and trimers (molecules formed of two or three identical atoms) was known, since they are undeflected in Stern–Gerlach experiments. It was understood that corrections for the presence of dimers should be applied when performing absolute measurements of atomic f values, and that this could be an important source of error. As a result, these molecules were mainly known for their deleterious effects, and atomic physicists sought to eliminate their influence from experiments.

In fact, dimers are often interesting and important. For example, alkali dimers are thought to undergo triplet-triplet electronic transitions which are excimer-like, in the sense that the lower state is repulsive (cf section 2.27). They are therefore potentially useful as broadband tunable laser sources [652, 653].

Early interest in the properties of clusters arose from the study of the absorption properties of interstellar matter. The development of cluster beams has been traced [654] to 1956, when experimenters attempted to produce intense molecular beams at low temperatures. This required the use of cooled nozzles, with the implication that numerous slow collisions would occur within them. Mass spectroscopy then revealed a far more complex distribution of masses than anticipated, and the study of free clusters was born. It is only when clusters could be formed as beams that they became suitable objects for spectroscopic study in the laboratory, and progress in the physics of mass-selected clusters was thereafter very rapid.

Since the properties of clusters have been more widely appreciated, their importance in the formation of laboratory discharges and plasmas has been studied [655], a subject of some practical importance. Clusters play a role in the photographic process, in surface chemistry and in catalysis. There are many different motivations for the study of clusters. One may merely be interested in how atomic properties are modified in this new environment. One may be concerned with the dynamics of this new quantum system, which is held together in a somewhat different way from the standard molecular pattern. One may wish to use clusters as building bricks to create further new 'supermolecules' of large size: clusters have a chemistry of their own, leading to a wide variety of potential applications. They also provide a new perspective in the study of condensed matter: they provide a bridge between free atoms and solids, which allows one in some sense to interpolate between the two, simply by changing the size of the cluster. Their volumes range from those of typical quantum systems to dimensions considered as macroscopic. Finally, they provide examples of many-body quantum systems dominated by correlations whose collective motions bear an interesting relation to those of nuclei. Their properties have been shown to match those of short pulse lasers, providing new targets for the investigation of dynamical effects [656].

The organisation of the present chapter is as follows: first, we describe different kinds of clusters which can be formed and the shapes which are commonly found. Next, we give an example of how clusters can be used to bridge the gap from the atom to the solid and, finally, we discuss the subject of giant dipole resonances in clusters.

12.2   General property of clusters

Clusters can be formed in a variety of ways,[1] but share a common property [658] which sets them apart from other forms of matter. They are held together by forces which never saturate, i.e. it should always be possible to add more atoms to a cluster.

The essential novel ingredient of cluster science is that the objects under study should retain self-similar properties as their size is increased. This is the concept of *stackability* [658]. Although clusters are finite in size, they may be made to grow indefinitely by stacking one more atom of the element from which they are composed. *This may be taken as the defining property of clusters* and distinguishes them from ordinary molecules.

Thus, we suppose that *any* number of atoms can be glued together to form a cluster, in such a manner that the sites on which the atoms are located remain nearly equivalent to each other. This has the interesting consequence that a cluster can grow to macroscopic size without losing fundamental characteristics established when it was born in the microscopic regime. We say that clusters bridge the gap from microscopic to macroscopic physics.

The view just presented contains an idealisation. In reality, these objects of study usually involve a complex interplay of properties, some of which properly belong to the science of clusters, while others are of the more traditional molecular nature. For example, at low temperatures, a metallic cluster may possess several isomers of rather different shape but closely similar energies. As the temperature is raised above a critical value T_c the cluster exhibits a random behaviour, 'flopping' constantly between its various isomeric forms, the effective shape being defined by an average. In this picture, molecular forces dominate at low temperatures, and the size alone is not enough to determine the state of the system. Above T_c, the situation is in fact simpler: a global description (the jellium model – see section 12.8.1) is then applicable. It extends through to the solid, and the situation above T_c is therefore the true cluster regime.

The clusters of greatest interest to us are those in which all the atoms are the same (homogeneous clusters). It is also possible to create mixed or heterogeneous clusters, in which two atoms of different types (which may or may not be chemically similar) are combined in indefinitely variable proportions.

As just noted, at the microscopic end of the range, clusters can be confused with molecules and, indeed, share many of their properties. To

---

[1] For examples of experiments involving clusters, there exists an extensive and fast growing literature. For example, the reader may consult conference proceedings, now available in book form [657] or review papers [682]

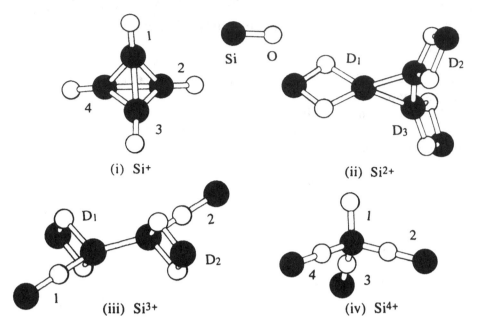

Fig. 12.1. Clustering of 4–6 SiO molecules to yield different bonding microstructures in which the Si atoms are tetrahedrally coordinated by other Si and O atoms. This leads to four different oxidation states of Si in the cluster (after A.M. Flanck *et al.* [659].

distinguish between molecules and clusters, stackability is the best criterion: it is not easy to change the number of atoms arbitrarily within a molecule without changing its structure in a rather fundamental way.

Of course, it is also possible to stack individual molecules together, and thereby form clusters of another kind: clusters are not necessarily built up from individual atoms of the same kind. If molecules rather than atoms are stacked, there is the additional possibility that the molecules on different sites may not be equivalent. For example, studies of the X-ray spectrum of SiO clusters [659] have revealed the presence of four different oxidation states of SiO, as shown in fig. 12.1. This kind of additional complication or internal chemistry cannot occur when the cluster is made up only from uncombined atoms.

Stackability is the property which makes atomic clusters similar to the corresponding solids when they become large enough to contain very many atoms. This has led researchers to ask the provocative question: how small is a solid [660]? This question should not be taken too literally: the properties of a cluster are usually quite different from those of the bulk. For a spherical cluster of about 1000 atoms, roughly one quarter of the atoms may be expected to lie on the surface. Furthermore, there can be several evolutions between the properties of the free atom and those of the solid as a function of cluster size, depending on the property under study. Thus different transitions may occur in different mass ranges. Finally, it must be borne in mind that the symmetry of a cluster is not necessarily that of the solid. Many clusters tend to be spherical, but this is not necessarily the case. Solids, on the other hand, stack uniformly to fill the whole of space.

Clusters may be held together by a variety of different forces, strong enough to hold the atoms together, but weak enough to allow some flexibility in the interatomic angles, which allows further atoms to be added to the system. If clusters are characterised by their stackability, not all binding forces are suitable to build up a cluster sequence.

12.3   Van der Waals clusters

One of the simplest types of cluster is formed by packing together atoms which are chemically inert. They are then held together by the van der Waals forces between neutral atoms, which impose no particular bond orientation and which do not change the electronic configuration on each site, so that infinite stackability is, in principle, readily achieved. The structure of van der Waals clusters is illustrated schematically in fig. 12.2.

As pointed out above, there are two extreme situations involving localised and delocalised valence electrons in cluster physics: at one end of the scale, the noble gases possess valence electrons which remain localised on individual atomic sites, while, at the other, the alkalis possess delocalised valence electrons which can wander over the cluster and resemble the conduction electrons of a metallic solid.

Between these two extremes lie clusters of atoms more stable than the alkalis, but less so than rare gases, and which may actually effect a transition from van der Waals to metallic behaviour as a function of the cluster size N. A good example is provided by Hg clusters: Hg atoms have a closed $6s^2$ subshell, and a resonably high ionisation potential. The small clusters (up to about 10 or 15 atoms) exhbit a van der Waals behaviour with quasiatomic $5d \rightarrow 6p$ transitions, while a conduction band appears for larger N [662] as the aggregates become metallic.

The structure of a van der Waals cluster is similar to that of an insulating solid in that the nature of the atoms remains unaltered, i.e. each atom inside the cluster carries the same number of electrons as the corresponding free atom. It is no surprise that rare gases, which have high ionisation energies, should form clusters of this type.

The lowest energy configurations of van der Waals clusters are regular shapes, corresponding to 'magic numbers' (see fig. 12.2), whose free

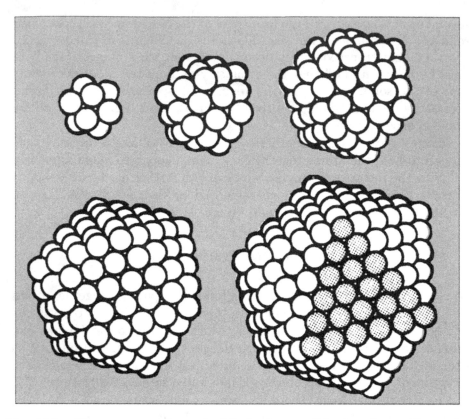

Fig. 12.2.   The structure of van der Waals clusters. The first five stable magic numbers are shown, containing 13, 55, 147, 309 and 561 atoms respectively (after J. Baggott [661]).

surface energy is small because they approach spherical shapes. These are called *geometric shell structures* to distinguish them from *electronic shell structures* like the shells of free atoms. Whether clusters are stable in this form will, of course, depend upon the temperature. Rare gases form two kinds of cluster: mostly, they are solids, although He tends to form liquid clusters. The fact that van der Waals clusters possess magic number configurations distinguishes them from a small 'piece' of a solid.

12.4   Clusters of molecular type: the fullerenes

Clusters of molecular type can come in many forms, but if we confine our attention to clusters formed from nearly identical atoms, a good example of a cluster of molecular type is afforded by the C_{60} molecule, known as buckminster-fullerene (because its shape is reminiscent of the domes built by the architect Buckminster-Fuller) [663]. The 'bucky-ball' is closer to a molecule than most entities described in this chapter because it owes its existence to a directional chemical bond which has rigidity, and confers stability to the symmetrical structure, illustrated in fig. 12.3. This bond is in fact the strongest of all known chemical bonds, and the spherical C_{60} structure, whose diameter is about 7 Å, is the strongest one can build.[2] The structure of C_{60} is related to an interesting problem in pure geometry. There exist five so-called *Platonic solids* [664], formed by requiring that all the apices of a regular solid should be entirely equivalent, i.e. they can be inscribed on a sphere, with identical connecting surfaces between them. It can be shown that the largest number of apices is 20 (the dodecahedron). If, however, one relaxes the condition that the connecting surfaces be identical, while still requiring that the solid be inscribed in a sphere, then another regular shape emerges with 60 apices, for which the connecting surfaces are only pentagons and hexagons. This is the structure of the 'bucky-ball', which belongs to a family of similarly constituted shapes known generically as the fullerenes. It is important to note that the fullerenes differ markedly in shape among themselves even when they do form closed hollow structures. This is a consequence of the geometrical principles just mentioned: there exists no Platonic solid with more than 20 apices, and the larger closed structures depart from the spherical shape.

Although less stackable than the other systems considered here, fullerenes, in other respects, also qualify as clusters, for several reasons. First, they do so because of their large size and symmetrical shapes. Second, there can exist similarly formed groupings of C atoms with diverse numbers (70, 120, etc) and related properties. Third the p bonds of C give rise to a delocalised electronic shell structure similar in many respects to the closed shells of metallic clusters: each carbon atom donates four electrons to a shell whose precise shape is determined by the fullerene structure, but which is closed, has a thickness of about one atom, and is very similar to the closed shells of metallic clusters.

The hollow cage structure of the fullerenes allows a spherical standing wave to be formed inside the cluster, and this has been advanced as the explanation for marked oscillations of the photoionisation cross section of C_{60} which persist in both the gaseous and the solid phase [666].

Clusters with hollow cage structures and more than 60 atoms are known as higher fullerenes. Their structures and modes of growth are a subject of study [667, 668, 669]. Experimentally, six different sizes (C_{60}, C_{70}, C_{76}, C_{78}, C_{82} and C_{84}) are well characterised as possessing hollow closed cages,

---

[2] It is in fact stronger than diamond, but does not fill space to form a uniform lattice, and so does not give rise to such a hard solid. The intercallation of foreign ions in solids formed from fullerenes is discussed in the next section.

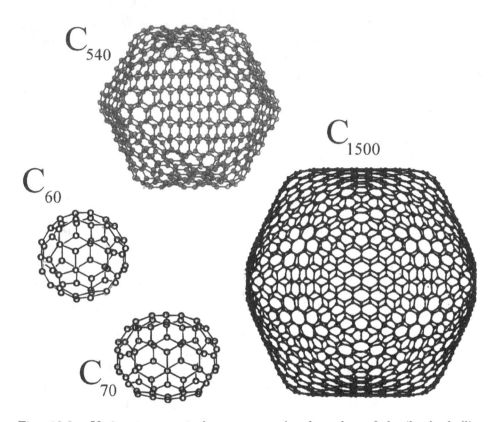

Fig. 12.3.   Various geometrical structures related to that of the 'bucky-ball': we show 60 atoms, which is spherical, 70 atoms, which becomes distorted from a sphere and 540 atoms, for which an icosahedral shape develops. Note that all the atomic sites are not equivalent. The structure of a giant fullerene, with 1500 atoms, is also shown (adapted in part from K. Prassides and H. Kroto [665]).

which are formed from five- and six-membered rings. Higher fullerenes, such as C_{90}, C_{96}, C_{102}, C_{106} have been isolated. In all cases, they avoid having adjacent pentagons in their networks, which has led to the proposal of the isolated pentagon rule (IPR) [670] and to the idea of pentagon migration during growth. C_{60} is then the smallest structure satisfying the IPR rule. For nonspherical structures, isomers of differing symmetries are observed, and their study is one way of inferring growth mechanisms [671].

An interesting property of the fullerenes is that, because they are completely hollow (i.e. all the atoms lie on the surface), fullerenes of different sizes can in principle be nested within each other like Russian dolls, in an arrangement sometimes referred to as a 'buckonion' [672].

Another interesting class of systems are the metallofullerenes, formed by intercallating one or more metal atoms amongst the C atoms during the preparation. The actual site occupied by the intercalated atoms has been much debated, and it seems that there are several different situations. For example, alkalis form metallofullerenes which tend to be superconducting at fairly high temperatures. Deciding whether an atom is internal or external to the fullerene cage (i.e. whether it is trapped inside or whether it is stuck to the outside of the cluster) is rather difficult experimentally: it may even occur that the question has no clear cut answer, because some of the electrons of the atom are inside, whilst the others are not: a notable example is La, which form objects in which it is fairly well established [673] that most of the metal atom is trapped inside the cage (endohedral capture, denoted as $La@C_{82}$), with a charge state close to $+3$ but the three remaining electrons are attached to the outside of the cage. This is perhaps related to the fact that La possesses a low triple ionisation potential as compared to an alkali. As yet, the three metals La, Y and Sc have been unambiguously observed to be trapped endohedrally in fullerenes [674], but other possibilities undoubtedly exist. Note that they all belong to the same homologous sequence and possess three electrons which are rather easily detached, leaving a very compact triply charged rare-gas configuration.

He atoms, being very small, are also prone to endohedral capture by C_{60}, and a single collision may suffice to form $He@C_{60}$ or $He@C_{70}$, depending on the temperature of the fullerene and the pressure of the surrounding gas [675]. An object of this type is potentially an interesting mixed target for high field laser physics to study the effects of laser pulse propagation and screening.

Finally, we note that there is much interest in so-called *nanotubes* made of C atoms. These appear to be essentially elongated giant fullerenes [676].

12.5   Intercallation of alkalis in fullerite

As remarked in section 11.8, materials formed from fullerenes are of interest to develop as anodes for 'rocking chair' batteries. We describe here some of the relevant properties, which depend not on endohedral capture, but rather on interstitial storage or reversible intercallation.

The fullerenes are very stable, and can be formed into a solid called fullerite by a compact stacking of C_{60} molecules which, for the phase of interest here, crystallises as a face-centred cubic structure of side $14.17\,\text{Å}$. Two interstitial sites are available: an octahedric site (one site per C_{60} of radius $2.06\,\text{Å}$) and a tetrahedral site (two sites per C_{60} of radius $1.12\,\text{Å}$ [677]). Pure fullerite is actually an insulator: states near the gap are primarily formed from π orbitals of the molecule. Since C_{60} has a very high electron affinity, an alkali atom can readily donate an electron to the host in order occupy interstitial sites. Also, since the fullerite structure is held together by weak van der Waals forces between the fullerenes, it is capable of partaking in the 'soft chemistry' described in the previous chapter: the host is able to expand without much change in its structure to accommodate intercalated alkali ions.

The actual mode of insertion and the sites occupied depend somewhat on the alkalis concerned, and this is mainly a question of size. In the case of Li, reduction has been found to proceed through a number of stages, with compositions ranging from 0.5 to 12 Li^+ ions per C_{60}. The first few stages are genuinely interstitial, while $Li_{12}C_{60}$ is a molecular compound with one Li on each pentagonal face, as has been demonstrated theoretically [678] and verified by mass spectrometry [679]. Reversibility can thus be expected mainly from the first stages.

The principal difference between alkali storage in graphite and in fullerite is that, in the former, the alkali ions are sandwiched between layers which move further apart (as in the layered oxides of section 11.8), while, in the latter, expansion is three-dimensional (as in the spinels also discussed in section 11.8). One way to study the relative effectiveness of storage in both materials is to mix alkali-saturated graphite with pure fullerite, and study the transfer of alkali between them [680]. It turns out that the heaviest alkali has the highest graphite affinity, while the lightest tends to transfer completely to fullerite. Fullerite has a higher electron affinity than graphite, but Cs is too large to fit into the tetrahedral site, and can only transfer to the octahedral sites.

When doped with an alkali A, fullerite becomes conducting over a range of concentrations close to A_3C_{60}, but becomes an insulator again around A_6C_{60}. There is also a moderately low temperature superconducting transition (it occurs at around 18 K for K_3C_{60}), which is found to vary fairly linearly with the size of the alkali, being highest for the largest intercallated ions. If, however, the lattice structure is made to change, then this transition disappears, which suggests that it is also connected with 'soft chemistry'.[3]

12.6   Shapes of cluster growth

Although it may seem from the other examples in the present chapter that most clusters should be nearly spherical, molecular clusters can grow in a variety of shapes, consistent with the symmetry of their bonds. Thus

---

[3] It is interesting to note in this connection that many of the so-called high temperature superconductors involve transition metals within their structure.

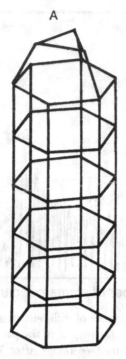

Fig. 12.4. The mode of growth of clusters of Si, according to the hypothesis described in the text (after J.C. Phillips [681]).

arborescent (i.e. 'tree-like') clusters exist. In fig. 12.4, we show the shape which is believed to be most stable for clusters of Si_n, according to the growth hypothesis of Phillips [681]: the ions pile up into a cylinder in layers of six atoms at a time, which can easily be cleaved in the range $6–10 \leq n \leq 100$, leading to reactivity minima for $\Delta n = 6$.

We give this example to stress that stackability does not necessarily imply a tendency for clusters to become spherical, or indeed to tend towards regular solids inscribed within a sphere (Platonic solids). Before making any such hypothesis, one must consider what is the nature of the forces holding the system together, and a variety of possibilities occur. The example given in the present section is still very regular, but a more disordered form of growth (arborescent growth) is also possible.

12.7 Metallic clusters

The clusters of greatest relevance in the present chapter are the so-called metallic clusters, i.e. the clusters formed from metallic elements (whether or not the clusters themselves should be considered as 'metallic' depends

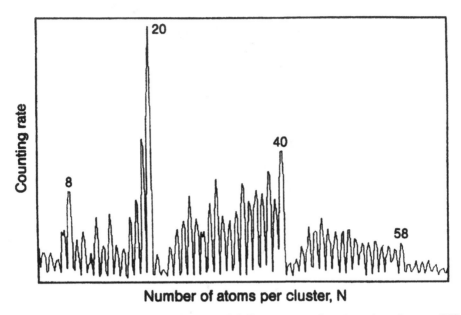

Number of atoms per cluster, N

Fig. 12.5. The relative abundances of different mass fractions in a beam of Na clusters. The discontinuities in the heights of the peaks at 8, 20, 40 and 58 atoms are attributed to a shell structure (see text – after W.D. Knight *et al.* [683]).

on the element concerned and on the criteria one uses). For example, if one forms clusters from alkali metals, then the outermost electron readily delocalises, so that an assembly of ions is formed inside a delocalised group of electrons rather similar to a conduction band in the solid [682]. As previously stressed, all metal atoms do not form clusters of this type.

When such clusters are formed in beams, essentially by many slow collisions of alkali atoms within a nozzle, it is found experimentally (by analysing the mass distribution using a time of flight spectrometer) that there are discontinuities in the intensity distribution of the peaks from which abundances can be computed for each cluster size. These discontinuities correspond to enhanced stability of metallic clusters around specific sizes (8, 20, 40, 58, etc). They are the same for all the different alkali metals, and are therefore referred to as 'magic numbers'; see fig. 12.5. They also turn out to be the same (at least, for the first few magic numbers) as those observed in nuclear physics. This similarity has led to an explanation based on the shell model and to the suggestion that the jellium model can be used to account for the properties of metallic clusters [683].

Metallic clusters are stable enough to be ionised (in some cases, several times, but see section 12.19 for a consideration of their stability) without breaking up. Cluster ions are extremely useful when one wishes to select a species of a specific size, since they can be accelerated and guided by external fields. The magic numbers persist for charged clusters: for a singly positively charged cluster, the number is increased by one with respect to the neutral species.

These facts are explained by the shell model, which is based on the following principles. First, one assumes that the electrons are confined by an approximately spherical effective potential. The simplest realistic potential for clusters is of the Woods–Saxon type (a hypothesis which can be further justified – see section 12.8.1 below), i.e. has the form

$$V(r) = \frac{-V_0}{1 + \exp\left(\dfrac{r - R_0}{a}\right)} \qquad (12.1)$$

where V_0 is the depth of the well, R_0 is the effective cluster radius and a is the depth of the surface layer. The Woods–Saxon potential is, in effect, a square well, or symmetrical short range potential with rounded edges, which is physically a reasonable assumption for an averaged inter-atomic potential. Using this potential, one then solves the Schrödinger equation (i.e. equation (2.2), with the potential in (12.1) substituted for the Coulomb potential) numerically, in order to calculate the energies of the closed shells and, hence, the magic numbers [684]. These are based on the same level-filling rules as the aufbau principle for atoms, but do not come in the same order, because the radial equation (which determines the energies of the levels) involves a long range potential, whose proper-ties are essentially different. Thus, for a Woods-Saxon well, the ordering of the subshells is $1s$, $1p$, $1d$, $2s$, $1f$, $2p$, $1g$, $2d$, $3s$, etc, giving as magic numbers 2, 8, 18, 20, 34, 40, 58, 68, 70, etc.

When the theoretical magic numbers are compared with the discontinu-ities measured in the abundance spectrum (fig. 12.5) or in the ionisation potentials (fig. 12.6), they line up pretty well: for K_n, the observed num-bers are 2, 8, 18, 20, 58, 92, and this is taken to imply that the original assumption of a spherical short range potential was a reasonable one. In fact, the result is not too critically dependent on the form of the poten-tial. Even a square well is a fair approximation, and one can also adapt a harmonic oscillator well, truncated at an appropriate radius, by adding to it a term proportional to ℓ^2, all of which generate very similar orderings of shells and magic numbers.

12.8 Theoretical models for the structure of clusters

Theoretical models to compute the structure and properties of clusters come in two types. First, we have models in which certain properties

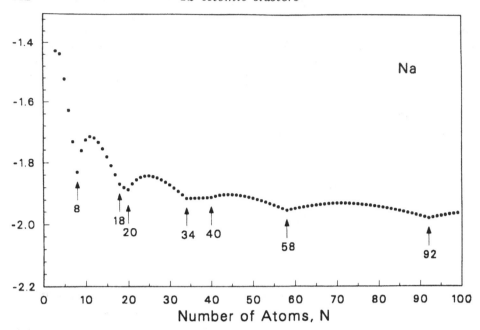

Fig. 12.6. Binding energies for K clusters, showing the discontinuities observed at the magic numbers (after W.D. Knight *et al.* [684]).

such as spherical symmetry are built in from the outset and incorporated as a simplification. Secondly, we have *ab initio* schemes, which set out to discover from first principles, using well-established methods taken over from quantum chemistry, what the optimum shape should be. We give one example of each type of approach.

12.8.1   The jellium model

One can develop a particularly simple scheme by using the assumption of spherical symmetry together with the jellium model of solid state or nuclear physics to compute the effective potential for clusters of different sizes. In this model, the electrons are treated as free particles by analogy with the conduction band of the solid and the ionic structure within the cluster is completely neglected. This obviously results in a great simplification of the problem, especially if the system is spherical, and might be thought too drastic an approximation. In fact, the jellium model only applies to a specific class of clusters (which we call metallic), but was of enormous importance to the history of the field as it revolutionised cluster physics.

For cluster physics to exist as a coherent subject, it is necessary that the properties of a cluster of one given size should not be totally different from those of another of a different size made up from the same atoms (this is

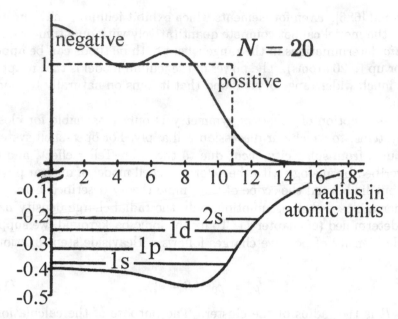

Fig. 12.7. Radial potential and charge density for an alkali cluster of 20 atoms, as obtained from the jellium model. Note the Friedel oscillations in the density of electronic charge. The ion density is assumed to be a 'top hat' function (after W. Ekardt [685]).

the concept of *stackability* referred to above). The Hückel model (see sub-section 12.8.2), which emphasises the ionic framework, does not provide such a situation: clusters appear essentially similar to molecules, each one with separate properties. The jellium model, on the other hand, by completely neglecting ionic structure, provides one approach to a genuine cluster theory. When it became apparent that a class of clusters exists which follows the predictions of the jellium model rather well, and in particular that the shell structure implicit within the jellium model persists from small to very large cluster sizes, experimental proof was provided that, under certain conditions, the ionic structure is indeed unimportant, and therefore that metallic forces allow a viable cluster sequence to exist.

Clearly, the validity of this approach will depend on the nature of the atoms involved. From the standpoint of a chemist, the obvious flaw of the jellium model is its total neglect of ionic structure. The model requires that the valence electrons should be strongly delocalised. This can only be true for certain metals which are very good conductors. It is also favoured if the ionic background is easily perturbed, in which case electronic single particle energies determine the structure. Finally, it tends to apply better when the wavefunctions have s character and when binding is non-directional [686]. Even for elements which exhibit jellium behaviour most clearly, the model cannot compete quantitatively with quantum-chemical *ab initio* determinations in the range where both methods can be applied (i.e. for up to 20 atoms). The power of the jellium model is that it applies over a much wider range of sizes and that it joins on naturally to models for solids.

The assumption of spherical symmetry is only reasonable for closed-shell systems, to which our discussion will apply. For open-shell systems, departures from sphericity occur due to the Jahn–Teller effect, and can be described by analogy with the deformed-shell model of nuclear physics [687], but lie beyond the scope of the simple theory described here.

Under a closed-shell assumption, only the radial charge density needs to be determined (cf chapter 1). To fix the ionic background, we suppose that the density of positive charges follows a Heaviside step function Θ thus:

$$\rho^+ = \rho_0 \Theta(R - r) \tag{12.2}$$

where R is the radius of the cluster. The purpose of the calculation is to determine the electronic density ρ^-, and this is achieved by a self-consistent procedure very similar to the one used for atoms.

We note that the potential $V(r)$ experienced by the electrons depends on *both* ρ^+ and ρ^- as follows:

$$V(r) = 2 \int_0^r \frac{\rho^-(r') - \rho^+(r')}{|r - r'|} + V_{exch}(r') 4\pi r'^2 dr' \tag{12.3}$$

in atomic units, where $V_{exch}(r)$ is the exchange term. It turn out that this term can be approximated as an explicit function of r:

$$V_{exch}(r) = -1.22 X(r) - 0.66 \ln\left(1 + \frac{11.4}{X(r)}\right) \tag{12.4}$$

where

$$X(r) = \left(\frac{3}{4\pi\rho^-(r)}\right)^{1/3}$$

Since $V(r)$ can be expressed explicitly as a function of the density, one can set up an iterative SCF procedure in which the radial Schrödinger equation is solved to determine the eigenfunctions $\psi_i(r)$, then the density $\rho^-(r)$ is determined from

$$\rho^-(r) = \sum_i |\psi_i(r)|^2 \tag{12.5}$$

where the sum is taken over all the occupied orbitals, and the process is repeated until convergence is obtained.

This was first done by Eckardt [685], and an example from his results is shown in fig. 12.7. Note that the potential curves have the expected shape, and extend well outside the radius R of the positive charge background (which is assumed to be fixed). Their form is most consistent with the original description of a Woods–Saxon potential (see previous section) and provides some further justification for using it. We also learn in this way that, for a sequence of clusters of different sizes, the depth of the well remains roughly constant, while the radius grows. Of course, such a potential cannot be completely realistic. In particular, it should not be 'flat' inside the well, but should exhibit local narrow wells near each ionic centre. The assumption of a 'top hat' ionic charge distribution with sharp edges is also unphysical.

The electronic charge distribution turns out (see fig.12.7) to exhibit oscillations, called Friedel oscillations, due to the sharp edge in the positive charge distribution at the surface of the cluster. It is possible to take the jellium model somewhat further than described here by allowing for the back reaction of the negative charge distribution on the positive charges in the self-consistent procedure. Note, however, that the jellium model is not appropriate for all clusters made from metal atoms.

12.8.2 *The Hückel model for alkali clusters*

An alternative to the spherical jellium approximation just described is to use the tried and tested methods of theoretical chemistry, namely the energy variational principle, to determine the most probable geometrical structure for atomic clusters. This is the basis of the Hückel method, a rough outline of which is as follows.

We suppose that there is only one delocalised electron per site (an appropriate assumption for an alkali cluster). The Hückel Hamiltonian for the cluster can then be written as:

$$\hat{H} = \underbrace{(E_{AT} + \omega) \sum_r |r><r|}_{1} + \underbrace{\beta \sum_{r,s} |r><s|}_{2} \qquad (12.6)$$

summed over sites. The first term is diagonal, and is responsible for most of the energy, while the second one is an off-diagonal term, whose evaluation is restricted to nearest neighbours in the tight-binding approximation. This Hamiltonian yields eigenvalues of the form

$$\epsilon_k = E_{AT} + \omega + \alpha_k \beta \qquad (12.7)$$

where the α_k are geometrical coefficients and ω, β are parameters. The total energy of the cluster of N atoms is then

$$E = N(\omega + E_{AT}) + \underbrace{\sum_{\text{all } k} \alpha_k}_{\text{occupied sites}} \beta \qquad (12.8)$$

The cohesion energy is defined as the difference between the total energy and the energies of the individual atoms, i.e.

$$E_{coh} = \mid E - N E_{AT} \mid = -N\omega + \mid \beta \mid \sum_k \alpha_k \qquad (12.9)$$

E_{coh} can then be minimised for clusters of one-electron atoms by starting from an arrangement with equal binding lengths and exploring all possible geometries. An example is shown in fig. 12.8.

   This has been done for $N \leq 5$ by Joyes [688], and for $N \leq 14$ by Wang and George [689] and Lindsay *et al.* [690]. As a byproduct of such calculations, one can discover whether *isomers*, i.e. different structures of nearly equivalent energy, occur, in which case one would expect the cluster beam to be a mixture, containing the different forms. If there is some difference in energy between them, then one may be able to vary the proportion of isomers in the beam by cooling it to a lower temperature There are of course much more elaborate models than the very simple one outlined above, and many more interactions may be included. All of them, however, make use of an energy variational scheme. By such methods, it is possible to compute stable structures and isomers for the alkali clusters [692]. An example of such calculations is shown in fig. 12.9. It is interesting to note that, while some clusters exhibit a high degree of symmetry, they are far from being spherical. This is visible for the isomers of Na_{20} in fig. 12.9, which would correspond to a closed-shell situation according to the model expounded in the previous section. One way of reconciling the models is to note that the actual observations are all performed at finite temperature, where the spectra of different isomers are superimposed (see section 12.17). Note that the Hückel method has wide applicability and is not confined to metallic clusters.

12.9   The transition from the atom to the solid

We have mentioned several times already that clusters should allow one to bridge the gap between the atom and the solid, and to track properties from one to the other, so that phase changes and transitions of various kinds can in principle be mapped out experimentally. This constitutes an important application of cluster physics, and we now give a concrete example.

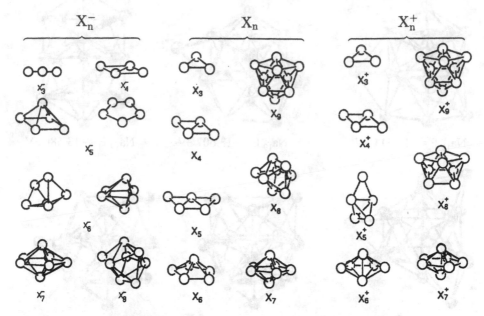

Fig. 12.8. The most stable isomers obtained by the Hückel method for the first few clusters of one-electron atoms and their ions (after P. Joyes [691]).

The problem of interest is one which was raised in chapter 11, namely instabilities of valence which can occur when measuring the $4f$ shell occupancy of lanthanide elements in the condensed phase. As commented previously, several elements in the sequence have different valences, and therefore different occupation numbers in the free atom and in the solid. Table 11.2 gives a detailed classification of lanthanide elements based on the number of $4f$ electrons in the free atoms and in the condensed phase, while table 11.3 gave the change Δn_f in the number of $4f$ electrons in going from the atom to the solid.

One can classify the rare earths into four distinct groups, according to the number q of $5d$ and $6s$ valence electrons:

group 1: $q = 3$ in atom, $q = 3$ in solid La, Gd;

group 2: $q = 2$ in atom, $q = 3$ in solid Nd, Pm, Dy, Ho, Er;

group 3: $q = 2$ in atom, $q = 2$ or 3 in solid Sm, Eu, Tm, Yb;

group 4: $q = 2$ in atom, $q = 3$ or 4 in solid Ce, Pr, Tb.

Groups 3 and 4 are of interest from the standpoint of valence transitions. For compounds containing elements from group 3, at least one of the valence states in the solid involves no change in the number of valence electrons from the free atom. Thus, we only expect to see an appreciable change in the core-level spectra on going from metal vapours to trivalent

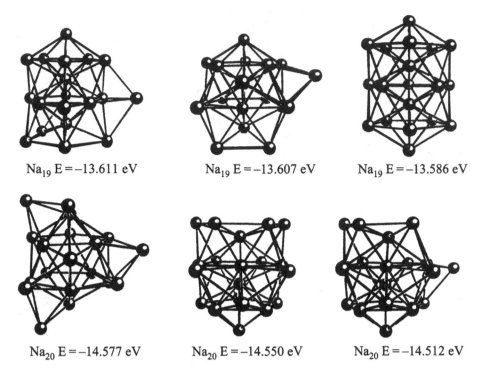

Na_{19} E = −13.611 eV Na_{19} E = −13.607 eV Na_{19} E = −13.586 eV

Na_{20} E = −14.577 eV Na_{20} E = −14.550 eV Na_{20} E = −14.512 eV

Fig. 12.9. Structures of Na clusters as obtained by *ab initio* calculations (see text – after R. Poteau and F. Spiegelmann [692]).

compounds.

Group 3 elements are therefore the most interesting to us, since a direct comparison with atomic spectra is fruitful. Because of the small exchange interaction for deep core hole excitation, the most favourable core hole spectra to study are those of the $3d$ subshell.

We now consider the $3d$ spectra of Sm and Tm. The simplest case is that of Tm, because it has 13 $4f$ electrons in the free atom, so that only one transition is allowed in photoabsorption from the ground state, *viz.*

$$3d^{10}\dots 4f^{13}6s^{2\,2}F_{7/2} \rightarrow 3d^{9}\dots 4f^{14}6s^{2\,2}D_{5/2}$$

which fills the valence shells of the atom. On the other hand, if instead of 13 outer electrons, the Tm atom has only 12 (see table 11.2), then more than one transition can occur. Thus, one can deduce by inspection of the core-excited spectrum if the valence of Tm has changed.

This situation is illustrated in fig. 12.10, which shows how the observed spectrum evolves for different clusters, interpolating between the spectra of the atom and of the solid (see [693]).

The first evidence that valence changes can be precipitated in this way for lanthanide clusters was obtained by Lübcke *et al.* [694] for clusters of

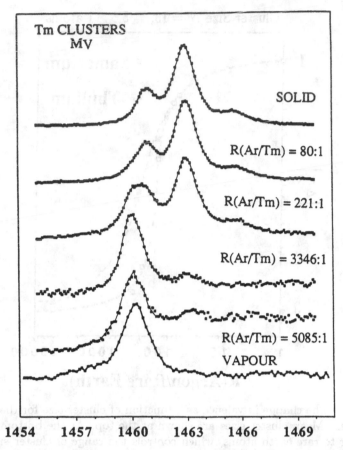

Fig. 12.10. Evolution of the $3d$ spectrum of Tm as a function of the cluster size, showing how the change in valence is observed (after J.-P. Connerade and R.C. Karnatak [637]).

Pr, Nd and Sm. However, they probed the $2p$ shell spectra, and so their data did not reveal the properties of the $4f$ electrons directly.

In all these experiments using rare-gas matrices, clusters of different sizes were produced by varying the relative concentrations of Tm and rare-gas atoms. In reality, such clusters are not mass-selected, but represent a narrow distribution over a range of sizes, but the peak in the distribution shifts according to the relative concentration, which is given alongside each spectrum in the figure. At the top and at the bottom of the figure are, respectively, the spectra of the solid and of the free atom. What is immediately apparent from the figure is that a change of valence occurs as a function of cluster size, with a state of more than one valence being present for the larger sizes and in the solid. It is possible, from a

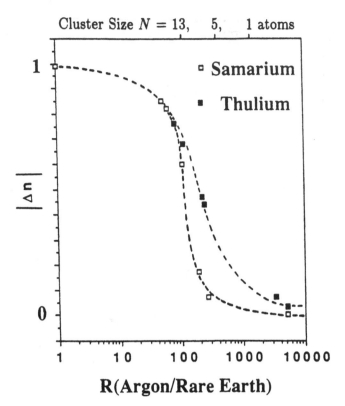

Fig. 12.11.   The change in valence as a function of cluster size for the elements Tm and Sm. Mean cluster sizes are shown at the top. At the bottom, R is the ratio of Ar to rare-earth atoms, which controls the range of cluster sizes (after C. Blancard *et al.* [693]).

study of the EXAFS spectrum of the sample (see section 11.14) to deduce the number of atoms in the first coordination cell, and one can plot the valence transition of table 11.2 as a function of cluster size. More recently, these plots have been checked against data obtained for free, mass-selected cluster beams by a molecular oxidation method, and the range of cluster sizes has been confirmed. Studies have also been conducted for Sm, for which a valence change was also observed, by using the kind of spectral analysis illustrated in fig. 11.4. The resulting curves are shown in fig. 12.11.

In such experiments, one should also note that the clusters were not free, but were deposited or frozen into a solid rare-gas matrix or *host*, so as to achieve a high enough density of clusters to observe the soft X-ray spectrum using a synchrotron radiation source. This technique, called *matrix isolation spectroscopy*, is often used to probe the interaction between single atoms and the host. Such experiments are open to the criticisms that: (i) the valence change might not be the same as for free clusters; and (ii) the number of atoms in the clusters is not quite exact, since they are not mass-selected, and is determined only indirectly by the EXAFS study.

In more recent experiments the properties of partially oxidised clusters of lanthanide atoms have been studied, in mass-selected beams of clusters [695]. What was studied in this case was not the optical spectrum but the mass spectrum, observed by time of flight spectroscopy, from which the degree of oxidation, and hence the valence, could be deduced as a function of the cluster size. The results are very similar to those of [693] and confirm not only the nature of the valence change, but the range of cluster sizes over which it occurs. Since not only the environment of the sample, but also the detection technique are different, this agreement provides important experimental confirmation. One can also account for the different reactivity with O of the atoms in the cluster according to valence by using a double-well model: the chemical activity of an electron is far greater when it is in the outer well.

The results this section are very relevant to the discussion in chapter 11 of the Anderson impurity model and the quasiatomic orbital collapse model. It is significant, in particular, that the transition occurs for rather small clusters, smaller than might perhaps be expected for the impurity model to be applicable. The quasiatomic model provides a straightforward explanation for the varying degrees of oxidation observed in [695]: the chemical activity of the lanthanide atoms is greater when the orbitals are in an expanded or outer-well state than when they are in a contracted or inner-well one. On the other hand, an important issue which needs to be determined is over what range of cluster sizes an effective conduction band actually appears, since its presence provides the hybridisation forces which play a crucial role in the impurity model.

12.10 Mass selection of clusters

In the previous section, we have remarked how important it is to achieve proper mass selection in cluster physics, since otherwise the experimental results are difficult to interpret. In an ideal experiment, one wishes to be sure that only clusters containing a definite number of atoms are present in the interaction region. It is a triumph of experimental technique that conditions very close to this ideal have actually been achieved.

In experiments on beams of mass-selected clusters, the difficulty is that beams already contain extremely low densities of clusters (typically 10^8 clusters per cm^3). After mass selection, the number of clusters will have

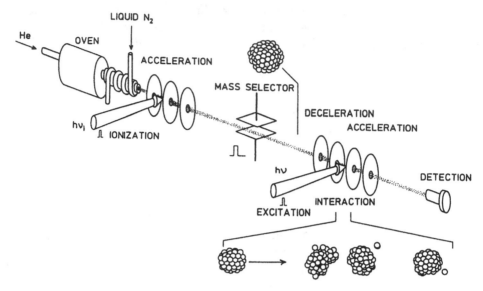

Fig. 12.12.   The double time of flight arrangement used to achieve mass selection of clusters by Bréchignac and collaborators (after C. Bréchignac and J.-P. Connerade [714]).

dropped by several orders of magnitude, and may be around 10^5 per cm^3, thus making detection extremely difficult.

The method used by Bréchignac and her collaborators, though simple in principle, requires a high detection efficiency. Its basic operation is as follows. First, clusters are produced by the usual techniques in a standard cluster source, shown at the left hand side of fig. 12.12. Next, these clusters are ionised by a short burst of radiation from an ultraviolet laser. This pulse provides a time marker, and subsequent events occur with a time delay which is referred to this marker. As previously noted, singly ionised clusters are stable, and can therefore be accelerated to a given velocity v, given by $\frac{1}{2}M_c v^2 = eV$, where M_c is the mass of the cluster and V the applied voltage. There follows a flight tube, about 1 m long, in which the clusters travel. Because of their different masses, different cluster fractions reach different velocities for a given applied voltage and can therefore be separated from each other in time. In practice, a drift tube about 1 m long allows good separation on a time scale of microseconds. Beyond this first tube, a pair of deflecting electrodes is placed, to which a pulse is applied in such a manner that *only* a given mass fraction is allowed through. The mass-selected beam is then probed by a second laser, and the resulting fragmentation products can then be analysed with a second drift tube of similar type.

This method of mass selection [714] has been used by Bréchignac and her colleagues for a number of important observations of the spectra of clusters, some of which are described in the following sections.

12.11 Collective oscillations in clusters

The shell model for metal clusters, described above, has an important implication which will not have escaped the reader: if electrons become delocalised from individual atoms and can roam freely over the whole cluster to form a closed shell, then this shell should be able to oscillate collectively, and should therefore exhibit giant dipole resonances analogous to those which were described in chapter 5 for free atoms.

There is, however, an important difference between the two. An atomic shell has a radius of the order of 1 Å so that the charge density is roughly 2.5×10^{24} cm^{-3} for a shell of ten electrons, whereas a cluster has a radius of the order of 10 Å, leading to a charge density of about 2×10^{21} cm^{-3}, for a closed shell of eight electrons, i.e. about 800 times smaller. According to the Drude model (see section 12.16), the frequency of collective oscillation scales as the square root of the electron density, and should therefore lie a factor of about $\sqrt{(800)} \sim 27$ lower than the atomic $4d \to f$ giant resonances, which occur around 100 eV. This is of course only a *very* crude estimate, because the Drude model does not really apply to atoms and the sizes quoted are very rough, ball park figures, but indicates that the collective resonance of the whole cluster shell should lie at around 3 eV, as compared with a much higher energy for the atomic giant resonances.

Of course, since clusters are made up of atoms, and since atomic giant resonances are localised excitations which survive even in solids, they must clearly be present in clusters also. We thus have the possibility of observing two kinds of collective resonance, one localised on individual atoms, and the other delocalised on the whole cluster, within the same sample.

12.12 Atomic giant resonances in clusters

Atomic giant resonances in clusters were observed by Bréchignac *et al.* [696] for clusters of Sb. Since the spectrum of the corresponding free atom was not available, the 'universal curve' deduced from the uncertainty principle (see section 5.16) was used for analysis. The resulting point (see fig. 12.13) shows that the observed resonance fits closely onto the line for free atoms.

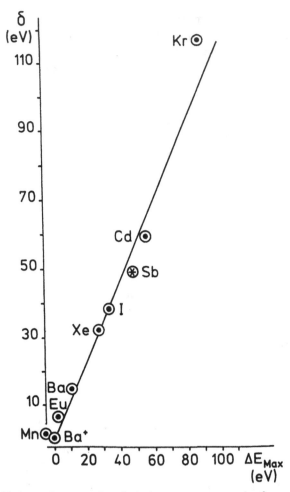

Fig. 12.13. Universal curve for the giant resonances in free atoms with the point for the quasiatomic resonance in clusters of Sb included (adapted from J.-P. Connerade [225]).

The resonances exhibit interesting variations as a function of atomic number, the origin of which is unexplained. As can be seen in fig. 12.14, their intensity relative to the rest of the spectrum fluctuates strongly, with the giant resonance actually disappearing from view for $n = 8, 12, 16$, etc atoms in the cluster. This behaviour seems in some way to be related to the fact that Sb_n clusters tend to be made up from smaller groups of four atoms stuck together.

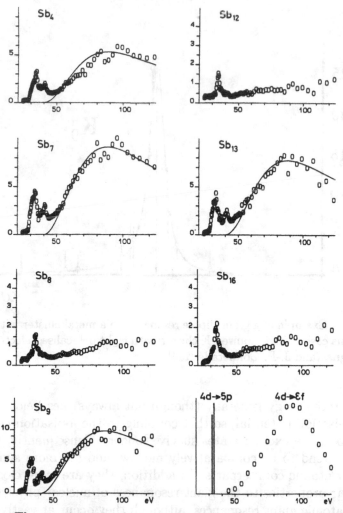

Fig. 12.14. The quasiatomic giant resonances in Sb clusters of different sizes. Note the marked dependence of the amplitude of the giant resonance on the number of atoms in the cluster (after C. Bréchignac *et al.* [696]).

12.13 Giant resonances intrinsic to metallic clusters

The second, and more important kind is the giant dipole resonance intrinsic to the delocalised closed shell of a metallic cluster. Such resonances have received a great deal of attention [684]. They occur at energies typically around 2–3 eV for alkali atoms, and have all the features characteristic of collective resonances. In particular, they exhaust the oscillator strength sum rule, and dominate the spectrum locally.

They have certain specific properties as compared with the atomic case:

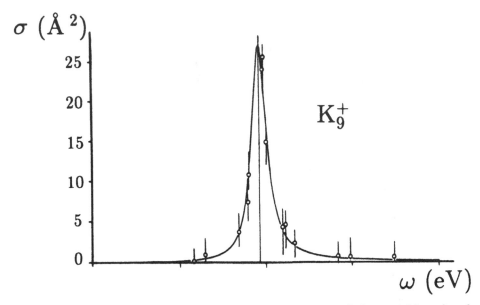

Fig. 12.15. Example of a giant dipole resonance in a metal cluster with a closed shell, in this case a singly ionised K cluster with eight delocalised electrons (after C. Bréchignac and J.-P. Connerade [714]).

the resonance energy generally (though not always) lies below the associated ionisation potential, so that coupling to the ionisation continuum is weak, and the excited states survive longer. Consequently, the giant resonances tend to be comparatively narrow, and are more symmetrical than their atomic counterparts. In addition, they are more closely similar in their properties to the giant resonances of nuclear physics than to the quasiatomic giant resonances, although they occur at vastly different resonance energies.

The fact that they tend to be fairly symmetrical (at least when they occur below the ionisation threshold) is related to their time characteristics: from the lifetime widths and resonance energies, one can deduce that the giant resonances in metallic clusters are many-body oscillations undergoing several periods. Giant resonances in metallic clusters can truly be considered as plasmons, and relate quite clearly to surface plasmons in solids.

In fig. 12.15, we show a typical resonance, as obtained for closed shell clusters. This similarity is believed to extend to their structure (although some experimental evidence indicates a more complex situation – see section 12.17). In particular, it is very interesting to consider departures from the purely spherical shape, when the shell has a number of electrons

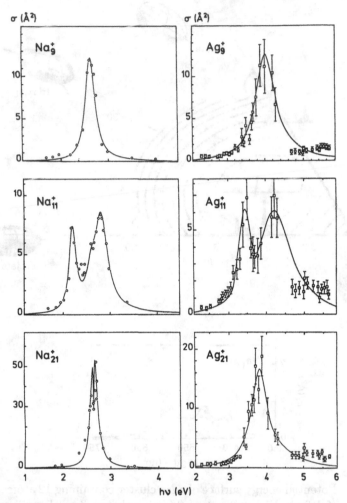

Fig. 12.16. Examples of giant resonances in clusters containing magic and nearly-magic numbers of delocalised electrons, showing how the giant resonances split into structures indicative of the symmetry of the system (after C. Bréchignac and J.-P. Connerade [714]).

close to but not equal to a magic number. For a spheroidal cluster, there are two distinct frequencies, corresponding to modes of oscillation, either parallel or perpendicular to the axis of deformation. Thus, the giant resonance splits into two resonances, one of which (the perpendicular case) has twice the spectral weight of the other because it contains two modes whereas the other (the parallel case) contains only one. This behaviour is clearly observed in the experimental data for clusters made up of atoms with one valence electron (see fig. 12.16). More generally, the indepen-

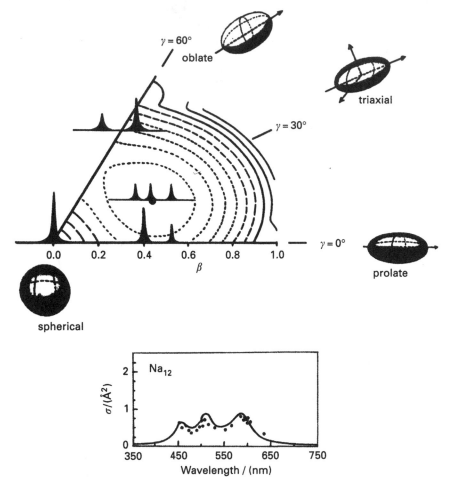

Fig. 12.17. Potential energy surface of a Na cluster containing 12 atoms, plotted in the plane of deformation parameters β (giving the quadrupole moment) and γ (the departure from axial symmetry). By averaging over an ensemble, the actual spectrum is obtained, and is compared with experimental data below the graph (after R.A. Broglia [702]).

dent particle model predicts that strongly deformed clusters possess three different radii along the principal axes. One then finds three different frequencies of the giant resonance. An example occurs for Na_{12}, and has been analysed by Pacheco and Broglia [701]: it is illustrated in fig. 12.17, which also contains a comparison with the data of [707].

The fullerenes also possess a shell of delocalised electrons: each C atom contributes four delocalised electrons, constrained to move on a sphere roughly one atom thick. This shell possesses a prominent giant resonance whose collective nature has been studied [703]. In this respect, fullerenes are similar to metallic clusters, although the energy range involved is different.

12.14 Electron attachment to clusters

All the methods and principles relating to electron attachment to atoms and molecules (see sections 2.19 and 2.24) are transferable to neutral clusters. A universal method of detection is mass spectroscopy of the negative ions produced. However, a cluster is a particularly fragile object, and so the most gentle experimental techniques for producing negative ions assume special significance. In particular, the method involving transfer of electrons by collisions with atoms in high Rydberg states (section 2.19) is especially suitable, and has been extensively applied, both to molecular [704] and to alkali clusters in which Jahn–Teller deformations of the cluster shape have been shown to result in an even–odd alternation in the distribution of cross section versus cluster size [705]. Collisions with low energy electron beams have also been used [706]. Since the electron energy is then much higher, it becomes necessary to consider how the system relaxes, both during and after the attachment process. Here, the distinction between van der Waals and metallic clusters reemerges. In a van der Waals cluster, the electron is first accommodated in an extended orbit, which then relaxes towards its ground state by localisation of the captured electron and an associated nuclear rearrangement [704]. Capture and relaxation are then treated as separate processes, occurring on different timescales. Metallic clusters are expected to behave in quite a different way, because of the high mobility of the delocalised electrons. This is considered in the next section.

12.15 Polarisation and negative ions of clusters

Clusters are very readily polarised, and, because of the resulting dipole (cf. section 2.26), can easily attach an extra electron to form negative ions. Giant dipole resonances involving the whole cluster are of course a measure of the dynamical polarisability of the whole system. When a charged particle approaches, a cluster is polarised, and the axis of the dipole points along the interaction line connecting the projectile and the target. As the charged particle moves past the cluster, the dipole rotates. A rotating dipole radiates, and the emitted light is called *polarisation radiation*. As a result of this radiation, the projectile slows down, and thus the polarisation of the target contributes to the crosssection for bremsstrahlung. Electronic relaxation thus occurs during the capture process: since the projectile is slowed down, the probability of capture is enhanced, and the complete expression for the cross section for radiative capture into a final state of zero angular momentum is [708]:

$$\sigma = \frac{4\omega^3}{3p^3c^3} \, |< n_f 0 \, | \, r \, | \, p1 > + \frac{2}{\pi} \int_0^\infty q\alpha(\omega, q) < n_f 0 \, | \, j_1(qr) \, | \, p1 >^2 dq$$

(12.10)

where p is the momentum of the incident electron, $\alpha(\omega, q)$ is the dynamical polarisability of the cluster, $\omega = | \, \varepsilon_f \, | + p^2/2m$, where ε_f is the binding energy of the electron after capture, $| \, p1 >$ and $| \, n_f 0 >$ are the radial wavefunctions of the initial (angular momentum 1) and final (angular momentum 0) states of the incident electron and $j_1(qr)$ is the Bessel function.

The first term represents the usual mechanism for radiative capture, while the second results from the polarisation mechanism. The ratio η of the partial cross sections of the two mechanisms of capture is approximately $\eta \sim 4 \, | \, \alpha_d(\omega) \, |^2 \, / \pi^2 R_f^3$, where R_f is the radius of the final state orbit of the electron. Within a giant resonance, the dipole dynamical polarisability is mainly determined by its imaginary part $\mathcal{J}m\alpha_d(\omega) = c\sigma_\gamma(\omega)/4\pi$, where $\sigma_\gamma(\omega)$ is the photoabsorption cross-section of the cluster.

In principle, this mechanism of enhanced radiative capture exists for atoms as well as for clusters. However, there are two reasons for which it is much more significant in the case of clusters. First, the energy of a giant resonance is much lower in a metallic cluster than in an atom, so that thermal electrons can more readily excite it. Second, the electron cloud in a cluster is more readily deformable, i.e. its dynamical polarisability is much larger. If typical values for metallic clusters are substituted, ratios of 10^4 or 10^5 are obtained, which show that the giant resonances play a very important role in the radiative capture of low energy electrons to form negative ions. Since the fullerenes possess giant resonances due to the polarisability of the shell of delocalised electrons, it is perhaps no surprise that they are also very efficient at capturing electrons [698].

This mechanism for capture is a very natural one for the formation of negative ions, since it provides a means of radiating the excess energy of the system which is directly associated with the dipole field of the target, and this dipole field is also responsible for attachment.

From equation (12.10), one finds that interference between the direct bremsstrahlung process and polarisation radiation results in an asymmetry of the giant resonance profile observed in fluorescence with electron excitation even when the corresponding photoabsorption profile is symmetrical. This is exactly analogous to the Fano resonances (chapter 6): bremsstrahlung plays the role of the continuum, while the resonant channel arises from dynamic polarisation of the giant resonance.

Before leaving this subject, it is worth noting that two oppositely charged clusters can also orbit around each other, giving rise to a new class of Rydberg system composed of two very massive 'particles.' This is an interesting problem because it lies at one of the limits between quantum and classical physics. Such a system (known as a Rydberg cluster) is claimed to be stable in a high angular momentum state [697], although effects due to polarisation radiation do not yet appear to have been considered.

12.16 The Mie solution and the Drude model

When a plane electromagnetic wave is incident on a spherical distribution of charge of radius R and dielectric constant ϵ_1 (which is in general complex), vibrations are set up, and an absorption cross section can be calculated from Maxwell's equations. The Mie solution for the cross section is

$$\sigma(R, \lambda) = \pi R^2 \left(\frac{2\pi}{\lambda}\right) \mathcal{I}m \left(\frac{\epsilon_1 - \epsilon_0}{\epsilon_1 + 2\epsilon_0}\right) \qquad (12.11)$$

This solution can be regarded as the classical limit for the response of a spherical cluster. A simple form for the dielectric constant is given by the Drude model:

$$\epsilon_1 = \epsilon_0 \left(1 - \frac{\omega_p^2}{\omega^2 + i\gamma\omega}\right) \qquad (12.12)$$

where ω_p is the plasmon frequency given by $\omega_p^2 = 4\pi n_e e^2/m$ and n_e is the electron density, while γ is a damping factor, which, for example in a metal, would be taken as the electron–phonon collision time. One thus obtains

$$\sigma(\omega) = \frac{4\pi n_e^2}{mc\epsilon_0} \frac{\gamma\omega^2}{\left\{\omega^2 - \left(\frac{\omega_0}{\sqrt{3}}\right)^2\right\}^2 + \gamma^2\omega^2} \qquad (12.13)$$

which represents a collective resonance, because the dielectric function of Drude describes the response of the whole sphere of charges. Under these conditions, a measurement of ω_0 is a determination of the electron density n_e. The fact that, to a first approximation, the frequency remains fairly constant as the cluster size increases shows that n_e does not change very significantly with size.

Small variations in ω_0 with size can be interpreted as a 'spillout', i.e. that some of the electronic cloud extends out from the cluster to a greater or lesser degree as a function of the size.

The Mie solution given here assumes a linear response: if the classical oscillator is driven hard, anharmonic terms appear which give rise to a nonlinear Mie scattering, similar to effects considered in chapter 9. The perturbative theory of nonlinear Mie scattering has been given for spherical metal clusters, and gives rise to second and third harmonic generation [699].

12.17   Influence of temperature

It was shown in section 12.8.2 that different isomers may occur for a given number of atoms in a cluster. For small differences in energy, the shape of a cluster may alter significantly. Bearing in mind that clusters are usually made by collisions in beams at fairly high temperature, many different structures can be present at the same time, so that the spherical form for clusters with closed shells is only an average over many other shapes.

This interpretation helps to reconcile the different models: on the one hand, we have those inspired by quantum chemistry, with clearly defined structures and, on the other, statistical mean field approaches such as the jellium model. Clearly, the two do not apply together, but, by averaging over many isomers, one can arrive at a nearly-spherical shape at finite temperature.

Experimental evidence supporting this view comes from measurements in which the cluster beam is cooled to low temperatures [709]: it has been found that the giant resonance splits up into more than one component even for clusters containing magic numbers of atoms.

There are two possible explanations. First, isomers can become resolved as the system is cooled down. Second, one may suppose that the shape is indeed spherical, but that the Drude model, being classical is not quite applicable: indeed a fully quantum-mechanical treatment shows that the mode structure should be more complex than suggested above.

One conceptually simple approach which has been used to represent temperature effects in metallic clusters is the random matrix model, developed by Akulin *et al.* [700]. The principles of the random matrix model, developed in the context of nuclear physics by Wigner and others, were outlined in chapter 10. The essential idea is to treat the cluster as a disordered piece of a solid. In the first approximation, the cluster is regarded as a Fermi gas of electrons, moving in an effective, spherically symmetric short range well. Without deformations, one-electron states then obey a Fermi distribution. As the temperature is raised, various scattering processes and perturbations arise, all of which lead to a random coupling between the states of the unperturbed system. One can thus introduce a mean-square interaction, i.e. a single parameter to represent the transformation between a 'cold' and an interacting situation. This perturbation is then treated by the random matrix method. The mean square interaction is a disorder parameter containing three contributions. The first depends on electron–electron and electron–ion contributions which are not included in the Fermi gas model. The second comes from scattering processes (phonons) and therefore depends on the temperature. The third is a shape effect, which depends on the magnitude of the deformation of the cluster.

The sum of the first two terms can be estimated by comparison with the known properties of metals, and thus one can study the influence of the shape. For each cluster size, it turns out that there is a critical temperature, which corresponds to the minimum energy. Below this temperature, the cluster is deformed (i.e. nonspherical) and its electronic energy is minimum. Above this temperature, the cluster becomes spherical, and its energy then begins to depend on the temperature. The implication is that there is a phase transition at a critical temperature, which will be zero for a closed shell (or naturally spherical cluster) and maximum near a half-filled shell.[4]

12.18   Applications of many-body theory to clusters

The appropriate theoretical framework for calculating giant resonances in clusters in the quantum limit is the random phase approximation (RPA) and its extension including exchange (RPAE), discussed in chapter 5. The many-body theory developed originally for nuclei (RPA), and then extended for giant resonances in atoms (RPAE) should contain all the necessary ingredients for a proper description of giant resonances in clusters, but the difficulty is that the potential is not completely known, and that it is more complex than the simple Woods–Saxon shape used in elementary calculations by the shell model. There is currently much activity in developing *ab initio* theoretical descriptions of spherical metal clusters. A full reference on this subject is the book by Bertsch and Broglia [713].

12.19   Coulombic fission of clusters

A cluster may become not only singly, but doubly charged, and the question then arises of whether the repulsive forces set up are large enough for fission to occur. The problem is similar (but not identical in detail) to that

---

[4] The theory neglects pertubations involving spins.

of Coulomb explosions in molecules, which was discussed in section 7.7. It is also related to the properties which clusters may exhibit when placed in a strong AC field, again by analogy with Coulomb explosions in strong fields, discussed in section 9.23.5.

It has been found experimentally that there exists a critical size of stability n_c below which doubly charged cluster ions do not exist. This is true for both heterogeneous (mixed) clusters and homogeneous clusters. Thus, for example, doubly ionised K clusters do not exist with fewer than 19 electrons ($n_c = 21$) [710]. The reason is intuitively obvious, since the repulsive Coulomb forces increase with decreasing cluster size. However, energy considerations alone do not suffice to achieve an understanding of the fission process. As pointed out by Bréchignac *et al.* [710], Coulombic fission competes with other fragmentation processes in metallic clusters, and the dynamics of the process must also be considered.

Staying with K as our example, the fission process

$$K_N^{2+} \rightarrow K_{N-p}^+ + K_p^+$$

and the 'evaporation' process in which a single neutral atom is ejected:

$$K_N^{2+} \rightarrow K_{N-1}^{2+} + K$$

coexist and compete. By observing the fragmentation products it is possible to study the relative contributions from each pathway.

In this way, a model has been developed to represent the competition between fission and fragmentation within the metallic drop approximation, analogous to that used in nuclear physics.

Within this model, the energy change Δ_N^{Z+} for the dissociation of a Z times ionised cluster containing N atoms is

$$\Delta_N^{Z+} = -a_S \left[N^{2/3} - (N-p)^{2/3} - p^{2/3} \right] - \frac{e^2}{r_S} \left[\frac{\alpha}{N^{1/3}} - \frac{\beta}{(N-p)^{1/3}} - \frac{\gamma}{p^{1/3}} \right]$$

where the first term is a surface term, corresponding to the change in energy due to the extra surface created by the dissociation process, while the second is a charge-dependent term representing the Coulomb repulsion.

If the dissociation process of the cluster A_N is written as:

$$A_N^{Z+} \rightarrow A_{N-p}^{(Z-q)+} + A_p^{q+}$$

The quantities α, β and γ are given by:

$$\alpha = \frac{Z^2}{2} - \frac{Z}{8}, \beta = \frac{(Z-q)^2}{2} - \frac{Z-q}{8}, \gamma = \frac{q^2}{2} - \frac{q}{8}$$

For clusters, there is no *a priori* relation between Z and N, except that, usually, $Z << N$. Thus, the surface term predominates. Since the surface variation is smallest when $p = 1$, there is a tendency for evaporation to

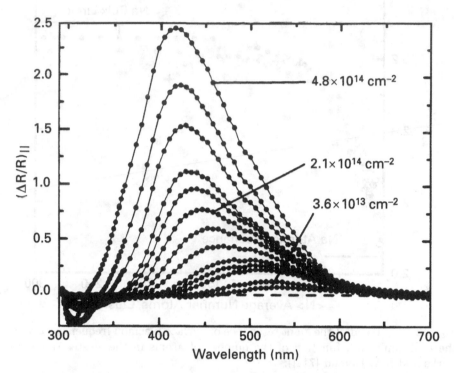

Fig. 12.18. Evolution of the giant resonances for adsorbed Na clusters on a surface at various surface densities (after J.H. Parks and S.A. Donald [712]).

predominate. On the other hand, for large doubly charged clusters, the negative contribution of the Coulomb term leads to assymmetric fission processes, corresponding to the ejection of a singly charged dimer.

In reality, the liquid drop model is oversimplified: one should also take into account the stabilisation effects due to shell formation, as well as the presence of an energy barrier in the fission process [711].

12.20   Conclusion: transition to the solid

As clusters become much larger in size, the collective mode can be tracked from the quantum limit to the bulk [712], so that the evolution of the many-body resonances is known over an enormous range. In the bulk limit, it has been shown that the resonances tend towards the surface plasmon of the solid, except that, since the solid does not have spherical symmetry, the $\sqrt{3}$ factor of the Mie–Drude theory does not appear and

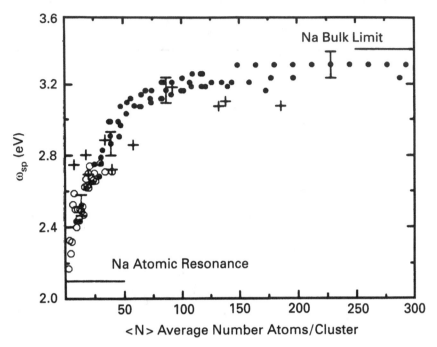

Fig. 12.19.   Trend of the giant resonance or surface plasmon frequencies towards the bulk limit as a function of the number of atoms in the cluster (after J.H. Parks and S.A. Donald [712]).

there is no dipole coupling to the radiation field.

The question of where metallic behaviour begins in a cluster as its size increases is an interesting one and relates to the question cited at the beginning of this chapter: how small is a solid? It turns out that, in fact, there is no single range of sizes where this transition takes place. The answer to the question depends on which physical variable is used as the criterion for determining solid state behaviour. An example is shown in figs. 12.18 and 12.19, which depict the connection between plasmon properties in clusters and in solids. Note that the size range for the transition is quite different from the one involved in the appearance of different valences (fig. 12.11). There is, in fact, no unique answer to the question: 'what determines the appropriate size range?' because it depends on which property is studied. Readers interested in the material just summarised will find a more detailed review in [714]. An interesting, but as yet incompletely resolved question concerns the properties of clusters in very intense laser fields, and, more generally, their nonlinear response [715]. This includes not only the generation of harmonics of the laser frequency, but also the possibility of observing harmonics of the giant resonance itself. As noted above, the fragmentation properties are strongly size-dependent, and the timescale for fragmentation also turns out to be well matched to the properties of short-pulse lasers. The study of clusters in strong laser fields is currently an active and rapidly evolving field of research.

12.21   The Parthian shot

Any reader persistent enough to have followed this book to its conclusion may wish to ponder an interesting remark by Richard Feynman:

> If, in some cataclysm, all of scientific knowledge were to be destroyed, and only one sentence passed on to the next generation of creatures, what statement would contain the most information in the fewest words? I believe it is *the atomic hypothesis* (or the atomic *fact*, or whatever you wish to call it) that *all things are made of atoms*.

References

Chapter 1

[1] N. Bohr (1922) *The Theory of Spectra and Atomic Constitution*
     Cambridge University Press, Cambridge

[2] E.C. Stoner (1924) Phil Mag **48** 719

[3] Sir H. Jeffreys and Lady B. Jeffreys (1982) *Methods of Mathematical Physics*
     Cambridge University Press, Cambridge.

[4] C.A. Coulson (1982) *The shape and Structure of Molecules*
     Clarendon Press, Oxford

[5] J.-P. Connerade R.C. Karnatak and J.M. Esteva (1987)
     Giant Resonances in Atoms Molecules and Solids
     NATO ASI Series B vol 151 Plenum Press, New York

[6] J.-P. Connerade (1991) Advances in Atomic Molecular and Optical Physics **29**
     325

[7] W.D. Knight K. Clemenger W.A. de Heer W.A. Saunders M.Y. Chou M.Y.
     and M.L. Cohen Phys. Rev. Lett **52** 2141

[8] J.-P. Connerade (1991) J. Phys. B: At. Mol. Opt. Phys. **24**, L109

[9] A. L'Huillier and Ph. Balcou (1993) Phys. Rev. Lett. **70** 774

[10] H.A. Bethe and R. Jackiw (1968) *Intermediate Quantum Mechanics*
     second edition W.A. Benjamin, New York

[11] T.A. Koopmans (1933) Physica **1** 104

[12] I.M. Band (1981) J. Phys. B: At. Mol. Phys. **14** 1649; I.M. Band V.I.
     Fomichev and M.B. Trhazkovskaya (1981) J. Phys. B: At. Mol. Phys. **14** 1103

[13] J.B. Mann and W.R. Johnson (1971) Phys Rev **A4** 41

[14] I.P. Grant (1970) Adv. Phys. **19** 747

[15] I.P. Grant B.J. McKenzie P.H. Norrington D.F. Mayers and N.C. Pyper
     (1980) Computer Phys. Commun. **21** 207

[16] C. Froese-Fischer (1977) *The Hartree-Fock Method for Atoms*
     John Wiley Interscience, New York

[17] J.C. Slater (1965) *The Quantum Theory of Atomic Structure* McGraw-Hill,
     New York

[18]   F.A. Parpia I.P. Grant K.G. Dyall and C. Froese-Fischer (private communication) Available from the Mathematical Institute St Giles Oxford OX1 3LB UK

[19]   J.-P. Connerade and K. Dietz (1987) Comm. At. Mol. Phys. **XIX** 283

[20]   R.D. Cowan (1981) *The Theory of Atomic Structure and Spectra* U. of California Press, Berkeley USA; G.E. Bromage (1978) *The Cowan-Zealot Suite of Computer Programs for Atomic Structure* Appleton Laboratory Report AL-R13 Rutherford-Appleton Laboratory, Chilton, Didcot Oxfordshire OX11 OQX UK

[21]   E.U. Condon and G.H. Shortley (1935) *The Theory of Atomic Spectra* Cambridge University Press (Cambridge)

Chapter 2

[22]   G.W. Series (1957) *The Spectrum of Atomic Hydrogen* (Oxford University Press, Oxford) A book under the same title, plus the word:Advances, also edited by G.W. Series, contains a reprint of the 1957 text, together with further, more recent articles on the spectrum of hydrogen. It was published by the World Scientific Press (Singapore) in 1988

[23]   C.E. Moore (1949) *Atomic Energy Levels* NBS Publications No. 467 US Govt. Printing Office, Washington

[24]   M.A. Baig and J.-P. Connerade (1984) J. Phys. B: At. Mol. Phys. **17** L271

[25]   M.A. Baig A. Rashid I. Ahmad M. Rafi J.-P. Connerade and J. Hormes (1990) J. Phys. B: At. Mol. Opt. Phys. **23** 3489

[26]   H.E. White (1934) *Introduction to Atomic Spectra* McGraw Hill Book Company, New York

[27]   J.-P. Connerade W.A. Farooq H. Ma M. Nawaz and N. Shen (1992) J. Phys. B **25** 1405

[28]   W.R.S. Garton and J.-P. Connerade (1969) Astrophys. J. **155** 667

[29]   K. Sommer (1986) Ph.D. thesis. Physikalisches Institut der Universität Bonn Nußallee 12 D5300 Bonn 1 Germany

[30]   M.A. Baig J.-P. Connerade C. Mayhew G. Noeldeke and M.J. Seaton (1984) J. Phys. B: At. Mol. Phys. **17** L383

[31]   M.J. Seaton (1966) Proc. Phys. Soc. **88** 815

[32]   M.A. Baig and J.-P. Connerade (1985) J.Phys.B **18**, 1101

[33]   A. Sommerfeld and H. Welker (1938) Ann. der Phys. **32** 56

[34]   J. Hormes and G. Happel (1983) J. Chem. Phys. **78** 1758

[35]   W. Jaskólski (1996) Physics Reports **271** June issue

[36]   J.-P. Connerade (1997) J. of Alloys and Compounds (In Press)

[37]   C. Fabre M. Gross J.M. Raimond and S. Haroche (1983) J. Phys. B. At. Mol. Phys. **16** L671

[38]   K.M. Müller-Dethlefs R. Frey and E.W. Schlag (1984) Ann. Isr. Phys. Soc. (Israel) **6** 449; K. Müller-Dethlefs (1984) Ann. Isr. Phys. Soc. (Israel) **6** 452

[39] A. Mank T. Nguyen J.D.D. Martin and J.W. Hepburn (1995) Phys. Rev. **A51** R1

[40] Xu Zhang J.M. Smith and J.C. Knee (1992) J. Chem. Phys. **97** 2843

[41] A.K. Dupree and L. Goldberg (1970) Ann. Rev. Astr. Astrophys. **8** 231

[42] D.R. Bates A.E. Kingston and R.W.P. McWhirter (1962) Proc. Roy. Soc. (London) **A 267** 297 and **A 270** 155

[43] T.W. Priest (1972) J. Chem. Soc. Farad. Trans. **68** 661

[44] E. Fermi (1934) Nuovo Cimento **11** 157

[45] M. Matsuzawa (1971) J. Chem. Phys. **55** 2685; (1972) J. Phys. Soc. Japan **32** 1088 and **33** 1108; (1973) J. Chem. Phys. **58** 2674; (1974) J. Elect. Spec. Rel. Phen. **4** 1

[46] E. Vogt and G.H. Wannier (1954) Phys. Rev. **95** 1190

[47] R.F. Stebbings (1977) CNRS Colloquium N^0 273 *Etats Atomiques et Moléculaires Couplés à un Continuum – Atomes et Molécules Hautement Excités* S. Feneuille and J.-C. Lehmann Eds. Société Française de Physique, Paris

[48] D. Smith and P. Spanel (1996) J. Phys. B: At. Mol. Opt. Phys. **29** 5199

[49] F.B. Dunning (1987) J. Phys. Chem. **91** 2244

[50] G. Vitrant J.M. Raimond M. Gross and S. Haroche (1982) J. Phys. B: At. Mol. Phys. **15** L49

[51] J. Parker and C.R. Stroud (1986) Phys. Rev. Lett. **56** 716

[52] J.A. Yeazell and C.R. Stroud Jr (1988) Phys. Rev. Lett. **60** 1494

[53] H.A. Lorentz (1987) *Selected Works* Nancy J. Nersessian Associate Ed. Floris Cohen Palm Publications Nieuwerkerk (the Netherlands)

[54] E. Schrödinger (1926) Naturwissens. **14** 664

[55] J.-C. Gay (1990) *Atoms in Strong Fields* C.A. Nicolaides C.W. Clark and M.H. Nayfeh Eds. NATO ASI Series B vol 212 Plenum, New York page 155

[56] V. Kostrykin and R. Schrader (1995) J. Phys. B: At. Mol. Opt. Phys. **28** L87

[57] E.H. Lieb I.M Sigal B. Simon and W. Thirring (1984) Phys. Rev. Lett **52** 994

[58] E.H. Lieb (1984) Phys. Rev. Lett. **52** 315

[59] R. Wildt (1939) Astrophys. J. **89** 295

[60] D.J. Pegg R.N. Thompson R.N. Compton and G.D. Alton (1987) Phys. Rev. Lett. **59** 2267

[61] C. Froese-Fischer J.B. Lagowski and S.H. Vosko (1987) Phys. Rev. Lett. **59** 2263

[62] S. Mannervik G. Astner and M. Kisielinski (1980) J. Phys. B: At. Mol. Phys. **13** L441

[63] J.O. Gaardsted and T.J. Andersen (1989) J. Phys. B: At. Mol. Phys. **22** L51

[64] S.E. Novick P.L. Jones T.J. Mulloney and W.C. Lineberger (1979) J. Chem Phys. **70** 2210

[65] C.F. Bunge M. Galan R. Jauregi and A.V. Bunge (1982) Nucl. Inst. Methods **202** 299

[66] T. Andersen (1990) Physica Scripta **34** 23

[67] A. Carrington (1996) Science **274** 1327

[68] A. Carrington I.R. McNab and C.A. Montgomerie (1989) J. Phys. B: At. Mol. Opt. Phys. **22** 3551

[69] A.S. Wightman (1950) Phys. Rev. **77** 521

[70] R.F. Wallis R. Herman and H.W. Milnes (1960) J. Mol. Spec. **4** 51

[71] M.H. Mittleman and V.P. Myerscough (1966) Phys. Lett. **23** 545

[72] J.F. Turner and K. Fox (1966) Phys. Lett. **23** 547

[73] J.M. Levy-Leblanc (1967) Phys. Rev. **153** 1

[74] O.H. Crawford (1967) Proc. Phys. Soc. (London) **91** 279

[75] J.L. Carlsten J.R. Peterson and W.C. Linneberger (1976) Chem. Phys. Lett. **37** 5

[76] K.D. Jordan and W. Luken (1976) J. Chem. Phys. **64** 2760

[77] H. Rinneberg J. Neukammer G. Jönsson H. Hieronymus A. König and K. Vietzke (1985) Phys. Rev. Lett. **55** 383

[78] H. Rinneberg J. Neukammer M. Kohl A. König K. Vietzke H. Hieronymus and H.-J. Grabka (1988) *Atomic Spectra and Collisions in External Fields* K.T. Taylor, M.H. Nayfeh and C.W. Clark Eds. Plenum Press, New York

[79] A.A. Konovalenko (1984) Sov. Astron. Lett. (USA) **10** 846

[80] X. Ling B.G. Lindsay K.A. Smith and F.B. Dunning (1992) Phys. Rev. A **45** 242

[81] M.T. Frey S.B. Hill X. Ling K.A. Smith and F.B. Dunning (1994) Phys. Rev. A **50** 3124

[82] R.F. Stebbings and F.B. Dunning (1983) *Rydberg States of Atoms and Molecules* Cambridge University Press, Cambridge

[83] R.S. Mulliken (1976) Acc. Chem. Res. **9** 7

[84] H. Lefebvre-Brion and R.W. Field (1986) *Perturbations in the Spectra of Diatomic Molecules* Academic Press, New York

[85] G. Herzberg (1969) Phys. Rev. Lett. **23** 1081

[86] G. Herzberg and Ch. Jungen (1972) J. Mol. Spec. **41** 425

[87] W.C. Price (1938) Proc. Roy. Soc. (London) **A167** 216

[88] J.-P. Connerade, M.A. Baig, S.P. McGlynn and W.R.S. Garton (1980) J. Phys. B: At. Mol. Phys. **13** L705

[89] M.A. Baig, J. Hormes, J.-P. Connerade and S.P. McGlynn (1981) J. Phys. B: At. Mol. Phys. **14** L25

[90] C. Mayhew, M.A. Baig and J.-P. Connerade (1983) J. Phys. B: At. Mol. Phys. **16** L757

[91] J.-P. Connerade and J. Hormes (1986) Zeit. fur Phys. D **4**, 3

[92]   J.-P. Connerade M.A. Baig C.A. Mayhew and J. Hormes
       (1995) J. Electron. Spectr. Rel. Phen. **73** 173

[93]   L. Karlsson L. Mattson R.G. Allbridge S. Pinchas T. Berkmann and
       K. Siegbahn (1975) J. Chem. Phys. **62** 4745

[94]   C.A. Mayhew and J.-P. Connerade (1986) J. Phys. B: At. Mol. Phys. **19** 3493

[95]   C.A. Mayhew, J.-P. Connerade, M.A. Baig M.N.R. Ashfold,
       J.M. Bayley, R.N. Dixon and J.D. Prince (1987)
       J. Chem. Soc Faraday Trans **83** 417

[96]   C.A. Mayhew and J.-P. Connerade (1986) J. Phys. B: At. Mol. Phys. **19** 3505

[97]   C.A. Mayhew J.-P. Connerade and M.A. Baig (1986)
       J. Phys. B: At. Mol. Phys. **19** 4149

[98]   M.A. Baig J.-P. Connerade J. Dagata and S.P. McGlynn (1981)
       J.Phys.B. **14**, L25

[99]   M.A. Baig, J.-P. Connerade and J. Hormes (1982) J. Phys. B. **15**, L5

[100] M.A. Baig J.-P. Connerade and J. Hormes (1986) J.Phys.B. **19**, L343

[101] J.-P. Connerade M.A. Baig and S.P. McGlynn (1981)
       J. Phys. B: At. Mol. Phys. **14** L67

[102] R.S. Knox (1963) Solid State Phys. Suppl. **5** 7

[103] B.I. Bleaney and B. Bleaney (1976) *Electricity and Magnetism* third edition
       Oxford University Press, Oxford page 543

[104] J.M. Shi F.M. Peters andJ.T. Devreese (1993) Phys. Rev. **B48** 5202

[105] J.A. Wheeler (1946) Ann. N. Y. Acad. Sci. **48** 221

[106] E.A. Hylleraas and A. Ore (1947) Phys. Rev. **71** 493

[107] M. A. Lampert (1958) Phys. Rev. Lett. **1** 450

[108] R.R. Sharma (1968) Phys. Rev. **170** 770 **171** 36

[109] O. Akimoto and E. Hanamura (1972) Solid State Commun. **10** 253

[110] J.R. Haynes (1968) Phys. Rev. Lett. **17** 860

Chapter 3

[111] M.J. Seaton (1966) Proc. Phys. Soc. **88** 801

[112] H. Friedrich (1990) *Theoretical Atomic Physics* Springer-Verlag, Berlin

[113] M.A. Baig and J.-P. Connerade (1984) J.Phys.B **17**, L469

[114] White H.E. (1934) *Introduction to Atomic Spectra* McGraw Hill Book
       Company, New York

[115] J. von Neumann and E.P. Wigner (1929) Z. Phys. **32** 467

[116] K.T. Lu and U. Fano (1970) Phys. Rev. **A2** 81

[117] J.A. Armstrong P. Esherick and J.J. Wynne (1977) Phys. Rev. **A 15** 180

[118] C.H. Greene (1985) Phys. Rev. **A32** 1880

[119] M. Aymar (1990) J. Phys. B: At. Mol. Opt. Phys. **23** 2697

[120] W.R. Johnson and K.T. Cheng (1979) J. Phys. B: At. Mol. Phys. **12** 863

[121] W.C. Price (1936) J. Chem. Phys. **4** 439

[122] J.A. Dagata, G.L. Findley, S.P. McGlynn, J.-P. Connerade and M.A. Baig (1981) Phys. Rev. **A24** 2485

[123] R.S. Mulliken (1935) Phys. Rev. **47** 413

[124] R.S. Mulliken (1942) Phys. Rev. **61** 277

[125] M.A. Baig, J.-P. Connerade and J. Hormes (1982) J. Phys. B: At. Mol. Opt. Phys. **15** L5

[126] M.J. Seaton (1983) Rep. Prog. Phys. **47** 167

Chapter 4

[127] N.B. Delone S.P. Goreslavsky and V.P. Krainov (1994) J. Phys. B: At. Mol. Opt. Phys **27** 4403

[128] P.A.M. Dirac (1927) Proc. Roy. Soc. (London) **A 114** 243

[129] R. Ladenburg (1921) Z. Phys. **4** 451

[130] H.A. Kramers (1924) Nature **113** 673 and **114** 310

[131] R. Ladenburg and H. Kopfermann (1928) Z. Phys. **48** 15

[132] I. Kimel (1982) Phys. Rev. **B25** 6561

[133] A.R.P. Rao (1988) *Fundamental Processes of Atomic Dynamics* J. Briggs, H. Kleinpoppen and H. Lutz Eds NATO ASI Series B vol 181 Plenum Press, New York page 51

[134] M.Y. Sugiura (1927) J. de Physique et le Radium Série VI **VIII** 8

[135] M. Gailitis (1963) Sov. Phys. JETP **17** 1328

[136] J.-P. Connerade H. Ma N. Shen and T.A. Stavrakas (1988) J. Phys. B **21**, L241

[137] M.J. Seaton (1966) Proc. Phys. Soc. **88** 801

[138] M. Rudkjøbing (1940) Det. Kgl. Danske Videnskabernes Selskab (Math-Fys Medd.) **XVIII**, 2

[139] M.J. Seaton (1951) Proc. Roy. Soc. (London) **A208**, 405

[140] J.W. Cooper (1962) Phys. Rev. **128** 681

[141] M. Rahman-Attia M. Jaouen G. Laplanche and A. Rachman (1986) J. Phys. B: At. Mol. Phys. **19** 897

[142] A.Z. Msezane and S.T. Manson (1975) Phys. Rev. Lett. **35** 364; (1982) Phys. Rev. Lett. **48** 473

[143] U. Fano (1969) Phys. Rev. **178** 131

[144] W. von Drachenfelds U.T. Koch R.D. Lapper T.M. Muller and W. Paul (1974) Z. Phys. **269** 387

[145] K.S. Bhatia J.-P. Connerade and Y. Makdisi (1990) J. Phys. B: At. Mol. Phys. **23** 3475

[146] E. Fermi (1934) Nuovo Cimento **11** 157

[147] J. Q. Sun E. Matthias K.D. Heber P.J. West and J. Güdde
       (1991) Phys. Rev. **43** 5956

[148] M. Nawaz W.A. Farooq and J.-P. Connerade
       1992 J. Phys. B **25** 1147

[149] A.P. Thorne *Spectrophysics* second edition Chapman and Hall, London

[150] E.P. Wigner (1959) *Group Theory* Academic Press, New York

[151] M.C.E. Huber and R.J. Sandeman (1986) Rep. Prog. Phys. **4** 397

[152] M. Faraday (1846) Phil. Mag. **153** 249

[153] Macaluso and Corbino (1899) Rend. Linc **8** 1 Sem. Ser. V Fasc. 3

[154] P. Zeeman (1896) Zitt. Akad. Amsterdam **5** 181, 242
       (Engl. transl. (1897) Phil. Mag. **43** 226)

[155] G.A. Schott (1912) *Electromagnetic Radiation and the Mechanical Reactions
       Arising from it; being an Adams Essay Prize in the University of Cambridge*
       Cambridge University Press, Cambridge

[156] P. Kapitza (1938) Proc. Roy. Soc. (London) **A167**, 1

[157] J.-P. Connerade C. Schmidt and M. Warken (1993) J. Phys. B **26** 3459

[158] A.F. Harvey (1970) *Coherent Light* Wiley Interscience, London

[159] J.-P. Connerade (1978) Nucl. Intrum. Methods **152** 271

[160] F.A. Jenkins and H.E. White (1957) *Fundamentals of Optics* third edition
       McGraw-Hill Book Company Inc, New York

[161] J.-P. Connerade (1990) in *Atoms in Strong Fields* C.A. Nicolaides C.W. Clark
       and M.H. Nayfeh Eds NATO ASI Series B: vol 212 Plenum Press, New York
       page 189

[162] P.W. Atkins and M.H. Miller M.H. (1968) Molecular Physics **15**, 503

[163] J.-P. Connerade W.R.S. Garton M.A. Baig J. Hormes T.A. Stavrakas and B.
       Alexa (1982) Journal de Physique **43**, C2-317

[164] J.-P. Connerade W.A. Farooq H. Ma M. Nawaz and N. Shen
       (1992) J. Phys. B **25** 1405

[165] W.H. Parkinson E.M. Reeves and F.S. Tomkins (1976)
       J. Phys. B: At. Mol. Phys. **9**, 157

[166] M. Nawaz W.A. Farooq and J.-P. Connerade (1992) J. Phys. B **25** 3283

[167] X.H. He and J.-P. Connerade (1993) J. Phys. B **26** L255

[168] A.C.G. Mitchell and M.W. Zemansky (1971) *Resonance Radiation and
       Excited Atoms* Cambridge University Press, Cambridge

[169] A.D. Buckingham and P.J. Stephens (1986) Ann. Rev. Phys. Chem. **17**, 399

[170] W. Gawlik J. Kowalski R. Neumann H. Wiegemann and K. Winkler
       (1979) J.Phys. B: At. Mol. Phys. **12**, 3873

[171] A.K. Hui B.H. Armstrong and A.A. Wray (1978)
       J. Quant. Spec. and Rad. Transfer **19**, 509

[172] J.-P. Connerade (1983) J. Phys. B: At. Mol. Phys. **16** 399

[173]   G.S. Agarwal P. Anantha Lakshmi J.-P. Connerade and S. West (1997) J.
        Phys. B: At. Mol. Opt. Phys. (In the Press)

[174]   A.J. Warry D.J. Heading and J.-P. Connerade (1994) J. Phys. B: At. Mol.
        Opt. Phys. **27** 2229

[175]   S. Giraud-Cotton and V.P. Kaftandjian (1985) Phys. Rev. **A 32**, 2211 and
        2223

[176]   T. Ya Karagodova V.A. Makarov and A.A. Karagodov (1990) Optika i Spektr.
        69 389; T. Ya. Karagodova A.A. Zakharov and A.V. Kolpakov (1992)
        Izvestiya Russ. Akad. Nauk. **56** 209; (1993) Optika i Spektr. **74** 1137

Chapter 5

[177]   N. Bohr (1922) *The Theory of Spectra and Atomic Constitution* Cambridge
        University Press, Cambridge

[178]   E.C. Stoner (1924) Phil Mag **48** 719

[179]   G. Herzberg (1937) used this phrase in his classic text *Atomic Spectra and
        Atomic Structure* Dover Publications, New York. He was convinced that *'the
        whole periodic system of the elements can be unambiguously derived by using
        the building-up principle in conjunction with the Pauli principle'*, but his own
        description for the long periods was unsatisfactory.

[180]   D.L. Ederer (1964) Phys. Rev. Lett. **13** 760

[181]   J.A.R. Samson (1966) J. Opt. Opt. Am **55** 935

[182]   D.L. Ederer and D.H. Tomboulian (1964) Phys. Rev. **A133** 1525

[183]   A.P. Lukirskii I.A. Brytov and T.M. Zimkina (1964) Opt. Spectrosc. **17** 234

[184]   D.J. Kennedy and S.T. Manson (1972) Phys. Rev. **A5** 227

[185]   J.W. Cooper (1964) Phys. Rev. Lett. **13** 762

[186]   S.T. Manson and J.W. Cooper (1968) Phys. Rev. **165** 126

[187]   F. Combet-Farnoux (1972) *Proc. Int. Conf. Inner Shell Ionisation Phenomena
        Atlanta* vol 2 University of Georgia Press, Atlanta, page 1130

[188]   T.M. Zimkina and S.A. Gribovskii (1971) J. Phys. (Paris) Colloq. **32** C4 282

[189]   J.-P. Connerade and M.W.D. Mansfield (1974) Proc. Roy. Soc. **A341**, 267

[190]   P. Rabe K. Radler and H.W. Wolf (1974) *Vacuum Ultraviolet Radiation
        Physics* Pergammon, Braunschweig, page 247

[191]   J.-P. Connerade J.-M. Esteva and R.C. Karnatak Eds. *Giant Resonances in
        Atoms Molecules and Solids* (1987) Plenum, New York

[192]   H.E. White (1934) *Introduction to Atomic Spectra* International Student
        Edition McGraw-Hill, Kogakusha, New York

[193]   G. Herzberg (1944) *Atomic Spectra and Atomic Structure*
        Dover Publications, New York

[194]   E. Fermi (1928) in *Quantentheorie und Chemie* H. Falkenhagen Ed.
        Hinzel-Verlag, Leipzig

[195]   M. Göppert-Mayer (1941) Phys. Rev. **60** 184

[196] R.I. Karaziya (1981) Sov Phys Usp **24**, 775

[197] D.M. Dennison and G.E.Uhlenbeck (1932) Phys. Rev. **41** 313

[198] N. Fröman (1966) Ark. f. Fysik **32** 79

[199] R.E. Langer (1937) Phys. Rev. **51** 669

[200] T.Y. Wu (1933) Phys. Rev. **44** 727

[201] G. Wendin and A.F. Starace (1978) J. Phys. B: At. Mol. Phys. **11** 4119

[202] J.J. Dehmer A.C. Parr and S.H. Southworth (1987) *Handbook on Synchrotron Radiation* Vol 2 Elsevier Science North-Holland, Amsterdam, page 241

[203] E. Shigemata J. Adachi M. Oura N. Watanabe K. Soejima and A. Yagishita (1996) Proceedings of the Oji Seminar *Atomic and Molecular Photoionisation* (Tsukuba) A. Yagishita and T. Sasaki Eds. Universal Academic Press, Tokyo, page 69

[204] A.R.P. Rau and U. Fano (1968) Phys. Rev. **167** 7

[205] J.-P. Connerade (1991) J. Phys. B: At Mol Opt Phys **24**, L109

[206] D. ter Haar (1946) Phys. Rev. **70** 222
(note that this ref. contains a minor misprint: the expression in line 11 must be set to zero to determine the bound states)

[207] M.A. Khan J.-P. Connerade and M. Rafique (1994) J. Phys. B: At. Mol. Phys. **27** L563

[208] D.C. Griffin K.L. Andrew and R.D. Cowan (1969) Phys. Rev. **177** 62

[209] D.C. Griffin R.D. Cowan and K.L. Andrew (1971) Phys. Rev. **A3** 1233

[210] J.-P. Connerade (1978) Contemp. Phys. **19** 415

[211] J.-P. Connerade (1984) *New Trends in Atomic Physics* Vol. II Les Houches session XXXVIII Elzevier, Amsterdam, page 645

[212] J.-P. Connerade and M.W.D. Mansfield (1975) Proc. Roy. Soc. (London) **A346** 565

[213] F.A Saunders E.G. Schneider and E. Buckingham (1934) Proc. Nat. Acad. Sci. (Washington) **20** 291

[214] R.A. Roig and G. Tondello (1975) J. Opt. Soc. Am. **65** 829

[215] M.S. Child (1974) *Molecular Scattering Theory* Academic Press, London

[216] A. Temkin and A.K. Bhatia (1985) in *Autoionisation* A. Temkin Ed. Plenum Press, New York, page 66

[217] J.-P. Connerade (1982) J.Phys.B. **15**, L881

[218] G.F. Drukarev (1981) Sov. Phys. JETP **53** 271

[219] L.I. Schiff (1968) *Quantum Mechanics* third edition McGraw-Hill Kogakusha, Tokyo

[220] M.V. Berry (1966) Proc. Phys. Soc. **88** 285

[221] L.P. Landau and E.M. Lifshitz (1965) *Quantum Mechanics* Addison-Wesley, Reading USA, page 514

[222] J.-P. Connerade (1986) Comm. At. Mol. Phys. **XVII**, 199

[223] N.F. Mott and H.S.W. Massey (1965) *The Theory of Atomic Collisions* third edition Clarendon Press, Oxford

[224] F. Combet- Farnoux (1972) *Proc. Int. Conf. on Inner Shell Ionisation Phenomena Atlanta* vol 2 University of Georgia Press, Atlanta p1130

[225] J.-P. Connerade (1984) J. Phys. B: At. Mol. Phys. **17** L165

[226] R.C. Greenhow J.A.D. Matthew R.M. Clark and G.A. Gates (1991) J. Phys. B: At. Mol. Phys **24** 4677; R.C. Greenhow and J.A.D. Matthew (1992) Am. J. Phys. **60** 655

[227] M. Richter (1993) *Proceedings of the International Workshop on Photoionization* Eds. U. Becker and U. Heinzmann AMS Press Inc. New York page 183

[228] M. Richter (1993) Vacuum Ultraviolet Radiation Physics (Paris July 27-31 1992) Eds. F.J. Wuilleumier Y. Petroff and I. Nenner World Scientific, Singapore, page 135

[229] J. Stöhr K. Baberschke R. Jaeger R. Treichler and S. Brennan (1981) Phys. Rev. Lett. **47** 381

[230] J.-P. Connerade and K. Dietz (1987) Comm. At. Mol. Phys. **XIX**, 283

[231] J.-P. Connerade, K. Dietz, M.W.D. Mansfield and G. Weymans (1984) J. Phys. B: At. Mol. Phys. **17**, 1211

[232] J. P. Connerade (1978) J. Phys. B: At. Mol. Phys. **11**, L409

[233] T.J. Lucatorto T. McIlrath J. Sugar and S.M. Younger (1981) Phys. Rev. Lett. **47** 1124

[234] J.-P. Connerade (1978) J. Phys. B: At. Mol. Phys. **11**, L381

[235] A.A. Maiste R.E. Ruus S.A. Ruchas R.I. Karaziya and M.A. Elango (1980) Sov. Phys. JETP **51** 474 and **52** 844b

[236] J.-P. Connerade, M.W.D. Mansfield, M. Cukier and M. Pantelouris (1980) J. Phys. B: At. Mol. Phys. **13**, L235; J.-P. Connerade, M. Pantelouris, M.A.P. Martin and M. Cukier (1980) J. Phys. B: At. Mol. Phys. **13**, L357

[237] M. Robin (1985) Chem. Phys. Lett. **119** 33

[238] G. O'Sullivan (1982) J. Phys. B: At. Mol. Phys. **15** L327

[239] A.N. Belsky G. Comtet G. Dujardin A.V. Gektin L. Hellner I.A. Kamenskikh P. Martin V.V. Mikhailin C. Pedrini and A.N. Vasil'ev (1996) J. of Electron Spectroscopy and Related Phenomena **80** 109

[240] A.N. Belsky R.A. Glukhov I.A. Kamenskikh P. Martin V.V. Mikhailin I.H. Munro C. Pedrini D.A. Shaw I.N. Shpinkov and A.N. Vasil'ev (1996) J. of Electron Spectroscopy and Related Phenomena **79** 147

[241] H.P. Kelly (1976) *Photoionization and Other Probes of Many-electron Interactions* F.J. Wuilleumier Ed. NATO ASI series B vol 18 Plenum Press, New York, page 83

[242] H.P. Kelly (1968) Advances in Theoretical Physics vol II Academic Press, New York, page 75

[243] G.F. Bertsch and R.A. Broglia (1994) *Oscillations in Finite Quantum Systems* Cambridge Monographs on Mathematical Physics Cambridge University Press, Cambridge

[244] K.A. Brueckner (1955) Phys. Rev. **97** 1353 and **100** 36

[245] J. Goldstone (1957) Proc. Roy. Soc. (London) **A239** 267

[246] Z. Altun S.L. Carter and H.P. Kelly (1982) J. Phys. B: At. Mol. Phys. **15** L709

[247] M. Ya. Amusia N. Cherepkov and L.V. Chernysheva
       (1971) Sov. Phys. JETP **33** 90

[248] G. Wendin (1984) *New Trends in Atomic Physics* Vol. II Les Houches Session
       XXXVIII North Holland, Amsterdam, page 558

[249] M. Ya. Amusia V.K. Ivanov S.A. Sheinerman and S.I. Sheftel
       (1980) Zh. Exp. Teor. Fiz. **78** 910

[250] M. Ya. Amusia (1990) *The Atomic Photoeffect* Plenum Press, New York

[251] H.P. Kelly S.L. Carter and B.E. Norum (1982) Phys. Rev. **25** 2052

[252] M. Ya. Amusia L.V. Chernysheva G.F. Gribakin and K.L. Tesemekhman
       (1990) J. Phys. B: At. Mol. Opt. Phys. **23** 393

[253] V.K. Ivanov and L.P. Kruskovskaya (1994) J. Phys. B: At. Mol. Opt. Phys. **27**
       4111

Chapter 6

[254] H. Beutler (1934) Zeit. Phys. **91** 132

[255] F. Paschen (1931) Sitz. Preuss. Akad. Wiss. **32** 709

[256] U. Fano (1961) Phys. Rev. **124** 1866

[257] J.-P. Connerade (1992) Journal de Physique (France) **2** 757

[258] A.G. Shenstone (1939) Rep. Prog. Phys. **5** 210

[259] Landau L P and Lifshitz E M (1965) *Quantum Mechanics* Addison-Wesley,
       Reading USA, page 514

[260] P.A.M. Dirac (1958) *The Principles of Quantum Mechanics* second edition
       The Clarendon Press, Oxford

[261] B. W. Shore (1967) J. Opt. Soc. Am. **57** 881

[262] F.H. Mies (1968) Phys. Rev. **175** 164

[263] S.M. Farooqi J.-P. Connerade C.H. Greene J. Marangos M.H.R. Hutchinson
       and N. Shen (1991) J. Phys. B: At. Mol. Opt. Phys. **24** L179

[264] A.G. Shenstone (1929) Phys. Rev. **28** 449

[265] W. Eberhardt G. Kalkoffen and C. Kunz (1978) Phys. Rev. Lett. **41** 156

[266] O.-P. Sairanen H. Aksela S. Aksela J. Mursu A. Kivimäki
       A. Naves de Brito E. Nömmiste S.J. Osborne A. Ausmees and S. Svensson
       (1995) J. Phys. B: At. Mol. Opt. Phys. **28** 4509

[267] H. Aksela O.-P. Sairanen S. Aksela A. Kivimäki A. Naves de Brito E.
       Nömmiste J. Tulkki A. Ausmees S.J. Osborne and S. Svensson
       (1995) Phys. Rev. **A 51** 1291

[268] P. Morin and I. Nenner (1986) Phys. Rev. Lett. **56** 1913

[269] S. Svensson L. Karlsson N. Mårtensson P. Baltzer and B. Wannberg
       (1990) J. Electron Spectr. Rel. Phen. **50**

[270] H. Aksela S. Aksela M. Ala-Korpela O.-P. Sairanen M. Hotokka G.M. Bancroft K.H. Tan and J. Tulkki (1990) Phys. Rev. **A41** 6000

[271] I. Nenner P. Morin P. Lablanquie M. Simon N. levasseur and P. Millie (1990) J. Electron Spect. Rel. Phen. **52** 623

[272] A. Naves de Brito N. Correira B. Wannberg P. Baltzer L. Karlsson S. Svensson M.Y. Adam H. Aksela and S. Aksela (1992) Phys. Rev. **A46** 6067

[273] H. Aksela S. Aksela A. Naves de Brito G.M. Bankroft and K.H. Tan (1992) Phys. Rev. **A45** 7948

[274] A. Naves de Brito and H. Ågren (1992) Phys. Rev. **A 45** 7953

[275] S. Svensson H. Aksela A. Kivimäki O.P. Sairanen A. Ausmees S.J. Osborne A. Naves de Brito E. Nõmmiste G. Bray and S. Aksela (1995) J. Phys. B: At. Mol. Opt. Phys. **28** L325

[276] J. Geiger (1979) J. Phys. B: At. Mol. Phys. **12** 2277

[277] J. Macek (1970) Phys. Rev. **A2** 1101

[278] A.M. Lane (1986) J. Phys. B: At. Mol. Phys. **19** L601

[279] I. Shimamura C.J. Noble and P.G. Burke (1990) Phys. Rev **A41** 3545

[280] C.A. Nicolaides and T. Mercouris (1996) J. Phys. B: At. Mol. Phys. **29** 1151

[281] C.A. Nicolaides and D.R. Beck (1977) Phys. Rev. Lett. **38** 683

[282] P.T. Matthews and A. Salam (1959) Phys. Rev. **115** 1079

[283] A.C.G. Mitchell and M.W. Zemanski (1971) *Resonance Radiation and Excited Atoms* Cambridge University Press, Cambridge

[284] H.A. Kramers (1927) Att. Congress. Int. Fis. Como **2** 545

[285] R. de L. Kronig (1927) J. Opt. Soc. Am. Rev. Sci. Instrum. **12** 947; (1946) Physica **12** 543

[286] H.M. Nussenzveig (1972) *Causality and Dispersion Relations* Academic Press, New York

[287] G. Titmarsh (1948) *Introduction to the Theory of Fourier Integrals* (1948) second edition Oxford University Press, London

[288] R. Feynman (1963) *Lecture Notes in Physics* Addison Wesley, Reading USA

[289] B.W. Shore (1967) Rev. Mod. Phys. **39** 439

[290] N. Bohr R. Peierls and G. Placzek (1939) Nature **144** 200

[291] J.D. Jackson (1960) *Dispersion Relations* Scottish Universities Summer Sschools G.R. Streaton Ed. Oliver and Boyd, Edinburgh

[292] W. Sellmeier 1871 Pogg. Ann. Chem. **143** 271 **147** 386, 399, 525

[293] M. Weingeroff (1931) Zeit. f. Phys. **67** 679

[294] J.-P. Connerade (1988) J. Phys. B: At. Mol. Opt. Phys **21** L551; (1991) *ibid.* **24** L51

[295] H.E. White (1931) Phys. Rev. **38** 2016

[296] J.-P. Connerade W.A. Farooq and M. Nawaz (1992) J. Phys. B: At. Mol. Opt Phys. **25** L175

[297] K. Yoshino D.E. Freeman and Y. Tanaka (1979) J. Mol. Spec. **76** 153

[298]   K. Dressler (1970) unpublished (cited by H. Lefebvre-Brion and R.W. Field (1986) in *Perturbations in the Spectra of Diatomic Molecules* Academic Press, New York)

[299]   H.P. Kelly (1968) *Advances in Theoretical Physics* vol II Academic Press, New York p75

Chapter 7

[300]   U. Fano and J.W. Cooper (1968) Rev. Mod. Phys. **40** 441

[301]   L.S. Cederbaum J. Schirmer W. Domcke and W. von Niessen (1977) J. Phys. B: At. Mol. Phys. **10** L549

[302]   J.-P. Connerade, M.A. Baig and M. Sweeney (1987) J. Phys. B **20**, L771

[303]   M.A. Baig, J.-P. Connerade and G.H. Newsom (1979) Proc. Roy. Soc. **A367**,381

[304]   H. Beutler (1934) Zeit. Phys. **91** 132

[305]   N. Mårtensson and B. Johansson (1981) J. Phys. B: At. Mol. Phys. **14** L37

[306]   J.-P. Connerade, M.A. Baig and G.H. Newsom (1981) Proc. Roy. Soc. **A378**, 445

[307]   T.W.B. Kibble (1966) *Classical Mechanics* McGraw-Hill European Physics Series, London

[308]   R.S. Berry (1986) in *The Lesson of Quantum Theory* J. de Boer E. Dal and O. Ulfbeck Eds. Elzevier Science Publishers, Amsterdam, page 241

[309]   E.U. Condon and G.H. Shortley (1935) *The Theory of Atomic Spectra* Cambridge University Press, Cambridge

[310]   S. Silverman and E. Lassettre (1964) J. Chem. Phys. **40** 1265

[311]   L.M. Kierman E.T. Kennedy J.-P. Mosnier J.T. Costello and B.F. Sonntag (1994) Phys. Rev. Lett. **72** 2359

[312]   K.T. Chung (1982) Phys. Rev. **A25** 1596

[313]   J.-P. Briand (1996) Comments At. Mol. Phys. **33** 9

[314]   I. Hughes (1995) Physics World April issue, p43

[315]   J.H.D. Eland and J.H. Sheahan (1994) Chem. Phys. Lett. **223** 531

[316]   H. Shiromaru M. Mizutani K. Koyabashi M. Yoshino T. Mizogawa Y. Achiba and N. Kobayashi (1995) *Proceedings 12th International Workshop on the ECR Ion Souce* Tokyo Metropolitan University Press, Tokyo

[317]   R.K. Yoo B. Ruscic and J. Berkowitz (1992) J. Chem. Phys. **96** 911

[318]   R.R. Jones P. Fu and T.F. Gallagher (1991) Phys. Rev. **A44** 4260

[319]   M. Aymar and E. Luc-Koenig (1995) J. Phys. B: At. Mol. Phys. **28** 1211

[320]   G. Wannier (1953) Phys. Rev. **90** 817

[321]   S.M. Silverman and E.N. Lassettre (1959) Ohio State University Report No. 9 page 12. See also Chem. Phys. **40** 1265

[322]   R.P. Madden and K. Codling (1965) Astrophys. J. **141** 364

[323]   J.W. Cooper U. Fano and F. Prats (1963) Phys. Rev. Lett. **10** 518

[324]   J. Macek (1968) J. Phys. B: At. Mol. Phys. **2** 831

[325]   J.M. Feagin (1988) in *Fundamental Processes of Atomic Dynamics* Eds J.S. Briggs H. Kleinpoppen and H.O. Lutz NATO ASI Series B: Physics vol 181 Plenum, New York page 275

[326]   C.D. Lin (1974) Phys. Rev. **A10** 1986

[327]   U. Fano (1976) in *Photoionization and Other Probes of Many-electron Interactions* F.J. Wuilleumier Ed. NATO ASI Series B vol 18 Plenum Press, New York, page 11

[328]   D.R. Herrick (1978) Phys. Rev. **17** 1

[329]   H.R. Sadeghpour and C.H. Greene (1990) Phys. Rev. Lett. **65** 313

[330]   J.-Z. Tang and I. Shimamura (1994) Phys. Rev. **50** 1321

[331]   M. Domke C. Xue A. Puschman T. Mandel E. Hudson D.A. Shirley G. Kaindl C.H. Greene H.R. Sadeghpour and H. Petersen (1991) Phys. Rev. Lett. **66** 1306

[332]   A. Menzel S.P. Frigo S.B. Whitfield C.D. Caldwell M.O. Krause J.-Z. Tang and I. Shimamura (1995) Phys. Rev. Lett. **75** 1479

[333]   D.R. Herrick and O. Sinanoglu (1975) Phys. Rev. **A11** 97; D.R. Herrick (1975) Phys. Rev. **12** 413

[334]   S. Watanabe and C.D. Lin (1986) Phys. Rev. **A34** 823

[335]   P.G. Harris H.C. Bryant A.H. Mohagheghi R.A. Reeder H. Sharifian C.Y. Tang H. Tootoonchi J.B. Donahue C.R. Quick D.C. Rislove W.W. Smith and J.E. Stewart (1990) Phys. Rev. Lett. **65** 309

[336]   J.-Z. Tang Y. Wakabayashi M. Matsuzawa S. Watanabe and I. Shimamura (1994) Phys. Rev. **A 49** 1021

[337]   C. Pan A.F. Starace and C.H. Greene (1994) J. Phys. B: At. Mol. Opt. Phys. **27** L137

[338]   H.N. Russell and F.A. Saunders (1925) Astrophys. J. **61** 38

[339]   N. Bohr and Wentzel (1923) Phys. Zeit. **24** 106

[340]   J.-P. Connerade and M.W.D. Mansfield (1976) Proc. Roy. Soc. **A348**, 539

[341]   H. Beutler (1933) Zeit. f. Phys. **86** 19, 495, 710; **87** 176

[342]   W.R.S. Garton and J.-P. Connerade (1969) Astrophys. J. **155**, 667

[343]   J.-P. Connerade and M.A. Baig (1987) Chapter IV of the *Handbook on Synchrotron Radiation* G.V. Marr Ed. Springer Verlag, Berlin, page 175

[344]   M.W.D. Mansfield and J.-P. Connerade (1978) Proc.Roy.Soc. **A359**, 389

[345]   R. Ding, W.G. Kaenders J.P. Marangos J.-P. Connerade and M.H.R. Hutchinson (1989) J. Phys. B **22**, L251

[346]   J.-P. Connerade (1972) Astrophys. J. **172**, 213

[347]   J.-P. Connerade, W.R.S. Garton, M.W.D. Mansfield and M.A.P. Martin (1976) Proc. Roy. Soc. **A350**, 47

[348]   J.-P. Connerade (1977) Proc. Roy. Soc. **A354**, 511

[349]   J.-P. Connerade (1977) Proc. Roy. Soc. **A352**, 561

[350]   J.-P. Connerade, W.R.S. Garton, M.W.D. Mansfield and M.A.P. Martin (1977) Proc. Roy. Soc. **A357**, 499

[351]   J.-P. Connerade (1977) J. Phys. B: At. Mol. Phys. **10**, L239

[352]   J.-P. Connerade and M.A.P. Martin (1977) Proc. Roy. Soc. **A357**, 103

[353]   P. Zimmermann (1989) Comm. At. Mol. Phys. **XXIII** 45

[354]   T. Nagata M. Yoshino T. Hayaishi Y. Itikawa Y. Itoh T. Koizumi T. Matsuo Y. Sato E. Shigemasa Y. Talizawa and A. Yagishita (1990) Physica Scripta **41** 47

[355]   J.-P. Connerade (1992) Adv. At. Mol. Opt. Phys. **29** 325

[356]   J.-P. Connerade (1970) Astrophys. J. **159** 685 and 695

[357]   J.-P. Connerade, M.A. Baig, W.R.S. Garton and G.H. Newsom (1980) Proc. Roy. Soc. **A371**, 295

[358]   J.-P. Connerade, M.W.D. Mansfield, G.H. Newsom, D.H. Tracy, M.A. Baig and K. Thimm (1979) Phil. Trans. Roy. Soc. **290**, 327; see also: M.A. Baig, J.-P. Connerade and C. Mayhew (1983) J. Phys. B **17**, 371

[359]   S.J. Rose, I.P. Grant and J.-P. Connerade (1980) Phil. Trans. Roy. Soc. **296**, 527

[360]   H. Hotop and D. Mahr (1975) J. Phys. B: At. Mol. Phys. **13** L301

[361]   J.E. Hansen (1975) J. Phys. B: At. Mol. Phys. **7** 1902

[362]   J.-P. Connerade (1996) in *VUV Spectroscopy* U. Becker and D.A. Shirley Eds. Plenum Press, New York

[363]   J.-P. Connerade, S.J. Rose and I.P. Grant (1979) J.Phys.B. **12**, L53

[364]   B. Lewandowski J. Ganz H. Hotop and M.W. Ruf (1981) J. Phys. B: At. Mol. Phys. **14** L803

[365]   J.-P. Connerade (1981) J. Phys. B: At. Mol. Phys. **14**, L141

[366]   R.A. Rosenberg M.G. White G. Thornton and D.A. Shirley (1979) Phys. Rev. Lett. **43** 1384

[367]   R.A. Rosenberg S.T. Lee and D.A. Shirley (1980) Phys. Rev. **21** 132

[368]   A.W. Potts and E.P.F. Lee (1979) J. Phys. B: At. Mol. Phys. **12** L413

[369]   Y. Itikawa T. Hayaishi Y. Itoh T. Koizumi J. Murakami T. Nagata Y. Sato A. Yagishita and M. Yoshino (1985) *Photon Factory Activity Report* KEK, Japan page 212

Chapter 8

[370]   E.P. Wigner (1946) Phys. Rev. **70**, 15; (1948) *ibid.* **73**, 1002

[371]   A.M. Lane and R.G. Thomas (1958) Rev. Mod. Phys. **30**, 257

[372]   Pearson (1988) *Quantum Scattering and Spectral Theory* Academic Press, New York, page 277

[373]   A.M. Lane (1986) J. Phys. B: At. Mol. Phys. **19**, 253

[374]   A.M. Lane (1986) J. Phys. B: At. Mol. Phys. **19** L601

[375]   J.-P. Connerade and A.M. Lane (1987) J. Phys. B: At Mol. Phys. **20** L181

[376]   J.-P. Connerade (1991) J. Phys. B: At. Mol. Phys. **24** L513

[377]   J.-P. Connerade and A.M. Lane (1987) J. Phys. B: At Mol. Phys. **20** 1757

[378]   Y. Komninos and C.A. Nicolaides (1987) Z. Phys. D **4**, 301

[379]   A.M. Lane (1985) J. Phys. B: At. Mol. Phys. **18**, 2339

[380]   J.-P. Connerade, A.M. Lane and M.A. Baig (1985) J. Phys. B: At. Mol. Phys. **18**, 3507

[381]   J.-P. Connerade and A.M. Lane (1988) Rep. Prog. Phys. **51**, 1439

[382]   J.-P. Connerade (1978) Proc. Roy. Soc. (London) A **362**, 361

[383]   J.-P. Connerade and A.M. Lane (1985) J. Phys. B: At. Mol. Phys. **18**, L605

[384]   J.-P. Connerade, H. Ma, N. Shen and T.A. Stavrakas (1988) J. Phys. B: At. Mol. Phys. **21**, L241

[385]   J.-P. Connerade and A.M. Lane (1987) J. Phys. B: At. Mol. Phys. **20**, L363

[386]   J.-P. Connerade (1978) Nucl. Instrum. and Methods **152** 271

[387]   Niemax K (1985) Appl. Phys. **B38** 147

[388]   W.H. Aldous and Sir Edward Appleton (1932) *Thermionic Vacuum Tubes* seventh edition Methuen Monographs on Physical Subjects Methuen & Co, London, reprinted 1961

[389]   W.G. Kaenders, N. Shen, J. Marangos, M.H.R. Hutchinson and J.-P. Connerade (1990) J. Mod. Opt. **37** Letter page 835

[390]   U. Griesmann, Shen-Ning, J.-P. Connerade, K. Summer K and J. Hormes (1988) J. Phys. B: At. Mol. Phys. **21** L53

[391]   U. Fano (1961) Phys. Rev. **124** 1866

[392]   F.H. Mies (1968) Phys. Rev. **175** 164

[393]   A.M. Lane (1984) J. Phys. B: At. Mol. Phys. **17** 2213

[394]   B.W. Shore (1967) Rev. Mod. Phys. **39** 440

[395]   J. Ganz M. Raab H. Hotop and J. Geiger (1984) Phys. Rev. lett. **53** 1547

[396]   E.A.J.M. Bente and W. Hogervorst, (1987) Phys. Rev. A **36** 4081

[397]   F. Paschen (1931) Sitz. Preuss. Akad. Wiss. **32** 709

[398]   H. F. White (1931) Phys. Rev. **38** 2016

[399]   A.G. Shenstone (1939) Rep. Prog. Phys. **5** 210

[400]   J.-P. Connerade (1978) Proc. R. Soc. A **362** 361

[401]   A. Safinya and T.F. Gallagher (1979) Phys. Rev. lett. **43** 1239

[402]   P.E. Coleman P.L. Knight and K. Burnett (1982) Opt. Comm. **42** 171

[403]   M.E. St.J. Dutton and B.J. Dalton (1990) J. Modern Optics **37** 53

[404]   M.E. St.J. Dutton and B.J. Dalton (1993) J. Modern Optics **40** 123

[405]   L.I. Pavlov S.S. Dimov D.I. Metchkov G.M. Mileva and K.V. Stamenov (1982) Phys. Lett. **89A** 441

[406]   S.S. Dimov L.I. Pavlov K.V. Stamenov (1983) App. Phys. **B30** 35

[407]   O. Faucher Y.L. Shao D. Charalambidis and C. Fotakis
(1994) Phys. Rev. **A 50** 641

[408]   P.L. Knight M.A. Lauder and B.J. Dalton (1990) Phys. Rep. **190** 1

[409]   Y.I. Heller V.K. Lukinikh A.K. Popov and V.V. Slabko
(1981) Phys. Lett. A **82** 4

[410]   Jian Zhang (1991) *Laser-induced Continuum Structure and Third Harmonic
Generation in One- and Two-valence Electron Atoms* PhD thesis. University
of Southern California

[411]   C. Cohen-Tannoudji and P. Avan (1977) *Etats Atomiques et Moléculaires
Couplés à un Continuum: Atomes et Molécules Hautement Excités* Editions
du CNRS, Paris, page 93

[412]   C.H. Greene (1989) in *Atomic Spectra and Collisions in External Fields* K.T.
Taylor M.H. Nayfeh and C.W. Clark Eds. Plenum Press, New York, page 233

[413]   J.-P. Connerade (1983) J. Phys. B: At. Mol. Phys. **16** L329

[414]   J.-P. Connerade (1985) J. Phys. B: **18** L367

[415]   J. Dubau and M.J. Seaton (1984) J. Phys. B: At. Mol. Phys. **17** 381

[416]   A. Giusti-Suzor and U. Fano (1984) J. Phys. B: At. Mol. Phys **17** 215

[417]   J. W. Cooper C.W. Clark C.L. Cromer T.B. Lucatorto B.F. Sonntag E.T.
Kennedy and J.T. Costello (1989) Phys. Rev. A **39** 6074

[418]   Baig M.A. Ahmad S. Connerade J.P. Dussa W. and Hormes J.
(1992) Phys. Rev. A **45** 7963

[419]   C. Bloch (1965) 36th *Enrico Fermi Summer School Varenna* Academic Press,
London, page 592

[420]   W.R.S. Garton and F.S. Tomkins (1969) Astrophys. J. **158** 1219

[421]   F. Gounand T.F. Gallagher W. Sandner K.A. Safinya and R. Kachru
(1983) Phys. Rev. **27** 1925

[422]   A. Giusti-Suzor and U. Fano (1984) J. Phys. B: At. Mol. Phys. **17** 215

[423]   M.N.R. Ashfold, J.M. Bailey and R.N. Dixon (1986) J. Chem. Phys. **79** 4080

[424]   A.H. Kung R.H. Page R.J. Larkin Y.R. Shen and Y.T. Lee (1987) Phys. Rev.
Lett. **56** 328

[425]   P. Richard C.F. Moore and J.D. Fox (1964) Phys. Rev. Lett. **13** 343

[426]   H. Rinneberg G. Jonsson J. Neukammer K. Vietzke H. Hieronymus G. Konig
and W.E. Cooke (1965) *Proceedings of the Second European Conference on
Atomic and Molecular Physics* A.E. de Vries and M.J. van der Wiel Eds.,
European Physical Society, Geneva, page 247

[427]   U. Heinzmann H. Heuer and J. Kessler (1976) Phys. Rev. Lett. **36** 1444

[428]   K. Ueda (1987) J. Opt. Soc. Am. **B4** 648

[429]   K. Ueda (1987) Phys. Rev. **A35** 2484

[430]   N.E. Karapanagioti D. Charalambidis C.J.G. Uiterwaal C. Fotakis H. Bachau
I. Sanchez and E. Cornier (1966) Phys. Rev. **A53** 2587

[431]   C.H. Greene and L. Kim (1987) Phys. Rev. **A36** 2706 and 4272

[432] F. Combet-Farnoux (1982) Phys. Rev. **A25** 287

[433] P.A. Moldauer (1963) Phys. Lett. **8**, 70

[434] B. Piraux and P.L. Knight P (1988) Phys. Rev. **A40** 712

[435] C.T.W. Lahaye and W.Hogervorst (1988) 'Electric field induced interference effects in bound Rydberg states of barium' Preprint

[436] A. Alyah (1988) Doctoral Dissertation Universität Bielefeld

[437] *Atoms in Strong Fields* C.A. Nicolaides C.W. Clark and M.H. Nayfeh Eds. (1990) NATO ASI Series B: vol 212 Plenum Press, New York

[438] *Atomic Spectra and Collisions in External Fields* K.T. Taylor M.H. Nayfeh and C.W. Clark Eds. (1988) Plenum Press, New York

[439] M.A. Baig J.-P. Connerade and M. Rafi (1987) J. Phys. B **20**, L741

[440] K.T. Lu and U. Fano (1979) Phys. Rev. **A 2** 81

[441] N.E. Karapanagioti G. Droungas and J.-P. Connerade (1995) J. Mod. Opt. Letter page 727

[442] J. P. Connerade and S.M. Farooqi (1991) J. Phys. B: At. Mol. Opt. Phys. **24**, L331

[443] S.M. Farooqi J.-P. Connerade and M. Aymar (1992) J. Phys. B: At. Mol. Opt. Phys. **25**, L219

[444] J.-P. Connerade (1993) J. Phys. B **26** 4041

[445] M.A. Baig and J.-P. Connerade (1984) J.Phys.B **17**, L469

[446] M.A. Baig S. Ahmad U. Griesmann J.-P. Connerade S.A. Bhatti and N. Ahmad (1992) J. Phys. B **25** 321

[447] J.-P. Connerade, M.A. Baig and M. Sweeney (1987) J. Phys. B **20**, L771

[448] C.M. Brown and M. Ginter (1987) J. Opt. Soc. Am. **68** 817

[449] J.-P. Connerade and V.K. Dolmatov (1996) J. Phys. B: At. Mol. Opt. Phys. **29** L831; (1997) *ibid.* **30** L181

Chapter 9

[450] M. Göppert-Mayer (1931) Ann. Phys. **9** 273

[451] F. Yergeau S.L. Chin and P. Lavigne (1987) J. Phys. B: At. Mol. Phys. **20** 723

[452] S. Augst D. Strickland D.D. Meyerhofer S.L. Chin and J.H. Eberly (1989) **63** 2212

[453] J.M. Worlock (1972) in *Laser Handbook* F.T. Arecchi and E.O. Shulz-DuBois Eds. vol II North-Holland, Amsterdam, page 1323

[454] N.A. Cherepkov and A. Yu. Elizarov (1991) J. Phys. B: At. Mol. Phys. **24** 4169

[455] B. Cagnac G. Grynberg and F. Biraben (1973) J. Phys. (Paris) **34** 56

[456] W. Perrie A.J. Duncan H.J. Beyer and H. Kleinpoppen (1985) Phys. Rev. lett. **54** 1790

[457] H. Kleinpoppen (1986) *Proceedings of the Second International Symposium on the Foundations of Quantum Mechanics* U. of Tokyo, Tokyo, page 59

[458]   N. Mason and R. Newell (1989) J. Phys. B: At. Mol. Phys. **22** L323, 777;
        (1990) *ibid.* **23** 2179

[459]   N. Bloembergen and M.D. Levenson (1976) in *High Resolution Laser
        Spectroscopy* K. Shimoda Ed. Springer Verlag, Berlin

[460]   K.D. Bonin and T.J. McIlrath (1984) J. Opt. Soc. Am **B1** 52

[461]   A. Javan (1957) Phys. Rev. **107** 1579; and A. Javan E.A. Ballik and W.L.
        Bond (1962) J. Opt. Soc. Am. **B52** 96

[462]   R.G. Brewer and E.L. Hahn (1975) Phys. Rev. **A11** 1641

[463]   K. Shimoda and T. Shimizu (1972) Prog. Quant. Elec. **2** 45

[464]   P.A.M. Dirac (1958) *The Principles of Quantum Mechanics* second edition
        The Clarendon Press, Oxford

[465]   R.P. Feynman F.L. Vernon and R.W. Hellwarth (1957) J. Appl. Phys. **28** 49

[466]   J.E. Bjorkholm and P.F. Liao (1974) Phys. Rev. Lett. **33** 128

[467]   E. Hanamura (1973) Solid State Commun. **12** 951

[468]   J.P. Marangos N. Shen H. Ma M.H.R Hutchinson and J.-P. Connerade
        (1990) J. Opt. Soc. Am. B **7** 1254

[469]   H. Puell and C.R. Vidal (1976) Opt. Commun. **19** 279

[470]   J.J. Wynne and R. Beigang Phys. Rev. **A23** 2736

[471]   J.-P. Connerade and A.M. Lane (1987) J. Phys. B. At Mol. Phys. **20** L363

[472]   M. Rahman-Attia G. Laplanche M. Jaouen and A. Rachman
        (1986) J. Phys. B: At. Mol. Phys. **19** 3669

[473]   S.H. Autler and C.H. Townes (1950) Phys. Rev. **A78** 340

[474]   B.R. Mollow (1969) Phys. Rev. **188** 1969

[475]   C. Cohen-Tannoudji and P. Avan (1977) *Etats Atomiques et Moléculaires
        Couplés à un Continuum: Atomes et Molécules Hautement Excités* (text in
        English) Editions du CNRS, Paris, page 93; the same material can also be
        found (in French) in the book *Processus d'intéraction entre photons et atomes*
        by C. Cohen-Tannoudji J. Dupont-Roc and G. Grynberg (1988) Editions du
        CNRS Paris, page 233

[476]   C. Cohen-Tannoudji (1977) *Frontiers in Laser Spectroscopy* Vol 1 North
        Holland, Amsterdam, page 3

[477]   R. Shimano and M. Kuwata-Gonokami (1994) Phys. Rev. Lett. **72** 530

[478]   P.L. Knight M.A. Lauder and B.J. Dalton (1990) Physics Reports **190** 3

[479]   N.E. Karapanagioti O. Faucher and D. Charalambidis (1995) *Fifth European
        Conference on Atomic and Molecular Physics (ECAMP5)* R. Thompson Ed.
        European Physical Society, Geneva page 661

[480]   P.A. Golovinskiy M.A. Dolgopolov and V.S. Rostovstev (1993) Laser Physics
        3 487

[481]   D.M. Volkov (1935) Zeit. Phys. **94**, 250

[482]   J.-P. Connerade K. Dietz and M.H.R. Hutchinson (1995) Physica Scripta **T58**
        23

[483]   W.F. Lamb (1995) Applied Physics B **60** 77

[484] Tisch J.W.G. Smith R.A. Ciarocca M. Muffett J.E. Marangos J.P. and Hutchinson M.H.R. (1994) Phys. Rev. **A49**, 28

[485] K. Boyer G. Gibson H. Jara T.S. Luk I.A. McIntyre A. McPherson R. Rosnan and C.K. Rhodes (1988) *Atoms in Strong Fields* C.A. Nicolaides C.W. Clark and M.H. Nayfeh Eds. NATO ASI series B 212 Plenum Press, London, page 283; see also A. Szöke and C.K. Rhodes (1986) Phys. Rev. Lett. **56** 720

[486] P. Lambropoulos (1985) Phys. Rev. Lett. **55** 2141

[487] L.J. Frasinski K. Codling P. Hatherly J. Barr I.N. Ross and W.T. Toner (1987) Phys. Rev. Lett. **58**, 2424

[488] J.H. Posthumus L.J. Frasinski A.J. Giles and K. Codling (1995) J. Phys. B: At. Mol. Opt. Phys. **28** L349

[489] J.-P. Connerade (1991) Adv. Atom. Mol. Opt. Phys. **29**, 325

[490] J.-P. Connerade and K. Dietz (1992) J. Phys. B: At. Mol. Opt. Phys. **25** 1185

[491] R.R. Freeman R.R. (1992) Physics World March issue vol 5 No 3 page 29

[492] C. Blondel M. Crance C. Delsart and A. Giraud (1991) J. Phys. B. **24** 3575

[493] T.J. McIlrath P.H. Bucksbaum R.R. Freeman M. and Bashkansky Phys. Rev. **A35**, 4611 (1987).

[494] L.V. Keldysh (1964) Zh. Eksp. Teor. Phys. (JETP) **47** 1945

[495] F.H.M. Faisal (1973) J. Phys. B: At. Mol. Phys. **6** L89

[496] H. Reiss (1980) Phys. Rev. **A22** 1786; (1987) J. Phys. B **20** L79

[497] P.H. Bucksbaum M. Bashkansky R.R. Freeman T.J. McIlrath and F. DiMauro (1986) Phys. Rev. Lett. **56**, 2590

[498] B. Yang K.J. Schafer B. Walker K.C. Kulander P. Agostini P. and L.F. DiMauro (1993) Phys. Rev. Lett. **71** 3770

[499] P. Agostini F. Fabre G. Mainfray G. Petite and N. Rahman(1979) Phys. Rev. Lett. **42** 1127

[500] Y. Liang S. Augst S.L. Chin Y. Beaudouin and M. Chacker (1994) J. Phys. B: At. Mol. Opt. Phys. **27** 5119

[501] P. Kruit J. Kimman and M. van der Wiel (1981) J. Phys. B: At. Mol. Phys. **14** L597

[502] D. Normand and J. Morellec (1988) J. Phys. B: At. Mol. Phys. **21** L625

[503] P.H. Bucksbaum A. Zavriyev H.G. Muller and D.W. Schumacher (1990) in *Multiphoton Processes* G. Mainfray and P. Agostini Eds. Service de Physique Atomique CE Saclay, Gif sur Yvette (France), page 181

[504] R.R. Freeman P.H. Bucksbaum H. Milchberg S. Darack D. Schumacher and M.E. Geusic (1987) Phys. Rev. Lett. **59** 1092; H. Rottke B. Wolff M. Brickwedde D. Feldmann and K. Welge (1990) Phys. Rev. Lett. **64** 404

[505] J.L. Krause K.J. Schafer and K.C. Kulander K.C. (1992) Phys. Rev. Lett. **68** 3535

[506] M.E. Faldon M.H.R. Hutchinson J.P. Marangos J.E. Muffett R.A. Smith J.W.G. Tisch and C.G. Walström (1992) J. Opt. Soc. Am. **B9**, 2094; T. Starczewski J. Larsson C.G. Walström J.W.G. Tisch R.A. Smith J.E. Muffett and M.H.R. Hutchinson (1994) J. Phys. B: At. Mol. Phys. **27** 3291

488 *References*

[507] Ya. B. Zeldovich (1967) Sov. Phys. JETP **24** 1006

[508] C. Zener (1932) Proc. Roy. Soc. (London) **A137** 696

[509] H. Breuer K. Dietz K. and M. Holthaus (1992) *Proceedings International Conference on Coherent Radiative Processes in Strong Fields Washington DC 1990* M. Jacobs Ed. Gordon and Breach, New York; see also J.-P. Connerade K. Conen K. Dietz and J. Henkel (1992) J. Phys. B: At. Mol. Opt. Phys. **25** 3771

[510] M.P. de Boer and H.G. Müller (1992) Phys. Rev. Lett. **68**, 2747

[511] A. L'Huillier K.J. Schäfer and K.C. Külander (1991) J. Phys. B: At. Mol. Opt. Phys. **24**, 3315

[512] J.E. Muffett C.G. Waldström and M.H.R. Hutchinson (1994) J. Phys. B: At. Mol. Opt. Phys. **27** 5693

[513] J.-P. Connerade and C. Keitel (1996) Phys. Rev. A **53** 2748

[514] W. Pauli and M. Fierz (1938) Nuovo Cimento **15** 167

[515] H. A. Kramers (1938) Nuovo Cimento **15** 108

[516] W.C. Henneberger (1968) Phys. Rev. Lett. **21** 838

[517] M. Gavrila (1992) in *Atoms in Intense Radiation Fields* M. Gavrila Ed. Academic Press, Orlando

[518] R.B. Vrijen J.H. Hoogenraad H.G. Müller and L.D. Noordam L.D. (1993) Phys. Rev. Lett. **70** 3017

Chapter 10

[519] H. Poincaré (1886) Acta Mathematica VIII 295; and (1892) *Les Méthodes Nouvelles de la Mécanique Céleste* Gauthier-Villars et Cie, Libraire du Bureau des Longitudes et de l'Ecole Polytechnique, Paris

[520] A. Hacinliyan (1990) Europhys. News **21** 7

[521] P. Cvitanović, I. Percival and A. Wirzba Eds. *'Quantum Chaos — Quantum Measurement'* (1992) NATO ASI Series C (Mathematical and Physical Sciences) vol 358 Kluwer Academic Publishers, Dordrecht

[522] W.-H. Steeb and J.A. Louw (1986) *Chaos and Quantum Chaos* World Scientific Co Singapore

[523] A.N. Kolmogoroff (1954) Dokl. Akad. Nauk. SSSR **98** 527

[524] V.I. Arnold V.I. (1963) Usp. Mat. Nauk. SSSR **18** 13

[525] J. Moser (1962) Nachr. Akad. Göttingen Wissenschaft 1

[526] J. Bauche-Arnoult J. Bauche and M. Klapisch (1982) Phys. Rev. **A25** 287

[527] D. Pfenniger (1990) Europhys. News **21** 6

[528] W.P. Reinhardt and D. Farrelly (1982) in *Atomic and Molecular Physics Close to Ionization Thresholds in High Fields* J.-P. Connerade J.-C. Gay and S. Liberman Eds. Les Editions de Physique (Les Ulis) France page C2-29

[529] A. Einstein (1917) Verh. Deutsche. Phys. Gesell. (Berlin) **19** 82

[530] M.L. Brillouin M.L. (1926) J. de Phys. **7** 353

[531] J.B. Keller J.B. (1958) Ann. Phys. **8** 180

[532] I.C. Percival (1973) J. Phys. B: At. Mol. Phys. **6** L229; J.G. Leopold and I.C. Percival (1979) ibid. **12** 709

[533] J.B. Garg J. Rainwater J.S. Petersen and W.W. Havens Jr (1964) Phys. Rev. **134** B985

[534] J.L. Rosen J.S. Desjardins J. Rainwater and W.W. Havens Jr (1960) Phys Rev **118** 687

[535] M. Göppert-Mayer and J.H.D. Jensen (1955) *Elementary Theory of Nuclear Shell Structure* John Wiley, New York

[536] L.S. Kisslinger and R.A. Sorensen (1960) Kgl. Danske Videnskab. Selskab. Mat.-fys. **32** No 9

[537] E.P. Wigner *Gatlinberg Conference on Neutron Physics* Oak Ridge National Laboratory Rept. No ORNL-2309, page 59

[538] L.D. Landau and Ya. Smorodinsky (1958) *Lectures on the Theory of the Atomic Nucleus'* Consultants Bureau, New York, page 55

[539] M.L. Mehta (1967) *Random Matrices and the Statistical Theory of Energy Levels* Academic Press, New York, second edition (1990)

[540] *Statistical Theories of Spectral Fluctuation* (1965) C.E. Porter Ed. Academic Press, New York

[541] M.V. Berry (1985) Proc. Roy. Soc. (London) **A400** 229

[542] R.G. Thomas and C.E. Porter (1956) Phys Rev **104** 483

[543] N. Rosenzweig (1963) *Statistical Physics* Benjamin, New York

[544] N. Rosenzweig and C.E. Porter 1960 Phys. Rev. **120** 1698

[545] H. Fröhlich (1937) Physica **4** 406

[546] J.-P. Connerade, M.A. Baig and M. Sweeney (1987) J. Phys. B **20**, L771

[547] J.-P. Connerade, M.A. Baig and M. Sweeney (1990) J. Phys. B **23**, 713

[548] F.J. Dyson and M.L. Mehta (1963) J. Math. Phys. **4** 701

[549] J.-P. Connerade (1997) J. Phys. B: At. Mol. Opt. Phys. **30** L31

[550] J.-P. Connerade I.P. Grant P. Marketos and J. Oberdisse (1995) J. Phys. B: At. Mol. Opt. Phys. **28** 2539

[551] T.A. Brody J. Flores J.B. French P.A. Mello A. Pandey and S.S.M. Wong (1981) Rev. Mod. Phys. **53** 385

[552] W.R.S. Garton and F.S. Tomkins (1969) Astrophys. J. **158** 839

[553] P. Zeeman (1913) *Researches in Magneto-Optics* MacMillan, London

[554] L. Landau (1930) Z. Phys. **64** 629

[555] R.J. Elliott G. Droungas and J.-P. Connerade (1995) J. Phys. B. **28** L537

[556] E.P. Wigner (1959) *Group Theory* Academic Press, New York

[557] A. Holle J. Main G. Wiebusch H. Rottke and K.H. Welge (1988) Phys. Rev. Lett. **61** 161; A. Holle G. Wiebusch J. Main K.H. Welge G. Zeller G. Wunner T. Ertl and H. Ruder (1987) Zeit. Phys. **5** 279

[558] P.F. O'Mahony M.A. Al-Laithy and K.T. Taylor (1989) in *Atomic Spectra and Collisions in External Fields* K.T. Taylor M.H. Nayfeh and C.W. Clark Eds. Plenum Press, New York, page 65

[559] G. Wunner G. Zeller U. Woelk W. Scheizer R. Niemeler F. Geyer H. Friedrich and H. Ruder (1988) in *Atomic Spectra and Collisions in External Fields* K.T. Taylor M.H. Nayfeh and C.W. Clark Eds. Plenum Press, New York, page 9

[560] H. Friedrich and D. Wintgen (1989) Phys. Reports **183** 39

[561] C.H. Lewenkopf (1990) Phys. Rev. A **42** 2431

[562] G. Casati B.V. Chirikov and I. Guarneri (1985) Phys. Rev. Lett. **54** 1350

[563] J. Zakrzewski K. Dupret and D. Delande (1995) Phys. Rev. Lett. **74** 522

[564] H. Hasegawa S. Adachi and H. Harada (1983) J. Phys. A **16** L503

[565] K.J. Drinkwater J. Hormes D.D. Burgess J.-P. Connerade and R.C.M. Learner (1984) J. Phys. B: At. Mol. Phys. **17** L439

[566] *Atomic and Molecular Physics Close to Ionization Thresholds in High Fields* (1982) J.-P. Connerade J.-C. Gay and S. Liberman Eds. J. de Physique **43** C-2

[567] C.A. Nicolaides C.W. Clark and M.H. Nayfeh (1990) *Atoms in Strong Fields* NATO ASI Series B vol 212 Plenum Press, New York

[568] Special Issue on Atoms in High Magnetic Fields, D. Delande and K.T. Taylor Eds. (1994) J. Phys. B: At. Mol. Opt. Phys. **27** issue No 13 (July)

[569] J. Pinard P. Cacciani S. Liberman E. Luc-Koenig and C. Thomas (1988) in *Atomic Spectra and Collisions in External Fields* Eds. K.T. Taylor M.H. Nayfeh and C.W. Clark Plenum Press, New York

[570] H. Crosswhite U. Fano K.T. Lu and A.R.P. Rau (1979) Phys. Rev. Lett. **42** 963

[571] W.P. Reinhardt (1983) J. Phys. B: At. Mol. Phys. **16** L635

[572] E.A. Soloviev (1981) JETP Lett. **34** 265

[573] P.F. O'Mahony and K.T. Taylor (1986) Phys. Rev. Lett. **57** 2931

[574] R.F. O'Connell (1974) Astrophys. J. **187** 275

[575] A.R.P. Rau (1979) J. Phys. B: At. Mol. Phys. **12** L193

[576] S. Feneuille S. Liberman E. Luc-Koenig J. Pinard and A. Taleb (1982) Phys. Rev. **A25** 2853

[577] M. Nayfeh D. Humm and K. Ng (1990) in *Atoms in Strong Fields* C.A. Nicolaides C.W. Clark and M.H. Nayfeh Eds. NATO ASI Series B vol 212 Plenum Press, New York, page 133

[578] D. Harmin (1981) Phys. Rev. **A24** 2491; (1982) *ibid.* **A26** 2696

[579] R.J. Damburg (1988) in *Atomic Spectra and Collisions in External Fields* K.T. Taylor M.H. Nayfeh and C.W. Clark Eds. Plenum Press, New York, page 141

[580] J.-P. Connerade (1990) in *Atoms in Strong Fields* C.A. Nicolaides C.W. Clark and M.H. Nayfeh Eds. NATO ASI Series B vol 212 Plenum Press, New York, page 1

[581] R.V. Jensen (1984) Phys. Rev. **A30** 386

[582] J.E. Bayfield and P. Koch (1974) Phys. Rev. Lett. **33** 258

[583] G. Casati (1990) in *Atoms in Strong Fields* C.A. Nicolaides C.W. Clark and M.H. Nayfeh Eds. NATO ASI Series B vol 212 Plenum Press, New York page 231

[584] J.-P. Connerade K. Conen K. Dietz and J. Henkel (1992) J. Phys. B: At. Mol. Opt. Phys. **25** 3771

[585] K. Dietz J. Henkel and M. Holthaus (1992) Phys. Rev. **A45** 4960

[586] D. Delande A. Bommier and J.-C. Gay (1991) Phys. Rev. Lett. **66** 141

[587] P.F. O'Mahoney and F. Mota-Furtado (1991) Comm. At. Mol. Phys. **25** 309; (1991) Phys. Rev. Lett. **67** 2283

[588] E. Korevaar and M.G. Littman (1983) J. Phys. B: At. Mol. Phys. **16** L437; J. Main and G. Wunner (1994) J. Phys. B: At. Mol. Phys. **27** 2835

[589] R.M Potvliege and P.H.G. Smith (1991) J. Phys. B: At. Mol. Opt. Phys. **24** L641 and **25** 2501; E. Charron A. Giusti-Suzor and F.H. Mies (1993) Phys. Rev. Lett. **71** 692; D.W. Schumacher F. Weihe H.G. Muller and P.H. Bucksbaum (1994) Phys. Rev. Lett. **73** 1344; S. Watanabe K. Kondo Y. Nabekawa A. Sagisaka and Y. Kobayashi (1994) Phys. Rev. Lett. **73** 2692

Chapter 11

[590] D. Pines (1962) *The Many-body Problem* Benjamin, Reading USA; D. Pines (1964) *Elementary Excitations in Solids* Benjamin, New York

[591] J.A.D. Matthew (1993) Contemp. Phys. **34** 89

[592] H.A. Kramers (1935) Physica **2** 483; F. Seitz (1935) Physica **2** 278

[593] E.W. Plummer and W. Eberhardt (1982) Adv. Chem. Phys. **49** 533

[594] F. J. Himpsel (1990) Surf. Sci. Rep. **12** 3

[595] P. Weightman (1995) Microsc. Microanal. Microstruct. **6** 263

[596] C.M. Teodorescu R.C. Karnatak J.-M. Esteva A. El Afif and J.-P. Connerade (1993) J. Phys. B: At. Mol. Opt. Phys. **26** 4019

[597] R.J. Cole J.A. Evans L.Duò A.D. Laine P.S. Fowles P. Weightman G. Mondio and D. Norman (1992) Phys. Rev. B **46** 3747

[598] R.J. Cole P. Weightman D.R. Jennison and E.B. Stechel (1994) J. Elec. Spec. Related Phen. **68** 139

[599] R.C. Karnatak, J.M. Esteva and J.-P. Connerade (1981) J. Phys. B: At. Mol. Opt. Phys. **14**, 4747

[600] N.F. Mott (1974) *Metal Insulator Transitions* Taylor and Francis, London

[601] A.M. Boring and J.L. Smith (1987) in *Giant Resonances in Atoms Molecules and Solids* J.-P. Connerade J.-M. Esteva and R.C. Karnatak Eds. NATO ASI Series B vol 151 Plenum, New York, page 311

[602] B.E. Koel G.M. Loubriel M.L. Knotek R.H. Stulen R.A. Rosenberg and C.C. Parks (1982) Phys. Rev. **B25** 5551

[603] M.N. Piancastelli *et al.* (1983) Appl. Phys. Lett. **42** 990; D.W. Lindle P.A. Herman T.A. Ferret M.N. Piancastelli and D.A. Shirley (1987) Phys. Rev. **B35** 4605; (1989) Solid State Commun. **72** 635

[604] C. Lamonier G. Wrobel and J.P. Bonnelle (1994) J. Mater. Chem. **4** 1927

[605] K. Brandt (1994) Solid State Ionics **69** 173

[606] L. Jourdaine and J.L. Souquet (1987) Rev. Solid State Sci. **1** 67

[607] B.C.H. Steele (1973) in *Fast Ion Transport in Solids* W. van Goo Ed. North-Holland, Amsterdam, page 103

[608] M. B. Armand (1973) in *Fast Ion Transport in Solids* W. van Goo Ed. North Holland, Amsterdam, page 665

[609] A. Kozawa (1980) in *Application of Solid Electrolytes* T. Takahashi and A. Kozawa Eds. JEC Press Inc. Cleveland Ohio, page 23

[610] M. Ménétrier A. Rougier C. Delmas C. Marichal J. Hirschwinger and P. Willmann (1994) *Groupe Français d'Etude des Composés d'Insertion (Montpellier)* Société Française d' électrochimie, Paris, page 67

[611] C. Delmas I. Saadoune A. Rougier and P. Willman (1993) *Piles et Accumulateurs au Lithium* G. Bronoel Ed. Société Françaiese des thermiciens et Société Française de Chimie, Paris

[612] L'Actualité Chimique (1992) n°1 numéro spécial: l'électrochimie Janvier/Février

[613] J. Morales J.L. Tirado M.L. Elidrissi Moubtassim J. Olivier-Fourcade and J.-C. Jumas (1995) J. Alloys and Compounds **217** 176

[614] M.P. Crosnier E. Delarue J. Choisnet and J.L. Fourquet (1992) J. Solid State Inorg. Chem. **39** 321

[615] J.C. Fuggle in *Giant Resonances in Atoms Molecules and Solids* J.-P. Connerade J.-M. Esteva and R.C. Karnatak Eds. NATO ASI Series B vol 151 Plenum, New York, page 381

[616] O. Gunnarsson and K. Schönhammer (1983) Phys Rev **B28**, 4315

[617] O. Gunnarsson O.K. Anderson O. Jepsen and J. Zaanen (1989) *Proc. Tenth Tanigushi Int. Symp.* Springer-Verlag Berlin

[618] T. Jo and A. Kotani (1988) Phys Rev B **38** 830; (1988) J Phys Soc Jap **57** 2288

[619] B.T. Thole G. van der Laan J.C. Fuggle G.A. Savatsky R.C. Karnatak and J.-M. Esteva (1985) Phys. Rev. **32** 5107

[620] D.W. Lynch and R.D. Cowan (1988) Phys Rev **B36** 9228

[621] J.M. Lawrence P.S. Riseborough and R.D. Parks (1981) Rep. Prog. Phys. **44** 1

[622] J.-P. Connerade (1982) J Phys C: Solid State **15**, L367; (1983) Journal of the Less Common Metals **93**, 171

[623] I.M. Band and V.I. Fomichev (1980) Phys. Lett. **A75** 178

[624] I.M. Band K.A. Kikoin M.B. Trzhavskovskaya D.I. and Khomskii (1988) Sov Phys JETP **67**, 1561

[625] M. Schlüter and C.M. Varma (1983) Helv. Phys. Acta **56** 147

[626] A. Bringer (1983) Solid State Commun. **46** 591

[627] B.K. Sarpal C. Blancard J.-P. Connerade J.-M. Esteva J. Hormes R.C. Karnatak and U. Kuetgens (1991) J. Phys. B: At. Mol. Opt. Phys. **24** 1593

[628] O. Gunnarsson and K. Schönhammer (1983) Phys. Rev. **B28** 4315; (1985) Phys. Rev. **B31** 4815

[629] J. C. Fuggle F.U. Hillebrecht J.M. Esteva R.C. Karnatak O. Gunnarsson and K. Schönhammer (1983) Phys. Rev. **27** 4637; O. Gunnarsson K. Schönhammer J. C. Fuggle F.U. Hillebrecht J.M. Esteva and R.C. Karnatak (1983) Phys. Rev. **28** 7330

[630] R. Monnier L. Degiorgi and D.D. Koelling (1986) Phys. Rev. Lett. **86** 2744

[631] F. Patthey J.M. Imer W.D. Schneider Y. Baer B. Delley and F. Hulliger (1987) Phys. Rev. **B36** 7697

[632] J.M. Esteva and R.C. Karnatak in *Giant Resonances in Atoms Molecules and Solids* J.-P. Connerade J.-M. Esteva and R.C. Karnatak Eds. NATO ASI Series B vol 151 Plenum, New York, page 361

[633] G. Kalkowski G. Kaindl G. Wortmann D. Lentz and S. Krause (1988) Phys. Rev. **37** 1376

[634] F. Marabelli and P. Wachter (1987) Phys. Rev. **B36** 1238

[635] R.C. Karnatak J.M. Esteva H. Dexpert M. Gasgnier P.E. Caro and L. Albert (1987) J. Magnetism and Magnetic Materials **63-64** 518

[636] R.C. Karnatak M. Gasgnier H. Dexpert J.-M. Esteva P.E. Caro and L. Albert (1985) J. Less Common Metals **110** 377; (1987) Phys. Rev. **B36** 1745

[637] J.-P. Connerade and R.C. Karnatak (1990) Comm. At. Mol. Phys. **XXIV** 1

[638] R.C. Karnatak J.-M. Esteva and J.-P. Connerade (1981) J. Phys. B: At. Mol. Phys. **14** 4747

[639] M. Lübcke B. Sonntag W. Niemann and P. Rabe (1986) Phys. Rev. **34** 5184 W. Niemann W. Malzfeldt P. Rabe R. Haensel and M. Lübcke (1987) Phys. Rev. **35** 1099

[640] H. Dexpert R.C. Karnatak J.-M. Esteva J.-P. Connerade M. Gasgnier P.E. Caro and L. Albert (1987) Phys. Rev. **B36** 1750

[641] J.W. Allen (1983) Journal of the Less Common Metals **93** 183

[642] B.T. Thole and G. van der Laan (1993) Phys. Rev. Lett. **70** 2499

[643] S.S. Hasnain (1991) Ed. *X-ray Absorption Fine Structure* Ellis Horwood, New York

[644] E. Schüle and W. Nakel (1982) J. Phys. B: At. Mol. Phys. **15** L639; (1987) J. Phys. B **20** 2299

[645] H. Ast S. Keller C.T. Whelan H.R.J. Walters and R.M. Dreizler (1994) Phys. Rev. **50** R1

[646] R.H. Garstang (1977) Rep. Prog. Phys. **40** 105

[647] D. Lai E. Salpeter and S.L. Shapiro (1992) Phys. Rev. **A45** 4832 D. Lai and E.E. Salpeter (1995) Phys. Rev. **A52** 2611 and (1996) *ibid.* **A53** 152

[648] T.O. Klaassen (1996) C. R. Acad. Sci. Paris **323** Série II 187

[649] P.W. Barmby J.L. Dunn C.A. Bates E.P. Pearl C.T. Foxon A.J. van der Sluijs K.K. Geerinck T.O. Klaassen A. van Klarenbosch and C.J.G.M. Langerak (1994) J. Phys: Condensed Matter **6** 7867

[650] H. Friedrich and M. Chu (1983) Phys. Rev. **A28** 1423

[651]   A. van Klarenbosch K.K. Geerinck T.O. Klaassen W.Th. Wenckebach and
        C.T. Foxon (1990) Europhys. Lett. **13** 237 A. van Klarenbosch T.O. Klaassen
        W. Th. Wenckebach and C.T. Foxon (1990) J. Appl. Phys. **67** 6323

                                Chapter 12

[652]   T.S. Vih and C.Y.R. Wu (1988) Opt. Comm. **68** 35

[653]   S. Shahdin B. Wellegetausen and Z.G. Ma (1982) Appl. Phys. B**29** 195

[654]   Becker E.W. (1986) Z. Phys. D Atoms Molecules and Clusters **3** 101

[655]   A. Garscadden and R. Nagpal (1994) Plasma Sources Sci. Tech. **3** 239

[656]   A. Ruff S. Rutz E. Schreiber and L. Wöste (1996) Zeit. Phys. D **37** 175

[657]   *Small Particles and Inorganic Clusters* C. Chapon M.F. Gillet and C.R.
        Henry Eds. (1989) Springer Verlag, Berlin; *Structures and Dynamics of
        Clusters* T. Kondow K. Kaya and A. Terasaki Eds. (1996) Universal Academy
        Press Inc. Tokyo

[658]   J.-P. Connerade (1996) in *Correlations in Clusters and Related Systems: New
        Perpectives on the Many-body Problem* J.-P. Connerade Ed. World Scientific
        Press, Singapore, page 5

[659]   A.M. Flanck R.C. Karnatak C. Blancard J.M. Esteva P. Lagarde P. and J.-P.
        Connerade (1991) Zeit. Phys. D - Atoms Molecules and Clusters **21** 357

[660]   A. Stace (1968) Nature **331** 116

[661]   J. Bagott (1990) New Scientist March issue

[662]   C. Bréchignac M. Broyer Ph. Cahuzac G. Delacretaz P. Labastie J.P. Wolf
        and L. Wöste (1988) Phys. Rev. Lett. **60** 275

[663]   H.W. Kroto J.R. Heath S.C. O'Brien R.F. Curl and R.E. Smalley
        (1985) Nature **318** 162

[664]   H.S.M. Coxeter (1963) *Regular Polytopes* second edition Macmillan, New York;
        (1969) *Introduction to Geometry* second edition John Wiley & Sons, New York

[665]   K. Prassides and H. Kroto (1992) Physics World **5** No4 44

[666]   Y.B. Xu M.Q. Tan and U. Becker (1996) Phys. Rev. Lett. **76** 3538

[667]   Y. Achiba and T. Wakabayashi (1993) Zeit. Phys. D **26** 69

[668]   T. Wakabayashi K. Kikuchi H. Shiromaru S. Suzuki and Y. Achiba (1993)
        Zeit. Phys. D **26** S258

[669]   Y. Achiba T. Wakabayashi T. Moriwaki S. Suzuki and H. Shiromaru (1993)
        Materials Science and Engineering B**19** 14

[670]   H.W. Kroto (1987) Nature **329** 529

[671]   T. Wakayabashi K. Kikuchi S. Suzuki H. Shiromaru and Y. Achiba (1994) J.
        Phys. Chem. **98** 3090

[672]   R.J. Tarento and P. Joyes (1996) Zeit. Phys. D **37** 165

[673]   S. Hino H. Takahashi K. Iwasaki K. Matsumoto T. Miyazaki S. Hasegawa K.
        Kikuchi and Y. Achiba (1993) Phys. Rev. Lett. **71** 4261

[674] N. Watanabe H. Shiromaru Y. Negishi Y. Achiba N. Kobayashi and Y. Kaneko (1993) Chem. Phys. Lett. **207** 493

[675] M. Saunders H.A. Jiménez-Vázquez R.J. Cross S. Mroczkowski D.I. Freedberg and F.A.L. Anet (1994) Nature **367** 256

[676] M. Endo and H.W. Kroto (1992) J. Phys. Chem. **96** 6941

[677] F. Béguin (1994) G.F.E.C.I. (Montpellier) page 23

[678] J. Kohanoff W. Andreoni and M. Parrinello (1992) Chem. Phys. Lett. **198** 472

[679] T.P. Martin *et al.* (1993) J. Chem. Phys. **99** 4210

[680] C. Hérold J.F Marêché and P. Lagrange (1994) G.F.E.C.I. (Montpellier) page 17

[681] J.C. Phillips *J. Chem. Phys.* **87** 1712 (1987); **88** 2090 (1988)

[682] W.D. de Heer (1993) Rev. Mod. Phys. **65** 611

[683] W.D. Knight W.A. de Heer and W.A. Saunders (1986) Z. Phys. D Atoms Molecules and Clusters **3** 109

[684] W.D. Knight K. Clemenger W.A. de Heer W.A. Saunders M.Y. Chou and M.L. Cohen (1984) Phys. Rev. Lett **52** 2141

[685] W. Ekardt (1984) Phys. Rev. **B29** 1558

[686] M. Brack (1993) Rev. Mod. Phys. **65** 677

[687] K. Clemenger (1985) Phys. Rev. **B32** 1359

[688] Joyes P. (1971) J. Phys. Chem. Solids **32** 1269

[689] Y. Wang and T.F. George (1987) J. Chem. Phys. **86** 3493

[690] D.M. Lindsay Y. Yang and T.F. George (1987) J. Chem. Phys. **86** 3500

[691] P. Joyes (1990) *Les Agrégats inorganiques Elémentaires* Les Editions de Physique, Les Ulis France

[692] R. Poteau and F. Spiegelmann (1993) J. Chem. Phys. **98** 6549

[693] C. Blancard, J.M. Esteva, R.C. Karnatak, J.-P. Connerade, U. Kuetgens and J. Hormes (1989) J. Phys. B **22**, L575

[694] M. Lübcke B. Sonntag W. Niemann and P. Rabe (1986) Phys. Rev. **B34** 5184

[695] C. Bréchignac Ph. Cahuzac F. Carlier and J. Leygner (1993) Zeit. Phys. D **28** 67

[696] C. Bréchignac M. Broyer Ph. Cahuzac M. de Frutos P. Labastie and J. Ph. Roux (1991) Phys. Rev. Lett. **67** 1222

[697] O. Knospe and R. Schmidt (1986) Zeit. Phys. D **37** 85

[698] D. Smith and P. Spanel (1996) J. Phys. B: At. Mol. Opt. Phys. **29** 5199

[699] J. Dewitz W. Hübner and K.H. Bennemann (1996) Zeit. Phys. D **37** 75

[700] V.M. Akulin C. Bréchignac and A. Sarfati (1995) Phys. Rev. Lett. **26** 220

[701] J.M. Pacheco and R.A. Broglia (1989) Phys. Rev. Lett. **62** 1400

[702] R.A. Broglia (1994) Contemp. Phys. **35** 95

[703] D. Östling S.P. Appell G. Mukhopadhyay and A. Rosen (1996) J. Phys. B: At. Mol. Opt. Phys. **29** 5115

[704]   T. Kondow (1987) J. Phys. Chem. **91** 1307

[705]   M. Nagaminé K. Someda and T. Kondow (1994) Chem. Phys. Lett. **229** 8

[706]   J. Huang H.S. Carman and R.N. Compton (1995) J. Phys. Chem. **99** 1719

[707]   K. Selby V. Kresin J. Masin M. Vollmer W.A. de Heer A. Scheidemann and W. Knight (1991) Phys. Rev. **B43** 4569

[708]   J.-P. Connerade and A.V. Solov'yov (1996) J. Phys. B **29** 365

[709]   H. Haberland B. von Issenforff J. Yufeng and T. Kolar (1992) Phys. Rev. Lett. **69** 3212

[710]   C. Bréchignac Ph. Cahuzac F. Carlier J. Leygnier and A. Sarfati (1991) Phys. Rev. **44B** 11386

[711]   C. Bréchignac Ph. Cahuzac F. Carlier M. de Frutos R.N. Barnett and U. Landman (1994) Phys. Rev. Lett. **72** 1636

[712]   J.H. Parks and S.A. McDonald (1989) Phys. Rev. Lett. **62** 2301

[713]   G.F. Bertsch and R.A. Broglia (1994) *Oscillations in Finite Quantum Systems* Cambridge Monographs on Mathematical Physics Cambridge University Press, Cambridge

[714]   C. Bréchignac and J.-P. Connerade (1994) J. Phys. B: At. Mol. Opt. Phys. **27** 3795 – topical review

[715]   F.C. Sparo J.R. Kuklinski and P. Mukamel (1991) J. Chem. Phys. **94** 7534

Index